Guide to Digital Home Technology Integration

QUENTIN WELLS

DELMAR
CENGAGE Learning

Australia • Brazil • Japan • Korea • Mexico • Singapore • Spain • United Kingdom • United States

DELMAR
CENGAGE Learning

Guide to Digital Home Technology Integration
Quentin Wells

Vice President, Career and Professional Editorial: Dave Garza

Director of Learning Solutions: Sandy Clark

Senior Acquisitions Editor: John Fedor

Managing Editor: Larry Main

Senior Product Manager: Michelle Ruelos Cannistraci

Senior Editorial Assistant: Dawn Daugherty

Vice President, Career and Professional Marketing: Jennifer McAvey

Executive Marketing Manager: Deborah S. Yarnell

Marketing Specialist: Mark Pierro

Production Director: Wendy Troeger

Production Manager: Mark Bernard

Content Project Manager: Christopher Chien

Senior Art Director: David Arsenault

Technology Project Manager: Christopher Catalina

Production Technology Analyst: Thomas Stover

For product information and technology assistance, contact us at
Professional & Career Group Customer Support, 1-800-648-7450

For permission to use material from this text or product,
submit all requests online at **cengage.com/permissions**
Further permissions questions can be e-mailed to
permissionrequest@cengage.com

Library of Congress Control Number: 2008930951

ISBN-13: 978-1-4354-0062-7

ISBN-10: 1-4354-0062-3

Delmar
5 Maxwell Drive
Clifton Park, NY 12065-2919
USA

Cengage Learning is a leading provider of customized learning solutions with office locations around the globe, including Singapore, the United Kingdom, Australia, Mexico, Brazil and Japan. Locate your local office at: **international.cengage.com/region**

Cengage Learning products are represented in Canada by Nelson Education, Ltd.

For your lifelong learning solutions, visit **delmar.cengage.com**

Visit our corporate website at **cengage.com**

Notice to the Reader

Publisher does not warrant or guarantee any of the products described herein or perform any independent analysis in connection with any of the product information contained herein. Publisher does not assume, and expressly disclaims, any obligation to obtain and include information other than that provided to it by the manufacturer. The reader is expressly warned to consider and adopt all safety precautions that might be indicated by the activities described herein and to avoid all potential hazards. By following the instructions contained herein, the reader willingly assumes all risks in connection with such instructions. The publisher makes no representations or warranties of any kind, including but not limited to, the warranties of fitness for particular purpose or merchantability, nor are any such representations implied with respect to the material set forth herein, and the publisher takes no responsibility with respect to such material. The publisher shall not be liable for any special, consequential, or exemplary damages resulting, in whole or part, from the readers' use of, or reliance upon, this material.

Printed in Canada
1 2 3 4 5 XX 10 09 08

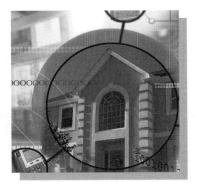

Contents

1

Introduction to DHTI   1

5
High-Voltage Wiring   138

6
Video and Audio Fundamentals   167

7

Audio and Video Installation and Setup 207

8

Security and Access System Fundamentals 249

9

Security System Installation and Setup 281

10

Telecommunications Fundamentals and Installation

312

11

Home Lighting Control

343

12

Heating, Air Conditioning, and Water System Control   376

13

Power Supply and Conditioning   417

14
Troubleshooting Integrated Technology Systems 448

Appendix A
DHTI Acronyms 477

Appendix B
CompTIA DHTI+ Examination Objectives 484

Series Preface

Y ou've heard these catch phrases many times. Home theaters. High-definition television (HDTV). Distributed audio and digital video. Lighting and home control systems. Broadband Internet access and wireless networking. What do all of these terms mean? Simply stated, it is cutting edge technology that is changing how we live—and will live—in our homes.

Why is this technology emerging? For one thing, the need for technical content and information is growing. But perhaps more importantly, people want the convenience and time- and cost-saving benefits of controlling their homes electronically as a way to spend more quality time with their families.

The residential integration industry is enjoying unprecedented growth in the installation of these new technologies. Due to the high demand, residential integration companies are seeking qualified individuals who understand the traditional fields such as new home construction, along with newer disciplines such as structured low-voltage cabling and data systems. Now is the time to jump on the bandwagon!

Recognizing the need for educational resources in this field, Delmar Cengage Learning is excited to present a four-textbook suite, called the *Residential Integration Series,* which addresses this exploding industry. These four texts encompass many of the aspects of the residential integration industry to include the basics of the business and how to get started in it, customer service skills, project management, and finally a text on the information required to prepare for the various certifications currently offered within the industry.

More specifically, these textbooks are:

Residential Integrator's Basics is the foundational text for this industry. Here you will find comprehensive information on computer networks, communications, home automation, cabling, wiring, and other topics. The final chapter addresses the Home of the Future.

Residential Integrator's Customer Relations pinpoints the types of customer service skills a residential integrator needs to be successful in this industry. These include working with both internal and external clients, working in teams (very important in this field), handling difficult client relationships, improving communications skills, training the client on both technology and equipment, and ensuring client satisfaction.

Residential Integrator's Project Management is based on *A Guide to the Project Management Body of Knowledge* (PMBOK), which is a widely accepted work published by the Project Management Institute (www.pmi.org). This text is divided into four sections, each covering a major phase of a residential integration project: The Foundation, Defining the Project, Planning the Project, and Executing, Monitoring/Controlling, and Closing the Project. This text is accompanied by a CD-ROM that contains project management templates specific to the industry that can be adapted to the needs of a particular company.

Residential Integrator's Certification provides in-depth coverage of the information required to prepare for the CompTIA DHTI+ exam. This includes low-voltage cabling, highvoltage wiring, computer networking, audio/video, security systems, home controls, and other industry-related topics. An appendix contains exam objectives for CompTIA's DHTI+ exam.

The textbooks are pedagogically rich with chapter objectives, critical thinking questions, study tips, and other suggested activities, along with chapter summaries and review questions to help you learn and retain the material.

Building Your Perfect Course Solution

It's your course, so why compromise? Now you can create a text that exactly matches your syllabus using **Cengage Learning Custom Solutions** online book-building application, **TextChoice.** TextChoice allows you to easily browse and select content from leading Cengage textbooks and custom collections—even include your own content—to create a text that is tailor-fit to the way you teach. Visit TextChoice at www.textchoice.com and learn how Cengage Learning Custom Solutions can help you teach your course, your way.

Preface

Data technologies and automation systems for homes and home businesses have expanded exponentially over the past few years. As is always the case when multiple independent developments occur in a free market economy, integration of the various hardware and software components to function seamlessly as an automated home network has lagged behind. This book is designed to provide basic knowledge of all the major home technologies and help the reader develop the necessary skills to install and configure these technologies so they function as a unified system to the maximum extent possible.

Some technologies, such as wired and wireless networks, work together easily; others, such as home entertainment systems, security systems, and home lighting and automation systems, require interfaces to connect and function with a home data network. To install and configure multiple technologies in a unified system, the home technology integrator (also known as the residential integrator) must understand not only the operation of each technology, but also the communication methods (wired or wireless) it uses, the protocols that govern its data flow, and the specific hardware and connection devices it employs. Home technology or residential integration is by definition a broad field, and one that is expanding daily. Every reader will need to study constantly in order to keep current with new developments.

Each section dealing with a specific technology provides basic knowledge about how the technology functions and the operation of its components, as well as practical information for connecting these components into a working system and integrating them to the maximum extent possible with the overall home technology network. The book starts with an overview of all the technologies and integration systems that can be installed and integrated in a home system; it then concentrates for several chapters on the design, installation, and configuration of high-speed, hardwired, and wireless local area networks in the home.

Successive chapters add detailed knowledge about other technologies that can be linked to, or controlled by, a home LAN. These include audio and video transmission systems, security systems, telecommunication systems, lighting systems, and numerous automation systems. The pedagogical design of the text provides an interactive learning experience that gives students sound theory and hands-on activities in each technology area to prepare them for the working world. Each chapter also provides case studies that challenge the student to develop solutions to real-life problems they are likely to find in the work environment.

One of the main objectives of this book is to prepare you to pass the CompTIA DHTI+ Certification exam (see http://www.comptia.com). Obtaining your certification will demonstrate to employers that you have the theoretical and practical knowledge needed to work in all areas of home technology and residential integration. Successfully completing the DHTI+ certification is a challenging task, but it qualifies you for wide employment opportunities in this growing area. Whether you work for one of the specialized firms that install and service only one or two home technologies (cable and satellite television hookups, for example, or security and fire alarm systems), or a broad-based systems integrator that supplies total network solutions, the knowledge in these chapters, confirmed by the CompTIA DHTI+ Certification, will assure you multiple career opportunities in an expanding technological field.

The Intended Audience

This book is intended for students and professionals who plan to install and configure home technology or residential integration systems. It is an ideal text for introducing students to all the residential integration systems and home technologies, and provides an excellent base of knowledge for more intensive study of individual specialties. It assumes that students have an elementary knowledge of electricity, some mechanical systems, and computer operation using Microsoft® Windows™, but the function of each integration system and home technology is explained in depth so those with no experience can quickly grasp both the conceptual and practical aspects of its operation.

Chapter Descriptions

The chapters in this book discuss the following topics:

Chapter 1   Introduction to DHTI, identifies the major data transmission and automation technologies that are part of home technology (residential integration) and explains how the integration of these technologies into a unified system can benefit the homeowner.

Chapter 2   Network Basics, identifies the major types of local area networks and explains how they function. This chapter also describes the various components in a computer network and how each item operates as part of an integrated system.

Chapter 3   Home Network Design and Configuration, describes the protocols used in computer networks, the wire types, connectors, and hardware pieces employed to build networks, and the standards that must be followed when wiring them. This chapter also describes how to set up and configure wired and wireless network hardware.

Chapter 4   Installing Network Components and Low-Voltage Wiring, explains how to plan a home network and install low-voltage wiring and other components. It also describes how to connect the components, test them, and configure them into a working network.

Chapter 5   High-Voltage Wiring, presents the safety standards for high-voltage AC wiring, and details how to calculate current loads. This chapter also explains how to install new AC circuits, including wiring, breakers, outlets, fixtures, and safety equipment.

Chapter 6   Video and Audio Fundamentals, includes analog and digital recording, broadcasting of both audio and video signals, and how each of these processes works. In addition, the chapter covers analog and digital standards for audio and video transmission, how these signals reach a home network, and how audio and video programming can be integrated in a home network.

Chapter 7   Audio and Video Installation and Setup, details how audio and video system components function and how they are combined into systems that display the various types of audio and video signals. The chapter also covers the setup and configuration of audio and video systems as part of a home network.

Chapter 8   Security and Access System Fundamentals, outlines the factors that must be considered in home security and access design, as well as the different types of home security systems available. Security and access system components are described in detail, along with the locations where each can be effectively used.

Chapter 9   Security System Installation and Setup, presents more information about installing low-voltage wiring for security and access systems and how to set up components for both wired and wireless systems. It also details how to configure and program security systems and access devices, the accessories that are available, and how these systems must be maintained and serviced.

Chapter 10   Telecommunications Fundamentals and Installation, explains the development of analog and digital telecommunications systems and the services these central systems provide. Local telephone systems and their internal services are described, along with their component parts. The chapter also covers telephone system wiring, connection, and configuration.

Chapter 11   Home Lighting Control, illustrates how home lighting is constructed and discusses the systems used to control lighting automatically. Wired and wireless lighting control components are detailed. The reader also learns how to install, configure, and program master controllers to direct them.

Chapter 12   Heating, Air Conditioning, and Water System Control, describes the design and operation of zoned and non-zoned HVAC systems, the components included in them, and the control systems that automate them. Also covered is how to wire, install, and program automation controls on the major types of HVAC systems. The second part of this chapter describes the components of water management and sprinkler systems and how they function. Sprinkler system automation is explained in detail, including wiring, component installation, accessories, and programming.

Chapter 13 Power Supply and Conditioning, describes electric power disturbances that affect home networks and digital data transmission, how to identify specific power disturbances, and the major power-conditioning components used to block or prevent power disturbances. It also describes how to install and configure power-conditioning equipment, and how to source and reduce electromagnetic interference (EMI) and radio frequency interference (RFI).

Chapter 14 Troubleshooting Integrated Technology Systems, describes how to analyze integrated system problems, read integrated system wiring diagrams and schematics, and how to use testing tools and equipment to trace system faults. It also describes how to identify major faults in integrated systems and correct or repair faults.

Appendix A DHTI Acronyms, contains a list of acronyms commonly used in the industry, along with their meanings.

Appendix B CompTIA DHTI+ Examination Objectives, lists certification objectives for the CompTIA DHTI+ examination with corresponding references to chapters and headings within this book where the related material is presented. Check **http://www.comptia.org** to see if updated certification objectives have been announced.

Features

To aid you in fully understanding networking concepts, this book includes many features designed to enhance your learning experience.

Chapter Objectives.   Each chapter begins with a detailed list of the concepts to be mastered within that chapter. This list provides you with both a quick reference to the chapter's contents and a useful study aid.

Illustrations and Tables.   Illustrations are provided to step you through various networking commands and installation procedures. Extra graphics are included to explain complex technology concepts and the use of specialized tools. Tables are provided to detail useful information about various technologies.

Chapter Summaries.   Each chapter's text is followed by a summary of the concepts introduced in that chapter. These summaries provide a helpful way to recap and revisit the ideas covered in each chapter.

Key Terms.   All of the terms within each chapter that were introduced with boldfaced text are gathered together in the Key Terms list at the end of the chapter. This provides you with a method of checking your understanding of all the terms introduced.

Review Questions.   End-of-chapter assessment begins with a set of review questions that reinforce the ideas introduced in each chapter. These questions ensure that you have mastered the concepts.

Hands-On Projects.   Hands-on projects are provided at the end of each chapter to give readers practical experience in the installation, cabling, and configuration of actual equipment used in each of the HTI technologies and residential integration systems.

Case Projects. Located at the end of each chapter are case projects. To complete these exercises, you must draw on real-world common sense as well as your knowledge of all the technical topics covered to that point in the book. Your goal for each project is to come up with answers to problems similar to those you will face as a working home technology integration technician.

Supplements

The following supplemental materials are available with the text.

Faculty Guide with CD-ROM. The printed faculty guide offers a complete teaching package. Components for each chapter include chapter objectives, key terms, classroom discussion questions, learning activities, teaching tips, chapter summaries, answers to the end-of-chapter review questions, a resource correlation grid, and an outline of the PowerPoint presentation.

CD-ROM. Each faculty guide has an accompanying CD-ROM that includes a PowerPoint™ presentation for each chapter, a computerized test bank, sample syllabi, and suggestions for additional resources.

CompTIA Authorized Quality Curriculum

The logo of the CompTIA Authorized Curriculum Program and the status of this or other training material as "Authorized" under the CompTIA Authorized Curriculum Program signifies that, in CompTIA's opinion, such training material covers the content of the CompTIA's related certification exam. CompTIA has not reviewed or approved the accuracy of the contents of this training material and specifically disclaims any warranties of merchantability or fitness for a particular purpose. CompTIA makes no guarantee concerning the success of persons using any such "Authorized" or other training material in order to prepare for any CompTIA certification exam.

The contents of this training material were created for the CompTIA HTI+ Exams covering CompTIA certification exam objectives that were current as of May 2008.

How to Become CompTIA Certified

This training material can help you prepare for and pass a related CompTIA certification exam or exams. In order to achieve CompTIA certification, you must register for and pass a CompTIA certification exam or exams.

In order to become CompTIA certified, you must:

1. Select a certification exam provider. For more information please visit http://www.comptia.org/.
   a. Click the "CompTIA Certifications" link at the right side of the home page.
   b. Click the "CEA-CompTIA DHTI+" link on the left side of the certification page.
   c. Click the "Find your test center" link in the center of the DHTI+ page.
2. Register for, and schedule a time to take, the CompTIA certification exam(s) at a convenient location.

3. Read and sign the Candidate Agreement, which will be presented at the time of the exam(s). To find the Candidate Agreement:

   a. Click the "exam policies" link on the left side of the DHTI+ page of the CompTIA Web site (see Step 1b).

   b. Click the "Candidate Agreement" link on the exam policies page to see the full agreement text.

4. Take and pass the CompTIA certification exam(s).

For more information about CompTIA's certifications, such as their industry acceptance, benefits, or program news, please visit the http://www.comptia.org home page's "CompTIA News and Special Information" links and the DHTI+ page's news links.

CompTIA is a nonprofit information technology (IT) trade association. CompTIA's certifications are designed by subject matter experts from across the IT industry. Each CompTIA certification is vendor-neutral, covers multiple technologies, and requires demonstration of skills and knowledge widely sought after by the IT industry.

To contact CompTIA with any questions or comments: Please call: (630) 678-8300 and press 5. Or e-mail: *questions@comptia.org*

Read This Before You Begin

The Hands-On Projects in this book help you apply what you have learned about Digital Home Technology Integration. The following section lists the minimum hardware requirements that allow you to complete all the Hands-On Projects in this book. Because DHTI covers many technologies, each of which requires hands-on experience as well as theoretical knowledge, the list is extensive.

In order to provide detailed and precise instructions in the Hands-On Projects, specific brands and models of hardware have been chosen for each project. Although the instructions are tailored for the specified equipment, you may use other similar equipment if it is more conveniently available. Remember, however, that if you choose to do this some of the instructions may need to be modified.

Note that programming instructions for automated devices, in particular, are usually very specific and must be followed precisely in order to obtain the desired results. If you choose to use other equipment, however, it will always have instructions provided by the manufacturer for programming and configuring it. These instructions should be used, as appropriate, in place of those provided in the Hands-On Projects. If the instructions for a piece of hardware are lost, you can usually obtain a printout from the manufacturer's Web site, or you can order an additional copy there.

Minimum Lab Requirements

Hardware: Many devices listed here are used in multiple hands-on activities. Numbers in parentheses indicate the number of activities for which each piece of equipment is used. If no number is shown, the equipment is used in only one activity. Those devices that are used in multiple activities are described with all the features they will require to complete every activity, even though all features will be not be used in every activity.

- (15) Two computers, each with Windows XP operating system, network interface card, CD burner/player drive, sound card, microphone, Internet access, and a Web browser
- Single color inkjet or laser printer and spare ink or toner cartridge for the printer
- Crossover patch cable to connect both computers in a network
- Laptop computer with wireless interface for public access wireless networks
- (3) Simulated stud wall section at least 8 feet wide and 6 feet high, or real wall with open studs
- (3) Three stud-mount outlet boxes
- (3) Simulated finished wall section at least 8 feet wide and 6 feet high, or a real wall with finished surfaces
- (2) Surface-mounted outlet box
- Cable raceway parts sufficient to run a finished raceway for a distance of 8 feet with at least one elbow or T junction extending one foot from the main raceway
- (2) Roll of Romex 2 conductor #14 cable
- Roll of Romex 3 conductor #14 cable
- (3) Standard AC duplex outlet
- Single-pole AC switch
- (4) Twenty twist-on electrical connectors for AC wiring
- Two double-pole, double-throw AC switches
- (7) Two plug-in lamps
- (5) Digital television with built-in digital tuner and decoder, antenna reception and input and output connections
- (3) Digital Video Disc (DVD) recorder/player
- (2) VHS Cassette recorder/player
- (4) Ten RCA-type video connector cables
- Blank DVDs
- (2) NTSC standard analog television set
- (4) Stereo audio system with amplifier, monaural select button, FM radio receiver, tape cassette recorder, and CD player
- (2) DVD player
- Commercially-recorded music cassette tape
- (4) Commercially-recorded music CDs
- Commercially-recorded DVD
- X10 wireless heat sensor and transmitter
- X10 security control panel
- (4) X10 mini-controller
- X10 Wireless Security Console (Model PS561 or similar)

- (2) X10 Wireless Camera System (model VR36A or equal) with power supply
- (8) X10 MS14A or MS13A Motion Sensor and transmitter
- X10 RF transceiver (Model RR501 or TM 751)
- (3) X10 wireless remote control module (Model 4000 remote and base or similar)
- (6) Two X10 plug-in lamp modules (Model 2000 or similar)
- X10 mini-timer controller module (Model 1100X or similar)
- PBX (DataLabUSA Model 308 or similar)
- Two telephone extension sets
- Cordless telephone set
- Telephone patch cable with RJ-11 or RJ-14 plugs on both ends
- One working telephone line accessible through an RJ-11 jack
- Fax machine
- Splitter
- (4) Roll of Category 5 cable
- (3) Roll of Category 3 two-pair telephone wire
- (4) Roll of single strand wire
- Twenty twist-on connectors for low-voltage wire
- Two RJ-14 plugs
- Two RJ-14 jacks
- Three RJ-11 or RJ-14 patch cables 8 feet in length
- Two RJ-45 jacks
- Four RJ-45 plugs
- Two surface-mounted RJ-14 hand-wired jacks
- Wall area or board on which to mount surface telephone jacks
- (2) Real or simulated three-way switched electric circuit in a classroom mockup
- Leviton 2202 WI DHC X10 500-watt Dimmer Switch with cover plate
- Leviton 22081 Slave Switch with switch cover plate
- Leviton DHC Split Receptacle with receptacle cover plate
- (5) Room with three or more live AC outlets
- (2) Thermostat (manual)
- 24-volt motorized damper
- (2) 24-volt AC transformer
- Drinking glass (glass, not plastic)
- Small plastic plate
- Cardboard box large enough to hold plate
- Ice cubes
- Automatic thermostat (Dayton Fuel Trimmer Model T-110 or similar)
- 9-volt electric light in a socket

- Automatic sprinkler controller (Rainbird Model ESP 6SI or similar)
- Four 24-volt electric lights in individual sockets that can be wired
- Multimeter (dial-face type, not digital)
- Small light socket suitable for wiring into a 110-volt circuit
- Light bulb for socket
- 110-volt two-wire extension cord (6 foot length)
- Four AA batteries
- Four AAA batteries

Tools: The only tools listed are specialized items. It is assumed that labs will have standard tools available.

- (2) RJ-45 punchdown tool
- (3) CatX combination cable stripper/crimping tool
- (4) Power drill and 1/2-inch speed bit for wood
- (4) Hair dryers

Introduction to DHTI

After studying this chapter, you should be able to: **OBJECTIVES**

- Understand how **Digital Home Technology Integration (DHTI)** is used to integrate, automate, and control an extensive variety of technology applications in today's consumer market.

- Describe the major technology areas that make up a complete DHTI system.

- List some of the important benefits that DHTI systems provide to users.

- Identify the main components from which all DHTI **subsystems** are constructed, plus the function of each.

- Define some of the knowledge and hands-on competencies needed by professional technicians who design, install, and maintain DHTI systems.

- Discuss how much DHTI is enough for a given situation.

Introduction

In this chapter, you will learn the meaning of Digital Home Technology Integration (DHTI) and what it can do. You will also learn the technology areas that form subsystems in DHTI and the basic categories of components that make up each of these subsystems. You will understand how the **network** enables

1

each part of the DHTI system to distribute and share data with the others so that all can provide better service to the user. You will also understand how the benefits of DHTI are making it increasingly popular and how the need for professionally qualified technicians who can design, install, and maintain DHTI systems is growing.

What Is Digital Home Technology Integration?

Digital Home Technology Integration (DHTI) is the concept of a connected home technology environment in which a central computer system, programmed or otherwise directed by the homeowner, manages and distributes incoming and outgoing Internet and audiovisual **digital data,** automates lighting, **appliances,** and utilities, and controls the local area network (LAN), security system, and other technologies of the home. The result is what is commonly referred to as a "smart-home" in which data and control systems are automated and integrated with one another to the maximum extent possible.

DHTI management and control can be minimal or extensive, depending on the preference of the individual and the amount of investment in equipment and **infrastructure.** A complete DHTI system requires a variety of subsystems, which are linked together and centrally controlled, but some degree of DHTI can be accomplished through individual control modules linked to various devices and appliances in the home. Figure 1-1 shows a diagram of a large DHTI system with arrows indicating the bidirectional flow of data to and from the various subsystems. Note that data flowing in from outside sources must pass through the gateway that protects the LAN from outside attacks. Likewise, information going out from any component of the DHTI system must also pass through the

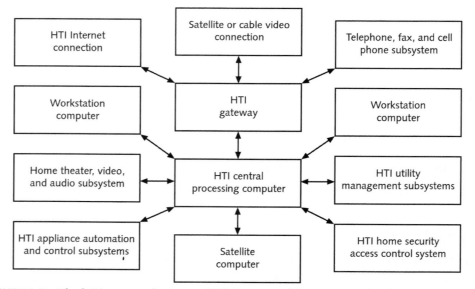

FIGURE 1-1 Block Diagram of a Large DHTI System with Arrows Indicating the Flow of Digital Data within the System, and with External Sources

gateway. This protects the home system from the inadvertent release of unauthorized information and from data theft by an outside intruder.

DHTI systems are generally used to provide, link, manage, and control six broad areas of the home technology environment:

- Internet and network (within the home) connectivity
- **Video** and audio signal reception and **distribution**
- **Telecommunications**
- Home security and **access control**
- Utility management (electricity, heating, air conditioning, and water)
- Appliance and lighting **automation** and control

Consumer demand for DHTI systems that incorporate some or all of these features is increasing. To provide the needed services, homes require a complete, automated home control network that integrates the computer-based systems with entertainment, heating and cooling control, lighting control, water management, home security, and other systems.

The complexity of networking a variety of electronic products into one home control network calls for trained home networking and integration professionals who are skilled in the design, installation, and troubleshooting of DHTI components, infrastructure, and subsystems. Consumers who employ DHTI professionals want to be assured of their technical qualifications in the field. To help provide this assurance of competence, the industry has developed a **credential** called DHTI Certified Professional.

The DHTI Certified Professional credential is designed for technicians who install and service residential LANs and other digital technology systems, including security, audio, and video reception and distribution, computer workstations, heating and air conditioning, automated lighting controls, cable and **satellite links,** and telecommunications systems. To earn a DHTI **certification,** professionals must pass a computer-based exam: DHTI+ Residential Systems Integration.

The DHTI certification program was developed jointly by **CompTIA** (http://www.comptia.org) and the **Internet Home Alliance (IHA)** (http://www.internethomealliance.com) in response to the growing need for qualified technicians to install and maintain networked home computing, entertainment, environment, and security systems, including the integration and distribution of the following:

- Communications network
- Entertainment network
- Security network
- Home control network
- Services and subscriptions
- Commercial wiring and cabling
- Ethernet, token ring, or other LAN topology

This book is designed to prepare technicians to pass the DHTI+ exam and become credentialed as a DHTI Certified Professional.

What Can a Home Technology Integration System Do?

This section takes a closer look at the particular data and services a DHTI system can provide for a home and its occupants. No single DHTI system would likely include all the features described here. Although each system's parameters reflect the individual needs and wants of its owner in the data and services it provides, all the functions noted here are presently available and new ones are being developed continuously.

Internet Connectivity

An Internet connection gives the home access to the **World Wide Web (WWW)** and to literally millions of sites on every imaginable, and some unimaginable, subjects. The first home connections to the Internet were provided by means of a **modem** that connected a home computer to a telephone line. The telephone line connected the home user's computer to a **service provider** who, in turn, connected it to the Web. This technology, a standard telephone line connected through a modem, is still widely used and serves adequately for many home users of the Internet, but it has limited data-carrying capacity. For this reason, most users now prefer a higher volume data link that can be set up in one of three ways: lease a larger capacity telephone line called a **digital subscriber line (DSL),** install a cable modem to provide Internet service from a cable company, or lease a satellite link. These types of connections each cost more than the regular telephone link, but they provide vastly greater data **transmission** capacity **(bandwidth).**

The Internet connection from the home computer to the service provider is an important part of the DHTI system's configuration because the type and capacity of that connection influences how the system distributes and uses Internet data. Whether the home LAN has Internet connectivity through a gateway hardware device or through gateway software installed directly on the DHTI system's central processing computer, either type of network security can perform the following specific tasks on the incoming and outgoing data:

- **Scan** incoming data to detect **viruses** and other forms of external data attacks, which can then be blocked from entering or affecting the system.
- Detect illegitimate attempts by outsiders to "hack" into the user's computer system and prevent them from accessing any information on the system behind its gateway firewall.
- Distribute and control Internet access to other computers within the home using the LAN setup.
- Block or **password**-protect access to some Internet sites, either on all computers on the network or on specific units designated by the system **administrator.**
- Manage all of the services available through Internet access, including e-mail, chat rooms, home or business Web sites, and others.
- Prevent unauthorized information (personal data, financial records, identity documents, medical records, etc.) from passing outside the DHTI system, whether sent by someone in the home, or requested by an outside source.

Within the home, the central processing computer can also manage network services for other computers and **peripherals** connected to the system. A home

LAN permits all connected computers to swap data directly, share files, use common printers, play games, and so on. Maximum use of computer resources is achieved along with maximum speed of communication within the network environment.

Audio and Video Reception and Distribution

Incoming digital video data usually comes to the home via a cable hookup or through satellite transmission, although broadcast television channels now also provide digital programming alongside their long-standing analog signal transmissions. Cable television requires a direct wire connection to the transmission facility. Most, but not all, areas of the country now have access to cable connection and those (mostly rural) areas that do not have cable are declining in number as the cable providers' wiring networks expand.

Satellite television access is available nationwide because the signal transmissions originate from several satellites positioned in geosynchronous (stationary in relation to the ground) orbits in space above North America. The satellites **downlink** (transmit) a coded audio/video signal to a small dish receiver mounted on the home and aimed directly at one or more satellites.

Both cable and satellite digital transmissions must pass through a **decoder** or **translator** unit before they can be displayed on screen. The decoder is set by the service provider to decode only those channels to which the user has subscribed.

Broadcast digital video and audio signals also require a decoder to render the digital data into a form that the television can display or the stereo system can play. Analog television and radio broadcasts do not require a decoder, but all analog television broadcasting in the U.S. will cease after February 17, 2009, leaving only digital television signals available, and analog radio signals may eventually follow the same course.

The type of incoming television signal selected by the user influences the configuration of the DHTI system that uses the signal. Cable hookups can be split and the incoming signal shared among several television displays or digital recorders. Each television can be set to a different channel and a recorder can record one channel while others are being viewed onscreen.

Satellite hookups require a separate decoder for each television display connected to the system. A digital recording device can be connected in tandem with each television so it can record a separate channel from the one being viewed. A maximum of four decoders can be connected to a single satellite dish, but more than one dish can be installed in a home video system if necessary.

The DHTI system that receives a digital television and/or audio signal can do any or all of the following with it:

- Distribute the signal to the various television and other receivers in the home.
- **Monitor** and control access to selected channels or programs on a channel for all television receivers or for selected receivers.
- Direct selected programs to recording devices for storage and later use.

Telecommunications

Telephones are the most universally installed technology in America, and a DHTI system can extend the utility and expand the functions this established service performs. The voice telephone line, which is standard in almost every home, can be made available in every room, either through hardwired extensions or through

wireless connections. The latter also allows telephone service to extend into the yard so a convenient receiver can be carried while gardening or be kept near the pool.

Standard telephone lines also provide the Internet access point in many homes. A separate line is preferable for this purpose, although one line can provide both data and voice transmission, but not simultaneously. When a line is used for data transmission, voice calls into or out of the home are blocked. The same is true of data transmission when a voice call is in progress.

A separate data line managed by the DHTI system provides much better Internet service and can also be used as a fax line with messages being routed to fax software on the home computer or to a separate stand-alone fax machine. DHTI systems can also use a data line as an emergency alarm line connected to the **home security system.** The system can be programmed to take control of the line in the event of an emergency and summon immediate assistance.

Voice and data transmission service can be expanded still further by the installation of a digital subscriber line (DSL), a high-speed data transmission telephone line that also has the ability to transmit voice and data simultaneously. A DSL has a transmission capacity of one megabyte of data per second or more, compared with a maximum of about 50 kilobytes capacity for a standard telephone line and modem.

Cell phones, which are actually miniature radio transmitters and receivers, are now found in most homes in addition to hardwired "land line" telephone service. Cell phones have now become integrated technology systems in themselves. In addition to wireless local and long distance telephone service, they provide instant messages, send and receive e-mail, record and display digital video images, connect wirelessly to the Internet, and supply extensive computer functions, as well as act as personal digital assistants (PDAs).

The home security system can be programmed to use a cell phone, which is not subject to localized power failures or cut wires, for outgoing emergency calls. This is just one of several ways that cell phone technology can be integrated into the DHTI system to provide better communication service to the home. Cell phones can also download and upload digital information from the home DHTI system and can remotely query the system for recorded messages, security status, and to activate automated systems.

Home Security and Access Control

A DHTI system can provide a wide variety of security and safety measures for the home. It allows access to the home and its surroundings and lets other security measures be centrally controlled and monitored for maximum protection of the occupants and their property. Security can be controlled from central locations within the home and also from remote locations by using a cell phone or a computer connected via the Internet to the DHTI central processing computer. Remote computer control can be accomplished from wired hardware, such as an office workstation, or using a wireless battery-operated laptop.

Security and monitoring features that can be included in a DHTI system include all of the following:

- Sensors on windows and other vulnerable entry points that trigger an alarm if intrusion occurs. These can be taped circuits on the glass, laser beams, or movement/vibration **detectors.**

- Heat and/or motion sensors that monitor the immediate area around the home or specific traffic corridors within it and trigger an alert if anyone enters the space.
- Video surveillance **cameras** that monitor both exterior and interior areas of the home for intruders.
- **Cipher locks** on exterior doors. These require **keypad** entry of a **numeric code** to open, and some types can be time-coded to block entry during specified hours.
- Sensors that respond to the presence or absence of specific conditions such as electric power to critical appliances (freezers, sump pumps, etc.), water in basement areas, or surface motion in a swimming pool.
- Strategically placed pressure-sensitive pads that respond either to pressure being applied (weight) or pressure being relieved (object picked up).

Any of these sensing and monitoring devices may trigger a variety of responses, including the following:

- Summon the police or other assistance, either by telephone or by radio.
- Activate an audible alarm.
- Activate additional lighting.
- Activate additional security procedures such as lockdown devices, circuit breakers, dropdown security doors, and others.
- Activate additional emergency equipment such as sprinkler systems, emergency generators, and transponders (battery-operated emergency radio transmitters).

Home Automation and Control

DHTI automation and control systems offer tremendous convenience and labor-saving comfort for most owners and truly essential technology assistance for certain groups, such as those with limited physical mobility and those with sensory impairment. For these groups, home automation can often be the key to living independently.

Home automation is accomplished by means of remote-controlled systems activated at will by the homeowner from a convenient location, or by systems that can be programmed on a preset schedule to self-activate at specified intervals or times. Many automation features are available, including the following:

- Centralized control of lighting throughout the house and outdoor lighting fixtures as well
- Control of window shades and curtains.
- Control of heating, air conditioning, and ventilation systems.
- Control of electric appliances, such as coffee makers, radios, alarm clocks, dishwashers, and others. Almost any appliance operating on standard house current can be controlled remotely.
- Control of safety systems such as swimming pool covers, driveway and sidewalk heating systems, and roof de-icing systems.
- Management of personal safety devices such as baby monitors, fall alarms, and vital sign monitors.

For persons with limited mobility, having as many home systems as possible automated and controlled from a central location is essential. If physical movement is severely restricted, control systems can often be adapted to function easily using whatever motion is available to the user or by using voice commands. Security and safety systems can be programmed to provide audible alarms to the visually impaired, as well as visual or vibration alarms to the hearing impaired.

Utility and Resource Management

A DHTI system can help minimize energy waste, increase efficiency, and reduce costs by continuously monitoring and managing utility usage. The system can turn off lights and other electrical devices when not in use. Lights left on in a room can be turned off automatically when a **sensor** detects that the room is unoccupied for a preset period of time. When anyone enters the room, the lights come on again, controlled by the same sensor.

Sensors, timers, and programmed controllers can manage heating and air-conditioning systems. This provides maximum energy efficiency while maintaining comfortable temperatures for the home's occupants. Heat can be turned down at night and when the home is unoccupied during the day, then turned back up just before the occupants awaken or return home. Air conditioning can be similarly adjusted to fit the schedules of those in the home.

Efficient home use of water includes the controlled watering of landscaped areas, which usually requires far more municipal water than any other consumer use. By adjusting sprinkler systems to provide only the needed amount of water for each part of the landscaping, and programming them to operate after sundown and for only the necessary time, a DHTI-managed system can use less water and be far more effective in maintaining the landscaping.

A DHTI system can also assist in managing a swimming pool. An automated pool cover that covers the pool when it is not in use prevents evaporation, helps hold heat in the pool, and acts as a safety barrier keeping people from falling into the water. Pool filtration systems can also be timed to clean and backflush (part of the cleaning process in most pool filtration systems) at night, and in some locations, the waste water used for these operations can be directed to the landscaping.

The Benefits of DHTI

The main benefits of DHTI are simple: They makes the user's life more convenient, safe, and fun. DHTI provides maximum use and enjoyment of electronic media, Internet, automation, and telecommunication systems. It makes technology available where and when the user wants it. Automated systems that function on demand or by programmed direction to meet predetermined needs in a timely manner often become essential services rather than mere conveniences. Similarly, DHTI's ability to make communication and data sharing easy to use throughout a home lets users make maximum use of technology to save time while accomplishing more. When that increased productivity is utilized in a home business setting, the payoff occurs not only in time efficiency, but also in increased profitability. When technology extends the senses and physical abilities of the physically challenged, while at the same time monitoring their safety and health, it becomes a critical, life-enhancing, if not life-saving, necessity.

Technology on Demand

Hardly anyone who has watched movies on a **home theater** system featuring a 60-inch screen with high-definition television viewing and surround sound audio reverberating from high-quality speakers will be content to go back to a small screen with standard resolution and a single 6-inch monaural audio speaker. Similarly, no one who has come home on a snowy evening to find the driveway and sidewalks clear of ice, the garage door opening as the car's front bumper approaches it, and the house warm and brightly lit, is anxious to return to the "good old-fashioned days" of snow shovels and manual lift doors. In other words, DHTI makes life a lot more pleasant and a lot less work.

Technology to Save Time

Users also save time with DHTI. Its automated features eliminate the need for many household tasks to be done manually and make services available on demand throughout the home with no need to move or alter equipment setups. With DHTI, each room in the house can be connected so that telephones, Internet, television, radio, and network services are all available wherever and whenever they are needed or wanted. Whole-house connectivity also means no waiting for multiple users, each of whom can log on simultaneously. The system apportions bandwidth and delivers data efficiently so everyone can perform their desired tasks using the shared resources.

DHTI saves time by allowing the user to control much of the timing of services to suit his or her own preferences. The system can be programmed to record audio and video programs for later listening or viewing (and eliminate commercial interruptions from them), warm the hot tub to perfect temperature just in time for planned use, or start the coffeemaker a few minutes before the alarm clock sounds each morning.

DHTI also saves the user time by automating the performance of routine tasks and functions using a set of programmed instructions. With DHTI, custom programming is available that permits a very sophisticated level of control in many automated systems simultaneously. There's no need for the user to think about or perform any of these daily tasks. Many maintenance functions, such as sprinkler activation and computer data backup, can be performed at night or when the home is unoccupied during the day. This timing makes more efficient use of resources and eliminates such inconveniences as computer downtime or children trying to play on wet grass.

Technology to Save Money

By taking over the management of many home functions, DHTI can make them more efficient and therefore more cost-effective. The automated system will not leave the water running too long or forget to turn off the lights when rooms are empty. It always turns down the heat or the air conditioning at night and activates the security system to make sure the occupants' rest is undisturbed. Standardizing household functions for time of performance and duration minimizes the amounts of energy and resources these operations use, which saves the homeowner money.

A sprinkler system, for example, can be programmed to water the yard from midnight until 3:00 am on Mondays, Wednesdays, and Fridays from April 15 to October 15, and also add a Saturday or Sunday watering whenever the preceding

5 days include two or more with daily high temperatures above 90 degrees. Operating on those instructions, the DHTI system can keep the yard green throughout the spring and summer. The homeowner may never see any evidence of the process and probably will not need to alter or adjust it unless there's a breakdown in the equipment.

DHTI security and safety systems can provide huge cost savings in the event of an emergency. The most dramatic of these may be a burglar frightened off by a blaring alarm before he or she can commit a robbery, but other warnings may actually save more dollars. Consider the loss from water damage caused by a burst washer hose or a sump pump with no power. A nearby water sensor will alert the owner in time to prevent harm and the DHTI system can even be programmed to take corrective action (shut off the water main or activate a generator) automatically. Similarly, the cost savings of quick fire detection and suppression can hardly be overstated.

Basic Components of DHTI Systems

Although DHTI can be used for an almost endless variety of data functions and to control devices and operations, all automated home technology systems consist of a combination of six basic components:

- Processors
- Communication links
- Sensors
- **Control devices**
- Display and monitoring devices
- Recording and storage devices

A desktop or laptop computer contains all six of these components and is, therefore, an example of an integrated technology system. Figure 1-2 shows a block diagram of a computer system with several input and output devices sending information to, and receiving information from, the central processing unit, which is the "brain" of the system and performs all the manipulation of its data.

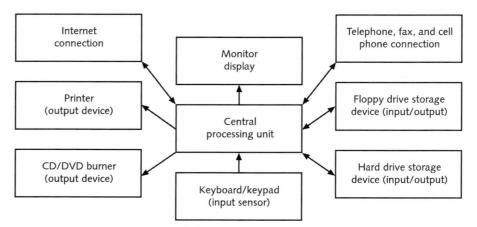

FIGURE 1-2   A Computer System Block Component Diagram

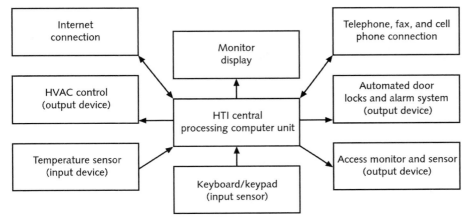

FIGURE 1-3   Block Diagram of a Small DHTI System

DHTI simply expands the scope and the scale of these components so their functions can be extended over a wider geographic area—the whole interior and exterior of a home and its surrounding yard—and to a much larger array of output and storage devices than just the computer's display screen and its hard drive. Figure 1-3 shows a small DHTI system that extends the input and output reach of a computer by providing it with input from the Internet and other segments of the home LAN and allowing the central computer to remotely control automated systems in distant segments of the LAN.

Processors

Processors are electronic devices that receive digital information (input), process this information according to a set of stored instructions **(program),** and send the results of their processing (output) to another device that can either display it, store it for future use, or perform a preprogrammed action using the output information or by executing its instructions.

A computer's central processing unit programmed with word-processing software constitutes a processor. It receives input data from the **keyboard,** processes the input according to its programmed instructions into a finished document, and sends the results to a monitor screen and then to a printer, which displays the data as a printed page.

In a DHTI system, a processor may receive many different types of input from a variety of input devices, such as cameras, sensors, **microphones,** telephone links, and others. The input data may be processed in a variety of ways, according to the manner in which the processor has been programmed, and the results of that processing may be output to various storage, display, and action devices in equally diverse ways.

Input to a processor, for example, could come from a digital thermometer, which is one type of temperature sensor. If the sensor sends data to the processor that the outside air temperature has fallen near the freezing point of water, the processor might compare the temperature information it receives with a preset schedule of instructions and, as a result, output instructions to subsystems throughout the home. It might send instructions to control devices to perform any or all of the following actions:

- Shut down the sprinkler system
- Cover the swimming pool

- Turn on the driveway heating system
- Turn on the rain gutter de-icing system
- Close vents and windows in the home
- Turn on the space heater in the doghouse

In DHTI, a single central processing unit is usually sufficient to operate and control the entire system, although some subsystems may have smaller processors to perform localized functions. Centralized processing permits the greatest degree of integration among all subsystems and helps ensure they are not acting in conflict with one another. It would be counterproductive, for instance, to have one automated function turn on the air conditioning system, while another opens windows and vents to the outside at the same time. If one processor controls all automated functions, its programming can be integrated for maximum beneficial effect.

Communication Links

Communication links are the infrastructure of an integrated technology system and work to tie that system together as a functioning unit. These links are the wires, cables, connectors, wireless transmitters, receivers, routers, **switches,** and other devices that connect the components of a system and permit them to transmit and receive data both among themselves and to the outside world.

These bits and pieces of hardware and lengths of wire are usually quite simple and mundane objects, but they are critically important parts of every integrated technology system. Correctly designing, installing, and maintaining communication links can do more to assure the reliability of an integrated system than any other facet of the system's creation.

In a computer, all of whose components are housed in a single case and closely wired together, communication links are made up of the integrated circuit boards, buses, sockets, and cables that connect the parts electronically so data can be transferred among them. DHTI systems use similar links, but because their components are usually farther apart, the communication links are longer and need to be more robust. Because they are usually built-in as part of the home's structure and are difficult and expensive to repair or replace, the cables, connectors, plugs, and other infrastructure of a DHTI system must be durable over time and able to resist the effects of weathering, outside interference, and other factors that can degrade their performance.

DHTI communication links can be hardwired using a variety of cable types and sizes and suitable end-point connecting devices. Hardwiring is performed most conveniently and at the least cost, if factored into the construction process of a home, but it can be done after the home is completed using a process called retrofitting. Retrofitting a home for DHTI may cost more than new construction wiring and may be less pleasing in appearance because of the difficulty of entirely concealing wiring and connectors after the finished surfaces of the home are complete.

Wireless connection of DHTI is often preferred for retrofitting existing homes and even for some new construction. Several existing wireless technologies can provide connectivity for a DHTI installation. Most require the installation of one or more radio transceivers within the home, which communicate with all the components of a DHTI system via high-frequency radio transmissions. No wiring is required for this type of installation except a power connection for the transceivers (hubs) and hardwired connections for Internet and television data coming into the home.

At least two other technologies, the **X10** system and the Home Plug system, use the existing electrical wiring in the home to carry digital data in addition to the standard electric current flowing through these circuits to power lights and appliances. Still another technology, called HomePNA, also requires no new wires to be installed in the home because it uses the existing telephone wires to carry data to devices on the home network.

Sensors

Sensors are devices that convert human activity or environmental conditions into data. A keyboard that reacts to finger pressure on each of its keys is a type of sensor. It senses pressure applied to each key and converts it into corresponding digital data, which goes to the computer's central processing unit. A microphone is another type of sensor that reacts to the air pressure of sound waves and converts voice commands, dictation, or music into analog electrical signals that can then be converted into digital data for the computer.

Sensors in a DHTI system can be the same keyboard, microphone, or mouse found in any computer system. They can also be other much more diverse devices, and can be separated by much greater distances from the **central processor** to which they send their data.

In a DHTI system, keypads can be located near the home's entrances where they serve as input devices for the codes to open cipher locks on doors. Other keypads can be located throughout the home and may control entertainment systems, lighting, curtains and shades, garage doors, and other functions.

Environmental sensors are devices that react to some aspect of conditions around them. A temperature sensor, for example, continually sends the current temperature to the processor, which then determines whether to adjust heating, air conditioning, or other systems based on the data it receives.

A light sensor measures the light level in a room and sends the data to the processor, which can turn the lights in the room off, if another sensor indicates the room is unoccupied. Alternatively, the processor may open window shades and curtains to increase a low light level during daylight hours. A light sensor placed inside a normally dark cabinet or case can even be used to shut down a system for safety reasons or trigger a security alarm if the door to the cabinet is opened, letting in light. Figure 1-4 shows a diagram of a lighting control segment of a home LAN, including sensors to detect interior and exterior light levels and control modules to adjust shades, curtains, and lighting both inside and outside the home.

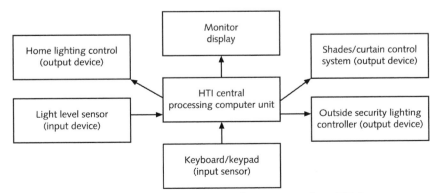

FIGURE 1-4   Block Diagram of a Lighting Control Segment of a DHTI System

Arrows show the direction of data flow to and from the various devices depicted in the diagram in Figure 1-4. From the keyboard, the LAN administrator can program the system parameters and set any of its features by time, as well as by sensor input.

Other types of sensors include the following:

- Heat (infrared) sensors
- Motion (vibration) detectors
- Pressure or weight pads
- Noise sensors (voice activators)
- Water level and/or pressure sensors
- Electric flow sensors (power on, power off)
- Temperature sensors
- Chemical sensors such as carbon monoxide monitors

All of these devices send data to the processor about the particular environmental condition they monitor. The processor compares this information to its programmed instruction set and directs other components of the DHTI system to respond accordingly.

Control Devices

Sensors collect information about conditions around the house and send it to the processor. The processor evaluates that information and issues commands to the appropriate control devices. These devices are the action components in a DHTI system. They act on the processor's instructions to turn off, turn on, or adjust the levels of subsystems in the home.

Control devices come in many forms:

- Switches, which turn electrical appliances and lights on or off
- Electromagnetic valve controls in water systems and gas lines
- Magnetic lock controls
- Electrical **relays,** which act as switches within circuits and route data to appropriate points in the DHTI system
- Electromagnetic mechanisms, called servos, that perform linear movements such as sliding a lock bolt open or closed or adjusting the set of a furnace damper
- Electric motors with remotely controlled switches that perform such functions as pulling curtain and shade lines, drawing a pool cover open or closed, or raising and lowering a garage door
- Hydraulic actuators that perform heavy duty functions such as opening and closing large driveway gates, or lifting heavy loads in an elevator or dumb waiter

All control devices in a DHTI system have the ability to receive digital instructions from a processor and convert those instructions into some form of action or movement. Many control devices can also be operated manually so the instructions they receive from the processor can be overridden, and the subsystems they control can be operated or shut down on site in the event of an emergency or a system failure.

Display and Monitoring Devices

Display and monitoring devices are usually computer screens or television screens that display video and audio signals brought into the home via the Internet, by cable, or from satellites. They also display software programs running on home computers, recorded programs from video recorders and DVDs, and data generated by the processor about the status of other systems throughout the home.

Speakers and earphones are audio devices that "display" sounds to our ears. They play the music of recorded CDs, allow users to speak with one another in separated parts of the home, and output the conversations of audio chat rooms. The processor can also use speakers to sound warnings or alarms if the DHTI system detects danger or other problems within its programmed parameters.

Other types of monitoring devices include continuity lights, which monitor circuits, and space monitors, which can use motion, heat, or pressure sensors to monitor a given area.

Recording and Storage Devices

Recording and storage devices save digital data so it can be retrieved and used again. Computers typically store their data on magnetic devices such as hard drives, floppy disks, zip disks, tapes, and memory sticks (flash drives).

CDs (compact discs) are storage devices that store data in a form readable by a beam of light reflecting off the disc as it spins in a player. CDs can store digital text data, still pictures, movies, and audio files (voice and music). They have largely replaced the earlier magnetic tape method of storing audio data.

Television programs and digital movies are usually stored on DVDs (Digital Versatile Discs) by recorders, which can also play these devices to recover and display the data on them. DVD technology, which uses much higher-capacity discs than CD technology, has mostly supplanted the former method of storing video data on magnetic tape in video cassettes, although tape cassettes are still used for some applications.

Video entertainment programs are nearly always stored on DVDs, large capacity hard drives, or videocassettes, all of which have the capacity to handle the large amount of data these programs contain. These media can also record images from security cameras and other monitors in the DHTI system. Storing images and data from system monitors, especially those devices in the home's security subsystem, allows any significant event to be reviewed after the fact, often a critical advantage in investigating crimes or in settling disputes. Home residents can use recorded data to determine the time and date of events that occurred while no one was at home.

CDs are the main method for storing audio files, but hard drives are also used extensively for music storage. Portable devices that use flash memory storage are now the preferred audio storage devices for people who want portability, small size, and instant access in their audio playback equipment.

The type and amount of data to be stored usually determines the type of device used. Video programs and high-resolution photographic images are the most data-intensive types of files, and so usually require the largest capacity storage devices. Next in size are sound files followed by graphic files. The smallest files are usually those composed of text or numerical data.

All data files can be compressed using mathematical computer algorithms to condense the data so it requires less storage space. Most compression methods result in some loss of quality in digital video, audio, and graphic files, but this is not true of text or numeric files, which retain complete fidelity to the original data when compressed.

Designing and Installing DHTI Systems—The Professionals

The need for professionals skilled in integrating home networking equipment is growing as a direct result of rising consumer demand for home networking products and Internet-enabled devices for entertainment, voice, and data systems. Homes with complete automated control networks that integrate computer-based systems together with entertainment, heating and cooling, water management, home security, and other systems are being constructed or retrofitted in growing numbers.

The DHTI credential was developed by the Computer Technology Industry Association (CompTIA). According to John Venator, president and CEO of CompTIA, the certification was deemed essential "in response to the growing need for qualified technicians to install and maintain networked home entertainment, environment, and security systems."

"During the next few years, as we're connecting products beyond the home computer—products like intrusion alarms, furnaces, air conditioners, and the next generation of smart-appliance, families will want to be assured that technicians are qualified."

"The goal of the DHTI certification program is to establish a recognized standard for the men and women who install and service home-networked products."

The certification is designed for technicians who install and troubleshoot integrated residential subsystems, beginning from the demarcation (d-mark) point, which is where the homeowner's network equipment and the external network service provider's equipment meet, to where the regulated signal stops.

The CompTIA DHTI+ Certificate is a cross-industry credential providing recognition that a Digital Home Technology Integrator (DHTI) professional has attained a standard of excellence in the integrated home network industry. The certification is based on a set of standards designed to measure the knowledge and understanding of core competencies regarding the installation, integration, and troubleshooting of the following subsystems:

- Home security
- Audio and video
- Computer networks
- **Heating, ventilation, and air conditioning (HVAC)** systems
- Cable and satellite
- **Broadband**
- Telecommunications
- Service provider wiring (external)
- AC electrical wiring

The required knowledge and skill sets addressed through the DHTI certification program are extensive. The DHTI+ certification exam focuses on specific content (tasks) and domain content.

The content of this book has been carefully coordinated with the knowledge required to pass the exam and obtain the DHTI certification.

How Much DHTI Is Enough?

The answer to this question is about the same as the answer to the question: How much equipment should mountain climbers carry? The answer is: Enough to reach the summit they are ascending, but not so much that carrying it is a burden. Similarly, a home technology system should provide the entertainment, communication, and automation the user needs and wants, but should not be so expensive that its cost outweighs its usefulness, nor so complex that the home-owner finds it difficult to use or maintain.

There is no ideal DHTI system. Each installation is a balance between features and costs. Each person has his or her own preferences about how many subsystems a DHTI design should have and what features and level of performance each of them should have.

That said, it is evident that the first major step in creating a DHTI system is to develop an overall design. The design should start with a list of objectives describing what the system should be able to deliver in data and services, and what it should be able to do in automation, monitoring, and control.

Specific detailed objectives allow DHTI technicians to design a system in the most economical way. Of course, when planning for wiring or wireless installation, whether in new construction or as a retrofit, it is a good idea to design for more capacity than you think you will need. Doubling the required number and capacity of communication links over what the design indicates are actually needed is generally a good investment in the future. The reasons for building in such excess capacity are threefold:

1. Installing wiring and other infrastructure for a DHTI system in the walls of a home is easily accomplished while the home is under construction, but difficult and expensive to do after the home's finish work is completed. Putting additional bandwidth (multiple wires with larger data capacity) in place during the initial installation requires very little additional effort and costs very little more. The wire and other hardware parts are not expensive. It is the labor of stringing and pulling them through walls (especially finished walls) that takes time and costs dearly. Hence, the axiom: never pull a single wire; always double it and double both wires' size, too, if possible. Even if you have no need for the second wire at present, chances are good you will later.

2. The second reason for installing increased capacity is a phenomenon known in the computer world as "Moore's Law." It states that computer (and some peripheral equipment) power and capacity doubles every 18 months. It has held true for at least the last 30 years. This law suggests that the bandwidth needed by DHTI systems today will increase by a factor of four within just 3 years.

3. Finally, when users sees how useful, how convenient, and how downright essential DHTI is in the American home, their systems as originally designed may seem inadequate. As they learn of new developments in technology integration, they will want to incorporate these into their existing systems. To many DHTI consumers, if some is good, more is better.

Summary

- DHTI is the concept of a connected home environment in which a computer system manages and distributes Internet and audiovisual digital data, and controls the network, appliances, security, and utilities of the home.
- HTI systems provide, manage, and control six areas of home technology environment:
  - Internet and network
  - Video and audio reception and distribution
  - Home security and access control
  - Telecommunications
  - Utility management
  - Appliance automation and control
- DHTI makes home life more convenient, safe, and fun.
- DHTI provides maximum use and enjoyment of electronic media, Internet, and telecommunication systems, and makes technology available on demand.
- DHTI systems consist of combinations of six categories of components:
  - Processors
  - Communication links
  - Sensors
  - Control devices
  - Display and monitoring devices
  - Recording and storage devices
- Trained professional technicians certified in DHTI design, installation, and maintenance are increasingly needed to meet the growing demand for home technology systems.
- DHTI systems should be carefully designed to accomplish the specific entertainment, data transmission, security, and automation objectives that the user desires. Provision should be made for expanding initial designs.

Key Terms

Access control   Restriction on who has the right to use a computer system or to enter a physical location.

Administrator   A person designated to maintain and control a computer system or subsystem.

Appliance   Any piece of equipment that performs a specified task.

Automation   Performing a task or function by means of a programmed device or system without the need for human supervision.

Bandwidth   The size of a digital data stream that determines the amount of data that can be transferred in a given amount of time.

Broadband   Any method of transmitting large amounts of data in a short time span. Usually accomplished by using multiple frequencies or data streams; a large (or wide) bandwidth technology.

Camera   A device that creates a video image in digital or analog form. Cameras can produce still images or moving images.

Central processor   The main computer in a DHTI system or, alternately, the component in a computer that performs calculations on data.

Certification   The process of attesting to the qualification or competence of a person to perform certain services.

Cipher lock   A door lock that opens only when a numeric code is entered on a keypad mounted near it. The code can also be placed on a memory card that is swiped through a reader to open the lock.

CompTIA   Computer Technology Industry Association, a global information technology (IT) trade association with more than 13,000 members in 89 countries. It works to advance the IT industry, promote IT public policy, and develop standards for the training of professionals in the industry.

Control device   Any device that initiates a process, directs equipment to perform a function, or responds in a preset manner to remote commands.

Credential   A written certification that an individual has specified qualifications, abilities, or expertise.

Decoder   A device for rendering data received in an encrypted form into a form that can be used by a computer or displayed by a video or audio system.

Detector   A sensor that is designed to react to a certain event such as pressure, motion, heat, or light.

Digital data   Data in the form used by computers and that consists only of the binary numeric digits 1 and 0. Nearly all forms of information can be rendered into digital form for processing or transmission and then converted back into a form that can be displayed or understood by people.

Digital Home Technology Integration (DHTI)   A connected home environment in which a computer system manages data and controls subsystems in the home.

Digital subscriber line (DSL)   A telephone line used for high-speed digital data and voice transmission and always available for the subscribing user's exclusive use.

Distribution   The process of transmitting data throughout a system so it is available to all components.

Downlink   Digital data being transferred into a ground-based system from a satellite.

Firewall   Hardware or software that controls the data entering or leaving a computer system. It is used to maintain the security of the system.

Heating, ventilation, and air conditioning (HVAC) systems   A specialized field in the construction industry that installs and maintains the equipment and infrastructure of these systems.

Home security system   A DHTI subsystem of hardware and software designed to prevent unauthorized persons from entering a home or yard, using any of its data systems, or removing anything from them.

Home theater   A DHTI subsystem for displaying television programs and recorded video programs. Similar to a home entertainment center, but usually specialized primarily for video viewing.

Infrastructure   Any of the wiring, conduit, connectors, wireless hubs, switches, routers, and other hardware that enable the subsystems of a DHTI system to communicate with one another and the outside world.

Internet Home Alliance (IHA)   An industry group whose objective is to develop the market for home technologies that require a broadband or persistent connection to the Internet.

Keyboard   An alphanumeric data input device for a computer. It may contain additional keys that input specific commands or run sequences of data.

Keypad   A numeric data input device for a computer or other hardware in a DHTI system.

Microphone   A device that converts sound waves into an electronic analog signal that can be converted to digital form, transmitted over distances, and replayed or stored.

Modem   An electronic device that converts digital data into a form that can be sent over a telephone line. Modems are the most common method of connecting a home computer to the Internet through a telephone line or cable connection to a service provider.

Monitor   A device for displaying data in text or picture (graphic) form. Also, a sensor that monitors one or more environmental conditions.

Network   A group of computers, information sources, and peripheral devices connected by cable or radio so they can share data and communicate with one another.

Numeric code   A group of digits that serves as a password and must be entered into a cipher lock to open it.

Password   A group of letters and numbers that must be entered into a security system in a home or into a computer system to gain access.

Peripheral   Any input or output device connected to a computer that sends or receives data from the processor. Examples of input peripherals are floppy drives and microphones. Examples of output peripherals are printers and control devices.

Program   A set of digital instructions that a computer executes in sequence to perform functions.

Relay   A remotely operated electric switch that is activated by a small current, but that controls a large current flow.

Remote-control   A hardware device or software program that allows a person or computer to direct the operation of a subsystem from a distance.

Satellite link   Digital television connection that beams signals to receivers on the ground from a stationary satellite orbiting Earth in space.

Scan   Survey of a data set, space, or sensors to determine if particular data or a set of conditions is present.

Sensor Any device that detects or measures human activity or environmental conditions and sends data regarding its measurement to a processor.

Service provider A company that provides data transmission service or other utility services to consumers.

Subsystem A group of hardware components and software set up to perform a specific task or function within a larger multiple-function DHTI system.

Switch An electrical device that completes (turns on) or opens (turns off) a circuit; also a network hardware device that directs a data signal on a specified path.

Telecommunications The general name for all communication and data functions carried on telephone lines or radio signals, or performed by telephone hardware and software.

Translator Another term for a decoder that converts encrypted data into a readable form for display by an output device.

Transmission The movement of data from one location to another. The data can be digital or analog and the locations close together or distant.

Video A digital display of pictures such as television programs or computer screens.

Virus A self-propagating program that is sent to a computer, remains resident in its storage, and can interfere with or disable its operation.

Wireless technology Any of several methods of communicating digital data by means of radio waves without the need for any wires connecting the sender and receiver.

World Wide Web (WWW) The global system of interconnected networks over which users can share data through the use of common protocols.

X10 A data-transmitting technology that uses the existing wiring of a home or building as a carrier for its signals.

Review Questions

1. What does the abbreviation DHTI stand for?
2. What are the major components found in a DHTI system?
3. What are some of the areas of home automation that can be addressed by a DHTI system?
4. Which of the following would not be purchased by a consumer from a service provider, as discussed in this chapter?
   a. Electric service
   b. Gas service
   c. A computer
   d. A DSL
   e. A decoder
5. What groups of people can benefit particularly from DHTI's automation features?
6. What are the six main areas of the home environment that DHTI systems provide, manage, or control?
7. What is the function of a sensor?
8. What device is usually used to connect a home network to the Internet through a telephone line?
9. What is one limitation of wireless technology in a home environment?
10. What is one advantage of using wireless technology in a DHTI system installed in an existing home?
11. What is HVAC?
12. How can DHTI reduce home utility costs?
13. What does a satellite digital television hookup require for each television set connected to it?
14. What is CompTIA, and what is its involvement in DHTI?
15. Why is a central DHTI processor usually better than independent processors in subsystems?
16. When installing infrastructure wiring, why is it usually better to install more than just the required capacity?
17. All of the following can be detected by a sensor except:
   a. Heat
   b. Smoke
   c. Time
   d. Motion
   e. Pressure

18. What is a firewall in a computer system?

19. Can a computer virus reside in a sensor?

20. If the outside temperature dropped suddenly to freezing, what actions might a complete DHTI system take in response?

Hands-On Projects

Project 1-1: Research DHTI Products on the Internet

In this project, you find several sources of DHTI hardware and locate specific items you want. You may wish to save the results of this project for future use. To complete the project, you should have a computer with access to the Internet and a functioning Web browser. Be sure your research includes a visit to at least one of these sites: http://www.smarthome.com, http://www.homeauto.com, or http://www.hometoys.com.

1. Open your Web browser and connect to the Internet.

2. Use your browser's search function or connect to another search engine you prefer, then do an Internet search for the phrase "home technology integration."

3. Browse the results of your search and find at least three companies that sell home technology integration products. Note or bookmark each company's site so you can return to it.

4. On each of the vendor sites you found, find a device that will remotely turn a lamp or other electrical device on or off as directed by a **remote control.**

5. Print out the information for each device and compare the cost and features of each device. Determine which one you would buy if you were installing your own DHTI system.

Project 1-2: Find Components for a Control System on the Internet

In this project, you find what other components are required to make the control devices you found in Project 1-1 functional. You also determine what the total cost for these additional components is and which of the three original devices now represents the best buy. You may wish to save the results of this project for future use. To complete the project, you should have a computer with access to the Internet and a functioning Web browser.

1. Open your Web browser and connect to the Internet.

2. Return to each of the three sites you noted in Project 1-1. Find the items you priced to control an electrical appliance and determine what additional components are needed to make the device fully functional.

3. Make a list of the needed items for each device and their prices. Print out the information on the subsystem parts if possible.

4. Add up the subsystem prices for each of the three devices. Would you change your mind about which one to buy for your own DHTI system? Why or why not?

Project 1-3: Design a Whole-House Lighting Control System

One of the advantages of DHTI is its ability to get multiple uses from many components. In this project, you explore a Web site and design a full DHTI light control subsystem. You may wish to save the results of this project for future use. To complete the project, you should have a computer with access to the Internet and a functioning Web browser.

1. Open your Web browser and connect to the Internet.
2. Return to one of the three sites you noted in Project 1-1. Explore this site to find the components for a complete home lighting control subsystem. Print out information on each component and determine how they will connect together.
3. Also determine how the components can interface with the home lighting. Be sure you plan for the necessary central control station and the wiring or wireless technology needed to connect it with the control devices on the home's lights.
4. Price your system as fully as you can.
5. Estimate the number of hours of your labor that will be needed to install the system. Determine the cost of your labor at $20.00 per hour.
6. Add your materials and components cost to your labor cost to determine what the completed lighting control subsystem will cost.

Project 1-4: Review the Requirements for Certification as a DHTI Professional

In this project, you will review the subjects covered in the examination you must pass to become a DHTI Professional. To complete the project, you need a computer with access to the Internet and a functioning Web browser.

1. Open your Web browser and connect to the Internet.
2. Go to the web site http://www.comptia.org.
3. From the home page, click the I'm Interested In list box, and then click DHTI.
4. On the DHTI page, click the DHTI+ exam and DHTI+ Objectives to learn about how you can become a DHTI Professional after completing study of this book.

Project 1-5: Determine the Availability and Cost of Broadband Internet Service in Your Area

In this project, you determine the availability and cost of broadband Internet service from providers in your area. To complete the project, you need a computer with access to the Internet and a functioning Web browser.

1. Using your browser's search engine, search for the phrase "Internet service provider." You will probably receive more than a million responses.
2. Refine your search by adding the state and town where you live to the search. For example: "Internet service provider" +Nebraska +Lincoln. (Note that the syntax given may work on some search engines, but not on others, because each search engine has its own rules for how to phrase searches.)

3. Note several of the Internet service providers (ISPs) for your area. Try to find two nationwide providers who offer service throughout the country, and two local providers who only offer service in a smaller area.

4. Visit the Web sites of the ISPs you selected or contact them by telephone to determine whether they can connect broadband (not just standard telephone line and modem) service to your home and what they charge for their service.

5. Compare the features offered by each service in relation to their cost. Determine which one you think offers the best broadband service for your needs.

Case Project

Case Project 1-1: Report on the Benefits of DHTI for a Physically Restricted User

You work for High Tech Home Tech, Inc. as a DHTI system designer and installer. A client calls you and requests information on what DHTI could do for her home environment. She tells you she lives in a fourth-floor apartment accessed by elevator and stairway. The home has two entrances, but she must use the elevator because she uses a wheelchair for mobility. She has full use of her hands and upper body and has no sensory limitations. She prefers to live alone, but wants to have some additional safety and security as well as convenience added to her home. Write a brief report summarizing what DHTI could do for this client's lifestyle. Divide your recommendations into categories that include safety devices, security devices, convenience devices, and entertainment systems.

Network Basics

Introduction

In Chapter 1 you learned what DHTI is, and about some of the data services, security, automation, and other benefits an integrated home network can provide. In this chapter, you will learn what a local area network (LAN) is and how it functions. You will also learn about the two basic types of network architecture, Ethernet and token ring, and how a LAN receives and sends information on the Internet. You will understand how a LAN connects to external information sources through the Internet and how internal information is transmitted within the LAN. Finally, you will learn about the different hardware components that can be connected as nodes in a LAN, some of the data services these devices provide, and how output devices, such as CRT monitors, **flat panel screens,** and printers, work.

What Is a LAN and How Does It Work?

A local area network (LAN) is a specifically designed configuration of computers and other devices, located within a confined area such as a home or office building, and connected by wires or radio waves that permit the devices to communicate with one another to share data and services. Computers and other devices connected on a LAN can send and receive information from one another without confusion. The network directs the communications passing through it and acts as a sort of electronic traffic cop to prevent collisions or mixing of data. The LAN can also be connected to the Internet and the World Wide Web, either with a direct cable connection or by a telephone link through a modem so that workstations on the LAN have access to all the networks and sites linked to the global Web.

Each individual hardware component, such as a computer or printer, that is connected to the LAN is called a node. Each node has a unique identification address so it can recognize information coming to it on the LAN and accept it, while rejecting information addressed to other nodes.

A LAN's connection to the Internet is sometimes accomplished by having one of the LAN's nodes connected as a node on a wide area network (WAN). A WAN is a network that extends over a larger geographic area than a LAN, and it is connected through one or more of its nodes directly to the Internet. Data from the LAN travels through its connecting node to the WAN and through the WAN's connecting nodes to the Internet. Data coming from the Internet first travels through the WAN and then to the LAN.

The OSI Model

The Open Systems Interconnect model (OSI model) is a standard means of describing a network operating system by defining it as a series of layers, each with specific input and output. The OSI model was developed by the International Standards Organization (ISO) and has seven layers that are numbered in order from the bottom layer (Layer 1) to the top layer (Layer 7). The names of the various layers, starting from the top layer, are as follows:

- Layer 7—Application layer (top layer), the layer in which applications on a network node (computer, printer, etc.) access network services such as file transfers, electronic mail, and database access.

- Layer 6—Presentation layer, the layer that translates Application layer data to an intermediate form called frames that provide security, encryption, and compression for the data.

- Layer 5—Session layer, the layer that establishes and controls data communication between applications operating on two different computers, regulating when each can send data and how much.

- Layer 4—Transport layer, the layer that divides long communications into smaller data packages called segments, handles error recognition and correction, and acknowledges the correct receipt of data.

- Layer 3—Network layer, the layer that adds addresses to the data messages, translates logical addresses into actual physical addresses, and routes the data to addresses on the network.

- Layer 2—Data Link layer, the layer that packages bits of data from the Physical layer into frames (logical, structured data packets), transfers them from one computer to another, and receives acknowledgment from the addressed computer.
- Layer 1—Physical layer (bottom layer), the layer that transmits bits (binary digits) from one computer to another and regulates the transmission stream over a medium (wire, fiber optics, or radio waves).

All parts of network operating systems discussed in this book function in one of these seven layers. If you can visualize the layer in which an operating system functions, you will have a clearer understanding of how it relates to the rest of the network operating system. Consider this example: A computer workstation is running a word processing program and the program is sending a document to a printer to be printed. The process might go like this:

1. The word processing application sends the document to its print driver component where it is translated into printing instructions that the specific printer can recognize and implement. This is a Layer 7 function.
2. The printer instructions are compressed into the smallest amount of data possible so they will require the least amount of time to transmit on the network. The data may also be encrypted to prevent it from being read by any unauthorized party. These are Layer 6 functions.
3. The compressed printer data is directed to the network, which establishes an electronic link between the data sender and the intended receiver. This link controls when the data can be sent and received. This is a Layer 5 function.
4. The data is divided into segments; error recognition and correction codes, as well as the confirmation of receipt codes, are attached, forming each segment into a packet. This is a Layer 4 function.
5. The data packets are addressed and routes are established over which they will be sent on the network to the appropriate node. This is a Layer 3 function.
6. The data packets are transformed into bits of electronic binary form (0 and 1 only) information and physically placed on the network. This is a Layer 2 function.
7. The electric current containing the binary data moves through wires, on fiber optic cables, or by radio transmission from the sender to the recipient node on the network. This is a Layer 1 function.

Data Packets

To maintain order and avoid loss of data through misdirection or collisions, all data that moves on a LAN is sent in packets. A packet consists of a variable-sized piece of data with information about what type of data it is, where it came from, and where it's going, attached at the beginning of the packet as a **header** and at the end of the packet as a **trailer.** Figure 2-1 shows the structure of a typical Ethernet data packet with header and trailer.

Each individual packet contains a small amount of data. The maximum amount of data a packet can contain depends on the type of network. Each network type has a protocol, a set of formatting guidelines and rules that allows

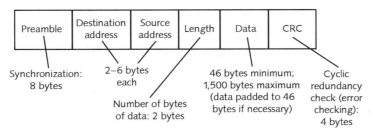

FIGURE 2-1   Diagram of a Typical Ethernet Data Packet

information sent by one computer or device to be accurately received and understood by another on the network. The protocol specifies how the data being sent is divided into packets of suitable size to travel on the network and adds the identifying information to each packet before it is transmitted on the network. Some protocols also determine if each data packet is correctly sent and then received by the destination device. If necessary, the protocol can resend a data packet that failed to arrive or arrived corrupted.

Sending the information on the network in packets provides several critical advantages. First among these advantages is that the information in each packet can easily be checked to be certain none is missing or corrupted. The receiving computer's network interface card (NIC), the device that connects the computer to the network, uses a system called checksum to verify the integrity of the data it receives. Checksum compares a value that is sent with the packet to a value calculated by the NIC using the data received in the packet. If the two values match, the data is deemed to be uncorrupted. If they do not match, the packet is assumed to be corrupted and must be sent again. Only the corrupted packet must be repeated, however, not the entire data stream. Using packets minimizes the amount of data that goes astray or is corrupted and must be repeated.

A second advantage to data packets is that sending information in these small pieces permits more data to quickly travel from many sources to many destinations than would be possible if the data were sent in large blocks. A large data block would fill the network's wires, preventing any other traffic from moving until the entire block was transmitted, checked, and its receipt confirmed. Data packets are small and are quickly delivered, checked, and confirmed to the sender. As soon as one packet is off the network, another can move onto it, either from the same sender to the same receiver or over a different network path. Data packets can be kept constantly moving on the network, and the aggregate of data they can contain is greater than what could be moved and verified in larger blocks.

The third advantage is that data packets make the information flow in a network easier to control by means of routers, switches, bridges, and other devices. Small packets are easily and quickly directed and they do not tie up the network for long at any critical point. Unlike large data blocks, packets from a single document can travel by multiple routes and be reassembled into their correct order after arriving at their destination. A delay or malfunction on one part of the network concerning a packet does not slow other packets down much: They are simply directed to another open route to their destinations. The blocked or corrupted packet is sent again by a different route to complete the transmission.

Each computer on a LAN must have a protocol installed on it for the computer to communicate with other computers on the network. Network BIOS Extended User Interface (NetBEUI) was, until recently, the most commonly used protocol. It is a proprietary Microsoft suite that can be run only on computers using the Windows® operating system. NetBEUI cannot be used with computers running the Linux operating system or with Macintosh computers using any version of the OS operating system. NetBEUI also cannot be used as an Internet protocol because it is not **routable** through many devices used on the Internet. TCP/IP protocol, which is routable and therefore can be used on the Internet, has now replaced NetBEUI as the dominant protocol for LANs.

Network Architecture

A network's architecture consists of the design of its wiring or radio wave connections, the configuration of its other physical components, its software (programming), and the protocols by which it operates. All of these parts must be tightly organized into a physical structure with a consistent operating methodology to establish a smoothly working communication system among all the devices connected to the network.

Each device connected to a network is called a node or **host.** A node can be a computer, a router, a printer, a sensing device, a video camera, a controller, or any number of other electronic devices. A host is always a computer.

The four common types of network architecture used today are Ethernet, token ring, FDDI, and wireless. Each has advantages and limitations, which you will examine in some detail in the following sections.

Ethernet

Ethernet is the most popular form of LAN in use today. 10-Mbps Ethernet, commonly known as 10BASE-T Ethernet, operates at a speed of 10 megabits per second (Mbps) of data. It is also referred to as Ethernet IEEE 802.3 after the industry standard with which it complies. Several variations of the 10-Mbps Ethernet exist, distinguished mainly by the type of cable and connectors used. Early Ethernet installations used either 50-ohm RG58/U coaxial cable, also known as thin Ethernet, or 50-ohm RG8/U coaxial cable, known as thick Ethernet, but these are both obsolete now.

By 2003, over 60% of installed networks were 10BASE-T. By 2007, Ethernet (the hardwired and wireless versions) accounted for almost 100% of home network connections. 10BASE-T Ethernet is still the dominant type of installation at present, but the need for greater data transmission capacity **(bandwidth)** has spurred the growth of still faster Ethernet technology for both business and home use.

100-Mbps Ethernet or Fast Ethernet operates at a speed of 100 Mbps. It can also handle data at 10 Mbps, and this feature allows devices running at the slower speed to operate on the same network along with those operating at 100 Mbps. As with 10-Mbps Ethernet, variations of 100-Mbps Ethernet are distinguished mainly by the types of cable used to transmit them. 100BaseTX, for example, uses two pairs of wire in a CAT5 twisted-pair cable that contains eight wires, while 100BaseFX uses fiber-optic cable.

1,000-Mbps Ethernet or Gigabit Ethernet (Gig-E) operates at a speed of 1,000 Mbps (1 gigabit per second). It is in wide use by businesses with large high-speed

LANs and heavy-traffic **server** connections. Many new home network users are opting for Gig-E to meet broadband Internet, video, and audio needs. Business users are also upgrading to Gig-E, or to InfiniBand, a still faster 10-Gigabit technology.

Gig-E is more expensive than Fast Ethernet, which is in turn more expensive than 10BASE-T, but some users may want to install these larger capacities to provide for future growth in their systems. Gig-E is currently the fastest version of Ethernet topology widely used for home LANs, but 10-Gigabit Ethernet is already available on the market and 60- and 120-Gigabit Ethernet is in development and scheduled to be marketed by the end of 2008. As use of larger capacity Ethernet technology increases, the relative cost of its components will decline.

Ethernets can be physically arranged in either of two configurations: a **bus topology** or a **star topology.** These configurations refer to how the nodes (devices) are connected to the Ethernet. In a bus configuration, each node is connected to the next by a direct line so the network forms one continuous line of nodes, each connected to the one on either side of it. There is no central point in this arrangement. When the end of the line is reached and there are no further nodes to be connected, the Ethernet is closed off with a **terminator** device specific to the infrastructure used—for example, some use 50 ohm and some 92 ohm. See Figure 2-2 for a diagram showing both the bus and star design Ethernet configurations. Note that the bus design connects nodes directly, while the star design connects each node to a hub, and none of the nodes are connected directly to one another.

In a star configuration, the Ethernet is set up with each node connected to a central hub, which serves as a distribution device. It passes all the information packets it receives from any device to every other device on the network. The hub does not use or even evaluate any data; it simply broadcasts all data it receives from any node to every other device on the network.

Because the hub generates a lot of data traffic that is not used, a star Ethernet may be slower than a bus Ethernet, especially if the network has many nodes. Since the hub replicates all the data it receives from any source and sends it to every node, the amount of data being sent increases for every node added to the network, even though most of the data sent to each node is not intended for that node and is discarded upon arrival. As the amount of data sent increases, more and more data packets from different nodes competing for bandwidth on the

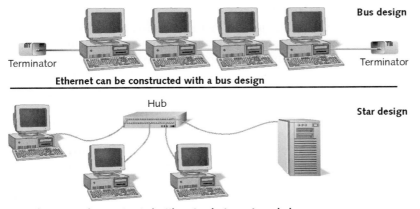

Terminator   Terminator
Ethernet can be constructed with a bus design

Hub   **Star design**

Ethernet can be constructed with a star design using a hub
FIGURE 2-2   Bus and Star Design Ethernet Configurations

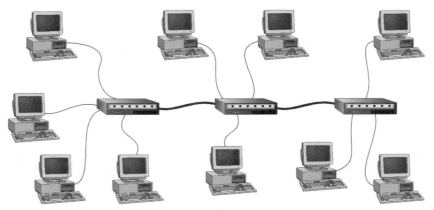

FIGURE 2-3   Combined Bus and Star Ethernet Design

LAN collide with one another. The network detects these collisions and resends the data packets involved, but the collisions and replication of data transmissions slow down the network.

In a bus Ethernet, data is sent on the network line in both directions from the source node. The data passes from one node to the next until it reaches the terminator at the end of the network. The terminator simply cancels the data signal, discarding the data so it cannot echo back on the network line and head back to the node it just came from. All information on the network passes through each node, but only once. There is no replication and broadcasting of data as in a star configuration. Each node determines if the data it receives is addressed to it. If it is, the data is read and receipt is confirmed. If it is not, the packet is passed on to the next node.

A large LAN might use a combination star and bus design with nodes connected to several hubs and the hubs connected in a bus configuration. This design is useful for constructing large networks with a minimum of wiring, but because all the hubs must still broadcast their data to the nodes, it tends to slow down as the amount of data flowing in it multiplies. Figure 2-3 shows a combined bus and star Ethernet design.

Ethernet Data Control

Ethernet networks are passive designs, which means they do not take any active role in controlling the movement of data on them. They just provide a means for data to move along the connecting wires. The nodes that receive and send the data must have some arrangement for controlling the flow. The protocol on an Ethernet causes any computer or device that is trying to send data on the network to first listen for silence. If the computer hears nothing on the network, it transmits a data packet, listening as it does so for any other data being transmitted at the same time. If it hears anything other than its own transmission, it stops transmitting and sends a collision signal indicating that two computers have attempted to send data at the same time. Each computer then waits a short, but random, amount of time and sends its data again.

This protocol ensures that two data packets are not sent at the same time, or if they are, that transmission is stopped and they are sent again. This type of network technology is called a contention-based system because each node on the network must contend for space on the network. This arrangement is known as the **Carrier Sense Multiple Access/Collision Detection (CSMA/CD)** method of data transmission.

Another, but somewhat slower, method is called **Carrier Sense Multiple Access/Collision Avoidance (CSMA/CA),** in which each node signals its intention to transmit data and waits to see if the line is clear before doing so. The AppleTalk protocol, used on networks with Macintosh computer nodes, uses CSMA/CA.

Thin and Thick Ethernet designs, wired with coaxial cable, are limited by the attenuation (weakening due to the distance traveled) of signals in the cable and can only support network segments up to 185 meters (Thin) or up to 500 meters (Thick) long. Fast Ethernet and Gigabit Ethernet can support network distances up to 1,000 meters with less attenuation of network signals.

Because data signals traveling in a wire get weaker over distance, an Ethernet that extends over a maximum allowed distance (as explained in the preceding paragraph), requires amplification of the data signals on the network. This is done with an amplifier **repeater** placed as a node on the network that amplifies the signal, including any noise or static it may have acquired, and sends it on.

Another type of repeater, called a signal-regenerating repeater, can also be used. This type of repeater reads the incoming signal and creates a full-strength duplicate of it, which it then sends on the network in place of the original weakened signal. A signal-regenerating repeater not only strengthens the signal, but also eliminates any noise and static that may have come into it. Figure 2-4 shows how a repeater is placed in a network to strengthen and clarify the signal.

A hub in a star Ethernet design can, if so configured, also act as a signal-regenerating repeater, which is sometimes a reason to use this configuration even though it's slower than the bus design. Even when they use repeaters, Ethernet LANs must be limited in size because the Ethernet protocol limits the length of time a signal can take on the network to reach its destination. Networks that traverse long distances and contain many repeaters have problems with this limitation because each repeater that the signal traverses adds a small amount of time to the transmission.

An Ethernet protocol rule known as the 5-4-3 rule requires that between any two nodes on a bus topology there can be no more than five segments, connected through four repeaters, and only three of the segments can include user nodes. Another rule, known as the 5-4 rule, limits the length of cable and number of hubs that can separate nodes on a 10-Mbps Ethernet. Somewhat different rules apply to Fast Ethernet and 10-Mbps Ethernet configurations wired with newer cable types.

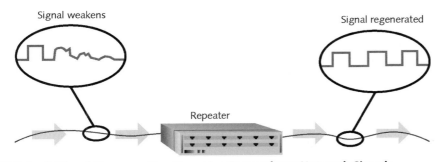

FIGURE 2-4   A Signal-Regenerating Repeater Strengthens Network Signals

Only very large home LANs will approach these limitations on overall size and cable length, but you should be aware that they exist and should examine them in more detail when working on a LAN that extends over more than a 500-foot-diameter circle.

Wireless Networks

As their name suggests, wireless LANs do not use wires to connect the nodes of the network. The nodes are not physically connected at all to one another or to a central device. They communicate with an access point or **wireless hub** using a wireless network interface card (NIC), which includes a transceiver (radio frequency transmitter and receiver) and an antenna. The **wireless NIC** allows the node to communicate over relatively short distances using radio waves, which it sends to the nearest hub and receives from the hub. Figure 2-5 shows how nodes on a wireless LAN connect by radio waves to access points or hubs wired to the network.

Because radio waves pass fairly easily through solid material such as wood, plaster, or even concrete, wireless connections can be made from a node to a hub through walls and other obstructions. This ability makes wireless LANs very useful and cost effective in already finished buildings where retrofitting wiring is both difficult and expensive. It is also an advantage where hardwiring a network may be impossible, such as on the beach at a summer home, or on a boat tied up at a private dock. Wireless LANs are limited, however, both by the low transmitting power of their NICs and hubs and by the fact that dense metals, especially ferrous metals, as well as heavy layers of concrete, stone, brick, or dirt, absorb radio waves. These factors restrict the distance over which a wireless network can be extended and may require more hubs than anticipated to obtain full-area coverage.

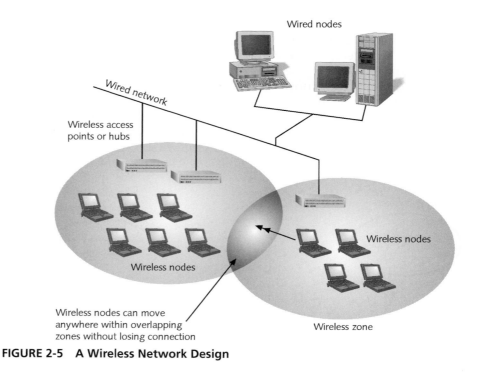

FIGURE 2-5   A Wireless Network Design

Hubs or access points must be placed so the wireless NICs of nodes can access at least one of them from any location within the LAN's defined area. Wireless networked nodes use star Ethernet topology and communicate with one another only through a hub rather than directly, node to node, as in a bus Ethernet. Hubs must be wired together or connected by wireless technology in a network that allows all hubs to communicate with one another and transmit the data they receive from their wireless nodes. A wireless LAN is thus not always entirely free of wired connections. It may be connected to a cable network by its hubs, which constitute nodes on a wired LAN.

The NICs used in wireless LANs have no way to detect whether collisions of data have occurred in the radio transmission process. For this reason, wireless LANs must use collision avoidance technology (CSMA/CA) in communicating from node to hub and back. Collision avoidance communication requires confirmation from the recipient that each packet of data was received accurately. This process slows down communication and makes wireless LANs less efficient, especially when multiple wireless nodes are trying to communicate to a hub simultaneously.

Wireless LANs are inherently less secure than hardwired LANs because data sent on a wireless LAN is broadcast as radio waves, which can be received by anyone tuned to the proper frequency if they are close enough to pick up transmissions from a node or a hub. The broadcast range of a wireless LAN is normally short, but it invariably extends out into some public space where unauthorized receivers can eavesdrop on the data being sent. Because all wireless network communication goes through the hubs, a single intercept receiver, properly placed, can eavesdrop on all the traffic traveling on the LAN. A laptop computer in a car parked curbside near a home with a wireless LAN can almost always pick up the wireless transmissions. For this reason, wireless LANs require strong internal security to prevent their data from being read by unauthorized users who may receive it. Figure 2-6 illustrates how a hacker can access a wireless LAN.

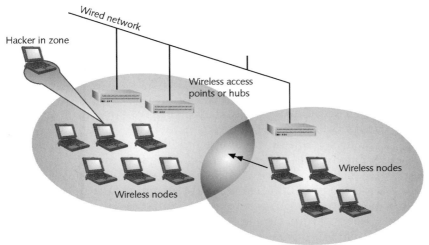

FIGURE 2-6   Any Hacker within Range of a Wireless LAN Can Receive Its Transmissions

Fiber-Optic Cable Networks

Fiber-Distributed Data Interface (FDDI) networks are fast networks that mainly use fiber-optic cable to communicate data at speeds of 100 Mbps or more. A variation of the FDDI topology uses copper wire in place of the fiber optic cable and is known as Copper Distributed Data Interface (CDDI). The network architecture of both FDDI and CDDI is essentially identical, except for the cable difference.

Because of the expense of their components, FDDI LANs were rarely installed in homes, but this has recently changed. Now that the majority of homes have broadband Internet access, the speed of FDDI technology can be fully utilized in a home LAN, and more users are installing fiber-optic systems.

FDDI topology utilizes a token-and-packet type of data transmission and a double token ring arrangement for the nodes. In a typical FDDI LAN, each node is connected by two cables to each of the nodes on either side of it, thus forming a circle layout for the network. Data normally travels on the primary ring in one direction and on the secondary ring in the opposite direction. Packets addressed to a node are accepted as they enter the node from the ring. Packets not addressed to a node are ignored and passed through to the next node, always in one direction.

An ISO standard for FDDI home networks is under development but has not yet been issued. In the absence of such a standard, several types of FDDI network structures have gained popularity in the consumer market. Among these are three main types, each of which is described in the following.

A fiber-optic cable network is wired throughout the home using FDDI technology for all data transmission. This type of system can accommodate video and audio data as well as computer data, but all data is transmitted using FDDI token ring and packet technology. For video and audio transmissions, FDDI technology somewhat limits the speed of data delivery, even using the broadband capability of FDDI cable. The data packet protocol of FDDI, which requires a high degree of transmission accuracy, confirmation of packet receipt, and the repeating of error packets, is the limiting factor rather than the capacity of the cable.

Another FDDI setup uses FDDI technology for part of the LAN in combination with Fast Ethernet technology for some segments of the LAN. The FDDI segment of the LAN is connected to the Ethernet portion through a router that can transform the data from one topology to the other in either direction. This allows the use of FDDI fiber-optic cable for high-speed video and audio data transmission, while other parts of the network use lower-cost Ethernet components. Figure 2-7 illustrates a combination fiber-optic and wired LAN.

Combination LANS may need to be wired for both FDDI and Ethernet at all nodes, even though only one topology is used at a given point. To accomplish this efficiently, two types of cable are bundled together in a single wrapped configuration known as Siamese cable, a cable type used throughout the home. The computer network data and all DHTI subsystems using Ethernet topology are carried on the Cat5 cables. Video, audio, and any other subsystem using FDDI topology are transmitted on the fiber-optic cable. Both segments of the network will be connected at one point so they can communicate with one another.

Finally, a home network can use a wireless Ethernet setup for computer data transmission and an FDDI network segment for video and audio data transmission. This eliminates the need for Siamese cable and still allows for the use of both Ethernet and FDDI data transmission methods for maximum speed and efficiency.

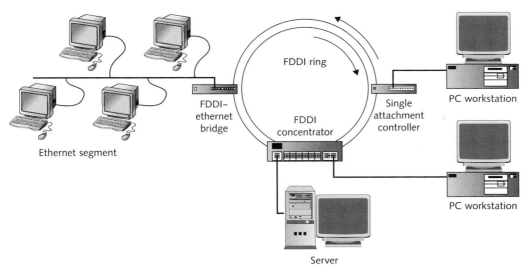

FIGURE 2-7   Combination FDDI and Fast Ethernet LAN

A variation of this arrangement is to use AC Power Line technology for the Ethernet portion of the network. See Chapter 3 for more details about this design.

Token Ring

The development of token ring LAN technology was dominated by IBM. Token ring operates at slower speeds than the newer versions of Ethernet, and its components are more expensive than Ethernet hardware. Primarily for these reasons, token ring LANs are now uncommon, especially in home applications, giving Ethernet almost 100% of the market. Some users maintain that token ring LANs have a better data control capability than Ethernet and are therefore superior for some applications, but this advantage has not overcome Ethernet's lower-cost appeal for home LANs. Few, if any, token ring LANs will be installed in future DHTI systems. The following information about them will be helpful in upgrading or maintaining an existing network.

A token ring is physically arranged in a star topology, one of the designs also used by Ethernet. Token ring LANs are so named because their protocol for data control uses a token, a small packet of data, to determine which node on the network can transmit data, and because all data actually travels in a circle or ring on the network.

In a token ring star design, each node is connected to a central device, referred to as a Multistation Access Unit (MAU), by two wires. The token packet (only one of these exists on a given network) is always present somewhere on the LAN. It travels from the MAU up one connecting wire to a node and back to the MAU through the other wire, then up one wire to the next node and back through the other. It passes through the MAU after each node and, after passing through all of them and returning to the MAU, travels back to its starting point at the other end of the MAU through the token ring's main ring cable. The token thus travels in a circle or ring on the network even though the nodes are arranged as a star. For this reason, the token ring topology is sometimes called a star ring. Figure 2-8 shows how a token ring network connects nodes through a MAU.

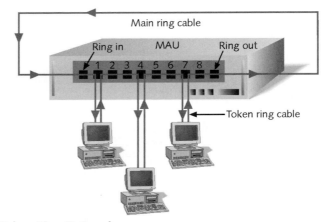

FIGURE 2-8   A Token Ring Network

Data control in the token ring is maintained by the simple rule that data can travel on the network only if it is attached to the one token data packet on the network. If the token is busy (already has data attached to it) when it reaches a node, the node cannot attach more data, and the token moves on until it reaches the node for which its attached information is intended. The receiving node reads the information from the token and attaches a code confirming receipt of the data. The token then travels back to the sending node, which reads the confirmation code and detaches the data it sent from the token, leaving it free for new data to be attached. If the node has more data it wants to communicate, it attaches that data to the token and sends it on. If not, the token moves to the next node that does have information to communicate, which attaches it to the token and sends it on. In this manner, the token continually circles the ring picking up and dropping off information as requested by the various nodes. The token method of data control is also used by some other LAN designs.

The MAU in a token ring can also serve as a repeater to boost and clean up the data signal if necessary. Token ring networks operate at a speed of either 4 Mbps or 16 Mbps. A token ring can theoretically have any number of nodes, but because data can travel only with the token, the efficiency of the network declines when the token must pass through many nodes and through the MAU many times for each packet of data communicated.

Token rings use either UTP or **STP cable,** which contains four wires (two twisted pairs) to connect the nodes with the MAU and to be used for the main ring cable. In Chapter 4, wiring and connectors are discussed in detail.

Receiving and Sending External Information on a LAN

Outside information comes into a home network from the Internet through a telephone line (standard or DSL), a cable connection, or from a satellite hookup. Information that originates in the LAN can also be sent out to other locations on the Internet. This outside data exchange is in addition to the data being sent and received among the various nodes on the LAN.

Communication between networks requires that they use a common protocol and that each computer or other device attached to each network have a unique address. The most common protocol currently used on LANs and to connect networks to one another is Transmission Control Protocol/Internet Protocol **(TCP/IP).** This protocol is used throughout the Internet. TCP/IP addresses each node on a LAN by a unique Internet Protocol address **(IP address),** which consists of four numbers separated by periods. Each of the four numbers in an IP address must consist of no more than eight bits of binary data, which means that each number cannot be higher than 256. This keeps the addresses short and standardized for consistency.

How Data Travels on the Internet

The Internet is really a vast collection of WANs, all interconnected to one another and able to communicate with each other. Data going from a node on a home LAN to a node on a distant network reaches its destination by traveling through the communication lines of other networks located between the originating point and the destination. Usually, the data can take many possible routes. Some are long and require more time for the data to arrive than others, but since all the WANs that make up the entire Internet are connected, it is possible to send data from a node on any LAN to a node on any other, provided the IP address of the destination is known and the data is directed toward that address by the Internet's data management devices, or routers.

Figure 2-9 shows the Internet as a web of interconnected, but independent, networks over which data travels from source to destination by various routes. A node on the California LAN A can send data to the node on New York LAN F by many different routes: router 1 to 3 to 8 to Network F, router 1 to 2 to 4 to 8 to Network F, or router 1 to 2 to 4 to 5 to 6 to 7 to 8 to Network F.

The Role of Routers

Routers are the hardware units responsible for directing data traffic between interconnected networks. They function as intelligent switches that have the ability to make decisions based on preprogrammed instructions and implement

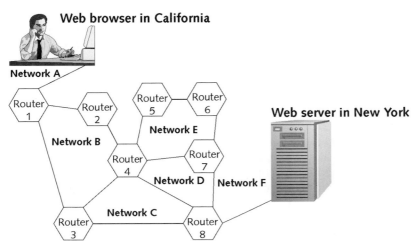

FIGURE 2-9   The Internet Is a Web of Interconnected Networks

those decisions by directing electronic data along specific routes. They read the IP addresses of each packet of data that comes to them, decide which is the most efficient path for the data to take to reach its destination quickly, and then switch the data packet onto that path.

A router is connected as a node in a network, but there is a significant difference between a router and any other node on a network. The router is always connected as a node in at least two networks and it may be connected as a node in many networks. For each network it is connected to, the router has a separate, unique IP address. Data from a node on network A to which a router is connected can be routed directly to any node on network B, to which the router is also connected, simply by switching the data packet from one network to the other so the addressee node can receive it on its network.

For data that is addressed to a node on a network to which the router is not directly connected, the router can only transmit the data to the remote network if the data packet was produced using a routable protocol such as TCP/IP or **Internetwork Packet Exchange/Sequenced Packet Exchange (IPX/SPX)**, which is the protocol used by Novell NetWare networks. The protocol used to produce data packets transmitted between nodes of a single LAN, NetBEUI, is not a routable protocol and data packets produced using it cannot pass outside the LAN through a router. This is a useful feature of NetBEUI, as you will see when other LAN components are discussed later in this chapter.

When a router receives a routable data packet, it reads the IP address on the packet and decides which of the other networks it is connected to should receive that data packet. It then switches the data packet to that network where another router connected to that network receives it, reads the IP address, and passes it on to another network closer to its destination, to which the new router is also connected. The process continues, perhaps through many networks and routers, until the data packet reaches the LAN on which its addressee node resides.

A router uses the IP address of a data packet and conditions existing on the network to determine how to route the packet. The route it chooses may not be the shortest, but is probably the fastest at the time the packet is transmitted. As routers send data packets, they "learn" the best (fastest) routes and remember them. If the first data packet sent to an IP address is slow to arrive, the next one to that address may be sent by a different route. When the router finds an efficient path to an address (not necessarily the absolute fastest route, but an acceptable one), it uses that path whenever data is sent to the same IP address, unless other conditions on the Internet slow the data down and require the router to seek a new route.

Connecting a Home Network to the Internet

More than half of American home computers and networks are now connected to the Internet via a high-speed DSL telephone connection, a cable connection, or a satellite hookup. Most of the rest connect to the Internet through an **Internet service provider (ISP)** by means of an ordinary telephone line using a modem. The modem converts the binary digital data of a computer into **analog data** (frequency modulation or FM) signals that can be communicated over a telephone line.

The amount of data that can travel over a communication line or a wireless connection in a given length of time is called bandwidth or line speed. The greater the bandwidth, the faster the communication can be because more data can move through the communication line in a given time. For digital data transmission, bandwidth is measured by the number of data bits per second (bps) that can be communicated over a given line or connection. One thousand bits per second is noted as one kilobit (Kbps), and one million bits per second is noted as one megabit (Mbps).

Home network Internet connections can have varying bandwidths, depending on the service provided by the ISP and the physical connection. It is important to remember that the useable bandwidth of any communication line is determined by the smallest capacity link in the line. If the home's central computer is connected to an ISP via a 1-Mbps DSL, but the modem in the computer has a capacity of 128 Kbps, then the connection has a maximum capacity of 128 Kbps, that of the smallest link in the system.

The most common methods of connecting a home LAN to an ISP are the following:

- The slowest, but least expensive, Internet connection to an ISP is affectionately known as Plain Old Telephone Service (POTS). This was formerly the most common method of home connection but now accounts for less than half of all Internet connections. It uses a dial-up system each time the connection to the ISP is made over the telephone line. The connection is not continuous, and when the line is not connected to an ISP, it can be used for regular telephone service, or any other telecommunications function. Data speed on a regular telephone line is a maximum of 56 Kbps.

- **Integrated Services Digital Network (ISDN)** technology also uses a telephone line to transmit data, but unlike POTS, the data is not converted to analog form. An ISDN line is digital and consists of two phone circuits, both carried on one pair of wires along with a slower third circuit used for control signals. Each data circuit can transmit data at up to 64 Kbps, and the two circuits can be combined to move data at a speed of 128 Kbps. This configuration of an ISDN line is known as the Basic Rate Interface (BRI) and is intended for home and small-business users. Another higher-cost ISDN level of service is called Primary Service Interface and is intended for larger users. It has 23 data channels and a control channel.

- A digital subscriber line (DSL) is a high-speed data and voice transmission line that still uses telephone wires for data transmission, but carries the digital data at frequencies well above those used for voice transmission. This makes possible the simultaneous transmission of voice and digital data on the same line. The regular voice telephone line must be dialed for each use, but the DSL part of the line is always connected to the computer. A DSL can transmit data at speeds up to 1.5 Mbps in both directions, or it can be set up as an asymmetric line (ADSL), which can transmit up to 640 Kbps upstream (to the ISP) and 7.1 Mbps downstream (from the ISP).

- A **cable modem** connects to the cable television line that is already installed or available in most homes. With a cable modem, digital data is converted to analog signals and placed on the cable at the same time as the incoming television signal. Incoming analog data signals are converted to digital for the computer by the modem. The data frequencies differ from the television signal frequencies, and the two signals do not interfere with one another on the cable. Depending on the individual configuration, a cable modem can transmit data at speeds from 500 Kbps up to 5 Mbps.

- A **satellite link** Internet connection to an ISP is now available nationwide. It is especially attractive in rural areas where telephone-based services may be limited and cable is sometimes not available. A satellite communication link uses a dish similar to a satellite television dish mounted on the home to communicate with a stationary satellite in orbit. The home's central computer is connected to the dish antenna. Incoming Internet data travels from the ISP to the satellite in orbit, then down to the home-mounted dish and into the central computer of the home LAN. The speed of the connection varies according to the ISP, but can go up to 1.5 Mbps. The uplink connection from the LAN to the ISP is usually by a telephone line/modem connection and is not as fast as the satellite downlink. A digital radio signal from the LAN up to the satellite, which in turn sends the signal to the ISP, is also available, but at a much higher cost than the telephone connection, which is usually adequate for sent data. Figure 2-10 illustrates how a satellite ISP sends data at high speed to home LANs via a stationary satellite and receives data from the LAN over a slower telephone/modem line.

- Wireless access using radio signals, infrared transmission, or beamed signals is possible in many areas, but has not proved very popular for two principal reasons. First, it is usually more expensive to install and use than other readily available technologies. Second, it is often subject to interference from other radio signal sources such as cell phones, radio transmitters, pagers, electric grids, large motors, and other devices. Wireless networks may themselves interfere with some devices such as pacemakers, aircraft control systems, and other sensitive electronic equipment. The wireless connection of nodes to a hub in a LAN works well where the required range of transmission is short (100 yards or less), but for more distant Internet connections through an ISP, wireless has not yet proved as economical or efficient as other methods.

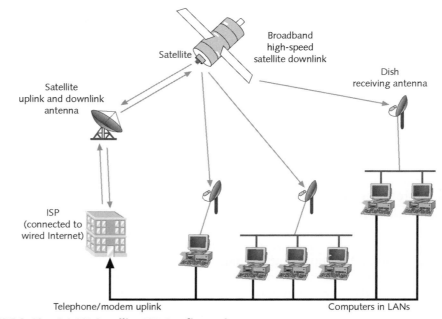

FIGURE 2-10   A LAN Satellite ISP Configuration

TABLE 2-1
Bandwidth Technologies and Speeds

Technology	Maximum Data Speed	Common Uses
POTS	Up to 56 Kbps	Home access through a modem to an ISP
ISDN or ISDL	64 to 128 Kbps	Medium-level home access through a modem to an ISP
DSL Digital Subscriber Line	1.544 Mbps	Home/business access; equal bandwidth in both directions
Asymetric DSL	640 Kbps upstream; up to 6.1 Mbps downsteam	Home/business access with the most bandwidth from the ISP to the user
Cable Modem	512 Kbps to 5 Mbps	Home/business access to ISP
802.11b Wireless	5.5 or 11 Mbps	Home/busines LANs
802.11a Wireless	Up to 54 Mbps	Home/business LANs
10BASE-T of Fast Ethernet	10 Mbps or 100 Mbps	Home/business LANs
Token Ring	4 or 16 Mbps	Home/business LANs
T1	1.544 Mbps	Business access to ISP
FDDI	100 Mbps	LANs and WANs
Gig Ethernet	1 to 10 Gbps	LANs and WANs

Table 2-1 compares a number of communication bandwidth technologies, their common uses, and their speeds.

Transmitting Internal Information on a LAN

Each node on a LAN can communicate with all the other nodes on the network. Depending on the LAN's architecture, each node may "talk" directly to another node (as in a bus Ethernet) or through a hub or intermediate device (as in a star Ethernet).

Every node has access to information stored on hard drives and other storage devices contained in each computer. This access may be limited by password protection or other security measures, but if it is not, then any node on the LAN can search the other nodes for information and retrieve the information off of any storage device in any computer on the LAN. A node can also receive information from sensors and other devices that are connected to the LAN and programmed to send their data on it.

In the same manner, any node can send information over the LAN to be stored on any storage device available on the system, or to any output device (printer, display monitor, etc.) that is connected to the LAN. A node can also send operating instructions and programs to control devices instructing them to take certain actions or perform particular functions.

Making all the information stored on every node of a LAN available to every other node is not always desirable. Control over which nodes have access, how much access they have, and which particular users of a node have that access is the responsibility of the LAN's administrator. This person determines each node's access and function using network management software installed on the LAN's central computer.

The network administrator also needs to control access to the LAN from the outside. This security control permits the administrator to prevent certain types or categories of information from reaching any node on the LAN by blocking it from entering the system. It also allows the administrator to prevent users from sending any unauthorized information stored on nodes in the LAN to any outside network node. Network administration software can define rights to send and receive data for the entire LAN, for each node on the LAN individually, and for each user on the LAN, as identified by their unique username and password.

A LAN is usually insulated and partially isolated from the Internet to which it is connected by means of a **firewall,** also referred to as a gateway. This is a software program (sometimes augmented by hardware devices) that controls information passing in either direction: from the Internet onto the LAN, or from the LAN onto the Internet. A firewall can be set up to block all information from some IP addresses, block information containing certain key words, require a password before allowing certain information to pass, or perform a variety of other restrictions. A firewall can also filter out viruses and data requests from unauthorized or unknown users. It can also prevent some or all data from being sent from the LAN to any outside user.

Major Hardware Components of a Home Network

A home LAN can have any number of nodes connected to it using one of the designs discussed. These nodes can consist of several types of hardware devices, each with a separate function or set of functions to perform in the system. In this section, you will examine pieces of hardware likely to be found in a home network

Computers

The heart of a home network is the central computer. This computer probably functions as a workstation and is multitasked to perform many other functions on the home network. In most designs, the central computer is also the access point for the home network's Internet connection, regardless of the type of connection used. The central computer is also a node on the home network, and as such it is able to communicate with other network nodes. Data coming into the home network from the Internet, which is addressed to other computers on the home network, is sent to them via the network from the central computer. The central computer controls access to and from the Internet for all nodes on the LAN.

The central computer's keyboard, mouse, microphone, and other input devices are set up to function as with any workstation, but additional software may permit the central computer to take control of other computers and devices on the network, as well as use its input devices to function as if they were attached directly to the controlled device. The same is true for some output devices such as a monitor, disk drive, or speakers: All can be programmed to function as if attached to another controlled device on the network.

The central computer may have a number of other tasks assigned to it in addition to its workstation and Internet data transmission functions. The central computer can direct the operation of other subsystems on the network, including the home security system, the utility control system, and the appliance control system, among others. Any of these functions may also be assigned to other workstations on the network.

Besides the central computer, a home network can include any number of other computers, either used as general workstations or having specialized functions. Some of these computers may have their own input and output devices, which function only for the computer to which they are attached. Others may receive data only from other devices on the network and output data only to other nodes on the network, which will be received and acted upon by other nodes. One or more of these additional computers may be programmed to function in the same manner as the central computer, either because of its convenient location in the system or to serve as a backup in case of a system failure in the central computer.

Monitors

A monitor is always connected to a computer, and the video display controller in the computer drives its display. The computer can be programmed to display on its monitor the data sent from any node on the network, including other computers, sensors, or control devices. Because a monitor lacks data-processing capability and can only display video data as directed by a video controller card in the computer, a monitor is never connected to a network as a separate node.

Some monitors are cathode ray tubes (CRTs), which display images on the flat front face of a glass cathode tube. This display is created by a beam of electrons shooting from the back of the cathode tube toward its front screen face. Plates on the top, bottom, and sides of the tube control the direction of the beam and direct it to continuously "paint" lines from left to right, top to bottom, across the screen. A grid placed in front of the filaments that generate the electron beam controls the intensity and color of the beam as it paints an electron stream across each line on the screen. Phosphor dots on the inside face of the screen light up when struck by the beam and produce dots of color on the display face. These dots of color are visible on the outside of the tube. The beam's color is controlled by the grid and can cause any combination of red, green, and blue phosphors on the screen to light at the appropriate level to produce the desired color for that area of the screen image. Figure 2-11 shows how a cathode ray tube (CRT) monitor works.

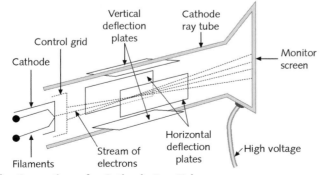

FIGURE 2-11   The Operation of a Cathode Ray Tube

Monitors come in many sizes, from 8-inch screens (screen size is always measured diagonally from one upper corner to the opposite lower corner) to 24-inch screens for computers and up to 42 inches for television displays. Screen size measures the actual distance across the screen. The image displayed on the screen is usually about two inches smaller than the measured size of the screen.

The **refresh rate** or **scan rate** is the number of times per second that the electron beam in the tube repaints the entire screen. The higher the refresh rate, the less flicker is apparent in the screen display and the more steady it appears to the viewer. Most monitors being sold today are set to refresh at a rate of 60Hz, which means the screen is completely repainted 60 times each second. This is adequate to prevent almost all flicker unless the display on the screen has an extreme amount of very rapid movement. In that case, the refresh rate can be set higher to eliminate the problem. The refresh rate of a PC monitor can be adjusted from the control panel in the Windows operating system.

Flicker in a monitor is also affected by whether the refresh method is **interlaced** or progressive (noninterlaced). Interlaced monitors paint their screens in two passes; they draw the odd-numbered lines on the first pass and the even-numbered lines on the second. This has the effect of changing only half the screen at a time and tends to minimize any flicker when the refresh rate is slow. Progressive monitors paint their screens completely in one pass. When the refresh rate is high (60Hz or more) their displays appear steadier and have less flicker than an interlaced display. Progressive monitors also appear to cause less eyestrain when viewed for long periods, though why this occurs is unclear. Computer monitors are progressive.

Television monitors are usually interlaced and nearly always have a refresh rate of 30Hz. This slow refresh rate, plus their limited **resolution,** makes television monitors unsuitable for use as computer displays and very few are used as such today. Table 2-2 defines some of the characteristics of monitors.

Flat panel monitors, while more expensive than CRT monitors, have now become the standard for computers, and their price difference is declining. They are available in sizes comparable to CRT monitors. In addition to providing sharper, less distorted displays than many CRT monitors, they have advantages such as being much lighter, being more compact in size, and consuming less electricity than CRTs. Flat panel monitors are known as liquid crystal display (LCD) panels. They form images by activating pixel-sized areas in a layer of liquid material sandwiched between two layers of glass or other transparent material. Two arrays of electrodes, one arranged in columns behind the liquid layer, and the other arranged in rows in front of it, produce the color display. By activating the electrodes on either side of the liquid layer, the controller allows only a particular color of light to pass through the layer at each pixel location. The lighted pixels together produce an image on the panel. Figure 2-12 illustrates how a flat panel (LCD) monitor works.

Early LCD panels were dim compared to CRT monitors. Dual-scan passive matrix displays improved performance by adding a second set of vertical electrodes to the sandwich, but these monitors still were not as bright as CRTs. Most new LCD panels are active matrix displays in which a transistor is added to enhance the brightness and color of each pixel. Active matrix displays can be used in lighted rooms and compare favorably with CRT monitors for brightness and resolution.

LCD panels can display only digital data. They can be used for analog (television) displays only if the data signal is first converted to digital form. **High-definition**

TABLE 2-2
Characteristics of Monitors

Characteristic	Description
Screen size	Diagonal (top left to bottom right) width of screen surface.
Refresh rate	The number of times per second an electron beam paints lines on a video screen from top to bottom.
Scan rate	The number of times per second an electron beam moves from the top to the bottom of a video screen; equal to the refresh rate in progressive screens, and to ½ the refresh rate in interlaced screens.
Interlaced	Screen refresh method in which the electron beam paints every other line during each top-to-bottom pass on the screen, odd-numbered lines on one pass and even-numbered on the second. Two passes are required to completely refresh the screen.
Progressive	Screen refresh method in which the electron beam paints every line during each top-to-bottom pass on the screen. Only one pass is required to completely refresh the screen.
Dot pitch	The distance between adjacent same-colored dots on a screen. Determines the resolution quality of the screen.
Resolution	The number of pixels (picture elements) that can be individually addressed on a screen by software.
Multiscan	A monitor that supports a variety of refresh rates and resolutions so that it can function with a variety of input devices.
Green monitor	A monitor that uses minimal electrical energy and thus supports the EPA Energy Star program.

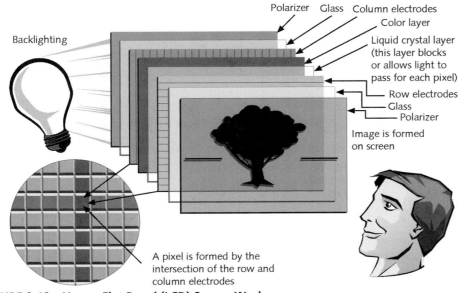

FIGURE 2-12 How a Flat Panel (LCD) Screen Works

TABLE 2-3
Monitor Resolution

Resolution	Horizontal Pixels	Vertical Pixels	Aspect Ratio
VGA	640	480	4:3
NTSC-TV	720	525	4:3
SVGA	800	600	4:3
XGA/SVGA2	1024	768	4:3
XGA+/SVGA3	1268	1024	4:3
High Def.	1680	1080	16:9

LCD panel television screens up to 60 inches in size (16:9 aspect ratio) are becoming more common in home theaters.

Monitors, both CRT and LCD, have several levels of resolution, which is a measure of the amount of detail that can be discerned on the screen. For a computer monitor, resolution is measured by the number of pixels on the screen that can be individually addressed by software and controlled by the computer's video card. The more pixels that appear on the screen, the finer the detail that can be seen clearly in the image. Resolution is controlled by the computer and does not depend on the number of dot **triads** that exist on the screen. A small monitor may use only a few triads to form each pixel. A large one can use many. Table 2-3 shows some of the common resolutions for computer monitors. The number of pixels noted for each resolution level is the same regardless of the monitor size.

Most monitors can display both analog and digital signals, and some computers have television signal tuners built into them. This allows the monitors to display standard television programs. If the resolution of the monitor is set to a higher level than 640 by 480 pixels, then the computer either resets the resolution to that level or the television display utilizes only part of the screen.

Printers

Printers are devices that output digital information on paper in text and graphic form. Several printer types are available, each using different technologies to produce printed images. Some printers print only black-and-white images, while others can print in color, even producing full-color photographs and other graphic images. The three major types of printers are **laser printers,** ink jet printers, and impact printers. Impact printers print only black and white, while the other two types can print in color or black and white.

Laser Printers

Laser printers are page printers. They form an image in much the same way as a photocopier and then print the entire page at one time. Laser printers have become so advanced and are now produced in such mass quantities that they are available at very reasonable cost and are very reliable. A small home-use laser printer can be purchased for about $125. Laser printers are the most economical type of printer due to the relatively low cost of the toner cartridges they use and the large number of pages (2,000 or more) that can be printed from a single cartridge.

Larger and heavier-duty laser models cost more and are intended to do a larger volume of printing over a period of time. They also print in color. A medium-range color laser printer costs around $500 and is easily capable of printing several thousand pages a month for several years without needing repair or service other than replacement of the toner (ink) cartridges.

Laser printers work by picking up toner on a charged image on the surface of a rotating drum. The charged image is created on the drum by laser light reflected off mirrors and onto the drum in a pattern controlled by data coming into the printer from the computer. Wherever the reflected laser light shines on the drum, it reduces or neutralizes the electrical charge of the drum in that area. The image is created a full page at a time without regard to whether it contains text, graphics, or both. The printer simply reduces the charge on areas of the drum according to its controlled instructions and these reduced charge areas form the image.

The areas with a reduced charge pick up toner from the cartridge as the drum rotates. The toner adheres only to the reduced charge areas on the drum because it is also negatively charged at a voltage below the white areas on the drum but above the image areas. The toner is thus negatively charged in relation to the white areas of the drum and is repelled from them (because like charges repel one another), but positively charged in relation to the dark image areas and is attracted to them (because opposite charges attract one another). As the drum surface rotates through the toner cavity in the cartridge, a toner image is deposited on its surface as formatted on the drum by the computer instructions the printer received.

Paper is moved through the printer at the same speed as the image drum rotates. The paper is also given a positive charge relative to the toner adhering to the drum. The toner is attracted to the paper as the paper passes near the drum. This transfers the toner image to the paper where it sits as loose bits of material until it is fused to the paper by hot rollers squeezing it onto the surface. The printed sheet then rolls out of the printer and into a tray. Figure 2-13 shows how a black-and-white laser printer works.

Color laser printing is done in a similar manner as black-and-white laser printing, except that the image-writing process is repeated four times, one for each color

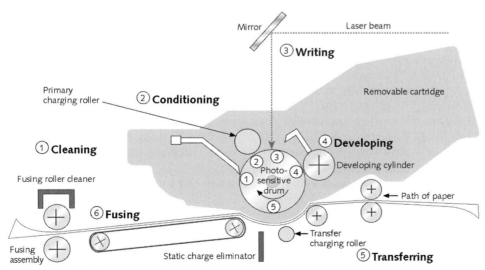

FIGURE 2-13   The Six Steps in Laser Printing Technology

toner needed for full-color printing (cyan, magenta, yellow, and black). A color laser printer must contain four separate writing, developing, and transferring systems that must be synchronized to work together and print their separate images in register (in the exact same position relative to one another). For this reason, it is a much more complex and expensive machine than a single-color printer.

In a color laser printer, the various colors of toner are deposited on the paper one after the other and then fused in place. The fusing process also turns the colors of toner somewhat transparent and blends them together in varying amounts to form the many colors possible. In color printing, digital instructions from the computer must not only specify the shape and size of images, but also the color of each dot in them. This can be done by identifying a specific color for an image area, or by having the computer separate the colors in a photograph or color graphic and send each of them to the printer to re-create the color image.

In a black-and-white laser printer and some color lasers, the removable toner cartridge not only contains toner but also the most wear-prone moving parts of the printer. These include the photosensitive drum, developing cylinder, and the primary charging roller. When a depleted toner cartridge is replaced, these parts are also replaced, thus keeping the printer in peak operating condition and eliminating the need for most common repairs caused by worn parts.

Some color laser printers do not combine any moving parts into the toner cartridges. Such printers require periodic replacement of the photosensitive drum and part or all of the fuser assembly to maintain high-quality printing. This maintenance involves expensive parts and makes color laser printers still more costly to use for high-quality printing.

Laser printers provide an excellent quality of printing with high resolution and good color control. Medium-quality printing for a laser printer is 300 dots per inch, about the same as the best dot matrix printer can achieve, but with much better ink coverage and much greater speed. High-quality laser printing is done at 600 or 1,200 dots per inch. Single-color printers run at speeds of eight pages or more per minute and color laser printers can often do four.

Ink Jet Printers

Ink jet printers print one line at a time using a print head that moves across the page. The print head consists of a row of tiny tubes that contain the ink. As the print head moves across the paper, it is directed by the controlling computer to squirt or jet ink onto the paper in minute quantities to form an image. The print head never actually touches the sheet. Only the ink is deposited as it passes, according to the copy design.

Because they print line by line, ink jet printers are slower than laser printers, but remain very popular for home use due to three factors: Their printing is high quality, they can print in color inexpensively; and their initial cost is lower than a comparable-quality color laser printer.

Most ink jet printers work by heating ink in the tubes of their print head until it boils. Tiny bubbles of ink form at the ends of the tubes and are ejected onto the paper. The droplets are charged (ionized) so they are attracted to the paper, and then the magnetic plates in the print head direct the ink placement to form characters and other shapes. Typical ink jet printers use 64 ink nozzles or 128, all of which can squirt ink according to instructions from the computer as the print head moves across the page.

Most ink jet printers have two print heads, one with a black ink cartridge for single-color printing and another with cyan, magenta, and yellow ink for color printing. This feature allows the black cartridge to be used independently of the color one, which conserves ink on single-color printing and produces brighter cleaner black areas in color printing than is possible by simply combining the three primary colors.

Early ink jet printers printed 300 dots per inch, but newer ones typically print up to 1,500 dots per inch and can produce color prints of photographic quality, especially on coated, high-quality paper that holds the ink, but that does not allow it to soak into the surface or bleed. A page from an ink jet printer is often somewhat wet when it emerges from the printer, especially if the print is in color and the ink coverage is heavy. Fast-drying inks reduce the time needed for ink jet prints to achieve full stability.

Impact Printers

The first computer printers were impact printers. Early impact printers were almost like automated typewriters. They had individual metal character keys, or a multiple-key daisy wheel, which struck the paper in the printer through a ribbon placed between the key face and the paper. Because its keys actually struck the paper, an impact printer could print multiple copies and forms using carbon paper or carbonless treated paper to create the copies. Where multiple-copy forms are still needed, some of these printers are still in use, but most impact printers today are dot matrix printers.

Dot matrix printers still strike the paper through a ribbon, but they use a matrix of small pins (nine pins in early models, twenty-four in later ones), which are used in combinations, to form characters. One matrix of pins forms all text characters and can also be directed to create graphics images one line at a time. The print quality is only fair and deteriorates as the ribbon continually wears out or is depleted of ink. Dot matrix printers are relatively slow because they print only one line at a time and their print head pins are moving mechanical parts that cannot move as fast as jetting ink or electronic data.

A printer can be connected directly to an individual computer and serve as its dedicated printer, but the same printer can also be connected to the LAN as a node with its own IP address. When connected to a network, a printer can receive data from any node in the network or even from a node in another network and print output for all of the computer workstations that have access to it. Such printers are designated as network printers, or shared printers, because the network connection allows several computers to share the output of a single printer. This greatly increases the amount of output a single printer can produce and often eliminates the need for dedicated printers at each workstation.

To function on a network, a printer requires a NIC and an IP address that identifies its location to other users who may wish to send it data for printing. The NIC is usually an internal component of the printer, but it can also be contained in an external box and connected to the printer through a USB or parallel port.

Web Pad

A web pad (also sometimes known as a web tablet) is a wireless single-purpose device designed to simplify and enhance Internet connectivity. A web pad consists of a wireless portable pad with an LCD screen and a base containing a wireless transceiver that sends and receives data from the portable pad. The base unit is connected to the Internet and through its wireless link, the webpad can send and receive data over the Internet.

Web pads have gained some acceptance in the marketplace, but their single function use and relatively high cost due to the wireless technology and LCD screens they use, has limited their sales. The best units include a serial port, a USB connection, a PC wireless slot, and often a SmartCard slot (more commonly used in Europe than in the U.S.), as well as a selection of wireless drivers and Internet software tools.

Network Interface Cards (NICs)

A network interface card (NIC) is an electronic board that plugs into one of the expansion slots that all PCs have on their motherboards or it attaches to the computer through an external port. The NIC has one or more ports built into it that are used to connect the NIC and its computer to a network using a cable that plugs into the port. A NIC can support Ethernet, token ring, or FDDI network architecture, but only one of the three. It may have ports that can accept more than one type of cable connection. Now that Ethernet has become the dominant type of network architecture, most computers have Ethernet NICs built into them as standard equipment. If a different type of NIC is needed, such as one for an FDDI network, it can be substituted for the original.

The function of the NIC is to send and receive information from the system bus in parallel and to send and receive information from the network in series. The NIC also converts the data that it receives from the system into a signal that is appropriate to the network. For an Ethernet card, this means converting the data from the 5-volt signal used on the computer's motherboard into the voltage used by twisted-pair cables. The component on the NIC that makes this conversion is called a transceiver (transmitter/receiver). Ethernet cards formerly had more than one transceiver to convert data into the appropriate voltage for coaxial cable and twisted-pair cable connections, but now the latter connection is standard. If a different connection is needed, a different NIC or an adapter can be installed. Figure 2-14 shows some examples of NICs.

Built-in NICs and those purchased separately all have identifying addresses coded into them by the manufacturer, which are used by the network to identify the computer (node) using the card. These addresses are called Media Access Control (MAC) addresses, physical or adapter addresses, or Ethernet addresses. They consist of six-byte (48-bit) hexadecimal codes, which are unique for each card. Part of the address contains the manufacturer identifier and the rest is a unique number. No two NICs have the same identifying code.

When selecting an external NIC, it is critical to match it with the network architecture to which it will connect, the specific type of cable connection it will use, and the type of slot in the computer (PCI or ISA) in which it will be installed.

Server

A server is simply a computer or similar device on a network that provides services or manages network resources. For example, a file server is a computer with a large storage device dedicated to storing digital files. Any user on the network can store files on the file server. A print server is a computer that manages one or more printers, and a network server is a computer that manages network traffic. A database server is a computer system that processes database queries.

FDDI

Token ring

Ethernet

FIGURE 2-14   Types of NICs

In large networks, servers are often dedicated, meaning that they perform no other tasks besides their server tasks. In a home network, however, a single central computer can execute several programs at once, including server functions. A server in this case generally refers to the software program that is managing resources rather than the entire computer, which may be performing other functions as well.

As its name suggests, a server is something like a waiter or food server. A client (node on the network) asks it for something—a file—and the server gets the file from storage and sends it to the client. In most cases, the server does not read or otherwise process this file, but simply hands it off to the client who asks for it.

The client may also request a service—printing, for example—and a server, which manages network-printing functions that provide the service to the requesting client. As stated earlier, in many networks, servers are highly specialized and often dedicated to a single function or service, but in a home LAN, a single server is likely to be multifunctional and sufficient for all required services.

Among the types of servers that a home network is likely to use are the following:

- Print server to control printer functions
- File server to store and retrieve files
- Video server to store and send video files to playing devices
- Audio server to store and send sound files to playing devices

- Backup server to keep backup copies of critical files updated on a scheduled basis
- Remote Access Server (RAS), which allows clients using Microsoft dial-up networking and other similar services to dial into the server from outside the LAN

Router

A router is an intelligent switch—in other words, a device that can make decisions according to preprogrammed instructions about the best routes for electronic data packets to travel, and direct the packets accordingly. Routers are the hardware units responsible for directing data traffic between interconnected networks. They read the IP address of each packet of data that comes to them, decide what is the most efficient path for the data to take to reach its destination quickly, and then switch the data packet onto that path.

A router is always connected as a node in at least two networks and may be connected as a node in many networks. For each network to which it is connected, the router has a separate unique IP address. A router routes the data packets it receives from one LAN onto another LAN, which either contains the data packet's destination node or is connected to another router closer to its destination node.

A home network will not require a router as one of its nodes unless it is connected by cable directly to another network. Most home networks connect to the Internet by DSL, cable, or by using a dial-up modem to connect to an ISP link. A router is not necessary for sending data to the Internet by these methods. Once the data reaches the ISP and is transmitted onto the Internet, routers in the system direct the data to its destination by choosing the most efficient path to send it, but the home user's connection to the Internet is solely through its ISP link and this is the only choice available for transmitting data out from the network or bringing data into the network.

Switches and Bridges

A switch or a bridge can sometimes be useful in a home network to reduce data traffic and thereby make the system more efficient. Both switches and bridges function only in a local network, and both make decisions about allowing data traffic to pass. These decisions are based on their address tables and the MAC address contained in each data packet that identifies the packet's destination address. Because bridges and switches do not send data to the Internet, but transmit it only within the LAN, the tables they maintain contain only local area addresses (as distinct from the Internet-routable IP addresses used by routers). The tables themselves are referred to as bridge tables, forwarding tables, or content addressable memory (CAM) tables, depending on the type of device.

When a hub receives data from a node in a LAN, it broadcasts the data to every other node in the LAN. Because the data is addressed to one node only, all the other nodes that receive a packet not addressed to them discard it. Each node receives all the data addressed to every other node, but discards everything except its own data. This means that the majority of data packet traffic on the LAN is discarded, but in a small network the volume of traffic is not great enough to warrant any improvement in this broadcast technique. The broadcast hub works fine.

On a larger or busier LAN, data packet traffic may be sufficient to justify using a switch or a bridge. Which to use depends on how traffic flows on the LAN and where it is heaviest.

If heavy traffic flows between two nodes on a combination bus and star Ethernet LAN, such as from a video server to a digital display device, and the other nodes have only lesser amounts, then a bridge may be called for. A bridge is a device that acts like a hub joining several segments of a LAN. Unlike a hub, however, a bridge does not broadcast the data it receives unless certain conditions are met. When the bridge receives a data packet from a segment of the LAN, it checks its routing table to see whether the destination node is located in the same segment of the network from which the packet arrived. If it is, the bridge refuses (discards) the packet because it can reach its destination by another route within the segment. If the destination address is not in the routing table for the source segment of the network, the bridge broadcasts the packet to all sections of the LAN to which it is connected, just like a hub.

Bridges work well in LANs where there is heavy traffic between nodes within segments of the network and not much traffic between nodes in different segments. They do not work well where traffic is more evenly distributed among nodes in different segments of the LAN. In this case, the bridge acts mostly like a hub and broadcasts too much unnecessary traffic on the network. The traditional 80/20 rule suggests that 80% of a node's data traffic is to another node within its own network segment. While this may hold true in some business LANs, the reverse is more likely to be the case in a home LAN. The majority of data traffic is coming from and going to locations outside the LAN in the form of e-mail, Internet data transfers, video programming, and similar data services. Only a relatively small portion of a home LAN's data traffic is likely to be directed to other nodes on the LAN.

Figure 2–15 shows how a bridge or switch connects two or more segments of a network in the same manner that a router connects two or more LANs. A device can function as either a bridge or a switch depending on how it is programmed to process the data packets it receives. A hub, by contrast, makes no data decisions, but merely broadcasts all data packets it receives.

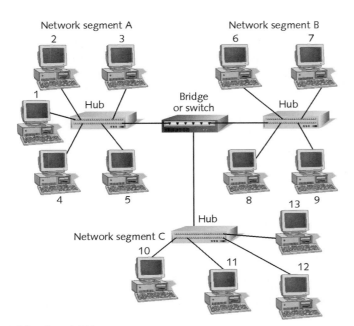

FIGURE 2-15   Bridge in a LAN

A switch, however, is useful to limit unnecessary traffic on a LAN, no matter where the traffic occurs within the network. Like a bridge, a switch has a routing table of MAC addresses for all the devices connected to it on the network. Also like a bridge, it refuses any packet it receives that is addressed to a node within its own segment because such a packet is received by another route on the network. If the address of the packet is not in the routing table of the switch's network segment, it does not broadcast the data as a hub or bridge would. It sends the packet only to that segment of the LAN where its addressed node is located. A switch connecting four segments of a network, each having an equal volume of traffic to the others, reduces network traffic by half over what would occur if the segments were connected by a hub or a bridge.

Wireless Access Hub

In a wireless network, an access hub receives data packets from all the nodes within its operating zone. The nodes may vary in identity and number because they are mobile and frequently move from the zone of one hub to that of another. The wireless hub does not know what nodes are within its range until each node's wireless NIC contacts the hub and associates with it by identifying itself and "signing on" to the hub. The hub broadcasts all the data packets it receives for all the nodes that are associated with it. The hub continues broadcasting data for each associated node until a node becomes associated with another hub. In a home LAN, only one wireless hub may be operating and all wireless nodes receive their data through that one hub. However, the same wireless node that operates on the home LAN may at times connect to a wireless hub at an office or other LAN located outside the home.

Note: The fact that any wireless node can associate with and receive data from any wireless hub broadcasting on its frequency means that every wireless network is potentially open to unauthorized users and should be made secure before any private information is sent on the network.

Wireless hubs are also connected to a wired network to which they also send all the data packets they receive. All data sent to a wireless hub from other networks and the Internet is usually sent via a wired connection. The hub then broadcasts these data packets to wireless receivers who are associated with it. A node on a wireless network can receive data only from other nodes on the network or from the Internet after the node has associated with a wireless hub on the network, and by doing so, announced its location. Until a wireless node associates with a hub and reveals its location, no other node on the network or the Internet can know where to send data intended for it.

Firewalls

The Internet now has more than 150 million users with more going on-line each day. Among that vast throng are many who use the Internet for malicious, immoral, unethical, or illegal purposes. A firewall prevents such users from sending inappropriate data to, or accessing data on, a home LAN. Placed at the connection point of the network with the Internet, a firewall can be a software program, hardware device, or a combination of both. A firewall enhances the security of networks by filtering incoming and outgoing traffic, thus protecting against unauthorized access to any node or stored data on the LAN and blocking unwanted data from entering the LAN.

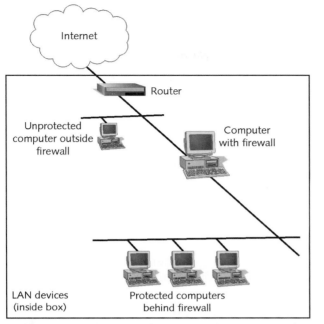

FIGURE 2-16   Firewall-Protected LAN with Unprotected Data Server for Internet Service

An effective firewall acts in both incoming and outgoing directions. It can be programmed to grant Internet access to the network only to selected "trusted" locations, or public access can be limited to selected "safe" services. On the other hand, outgoing traffic can also be filtered to ensure that valuable data is sent only to approved locations. Figure 2-16 shows a LAN with firewall protection on its central computer and an unprotected computer set up for Internet access to cleared data.

A firewall with full ingress/egress packet filtering can support dynamic packet processing for all of these filtering parameters:

- Internet protocol—TCP, UDP, ICMP, and others
- Source IP address
- Destination IP address
- Source TCP/UDP port
- Destination TCP/UDP port
- SYN packet

The firewall module should also provide an alerting, logging, and reporting system, which permits easy inspection of firewall activity as well as timely warnings for suspicious traffic activity. Firewalls also protect against virus programs, which can be sent to a node via the Internet and reside in its storage indefinitely. Such programs may harm a computer's data or permit access by a hacker. Because new and more sophisticated virus programs are continually being created, it is important that firewalls and other security devices be upgraded frequently to incorporate defenses against the newest attacks.

Firewall software is often purchased as a "fit-and-forget" antihacking program that can be easily installed and that offers straightforward LAN protection.

Most systems can also be customized to control which programs may access the Internet. In addition, they may have an IP-filtering tool that the user can set to specify which IP addresses can connect inwards or outwards. If, for example, a rogue program hidden within an unspecified e-mail attempts to connect to the network, the firewall can detect it and prevent it from being allowed to connect.

Summary

- A local area network (LAN) is a specifically designed configuration of computers and other devices, located within a confined area such as a home or office building, and connected by wires or radio waves that permit them to communicate with one another to share data and services. A wide area network (WAN) covers a large geographic area and is connected to the Internet.

- All data on a LAN is sent as packets, which consist of a small piece of data with information attached about what type of data it is, where it came from, and where it is going.

- Ethernet is the most common network topology and can be physically arranged in either a bus topology or a star. Other network topologies include wireless, FDDI, and token ring.

- Wireless LAN nodes are not physically connected at all to one another or to a central device. They communicate with an access point or wireless hub using a wireless network interface card (NIC) that includes a transceiver and an antenna.

- The Internet is a collection of LANs and WANs all interconnected to one another and able to communicate with each other. Data going from a node on a home LAN to a node on a distant network reaches its destination by traveling though the communication lines of other networks located between the originating point and the destination.

- Over half of home networks connect to the Internet through a high-speed (broadband) connection: a DSL, a cable link, or a satellite connection. The rest use a slower connection through an Internet service provider (ISP) by means of an ordinary telephone line and a modem.

- The heart of a home network is the central computer, which performs multiple functions. Each device connected to a network is called a node, or host. A node can be a computer, router, printer, sensing device, video camera, controller, or any number of other electronic devices. A host is always a computer.

- Monitors are CRTs that display images on the flat front face of a glass cathode tube or flat panel LCD. Their defining characteristics include screen size, refresh rate, resolution, and interlaced or progressive scanning.

- The three main types of printers in use on networks today are laser printers that operate similar to a photocopier; ink jet printers that squirt or jet ink onto the paper from the print head, and dot matrix printers that strike the paper through a ribbon using a matrix of small pins.

- A NIC is an electronic board that is mounted on a computer's motherboard, plugged into a PC expansion slot, or connected through an external port. The NIC connects the computer to a network using a cable that plugs into a port on the NIC.

- Hubs, routers, switches, and bridges are all devices that are used for transmitting or directing data traffic between interconnected networks and nodes. A server provides network services and manages network resources.

- A firewall is a software program, hardware device, or combination of both that filters incoming and outgoing traffic, thus protecting against unauthorized access to any node or stored data on the LAN and blocking unwanted data from entering the LAN.

Key Terms

Analog data Data in nonnumeric form such as radio waves, sound waves, and so on.

Bandwidth The amount of data that can travel over a communication line or wireless connection in a given length of time.

Bridge An intelligent switch that limits data flow on a LAN.

Bus topology One form of network architecture for Ethernet.

Cable modem A device that converts digital data to analog signals and connects a LAN to an ISP via the cable television connection.

Carrier Sense Multiple Access/Collision Avoidance (CSMA/CA) A method of data transmission in which nodes avoid data packet collisions through use of a token or other device controlling the movement of data.

Carrier Sense Multiple Access/Collision Detection (CSMA/CD) A method of data transmission in which data packets contend for space on the network, and nodes sense packet collisions that require resending.

Cathode ray tube (CRT) A monitor or TV screen.

Checksum A mathematical method for a receiver to determine if a data packet has been corrupted.

Digital data Data in the form of binary numeric code or a derivative of binary.

Digital subscriber line (DSL) An advanced form of ISDN line that makes possible the transmission of voice and high-speed digital data on the same line at the same time.

Dot matrix printer A printer that prints using a matrix of small pins that strike the paper through a ribbon and combine to form characters.

Ethernet The most common form of LAN architecture. It uses bus or star topology and employs CSMA/CD to manage the flow of data on the network.

Fiber-Distributed Data Interface (FDDI) Large fast networks that are constructed almost entirely with fiber-optic cable.

Fiber-optic cable A very high-speed means of transmitting data using light beams through glass or plastic threads or fibers.

Firewall A software program or hardware device that controls information passing from the Internet onto a LAN and from a LAN onto the Internet.

Flat panel screen An LCD monitor or television. *See* liquid crystal display (LCD).

Header Data at the beginning of a data packet identifying its source and destination.

High-definition TV New television standard with higher resolution, improved picture quality, and a 16:9 width-to-height ratio.

Host Another name for a computer on a LAN.

Hub A device that connects nodes on a LAN and broadcasts data received from any node to all other nodes.

Integrated Services Digital Network (ISDN) A technology that uses a telephone line to transmit digital data at high speed.

Interlaced A type of CRT on which the screen is refreshed alternately on its odd and even lines.

Internet A worldwide web of interconnected, but independent, networks over which data travels from source to destination by various routes.

Internet service provider (ISP) A company that provides Internet connections to home and business LANs.

Internetwork Packet Exchange/Sequenced Packet Exchange (IPX/SPX) The protocol used by Novell NetWare networks.

IP address An Internet Protocol address is a unique node address that consists of four numbers separated by periods.

Laser printer A printer that works by picking up toner on a charged image on the surface of a rotating drum and depositing it on paper where the image is fused in place.

Line speed The amount of data that can travel over a communication line or wireless connection in a given length of time.

Liquid crystal display (LCD) A thin lightweight type of video display that uses liquid crystal material sandwiched between two layers of electrodes to create a color image.

Local area network (LAN) A regionally confined network consisting of computers that communicate and share data and services.

Monitor   A screen output display for a computer or TV signal. Can be CRT or LCD.

Multistation Access Unit (MAU)   Device used in a token ring star design to which all nodes are connected.

Network BIOS Extended User Interface (NetBEUI)   A Microsoft proprietary protocol commonly used for LANs.

Network interface card (NIC)   A device for connecting a node to a LAN.

Network printer   A printer of any type connected to a LAN as a node with its own IP address.

Node   A computer or other device connected to a LAN by a NIC.

Non-interlaced   A type of screen refresh that renews each line of the screen in order. This method of refreshing is also called progressive.

Packet   A small segment into which data is divided and packaged with a header and trailer for transmission on a network.

Plain Old Telephone Service (POTS)   The most common method of home Internet connection.

Progressive   A type of screen refresh that renews each line of the screen in order. The other type of refresh scheme is interlaced, which renews every other line.

Protocol   A set of rules and standards that a network uses to communicate among its nodes.

Refresh rate   The number of times per second that the electron beam in a CRT repaints the entire screen.

Repeater   A device in a LAN that receives and strengthens the data signal to offset its attenuation over distance.

Resolution   The number of pixels on a monitor that are individually addressable by software.

Routable   A protocol that allows data to be sent to interconnected networks on the Internet.

Router   A device that connects two or more networks and directs the data traffic passing between them.

Satellite link   An Internet connection to an ISP via a satellite through a receiver dish antenna.

Scan rate   The number of times per second that the electron beam in a CRT repaints the screen from top to bottom. Same as refresh rate in progressive screens, equal to twice the refresh rate in interlaced screens.

Server   A computer or device on a network that provides network services or manages network resources.

Star topology   One type of network topology in which nodes are arranged in a star pattern.

STP cable   Shielded twisted-pair cable used for LANs.

Switch   A device used in a LAN to direct data traffic among the nodes.

Terminator   A device on an Ethernet that ends the data flow in a bus topology.

Token ring   A type of network in which data flows in a circular pattern and is controlled by a token.

Trailer   Data attached to the end of a data packet.

Transmission Control Protocol/Internet Protocol (TCP/IP)   The most common protocol used to connect networks.

Triad   A set of three dots in a color monitor which in combination can produce all colors.

Wireless hub   A device to which nodes in a wireless LAN can connect using radio waves.

Wireless NIC   A device in a wireless node that connects it to a hub using radio waves.

Zone   The area around a wireless hub which its transmission reaches and from which it can receive data from wireless nodes.

Review Questions

1. What are the three types of Ethernet and what distinguishes them from one another?
2. The device on a LAN that broadcasts the information it receives is called a _____.
   a. switch
   b. router
   c. bridge
   d. repeater
3. What is the difference between a switch and a bridge?
4. The header of a data packet contains a destination address. True or false?

5. How does a token ring network differ from an Ethernet network?

6. CSMA/CA is a protocol procedure for _____.
   a. detecting data packet collisions
   b. detecting a line fault
   c. avoiding data packet collisions
   d. avoiding line interference

7. A device that strengthens the data signal on a network and also amplifies any noise on the line is called a(n) _____.

8. Describe how the token controls the flow of data on a token ring.

9. A wireless NIC connects to an RJ-45 port. True or false?

10. An inkjet printer uses powdered toner to print. True or false?

11. The term CRT in a computer monitor refers to _____.
    a. certified receiving terminal
    b. central route tray
    c. cathode ray tube
    d. cyan, red, turquoise (colors)

12. An interlaced display differs from a non-interlaced display because it scans (refreshes) its screen in a progressive manner. True or false?

13. How does a router function in networks?

14. An IP address is used on the Internet to direct _____ to their destination.

15. What is the function of a firewall on a network?

16. The abbreviation "Mbps" stands for _____ and refers to the speed at which data travels on a network.

17. A Multistation Access Unit (MAU) is used on a _____ network.

18. Can a wireless node outside the zones of all access points still connect to a wireless LAN? Why or why not?

19. A wireless LAN is inherently less secure than wired LANs. True or false?

20. In a home LAN, an Internet service provider (ISP) offers a connection for the LAN to the _____.

21. What is a DSL line, and does it carry digital or analog data?

22. In most satellite ISP connections, data is sent from the LAN to the ISP by _____.

23. The aspect ratio of most computer monitors, television screens, and high-definition television monitors is 16:9. True or false?

24. What does a laser printer toner cartridge typically contain besides toner?

25. If a router receives data from a node that is using only the NetBEUI protocol, it discards the data packet. True or false?

Hands-On Projects

Project 2-1: Research Installation of a DSL

In this project you will research installing a DSL in a home. For this project, you need access to a computer connected to the Internet or a telephone yellow pages directory. You will also need access to a local telephone.

1. Find out what companies provide DSL service in your area and the specific neighborhoods they cover. You can find this information in either of two ways:
   a. Look in your local telephone directory under the heading "Internet service providers."
   b. On the Internet, do a search for the phrase "Internet service providers," adding in the name of your city and state.

2. Select three companies whose phone directory advertisements or Web sites indicate that they provide DSL service.

3. Determine from the Web site or call the company to find out if your home can be served with a DSL line. If it can, continue with the project as if it were

for your own home. If it cannot be served, select an address (such as for your school) that can be served and continue the project as if for that address.

4. Contact each provider by phone and determine the following information:

    a. What is the cost for DSL service from each provider?

    b. What additional services does each offer?

    c. Are you required to use the same company that provides the DSL line as your ISP? If not, what other ISPs are available?

    d. What equipment are you required to buy or lease to use the service?

    e. What is the installation cost for a DSL line?

    f. What is the monthly cost?

5. Write a short report noting the answers to the preceding questions for each of the three providers. Indicate which of the three you would choose as a DSL line provider. If you choose one other than the lowest cost provider, explain why you selected it over the others.

Project 2-2: Compare Cable Modem and Satellite ISPs

In this project, you compare the cost and availability of cable modem and satellite service to your home. For this project, you will need access to a computer connected to the Internet or a telephone directory. You will also need access to a local telephone.

1. Find out what companies can provide cable modem and/or satellite service in your area and the specific neighborhoods they cover. You can find this information in one of the following ways:

    a. Look in the yellow pages listings of your local telephone directory under the heading "Internet service providers."

    b. On the Internet, do a search for the phrase "satellite Internet service providers," adding in the name of your city and state.

    c. Do another search for the phrase "cable Internet service providers," again adding in the name of your city and state.

    d. There may be only one service provider for each type of service in your area. If there's more than one, select one provider of each type of service and continue the project with those providers.

2. Determine from the Web site or call the company to find out if your home is within their service area. If it is, continue with the project as if it were for your own home. If it is not, select an address (such as for your school) that is within the service area and continue the project as if for that address.

3. Contact each provider by phone and determine the following information:

    a. Is the firm's service two way or downstream only?

    b. If the service is one way, how can you obtain an upstream connection to go with it?

    c. What are the startup costs for each service?

    d. What is the monthly fee?

    e. What services are included?

    f. Do you have any choice of ISP other than the cable or satellite provider?

    g. Does the service require a long-term contract or other unusual conditions?

4. Write a short report comparing these two services and the DSL line provider you selected in Project 2-1. Note which service you would choose and your reasons for doing so.

Project 2-3: Show How Bridges, Switches, and Routers Work

In this project, use Figure 2-17 to answer the subsequent questions about the functions and effects of network devices.

1. If a data packet is sent from Computer 3 to Computer 12, what routes does it travel if Device A is a router, Device E is a switch, and all other connecting devices are hubs?

2. If a data packet is sent from Computer 3 to Computer 12, what routes does it travel if Device A is a router, Device E is a bridge, and all other connecting devices are hubs?

3. If a data packet is sent from Computer 9 to Computer 6, what routes does it travel if Device A is a router, Device E is a switch, and all other connecting devices are hubs?

4. If a data packet is sent from Computer 9 to Computer 6, what routes does it travel if Device A is a router, Device E is a router, and all other connecting devices are hubs?

5. If a data packet is sent from Computer 10 to Computer 1, what routes does it travel if Device A is a switch and all other connecting devices are hubs?

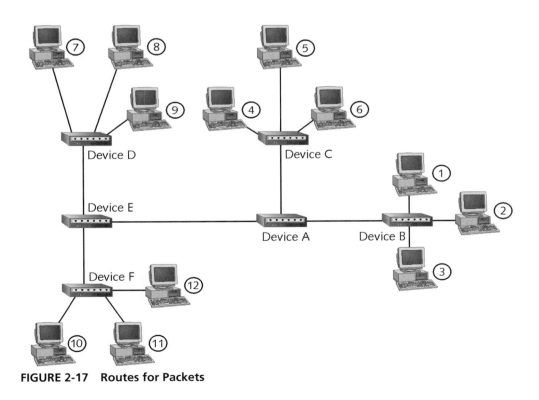

FIGURE 2-17   Routes for Packets

Project 2-4: Find a Firewall Protection System

In this project, you find a suitable firewall software package for use on your own LAN. For this project, you need access to a computer connected to the Internet. The information you obtain in this project could be useful in the future should you ever need to make a related purchase.

1. Open your Web browser and connect to the Internet.

2. Connect to a search engine that you prefer, or go to http://www.yahoo.com.

3. Do an Internet search for "firewall software protection" or another phrase that you think will locate firewall programs available on the Internet.

4. Browse the results of your search and find at least three companies that sell firewall software. Bookmark each company's site so you can return to it.

5. Evaluate each firewall product for the protection features it offers, how easy (or complex) its installation process is, what additional hardware, if any, it requires, and its cost.

6. Select the firewall product you prefer. Write a short report noting why you selected it over its competitors.

Project 2-5: Determine the Resolution of a Computer Monitor

In this project, you will use the Windows operating system of a computer to determine the resolution of its monitor.

1. Press the Windows key on your computer keyboard, or click the Start button in the lower-left corner of the screen.

2. From the Start menu, point to Settings, and click Control Panel.

3. In the Control Panel window, double-click Display.

4. In the Display window, click the Settings tab at the top of the window.

5. In the Settings screen, find the display resolution of the monitor.

6. Click the resolution setting and try to move it to a lower resolution. Do not attempt to increase the resolution to a higher level. The computer's video card or monitor may not be able to display a higher resolution.

7. If you can reset the resolution lower, see how the detail visible on the screen decreases. When you have observed the lower resolution, reset the display to its normal resolution.

Project 2-6: Change the Toner or Liquid Ink Cartridge in a Printer

In this project, you will change the black toner cartridge in a laser printer or the black liquid ink cartridge in an ink jet printer. Because the old cartridge will probably not be empty when you remove it, you will insert the same cartridge back into the printer as if it were a new one.

1. Find the user manual of the printer you plan to work on and review the instructions for changing its black toner or ink cartridge. If the printer is only one color, black is the only cartridge it contains. If it is a color printer, be sure you are changing the black cartridge and not one of the color ones.

2. Follow the instructions in the user manual to remove the old toner or ink cartridge from the printer. Most printers do not have to be turned off to change their cartridges, but it is a good idea to do so, especially if the printer is connected to a network or another computer that you do not control. If it is, someone could attempt to use the printer while you are working on it, so turning off the power is the safest procedure.

3. Perform any cleaning or other maintenance called for in the user manual, even though the cartridge may not be empty yet.

4. Return the cartridge to the printer as if it were a new one. Be sure the cartridge is seated correctly in the printer and that you have closed any moving parts you opened in removing it.

5. Turn the printer on again and print a test page to be sure the cartridge is functioning correctly.

Case Projects

Case Project 2-1: Design a Two-Family LAN

Two families who live next door to one another, in houses whose sidewalls are only 30 feet apart, both plan to install LANs in their homes. They would like the two LANs to be connected to one another so they can share data easily and so the two families can share the use and cost of a high-speed Internet connection. Write a short report or draw a diagram showing how you would set up these two LANs so they are connected directly to one another and can share an Internet connection.

Case Project 2-2: Recommend a Simple Network Design

A client asks you to recommend a network topology for her home LAN. She has only three computers and one printer to network. One computer has an Internet connection, which the others need to share. She might want to expand the LAN in the future, but for now she wants to get it working with a minimum of expense and new equipment. Write a short recommendation about the topology you recommend as being least costly to install, but that is expandable for later. Give the reasons why you chose it over other topologies.

Case Project 2-3: Research a Printer to Meet a Client's Needs

A client wants to purchase a new printer for home LAN use. All members of the family will use the printer. Most of the printing they do will be color photographs taken by their digital cameras and downloaded onto a computer. They also want to print their own color holiday cards. They will print some school reports with color illustrations as well as some business reports with graphs and diagrams, also in color. Finally, they will print letters, reports, and other documents in black and white, but these probably will not amount to more than 10% of their total printing output. Research at least three brands of printers and recommend a cost-effective quality printer for the type of printing the client will be doing. Specify the type of printer (laser, ink jet, or dot matrix) and the model you chose. Give a first, second, and third choice. (Three large printer manufacturers are Hewlett Packard at http://www.hp.com; Epson at http://www.epson.com; and Canon USA at http://www.canon.com. You can find printer information at these sites, but you may need to check retail dealers for prices.)

Case Project 2-4: Research Possible Displays for a Visually Impaired Client

A client who suffers from glaucoma, a condition that severely impairs sight, asks you to recommend a display for his computer that can enable him to read the screen accurately. To do this, he needs the ability to display the screen at a size at least three times larger than a 17-inch monitor. Four times larger would be even better. He does not need a resolution greater than 1024 × 768, because he cannot see fine detail, but he does need the large-size display. Research what type of display he should buy. Can you find a monitor or flat screen display three times as large as a 17-inch screen? What about an LCD projector? (Start your search at http://www.canon.com and http://www.proxima.com.) Do not forget to check on a video magnifier (like VideoEye! at http://www.videoeye.com or Aladdin Genie Pro Video Magnifier at http://www.optelec.com/Catalog.asp). Write a short report on what is available and the cost of each solution.

Home Network Design and Configuration

Introduction

In Chapter 2 you learned how networks are structured, how data is transmitted on a network, and the various protocols used to assure the uncorrupted movement of information. In this chapter, you will learn the varied information needs and installation requirements of home LANs and about other networking technologies that use power lines or telephone lines, already installed in a home, to carry data. You will also learn about several types of wireless technologies that can create a home LAN without any wiring at all. Lastly, you will find out about the network configurations and settings for a simple network.

Home Information Distribution Requirements

The definition of a home network has expanded with advances in technology. Initially, a home network was limited to connected computers and home automation (control of home appliances, security, lighting, temperature, etc.). Now, home

LANs incorporate not only these elements, but also full home video and audio distribution, video gaming capability, wireless networking, and connectivity to remote devices such as cell phones and controllers. The exponential growth of high-speed Internet subscribers (DSL, cable modem, etc.) has greatly expanded the requirements of home network users and the capability of the home LAN to meet those requirements. Today's home LAN is a network design providing intelligent communication and mutual data transfer between computers and peripherals, video and audio systems, digital home appliances, automated utilities, security systems, and other devices. It also supplies a gateway and broadband access to the Internet.

Data transmission requirements both to and from home LANs have multiplied by a factor of ten or even a hundred. Five years ago, a 10BASE-T Ethernet was the fastest technology likely to be found in a home LAN. At present, Fast Ethernet (100 Mbps) is the norm for new LAN installations, and Gigabyte Ethernet is increasingly common. Still higher speeds of 10 gigabits per second and more will be available soon.

The home network can be a hardwired system, wireless, or a combination of both, but it must provide a far greater degree of compatibility and adaptability to individual components and systems operating on different protocols and standards than is usually required in business systems.

Home technology systems are usually engineered to give adequate performance in specific consumer applications at a minimum cost. This design standard makes such systems affordable for the average homeowner, but often requires some limitations in the hardware, protocols, or other design parameters as compared to business-oriented networks.

Wired Network Types

Chapter 2 examined the structure of Ethernet LANs, which are by far the most widely installed type for home use. Many Ethernet LANs are hardwired, but the technology is equally well suited to wireless setups. Ethernet LANs can also be wired using FDDI fiber-optic technology to achieve still higher data transmission speeds. FDDI technology can also use **firewire** data transmission, which is particularly well suited to video and audio data where some tolerance for error in the compressed data is acceptable.

Token ring technology has been little used for home LANs, and this type of network has now almost passed into obscurity. While a few legacy token ring networks remain, new installations are almost exclusively Ethernet and this is the technology we will concentrate on in this chapter.

The cost of wiring and installing Ethernet may be high for some homeowners. Much of that cost centers around the expense of installing new wiring dedicated to the network. To reduce or entirely avoid this expense, several technologies have been developed that use wires already existing in the home to transmit data and require no new wires to be installed. Still other technologies do not use wires at all, but rely on radio waves to transmit network data. This chapter looks at several of these "no new wires" and wireless technologies in the subsections that follow. But first, the text takes a look at what is required to physically wire an Ethernet network in a home.

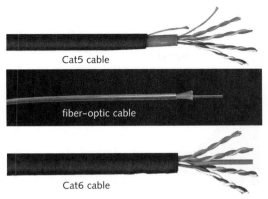

Cat5 cable

fiber–optic cable

Cat6 cable

FIGURE 3–1   Composition of Network Cables

Wiring an Ethernet Network

Hardwired 10BASE-T Ethernet and Fast Ethernet topologies both use Category 5 (Cat5) wiring for all the connections in the network. They can also use fiber-optic cable, but this is more expensive than Cat5 cable and also uses more costly plugs, jacks, and other connectors, which further increases its cost differential.

The vast majority of home LANs using Ethernet or token ring topology are wired with **Category 5 (Cat5)** or **Category 6 (Cat6) cable.** Figure 3-1 shows the composition of Cat5 and Cat6 cable, as well as that of fiber-optic cable. Table 3-1 compares how the different cables are used in the various Ethernet configurations and the maximum length the cables can extend. Coaxial cable is included in this table for comparison, although it is now rarely used for LANs.

In many new homes, Cat5 or Cat6 wiring is installed during construction so every room in the home is wired for network service in much the same manner as it is wired for electrical service. The contractor rarely installs the network equipment, but simply puts the wiring in place, adds standard jacks, and finishes the connecting point boxes in the walls with cover plates. When the homeowner

TABLE 3-1
Network Types, Speeds, and Cable Types

Network Type	Transmission Speed	Cable Type	Connectors	Maximum Cable Length
10BASE2	10 Mbps	Coaxial	BNC Connector	185 meters/ 600 feet
10BASET	10 Mbps	Cat5	RJ-45 Connector	100 meters/ 328 feet
100BASET	100 Mbps	Cat5 or Cat6	RJ-45 Connector	500 meters/ 1,640 feet
100/1,000BASEFX	100/1,000 Mbps	Fiber-Optic Cable	ST or SC Fiber-Optic Connector	500 to 2,000 meters/ 6,560 feet

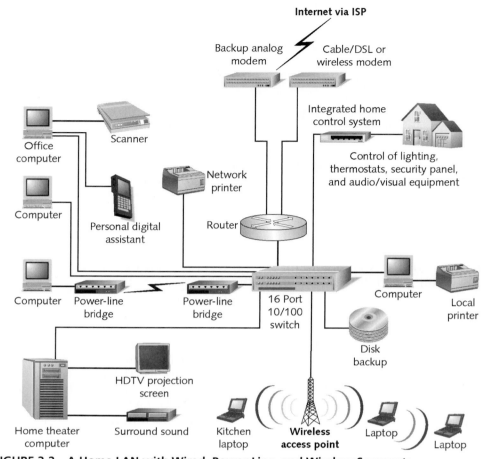

FIGURE 3-2 A Home LAN with Wired, Power-Line, and Wireless Segments

is ready to install a network, he or she can refer to the wiring diagram provided by the contractor and connect the network components directly into the wall-mounted points without the need to install any wiring except the patch cables from the wall connectors to the nodes. Figure 3-2 shows a home network with wired, power-line, and wireless segments.

If a home has installed Cat5 or Cat6 wiring, it's definitely the best option to use for a network. All the other wired and wireless network technologies have some potential drawbacks (limited capacity, potential for **interference,** slower speeds, etc.) that usually make them poorer choices for network infrastructure even under ideal conditions.

Most home LANs, however, are not installed under ideal conditions. Network wiring as a built-in feature has only been around for a few years, and not all the homes constructed even during the last decade have it. It is estimated that less than 10% of current U.S. homes have Cat5 or equal wiring that was installed at the time of construction. This figure is rising, however, as over half of new homes currently under construction are being wired for a LAN. For those that are not, and for those previously built without LAN wiring, installation of wiring after the home is completed is more difficult and more expensive. For many of them, other alternatives may be preferable.

In existing homes, network wiring must be installed by running or pulling it through the walls from access opening to access opening in order to provide connecting points in all the rooms where network service is desired. Pulling wires, also known as fishing wires, through existing walls often means taking the easiest route for running the wires, rather than the shortest. This tends to increase the amount of cable used over the amount used in a comparable new home where the wiring can be run more economically via the shortest routes. Longer cable runs, which out of necessity are sometimes pulled through tight obstacles and around sharp corners, may slow the LAN's operating speed somewhat.

Pulling wire for an installation in an existing home nearly always uses more wire, takes more time, and does some damage to the home (holes in the walls, moldings removed, etc.) that must be repaired. As a general rule, the older the home is and the more often it has been renovated in the past, the more difficult and expensive it is to retrofit wire it for a network. Heavy construction (brick, concrete block, plaster over wood lath, or heavy sheetrock) is more difficult to work with than lighter construction (wood siding, stucco, or plasterboard interior walls).

As you consider whether to install new wiring for a home LAN, carefully evaluate how long the job will take and how much damage will require repairing. Substantial homes built before the 1960s may well be exceedingly difficult to retrofit with new wiring. And some built before the 1930s may be virtually impossible without literally destroying a substantial part of the structure.

As the potential cost and difficulty of retrofit wiring increase, the use of other wired or wireless technologies for a home network becomes more attractive. Several of these technologies are less expensive to begin with than dedicated network wiring. If the cost differential mounts in a particular application, it will not be difficult for the network designer to see that another choice of technology is best for the job.

HomePNA or Phone-Line Network

Phone-line networking is a technology based on sending network data over the same wires that carry voice telephone conversations in the home. The Home Phone Networking Alliance (HPNA), a group of networking technology companies, developed the specifications for this technology. The phone-line data transmission standard that HPNA developed is commonly referred to as **HomePNA.**

HomePNA 1.0

The first version of the standard was designated HPNA 1.0 and transmitted data at up to 1 Mbps. It used frequencies slightly higher than the frequencies used for voice transmission on telephone lines and could thus be transmitted over the same wires simultaneously without interfering with voice transmission. HPNA 1.0 network systems are low cost and easy to install using existing telephone wiring. They are not much used in the United States now, having been superseded by newer HPNA standards (described in the following), but HPNA 1.0 is very popular in other parts of the world, especially in Asia where systems can be found in thousands of residential applications.

HomePNA 2.0

HPNA 2.0 can transmit data at speeds up to 10 Mbps, and is based on technology developed by Broadcom. In addition, it improved on HPNA 1.0 by permitting the prioritization of different types of data on the network and an enhanced data transmission protocol that overcomes interference problems on the telephone lines.

HPNA 2.0 allows IP telephone data and video data packets to be sent with a higher priority on the network than other data packets. Long delays in transmitting these types of data cannot be tolerated because they cause a great reduction in quality of the transmitted data when heard or viewed by the end user. HPNA 2.0 uses Quadrature Amplitude Modulation (QAM) protocol to transmit data, but can switch to Frequency Diversity Quadrature Amplitude Modulation (FDQAM) if interference is encountered on the telephone lines during operation of the network. FDQAM permits data to be transmitted on several frequencies (multiples of the primary frequency) simultaneously, thereby bypassing interference, at frequencies above or below it, and providing a cleaner, more accurate signal on the network.

HomePNA 3.x

HPNA 3.0 is an even faster specification that appeared on the market in 2004. It can transmit data at speeds up to 200 Mbps and uses both QAM and FDQAM data transmission protocols. This HPNA technology varies its data signals for maximum transmission efficiency by adjusting both to the different types of data it is transmitting and the characteristics of the wiring over which it is moving. It is also backward-compatible with HPNA 2.0, so it can be used on legacy systems without replacing all the components.

HPNA 3.0 controls data transmission on the entire network by means of a master control unit that assigns transmission schedules and priorities to all the devices on the network. The master control unit periodically sends its control message, called a media access plan, to all network devices. This coordinates the data transmission of each node. All other nodes on the network are slaves to the master controller, which provides for both synchronized (guaranteed priority) data transmission and random access (unscheduled priority) data transmission.

In HPNA 3.0 systems, most devices are capable of being set up as the master control unit or as a slave unit. The system can set up the master control unit automatically or it can be designated manually by the user. If a master control unit is shut down in a HPNA 3.0 network, the system will automatically designate a new control unit and restructure the network accordingly.

HPNA 3.1 is the newest HPNA technology and provides for data transmission rates of 250 Mbps using both telephone lines and coaxial cables. It uses the same frequency band (above DSL line frequencies and below television channel frequencies) on both types of cables. It is compatible with both HPNA 3.0 and HPNA 2.0 devices. HPNA 3.0 uses only FDQAM protocol when transmitting on coaxial cable (which is shielded from interference) and both QAM and FDQAM on unshielded telephone lines.

HomePNA Advantages and Disadvantages

As a network technology for home LANs, HomePNA 2.0 or 3.x offers some attractive features, particularly for installations in homes that are already extensively wired for telephone service. Consider the following:

• The system can be installed one unit at a time wherever a telephone line connection is available. The cost of individual transmission units is low.

• A large variety of networking devices are available from many manufacturers, all compatible with one another and all operating on the same HomePNA standard.

- HomePNA can accommodate both PCs and Macs on the network.
- The network functions as a peer-to-peer system that requires no hubs or routers. All data is sent to all nodes. Each node reads only what is addressed to it.
- The HomePNA standard transmits data at frequencies that do not interfere with analog voice traffic, so the telephone lines can also be used for voice transmission while the network is operating.
- The HomePNA 2.0 and 3.x standards are fast enough that they can meet almost all home network requirements including video transmission.
- As many as 25 devices can be networked using HomePNA, and the HomePNA system can also be connected as a segment to an Ethernet LAN.

The limitations of a HomePNA system mostly derive from limitations of the telephone wire system over which it operates:

- Each device must connect to the network through a telephone jack (or a BNC connector for HPNA 3.1 coaxial cable). The network is limited in scope by the number of available connection points. New jacks can be added, but wiring these is as difficult as wiring new jacks for an Ethernet.
- The system has some physical size limits. The maximum length of wiring between all devices on the network is 200 meters or about 650 feet (some HomePNA 3.1 systems can be "power boosted" to achieve longer distances up to 1,500 meters), and the area encompassed by the network cannot be more than 10,000 square feet, which is slightly less than the area of a quarter-acre lot.
- HomePNA will not work in some large homes that have old and extended telephone wiring.
- Once installed, HomePNA systems sometimes increase noise levels or cause distortion on the voice side of the telephone lines. This problem can often be solved with filtering devices, but doing so adds to the cost of the network.

How HomePNA Works

To transmit digital data over wires that are simultaneously carrying voice conversations, the HomePNA technology uses frequency-division multiplexing (FDM). FDM uses different frequencies from those used for voice transmission and sends each type of signal over the wire in its own separate channel or segment of bandwidth. By dividing the available bandwidth on a telephone line into uniform segments and sending data on each segment at a distinct frequency, FDM can actually increase the data capacity of a telephone line so it can simultaneously handle analog voice transmission, a high-speed DSL modem, and the 10-Mbps or higher data transmission of a HomePNA network. Figure 3-3 diagrams how network devices and voice telephones can be combined on a HomePNA network and used simultaneously. The network nodes require adaptors to transmit and receive data on the telephone lines. The telephones can transmit and receive voice traffic without adaptors. One ISP connection serves for both data and voice transmission.

To begin a HomePNA network, which can be expanded by adding new devices later, requires only a starter kit (available for about $100). These kits are available online from several U.S. suppliers, including http://www.homepna.com, http://www.netsys-direct.com, and http://www.superwarehouse.com, as well

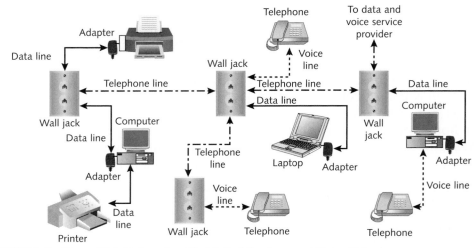

FIGURE 3-3   HomePNA Network with Digital Devices and Voice Telephones

as international firms in Europe and the Far East. The kits typically include HomePNA adapter cards for two computers, the cables needed to connect them to the telephone line, and an installation CD. Two versions of HomePNA adapters are available: an internal Peripheral Component Interconnect (PCI) card and an external Universal Serial Bus (USB) adapter, which connects to the computer through a **Universal Service Bus (USB)** port.

Additional individual PCI cards or USB adapters can be purchased to increase the number of nodes on the network. No other hardware is required. Laptop computers cannot accommodate the internal PCI cards, but work well with USB adapters. If the computer does not have a USB port, it can still be connected using a USB-to-PCMCIA adapter or USB-to-parallel port adapter.

The HomePNA architecture consists of adapters and bridges. One of these must be connected to each node of a HomePNA network and to the telephone or coaxial lines of the network. The adapter or bridge translates the data from the node into HomePNA protocol and sends it to the telephone lines or coaxial cables for transmission to other nodes on the network. If the HomePNA network connects to an Ethernet LAN, this must be done through a bridge that can translate data coming to and from each of the two network segments.

Both HomePNA adapters and bridges include two chips. The smaller 4100 chip is actually a transceiver that translates the analog data signals traveling over the telephone wires into digital data and sends it to a second chip, the 4210, to be interpreted. Conversely, the 4100 chip translates digital data received from the 4210 chip into analog signals and sends them out on the network telephone lines. The chip's signal is strong enough to carry up to 1,000 feet on a typical phone line. The translator chip does not process the data it receives or sends; it only converts it from analog to digital form, or the reverse, and sends it on to the network or the controller chip.

The controller chip receives data in digital form from the translator and filters it to remove noise that may have been picked up during transmission on the network phone line. The filtered data is then sent to the computer or other device where it is processed. For each packet of data received, the computer sends an acknowledgment back to the sender noting that the data was received. The computer processes all the data it receives, but only acknowledges the data packets addressed to it. Everything else is discarded. On a network with many nodes, this peer-to-peer arrangement means most data is not used, but because the computers communicate at a maximum speed of 10 to 200 Mbps on the LAN, the transfer of data is still fast enough for every application.

The two electronic chips that are the heart of devices using the HomePNA standard were both developed by Broadcom, but many companies now manufacture devices for HomePNA networks. These include 3Com, D-Link, Diamond, Intel, Linksys, and SMC.

Installing HomePNA

Installing a HomePNA network in a home where telephone jacks are in place next to or near each node is simple. Kits containing the necessary adapters, cables, and software for a two-computer system are readily available and additional nodes can be added using either PCI or USB adapters, each for $75 or less. When each node's hardware is in place, the cables are connected to a phone jack, and the software is installed on the computers. After this, the network is operational.

More than one device can be connected through a single jack if the connection point is fitted with a splitter such as those used for supplying multiple phone extensions in a room. Be sure to buy high-quality fittings to avoid imperfect connections that can introduce noise or interference on the line.

The signals transmitted over the phone lines by a HomePNA network are analog and not digital, and they are subject to degradation by noise on the line. Many of the analog devices that send data over phone lines (lines that are shared by the HomePNA network) introduce noise on the line along with their signal. The noise level is usually very low and does not affect the relatively slow signals of telephones, fax machines, or most other telephone line devices. Because HomePNA sends data at high speed and low power, noise can seriously disrupt the signals.

A slow rate of data transfer on a HomePNA network, or complete failure to communicate, is most often caused by noise from other equipment operating on the line. The noise corrupts some of the data packet transmissions, requiring them to be re-sent and slowing down the network, or it blocks so much of the data that the system cannot function at all. The solution to such noise problems is to remove the offending device from the network lines or place a low-pass filter between the device and the jack so its signals must pass through the filter. The filter blocks noise coming from the device, but still permits its legitimate signals to pass. Filters only operate one way: from the device to the line. A filter placed on the line from one device has no effect on noise originating in other devices. This means that a filter must be inserted between each noise-producing device and the telephone line to which it is connected.

Some fax machines, cordless phones, and other AC-powered communications devices also can produce another type of signal noise that must be filtered out of the AC power line to prevent it from affecting the telephone line network. To accomplish this, a different type of low-pass filter is placed between the AC wall outlet and the power cord connecting the device to it.

Because some resistance exists in the wire when passing electrical signals, all signals lose some of their energy and get weaker as they travel over long distances in it. This weakening process is called attenuation and affects all signals, both digital and analog. HomePNA signals are low power to begin with and are generally limited to travel of 650 feet in phone lines. This is adequate for the vast majority of homes, but not all. A few homes have wire runs that exceed this limit between two or more nodes on the network, and they will not be able to communicate with one another. HomePNA is a peer-to-peer system, so most nodes can function normally on the network, even though the two or three most distant from one another cannot.

When using HPNA 2.0 or 3.0 devices, there's no cure for this problem except to relocate the nodes closer to one another. HomePNA 3.1 networks that use coaxial cable can be "power boosted" to extend their range by several hundred feet. Using a 3.1 system costs more than a 2.0 or 3.0 network, but may still be preferable to rewiring nodes so the connections between them are more direct and shorter. The latter option may be so expensive that a wireless network or another technology is a better option.

AC Power-Line Networks

Like HomePNA, a **power-line network** uses existing wiring to carry data on the network instead of using telephone wires; however, this technology achieves the "no new wires" objective by transmitting network data signals on the regular AC electric wiring that also supplies power throughout the home. Two competing power-line network technologies are available. The original technology was called **Passport,** and was manufactured by a company named Intelogis.

The HomePlug Alliance, an industry group established by companies that manufacture products for the power-line networking technology, selected a newer technology called **PowerPacket** as the standard for power-line networking. Intellon developed this technology, and several additional companies, including Enikia, Adaptive Networks, and DS2, are now producing products based on it.

Power-line networks offer many of the same advantages of HomePNA networks because both use existing wiring rather than requiring new installation. Among the features that power-line technologies offer are the following:

• The system can be installed one unit at a time wherever an electric outlet is available. Because power outlets are usually more numerous and more widely distributed in a home than telephone jacks, power-line networks can often reach areas where HomePNA does not go.

• Many manufacturers are now producing PowerPacket-compatible devices, all of which operate on the same standard. The older Passport standard has fewer manufacturers.

• Power-line networks can accommodate both PCs and Macs.

• Like HomePNA, power-line networks function as peer-to-peer systems, which require no hubs or routers. All data is sent to all nodes. Each node reads only what is addressed to it and discards the rest.

• Power-line technology transmits data at frequencies that do not interfere with the low-frequency AC electric current flowing in the same wire. The network can function without regard to whether electric current is flowing or how much.

• Power-line technology allows data transmission at rates up to 14 Mbps, somewhat faster than 10BASET Ethernet or HomePNA 2.0 and easily able to meet all home network requirements.

• Unlike HomePNA technology, power-line systems do not require a card to be installed in the computer (although there are companies working on PCI-based systems).

Most of the disadvantages of networking through power lines apply only to the older Passport technology, which is now being phased out. The disadvantages are as follows:

- The network speed is much slower, about 50 Kbps to 350 Kbps.
- Performance can be reduced, or in some cases entirely blocked, by increased home power usage and older wiring.
- Passport works only with Windows-based computers and limits printing features.
- The access devices used in electrical outlets are rather large and unsightly.

Products based on PowerPacket technology have overcome most of these problems, but networks built with them may still be subject to interference from other electrical devices on the power lines. As with HomePNA, this problem can be solved with filters on the offending devices.

PowerPacket Network

PowerPacket technology is derived from the same FDM technology used for phone-line networking. The range of frequencies available for use on the electrical subsystem (4.3 MHz to 20.9 MHz) is divided into 84 separate carriers by a system called orthogonal frequency-division multiplexing (OFDM), which is similar to the technology used in DSL modems. These OFDM-generated frequencies can all be used for data transmission without interference from AC power running on the same lines. Figure 3-4 illustrates how a power-line network connects nodes through adapters to the home's AC power lines. Only one of the nodes has a data connection to an ISP; this connects all nodes to the ISP via the AC power lines.

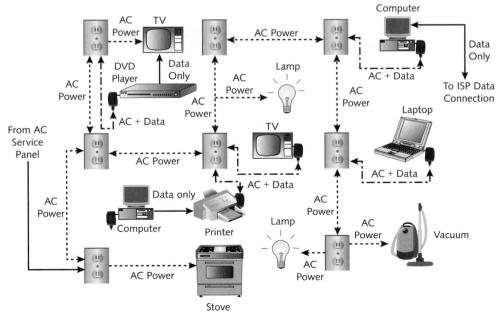

FIGURE 3-4 Power-Line Network Connected through AC Wires

OFDM sends packets of data simultaneously along several of the carrier frequencies, which allows the network to operate at speeds comparable to Ethernet systems. Current power-line networks transmit data at about 14 Mbps and new designs are increasing the technology's speed potential toward 100 Mbps.

If noise or a surge in power usage corrupts a data packet being sent on one frequency, the chip in the NIC that sent the data detects the problem and resends the packet on another frequency. This multiple frequency design, in which only some frequencies are used at any given time, allows PowerPacket to switch data packets onto clear frequencies as often as necessary to maintain an Ethernet-speed connection without losing any data.

Installing a PowerPacket Network

A small wall adapter, which fits in the electrical outlet supplying the computer's power, connects the device on a PowerPacket network to the AC wall outlet by means of a USB or Ethernet cord. This is a second low-voltage data connection that is in addition to the high-voltage AC power supply connection required for each piece of equipment. Manufacturers are now developing AC power cords with PowerPacket technology built in so that in the future only one connection will be needed: the combination PowerPacket/AC power cord.

Kits for setting up two computers on a PowerPacket network are available for less than $75 and additional adapters and connecter cables cost about $50 each. A different NIC adapter is required for a printer than that used for a computer, so you must specify the device to be connected to the network when buying. PowerPacket networks are compatible with HomePNA systems, wired Ethernet LANs, and wireless LANs. They can be connected to any of these other technologies through a suitable adapter on one node of the PowerPacket network. For HomePNA connections, a bridge is used that functions as a node on both types of network. For Ethernet and wireless networks, a network computer or a wireless hub can be connected to the power-line system through a regular adapter.

Software available with the PowerPacket adapters automatically detects all nodes (computers and printers) on the network and configures each device for the presence of all the others on the network. Like HomePNA, PowerPacket is a peer-to-peer network system in which all data is sent to all nodes, but individual data packets are accepted only by the device to which they are addressed. Other nodes simply discard packets not intended for them.

The network's Internet connection can be by cable modem, DSL, or normal modem. In each case, the **proxy server** software included with the system allows all computers to share the Internet connection.

Computers can be added to the network by simply plugging in a new outlet adapter, connecting the computer to it, and installing the software on the new node. Existing nodes detect the presence of the additional node and configure for it. Additional printers can be added using the printer plug-in adapter. File and printer sharing is done through Windows.

Passport Networks

Passport, the older Power-line technology, used frequency-shift keying (FSK) to send data over the electrical wires. FSK used only two frequencies, one for 1's and the other for 0's, to transmit digital information between the computers on the network. This method of data transmission proved to be too slow and too prone to interference to compete with other network technologies. Consequently, the manufacture of Passport AC power-line equipment has been discontinued.

Ethernet Power-Line Networks

Netgear and Panasonic have introduced high-speed Ethernet network hardware that uses power-line data transmission technology. The Panasonic HD-PLC power-line system uses a minimum of two identical adapter units, one of which is designated the master unit and the other a slave unit. The master unit controls the slave unit (also called a terminal unit) and additional terminal units (up to a recommended maximum of eight) can be placed on the network with it.

To set up the network, the master unit is plugged into a wall outlet and connected to a computer or a router by an RJ-45 Ethernet connector cable. The terminal unit is plugged into another wall outlet and connected to a second computer or router in the same manner. The two units are paired manually to communicate with one another. The two computers are then connected in a network and can exchange data at a speed up to 100 Mbps.

Netgear's Ethernet equipment is similar to Panasonic's but smaller in size. Both systems are very easy to install and set up, but both are more costly than PowerPacket components. A starter kit from Panasonic provides the adaptors needed to connect two computers on a Power-line network and costs about $200. Nothing else is required for the basic network. The HD-PLC power-line technology permits easy networking of computers connected to the same AC home wiring. Specific adapters can also connect printers and other peripherals to the network. The system is easily compatible with wired or wireless Ethernet segments. Additional adapters sell for about $125 each. Netgear's equipment is slightly less and, unlike Panasonic's package, includes Ethernet connector cables to connect each computer or peripheral to the adapter.

By connecting one or more wireless hubs through terminal units to the master unit, the Ethernet power-line system can serve as the wired portion of a wireless Ethernet network. It joins all the hubs and their associated wireless nodes to the master unit computer at data transmission speeds faster than that of most wireless LANs.

The Ethernet power-line technologies are fast and reliable, but are still subject to some interference problems, especially from hair dryers, vacuum cleaners, and other high-speed motors operating on power from the AC wiring. Touch-sensitive switches also seem to affect these systems adversely. Also, the adapters cannot be plugged into a power strip with a surge protector or noise filter because either of these devices will strip out the network data from the AC wires. AC power variations, such as surges, sags, interruptions, and so on (discussed in Chapter 13), do not affect power-line network data transmission, but lightning strikes and other severe power disturbances may cause problems.

Security for Ethernet power-line systems is better than for wireless networks because the adapters are individually set up, before they are first used, to communicate only with one another. Another adapter, even one plugged into the same wiring, cannot communicate with either the master or slave adapters in an operating network. Data on the network is also encrypted for further security.

X10

The **X10** technology started as a simple remote control system and has been expanded to perform other functions as well. Like PowerPacket and HD-PLC power-line, it uses the AC wiring in a home to transmit signals, but unlike the other systems, it is not a high-speed or high-capacity network system. While the other network technologies move vast amounts of data rapidly, X10 moves small data packets at ordinary speeds. Its major function is not the transmission of data files, but the control of home automation devices, and in this area it is the leading technology.

X10 is older than the other power-line technologies and, though equipment for it was originally produced by only one company, many firms now manufacture and market X10-compatible equipment, including GE, RadioShack, Sears, Stanley, IBM, Zenith, and others. Although sometimes used to control automation systems in business environments, it was designed for home use and has remained primarily a residential application. X10 equipment is inexpensive and designed to be easily installed by the average homeowner.

Basic X10 equipment consists of modules, which are plugged into electrical outlets and receive data on the AC wiring, and control panels, which are also plugged into AC outlet receptacles and send data over the power lines. In the basic X10 protocol, data packets are sent only in one direction: from control panels (transmitters) to modules (receivers). Each receiving module has a unit number by which it identifies data being sent to it. Control panels need no identifying number as they do not receive data, but only send it.

Basic X10 technology has been greatly expanded over the years and now includes two-way communication between control panels and modules, wireless communication between modules and controllers, and a much greater data transmission capacity (including some video and audio signals) than the original equipment could handle. Figure 3-5 shows an extensive X10 network with connections to lighting and appliances and various control panels (wired, telephone, wireless, and infrared remote) located throughout the home.

After the success of the original control devices, manufacturers of X10 hardware added new products to the technology along with new commands and bidirectional communication capability for some devices. Radio-based X10 devices and computer interfaces have also been developed to permit automated home lighting and appliance control in place of the former manual remote control. An X10-compatible telephone answering machine allows control of the system by calling in commands, and timers can send preset X10 signals as programmed. The line of X10-compatible equipment is growing continuously.

How X10 Works

The X10 protocol begins a data packet with a destination address consisting of a house code and a unit number of a module, or a group of destination unit numbers if the same command is to be sent to several modules. After the address, the packet includes a command signal. The data packet is sent to all modules on the network. Addressed modules accept and execute the command. Modules not addressed also receive the data packet, but discard it.

To control a lamp, for example, using an X10 module, the lamp is plugged into a module, which has a preset unit number, and the module is plugged into a wall outlet. The lamp's switch is then turned on and left on because the lamp can now be controlled from an X10 control panel located anywhere in the house.

A control panel can communicate through the AC wiring with every module in the home's X10 network simply by being plugged into an outlet. Additional control panels can be added wherever they are needed. Each control panel is an independent unit and functions without regard for what any of the others may be

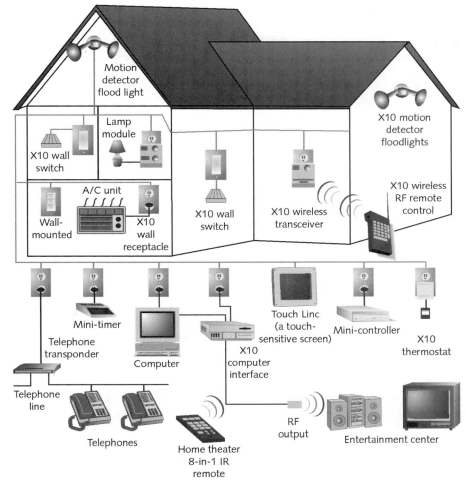

FIGURE 3-5   An X10 Network with Multiple Controllers and Modules

doing. A control panel requires no address or unit number because the modules that receive data from it do not need to know where the commands come from, only that they are addressed to the module.

The only control panel that is active on the network at a given time is the one that is sending a command signal. Commands travel only one way: from a control device into the AC wiring and thence to all modules on the network. Each signal is sent throughout the home's wiring and is eventually received by the addressee for which it is intended. The system is very flexible. Several modules can be assigned the same unit number, and they will always respond to the same commands. A command can be sent to several unit numbers at the same time.

X10 technology distinguishes between lamp modules and appliance modules, which are capable of handling larger AC power loads. Lamp modules can be turned on and off or dimmed and are limited to loads of 300 watts or less. Appliance modules simply turn on and off, but can handle up to 1,000 watts of power. To remotely turn on your coffeemaker in the morning using an X10 network requires that it be connected through an appliance module.

Because X10 signals travel on power lines, they can travel outside the home where the control panel is installed and into other homes that share a common AC power grid. For this reason, X10 includes in its protocol a house code, represented by one of the letters A through P. X10 modules respond only to control signals that carry the correct house code. The X10 protocol allows for 16 house codes and 16 unit codes (module numbers). More than one module can share a unit code. If 16 unit codes are not enough, even with sharing unit codes, a second or even third house code can be used for one network.

Advantages of X10

Compared to wireless, HomePNA, or power-line technologies, X10 is old, simple, and slow, but it still has many uses and is particularly well- suited to automated or remote lighting and appliance control. In addition to now being capable of radio or telephone control, X10 devices are also available that can interface with Ethernet, thus allowing X10 networks to be computer controlled.

For managing such automation tasks as lighting and appliance control, activating mechanical functions, and managing simple control sequences, X10 technology offers one of the most cost-effective and easy-to-install network systems available. Despite some limitations, it continues to be popular worldwide in the face of heavy competition from newer and more sophisticated network technologies. Part of the system's appeal is that its components are less expensive than any other networking solution. Another plus is that, in simple configurations, it requires very little technical knowledge to install.

Disadvantages of X10

The weaknesses of X10 as a home network system are mainly the result of the same simplicity that makes it an attractive automation choice. The most significant of X10's shortcomings is that the protocol has no signal verification. If a sent command is not received, the control panel that sent it does not know. Moreover, X10 is slow, which makes its lack of signal verification worse. It takes about a second to send an X10 command. While that command is being sent, another cannot be sent, or both data packets (commands) will collide and be lost. Data packets may fail to arrive because most houses are wired with two separate 110-volt circuit legs. X10 signals sent from a control panel connected to one leg may not reach a module connected to the other. A signal bridge can correct this problem.

X10 data packets can be blocked by other carrier-current devices, including wireless intercoms. They can also be blocked by power-conditioning equipment such as power strips, computer power supplies, and other devices. An X10 device called a "choke" can overcome this problem, but finding where to place the choke can be difficult. Interference from an outside source is also a potential problem for X10 signals. An expensive signal block may be required to filter such outside signals. Signal block installation is also costly and must be done at the main service panel by a professional electrician. Finally, appliances in the home can generate interference or block X10 signals entirely. Such devices require noise filters to correct the problem.

Newer X10 devices have largely overcome or even completely eliminated many of the disadvantages of the original X10 technology. Some X10 devices repeat commands to increase reliability. Others, especially those that use wireless transmission, can confirm the receipt of commands or their current status following a command signal. Improved filters have greatly reduced the noise interference problem.

Wireless Protocols and Standards

Wireless technology is becoming increasingly popular for home LANs, especially those installed in existing structures. The primary reason for this popularity is that wireless networks require few new wires or no new wires (if HomePNA or power-line technology is used) to be installed. The corollary reason is that, because it requires no new wires to be installed, wireless technology is usually less expensive than hardwired systems.

Wireless popularity has also been enhanced by the appearance of numerous wireless products that offer data transmission speeds and reliability levels comparable to hardwired systems. Many of these products are custom-designed specifically for home LAN applications and, for the trade-off of more limited application, offer easier installation, quicker configuration, and lower cost than hardwired devices adapted from business LANs.

In a wired network, each node's NIC sends digital data down a cable by changing the voltage on the wires from +5 volts to –5 volts using a prearranged protocol. Wi-Fi, **HomeRF,** and other wireless technologies simply replace the cable transmissions with short-range broadcasts of the data from small low-powered two-way radios. Instead of changing voltage on a wire, the wireless NIC encodes the zeros and ones by laying an alternating radio signal over a constant existing carrier signal, again using a prearranged protocol. The alternating signal encodes zeros and ones on the radio waves. Figure 3-6 shows a network configuration with several wireless segments. Wireless access points, called hubs, can be connected directly to the wired network or connected through a wireless bridge to another wireless access point.

At least four major wireless protocols are available:

- **Wi-Fi (Wireless Fidelity)** is the most widely used wireless technology at present. It is an **IEEE 802.11b** wireless standard and can transmit data at up to 11 Mbps.
- **WiFi5** (also known as WiFi/a) is an improved version of the original Wi-Fi technology and is also based on the same IEEE 802 standard. It is designated as IEEE 802a and can transmit data at speeds up to 54 Mbps.
- **WiFi/g (IEEE 802.11g)** is another version of the IEEE 802 standard that has a fast transmission speed of 54 Mbps. Unlike WiFi5, it is compatible with older and slower 802.11b hardware.
- **HomeRF (Home Radio Frequency)** is a wireless technology that was developed for the home market. It uses the Shared Wireless Access Protocol (SWAP) to transmit data on multiple frequencies in the 2.4 GHz radio band at rates up to 5.4 Mbps.
- **Bluetooth** is a short-range wireless technology limited to transmission distances of about 100 meters or less, which generally confines it to connecting nodes within a single room, or adjacent rooms.

Unlicensed Shared Use of Wireless Frequencies

All the major home LAN wireless technologies operate in the range of radio frequencies for which no transmitting license is required by the Federal Communications Commission (FCC) in the United States or by a similar agency in any foreign country.

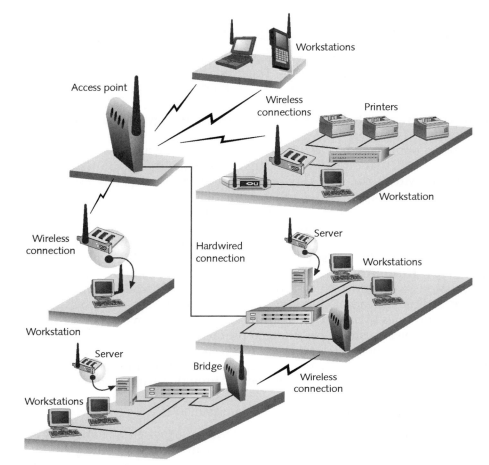

FIGURE 3-6   Wireless Network with Wireless and Wired Segments

The 2.4 GHz industrial, scientific, and medical (ISM) radio band is 83 MHz wide. This ISM band is used by all the main wireless networking technologies, cordless analog and digital phones, microwave ovens, and some medical equipment. As is the case with most unlicensed radio bands, no one owns or has exclusive use rights to any particular frequency in the band, so all users must share the radio spectrum.

Interference Potential

Because millions of users now broadcast wireless transmissions on the ISM band, the potential for interference of any one signal by others is great. This potential is reduced by the requirement that the transmitting power of an ISM device remain low and its broadcast range limited.

But even with limited range, wireless broadcasts are virtually certain to encounter some interference. The Federal Communications Commission (FCC) has rules that define how the ISM band can be used. They contain two key provisions that make spectrum sharing possible by reducing and mitigating interference. The first provision requires that all ISM devices must include some type of interference mitigation so that when interference does occur, the equipment automatically adjusts itself to reduce the effects of the interference.

The second provision requires that ISM devices must employ some sort of **spread spectrum** radio technology. Unlike conventional radio signals (broadcast AM or FM stations, for example) that occupy a fixed spot in the radio spectrum, spread spectrum transmitters constantly change the frequency at which they broadcast. This process is known as "hopping," and it significantly reduces the amount of power needed to broadcast a wideband radio signal through the air. This power reduction also minimizes any potential radiation hazard from the signals.

The two basic types of spread spectrum are **Frequency Hopping Spread Spectrum (FHSS),** which is used by Bluetooth and the HomeRF wireless networking system, and **Direct Sequence Spread Spectrum (DSSS),** which is used by 802.11b. As their names imply, FHSS systems hop around the radio band in a pseudo-random fashion, while DSSS systems step through a range of frequencies in a sequential order. DSSS provides higher data throughput than FHSS, but it is also more susceptible to interference.

Wi-Fi (Wireless Fidelity)

Wi-Fi (Wireless Fidelity) is the marketing name for an 802.11b wireless Ethernet specification. The Wi-Fi specification is still a young technology. It was introduced in 1999 by Apple Computer in its AirPort components. New equipment using the specification is continually being developed. 802.11b is an extension of the Ethernet protocol to wireless communication. It can handle many kinds of data. It is primarily used for TCP/IP, but can also handle other forms of networking traffic, such as AppleTalk or PC file-sharing standards.

The 802.11b standard was originally designed to enable high-performance radio to support roaming devices in large offices or business campus environments. It has now been expanded for use in LANs and is the most widely used home wireless LAN technology.

The 802.11b specification allows for the wireless transmission of data over the 2.4 GHz unlicensed band. Up to 11 Mbps of data can be sent over distances ranging up to several hundred feet. The range of any specific transmitter depends on what type of obstructions its signal encounters between transmitter and receiver. A clear line of sight between these devices enables the network to function at near top speed. Thick stone or concrete walls slow down data transmission dramatically

Each Wi-Fi transceiver can connect, with appropriate software, to any other similarly equipped device for ad-hoc computer-to-computer (peer-to-peer) transmission, but this is not the usual configuration. Most Wi-Fi LANs (and other wireless networks) use one or more access points or hubs, which are dedicated stand-alone hardware units with more powerful antennae than are found on any wireless NIC. The Wi-Fi access point often includes routing to the Internet, a Dynamic Host Configuration Protocol (DHCP) server, NAT, and other features necessary for a home network that includes a residential business operation.

Residential gateway devices are available that offer features similar to access points without the advanced management required for corporate networks or high-traffic installations.

The Wi-Fi standard is backward-compatible to an earlier specification, 802.11, but as a general rule, wireless networks adjust their data transmission speed downward to accommodate the slowest device on the network. This means that using a legacy device in an 11 Mbps WiFi network will cause all the data on the network to move at that device's slower speed capability: 1, 2, or 5.5 Mbps to and from the transceivers.

The WiFi standard is also compatible with the newer 802.11g standard (WiFi/g) that operates at 54 Mbps over the same 2.4 GHz band that WiFi uses, but again, the network will operate at the speed of its slowest connected component. WiFi/g components are more expensive than WiFi equipment and any benefit from their higher speed is lost if they are combined with the lower-cost but slower technology.

Because they operate on different frequency bands, WiFi and WiFi/g are not compatible with WiFi/a technology, which offers data speeds of 54 Mbps over the 5 GHz band. Several new combination 802.11a/b chipsets have been released, which make these two standards compatible with one another, but they have not achieved wide acceptance.

An industry group known as the Wireless Ethernet Compatibility Alliance (WECA) certifies its members' equipment as conforming to the 802.11b standard and allows approved hardware to be stamped Wi-Fi compatible. The Wi-Fi seal of approval is an attempt to guarantee compatibility between hundreds of vendors and thousands of devices. (The IEEE does not have such a mechanism, as it only promulgates standards.)

Wi-Fi is the only wireless standard now being used for public short-range wireless networks, such as those found at airports, schools, hotels, conference centers, coffee shops, and restaurants. Several companies currently offer hourly, monthly, session-based, or unlimited paid access via these networks around the U.S. and internationally. The Wi-Fi standard's wide acceptance has made it the dominant wireless protocol for business, and it has also surpassed HomeRF technology's early lead in the home network arena. WiFi networks account for about 70% of the currently installed wireless home networks, and their popularity is still growing.

All wireless networks have serious security risks associated with them because they broadcast digital data over open radio frequencies. The data can be picked up by any suitably tuned receiver located within range of the wireless transmitter.

Wired Equivalent Privacy (WEP) is an encryption standard that wireless networks use to improve security. WEP encrypts network traffic mathematically to prevent unwelcome eavesdropping by wireless receivers outside the network. Nearly all wireless equipment use this encryption method, but it is usually a built-in feature that can be activated or left off by the user. Even though it does not make a wireless network entirely secure, WEP should always be used to increase network privacy.

Additional wireless network security can be achieved by using an Access Control List for the network. All wireless NICs have a unique string of numbers called a MAC address installed at the time of manufacture. When a wireless network is set up, its access points can be programmed to permit only specified MAC addresses to connect to the network, thus preventing any outside nodes from accessing the system.

WiFi5

WiFi5 is an enhanced wireless networking technology that was introduced as a high-speed successor to Wi-Fi (802.11b). While Wi-Fi transmits at speeds of 11 Mbps, WiFi5 rams data through the airwaves at up to 54 Mbps. WiFi5, however, transmits on the 5 GHz band rather than the 2.4 GHz band that WiFi uses. This makes the two technologies incompatible with one another. A user planning to upgrade from WiFi to WiFi5 must start over with entirely new WiFi5 hardware which, though much faster, is also more expensive.

Dual-mode access points that bridge communication between the two technologies are now available, but they are even more expensive than the WiFi5 hardware. These cards also suffer from the same disadvantage as backward-compatible equipment in other technologies: When dual system hardware is used, the network speed slows to the capability of the slower technology. This means that the user pays the cost of WiFi5 equipment, but the network has only the speed of regular WiFi.

Still another WiFi technology, WiFi/g is also now available. It claims the speed of WiFi5 (up to 54 Mbps) and operates on the 2.4 GHz band, making it compatible with WiFi. WiFi/g, has not been around long enough to see if its performance matches WiFi5, but even if it does, it will probably still face the same noise limitations as WiFi. Its higher speed on the lower range band may make it vulnerable to interference from cordless phones and microwave ovens, which operate at the same frequencies.

All the WiFi technologies can function either by routing signals through hardwired access points (hubs) or in ad hoc peer-to-peer mode in which nodes communicate directly with one another. The hub infrastructure allows the wireless network to reach a wider area since multiple hubs can be positioned around the home and each will extend the network to the maximum limit of its broadcast range. Internet access is also better with a hub infrastructure that can distribute incoming data equally to all wireless nodes and relay outgoing data from several simultaneously. Peer-to-peer networks must have one node connected to the user's ISP and other nodes can only access the Internet through the connected node.

Both WiFi5 and WiFi/g use WEP encryption for security, but as with regular WiFi, this protection alone is insufficient to prevent unauthorized access to the wireless network. Access control accomplished by limiting network users to specified MAC addresses should always be employed with these networks.

Home Radio Frequency

Home Radio Frequency (HomeRF) is a localized wireless technology designed for home LAN use. It was the most popular wireless home LAN technology in the 1990s, but has since been overtaken by WiFi. HomeRF is now used in about 30% of installed home wireless LANs, and that percentage has been dropping for several years. It operates in the same 2.4 GHz band as WiFi and uses a digital frequency hopping spread spectrum radio transmission technique called Shared Wireless Access Protocol (SWAP). This protocol can carry both voice and data traffic and can interoperate with the **Public Switched Telephone Network (PSTN)** and the Internet. It was originally designed to provide data networking and dial tones between devices, such as PCs, cordless phones, Web tablets, and a broadband cable or DSL modem.

The original HomeRF technology was capable of transmitting data at a maximum 1.6 Mbps, which was far slower than the 11 Mbps speed of WiFi. HomeRF 2.0, the current version of the technology, has now been on the market for 5 years and is much faster than its predecessor, reaching transmission speeds of about 10 Mbps.

2.0 hardware is not only nearly as fast as WiFi, it's less expensive and inherently more secure. Instead of sending data at one frequency continuously as WiFi does, HomeRF hops between frequencies. Each time an access point and a node on the network exchange data, they hop to a different frequency. This transmission method avoids interference that may be present on any single frequency and makes eavesdropping by any outside device extremely difficult.

WiFi networks can reach a distance of about 100 meters or 330 feet, while HomeRF equipment is limited to a range of about 150 feet. This is adequate for most home LANs and will also allow the system to function well in many yards. HomeRF manufacturers recommend a limit of ten modules on a network, even though the technology will theoretically support more than 100. A broadband Internet connection can be shared on a HomeRF network, but multiple simultaneous users, especially if they are doing heavy downloading or file sharing, can quickly overload the system, causing it to slow down dramatically or even stop functioning.

HomeRF technology differentiates between types of data, in contrast to WiFi which handles all data in the same way. HomeRF supports eight voice connections and eight multimedia connections simultaneously, and the bandwidth is automatically divided among them. If no voice or multimedia connection is operating, the whole bandwidth can be used for data. If any voice or multimedia signals are detected, the system apportions half the bandwidth to voice and the other half to data. The data is prioritized to obtain maximum transmission speed.

Because HomeRF does not require access points, but functions as a peer-to-peer network, it is less expensive to install than any other wireless technology. Part of this cost difference is negated by the fact that many computers and peripheral devices now have built-in WiFi hardware while none have HomeRF technology.

Bluetooth

Bluetooth technology is based on short-range radio transmitters and receivers that are built into individual devices and that broadcast signals in the 2.0 to 2.4 MHz radio frequency band. The transmitters and receivers are application specific integrated circuits (ASICs) and can transmit data at rates as high as 721 Kbps with up to three voice channels also available. The data rate and additional voice channels are achieved through a very high rate of frequency hopping (1,600 hops per second as compared to the 50 hops per second used by HomeRF).

The range of the original technology is only about 10 meters, however, making it practical for use only in small areas. Bluetooth technology allows wireless connections between computers, printers, fax machines, and other peripherals, but does not have the range to serve as a wireless connection between an access point and laptops dispersed throughout a home and yard. Some vendors (Trendnet, Asus, LevelOne, Epox, and Air Live, among others) have developed Bluetooth devices with a higher transmitting power, which increases the range of the technology up to 100 meters, but these higher-power devices have not yet gained wide acceptance for home network use.

Bluetooth technology is most effective when used in wireless personal area networks (WPANs) that consist of electronic devices in a single room or in otherwise close proximity to one another. If increased power in new Bluetooth devices gives them the range to encompass a full home and yard area, this technology may well become a first choice in home networking. It is supported by many large companies that are already producing the lower-power devices for it. Bluetooth advocates are also encouraging its use by waiving payments for rights to use the technology.

Which Wireless Technology Is Best?

The general answer is this: All of them. The specific answer for a user depends on factors in the individual application. Each technology has its strong points and each has some limitations. Which one should be used in a particular home network design depends on the design, its physical location, what objectives it is trying to accomplish, and cost. In many cases, a combination of wireless technologies, which utilizes the advantages of each, and a wired technology may be the best solution.

The following are a few general statements about wireless technologies. Remember that, like all generalizations, they have exceptions in certain situations.

- Wi-Fi and HomeRF provide data transfer rates about equal to Ethernet and faster than Bluetooth.
- WiFi5 and WiFi/g are both faster than WiFi and HomeRF, but are also more expensive to install in a home network.
- Wi-Fi consumes more power than Home RF and much more than Bluetooth. This is only important where network devices are battery-powered and must be recharged more often when power consumption is high.
- Wi-Fi and WiFi5 have about twice the wireless range of HomeRF and all of these technologies have a larger coverage area than Bluetooth.
- Wi-Fi and WiFi5 can be used to connect to the Internet, network computers, and PDAs, as well as a wide range of home appliances and devices. They do not connect to voice lines.
- HomeRF can connect to all the devices that Wi-Fi can and also connect to voice telephone lines.
- Bluetooth can connect even more types of devices such as telephones, PDAs, computers, printers, headsets, GPS devices, storage devices, and others, but only if they are physically close to one another (typically no more than 20 feet apart).
- Bluetooth devices are more expensive than Wi-Fi, but lately have come down in price.
- Not all Bluetooth devices support the same profiles, and each network using this technology must be designed with products that have a common profile.
- It is easier to configure Wi-Fi and HomeRF equipment than Bluetooth equipment.
- Wi-Fi and HomeRF are in direct competition. Bluetooth is not in direct competition with either.

Much overlap exists in the capabilities of wireless technologies. WiFi, WiFi/g, and WiFi5 have about three quarters of the home wireless LAN market now, with HomeRF accounting for most of the rest. Bluetooth technology in home applications is mainly used to link devices in a single room while the remainder of the home network is hardwired or power-line based. Each wireless technology has strengths that make it the preferred choice for some applications. Each also has weaknesses that may disqualify it for some uses. Some advocates of each technology are so enthusiastic about their cause that they cannot see the benefits of the other alternatives, but each wireless protocol has a valid place in the home LAN arena.

Network Configuration and Settings for a Home Network

The basic architecture of a home network is the same whether the home LAN directs its data over its own dedicated hard wires, sends it as overlaid signals on wires already carrying other electrical current and signals, or transmits it wirelessly by radio waves over a limited area. The components needed to accomplish the movement of data are similar for each type of network. A hardwired network requires wiring to be installed in the home and connected to each node through its NIC. A power-line-based network also uses NICs to connect each node, but to the AC wiring or telephone circuitry already present in the home. A wireless network requires the installation of one or more wireless access points or hubs, which send and receive data by radio to and from the nodes, each of which processes the data through its own transceiver, a wireless NIC. Once you've decided on the type of LAN you want, installation is a straightforward process. Chapter 4 discusses the design and installation of a home LAN in detail.

The setup and configuration process for a LAN is very similar for computers running Windows 2000, WindowsXP, or Linux operating systems. Installation and configuration of a wireless hubs and NICs is also similar. For detailed instructions on component setup using these systems, refer to the following sources:

- Windows 9x Installation: *A+ Guide to Hardware,* by Jean Andrews, Ph.D., Boston, MA, Course Technology, 2006
- Linux NIC Installation: *Guide to Linux Installation and Administration Second Edition,* by Nicholas Wells, Boston, MA, Course Technology, 2003

Some of the configuration parameters for a LAN and its connection to the Internet are discussed in the following text.

TCP/IP Addressing

Transmission Control Protocol/Internet Protocol (TCP/IP) is the primary protocol used on the Internet. A computer must have TCP/IP loaded in order to access the Internet, and even most networks that do not connect to the Internet still use TCP/IP. It is important to understand the basis of TCP/IP so you can accurately set up a home LAN to use it correctly.

In a TCP/IP network, nodes do not have names, they have **IP addresses** that consist of four sets of eight-bit numbers separated by periods (dots). As was noted earlier, the eight-bit limitation on the size of each number means that no IP address number set can be higher than 255, which is the maximum number that can be expressed in binary using eight bits (11111111). TCP/IP rules forbid the use of some combinations of numbers and reserve others for special functions.

The total number of IP addresses available using four eight-bit numbers is less than 3.3 billion and the Internet has now grown so large that a serious shortage of available IP addresses has developed. (Three billion Internet-connected computers in the world are not enough! What to do?) This shortage of IP addresses has now resulted in the creation of a means to share IP addresses, which is discussed a little later in this chapter.

The TCP/IP protocol supports file sharing and printer sharing and a number of other services that are also available in different protocols. TCP/IP also has some functions—grouped under the umbrella term TCP/IP services—that are unique to this protocol and not available in others. The most important of these are Hypertext Transfer Protocol (HTTP), **TELNET, PING, IPCONFIG,** and **TRACERT.**

- Hypertext Transfer Protocol (HTTP) is the protocol of the World Wide Web. Web sites are constructed using Hypertext Markup Language (HTML), which HTTP can read and interpret.
- TELNET allows a user with a computer in one location to access a computer in a remote location as if the user were physically sitting in front of the remote machine, looking at its screen, and entering data on its keyboard.

- PING is a utility that enables a user at one computer on a network to determine if that node can communicate with another node also connected to a network. **PATHPING,** an improved version of PING, displays the route followed by an inquiry through every IP address to the destination of the queried computer and back.

- IPCONFIG is a utility that displays the computer's adapter address, IP address, subnet mask, and default gateway, and allows the Dynamic Host Configuration Protocol (DHCP) lease to be renewed or released by the user. WINIPCFG is the Windows 9*x* utility for this function.

- TRACERT shows the complete path that data packets are taking from the computer to reach any given destination.

The function of TCP/IP is to link together a whole array of LANs to form a wide area network (WAN) which is, in turn, linked via telephone line, cable, satellite, or other means, to all the other WANS around the world. This linked array of WANs is what we know as the Internet.

To make data transmission on the Web fast and efficient, TCP/IP is designed to use the high-speed expensive transmission links as little as possible, leaving them free to transmit long-distance traffic quickly. As discussed in the last chapter, routers are special network devices that direct data packets to specific IP addresses. Only packets with an IP address are passed by a router onto the WAN and perhaps the Web. All non-IP-addressed packets travel only in the LAN. Routers are by far most commonly used in TCP/IP protocol networks, although they can be used by some other protocols, such as IPX/SPX.

A computer using TCP/IP has a number of settings that must be correctly configured for it to operate on the network. TCP/IP settings can be set for both dial-up connections (modems) and direct (NIC) connections. Windows 2000 makes the configuration easy to access if the user right-clicks My Network Places, and then clicks Properties. Simply select the type of connection you want to configure, then set its TCP/IP properties.

A node in a LAN that wants to send data to another node located in another LAN via the Internet addresses the data packet to the IP address of the computer with which it is trying to communicate. The only IP address it can send the packet to, however, is the default gateway in its LAN, which is the computer node connected to the WAN. This machine serves as a router and, like all routers, is connected to at least two networks—in this case, the LAN where the node desiring to communicate is located, and the WAN, which is part of the Internet. The router sends the data packet to the WAN where another router will pass it on to its destination.

Domain Names

Domain names were developed as a means of making IP addresses more friendly to users. On the Internet, special computers called **Domain Name Service (DNS)** servers keep databases of IP addresses and their corresponding domain names. A node seeking data from another node with the domain name of \\new-horizons-college.edu can query a **DNS server** and obtain the IP address of \\new-horizons-college.edu. It can then use that address to find the correct node.

Because they are user-friendly and easily identified with commercial and other Web sites (imagine trying to remember or even look up a numeric IP address for every Web site you want to visit!), Internet domain names are highly regulated by the **Internet Corporation for Assigned Names and Numbers (ICANN).** Those who want a DNS name that can be accessed on the Internet must register the unique name with an ICANN-accredited registrar and pay a yearly fee.

DNS names used to end with one of these seven domain name qualifiers:

- .com for general business
- .org for nonprofit organizations
- .edu for educational organizations
- .gov for government organizations
- .mil for military organizations
- .net for Internet organizations
- .int for international

However, this is no longer true. As more countries joined the Internet, new endings were added for each country. A few examples of these two letter (digraph) endings are in the following list, and a complete list can be found at www.iana.org/cctld/cctld-whois.htm:

- .ar for Argentina
- .be for Belgium
- .ca for Canada
- .de for Germany
- .cn for China
- .ve for Venezuela

Recently, ICANN announced the creation of several new domain endings to keep pace with the demands of the growing Internet. These include:

- .biz for businesses
- .name for individuals
- .museum for museums
- .pro for professionals
- .aero for aviation
- .coop for cooperatives
- .info for general information

Dynamic Host Configuration Protocol

Dynamic Host Configuration Protocol (DHCP) enables a gateway computer in a LAN to create a pool of IP addresses that can be assigned to other nodes on the LAN when they need them. When a node no longer needs the IP address assigned to it, because it's no longer communicating on the Internet, that IP address is returned to the pool so it can be assigned to another node as needed. The group of addresses in a computer's pool is called a DHCP scope, and when a node is assigned one address, the duration of the use is called a lease. Leases from the DHCP pool can also include the names of the LAN's gateway and the DNS server, which helps speed data transmissions for the leasing node.

Most home LANs are not large enough to need this feature, but it is a good idea to set it up in case future expansion creates a need. It is especially useful where several LAN nodes use a dial-up Internet connection, but each machine is only on the network for short periods. When each is on line, it needs an IP address; the rest of the time, it does not. If you accept the automatic TCP/IP settings for Windows 2000 or Windows XP, these systems will use DHCP.

Dial-up connections to the Internet use a different hardware protocol called Point-to-Point Protocol (PPP). This is a streaming protocol developed to permit access over a telephone line. Windows simply regards the dial-up modem as a specialized NIC that has its own configuration in the network settings. PPP is required in addition to, not in place of, TCP/IP, which the network must still use for Internet traffic. PPP is only used for the telephone line portion of the connection. As soon as the data reaches the Internet through the ISP, it must use TCP/IP to find its destination.

Network Addressing Translation

The Internet has grown larger than anyone ever imagined. Although the exact size is unknown, the current estimate is that there are over 300 million hosts and more than a billion users actively on the Internet. That is more than the entire population of North and South America! In fact, the rate of growth has been such that the Internet is effectively doubling in size each year.

When IP addressing first came out, the system was designed to provide enough available addresses to cover any need. Theoretically, there are 4,294,967,296 unique IP addresses, but the actual number of usable addresses is smaller (somewhere between 3.2 and 3.3 billion) because of the way addresses are separated into classes, and because some addresses are set aside for multicasting, testing, and other special uses.

With the exponential expansion of the Internet and the huge increase in home networks and business networks, the number of available IP addresses is now simply not enough. The obvious solution is to redesign the IP address format to allow for more possible addresses. This is being done (in a development called IPv6), but the new system will take several years to implement because it requires modification of the entire infrastructure of the Internet.

Network Address Translation (NAT) is a temporary interim solution to the IP address shortage. NAT allows a single device, such as a router, to act as an agent between the Internet (or "public network") and an entire LAN. This means that only a single unique IP address, the one assigned to the agent router, can represent all the nodes on the LAN to which the router is connected.

When a LAN connects to the Internet through an agent device, the connected LAN is called a stub domain. The data traffic on a stub domain is mainly local, within the domain itself, and is not addressed to any node outside the LAN. Nodes in a stub domain can use both registered and unregistered IP addresses for addressing local traffic to other nodes within the LAN.

A registered IP address is one that is registered through an ICANN-approved registrar to only one node on the Internet. An unregistered IP address cannot be used on the Internet at all because it may conflict with a registered IP address assigned to another node.

Unregistered addresses are locally assigned by a LAN's agent device to other nodes and they can be used only within the LAN for locally addressed traffic. If a node with an unregistered IP address needs to communicate on the Internet, it must use NAT through the LAN's agent device to do so.

The agent device (firewall, router, or computer) that connects the LAN to the rest of the network world uses NAT in varied ways to direct Internet traffic to the correct node in the LAN:

- Static NAT maps an unregistered IP address in the LAN to a registered IP address (the one assigned to the LAN's agent device) on a one-to-one basis. It is used when a node on the LAN needs to be accessible from outside the local network.

- Dynamic NAT maps an unregistered IP address to a registered IP address from a group of registered IP addresses.

- Overloading maps multiple unregistered IP addresses to a single registered IP address by using different ports. This is known also as PAT (Port Address Translation), single address NAT, or port-level multiplexed NAT.

- Overlapping occurs when the IP addresses used on a stub domain LAN are registered IP addresses that are already in use on another network. The agent router must maintain a lookup table of these addresses so it can intercept them and replace them with registered unique IP addresses. The NAT agent router translates the "internal" addresses into registered unique addresses, and translates the "external" registered addresses to addresses that are unique to the stub domain LAN.

Security Firewall Configuration and Filtering

The network security concerns of a home LAN are smaller in scale, but otherwise much the same in content as those that occur on larger business networks. The influx of offensive material coming into the LAN and the potential for theft of data from, or damage to, the LAN are the main dangers against which a **firewall** protects. The word "firewall" for these hardware devices or software programs is derived from their function as a barrier. A firewall in a building prevents fire in a high-risk area from spreading to the rest of the structure. A network firewall performs the same function: It prevents unwanted data and unauthorized persons on the Internet from gaining access to the LAN, either to send offensive data into it or steal data from it.

A network firewall does not block all access to the Internet, but it **filters** incoming data and access requests that arrive through the outside connection. Any packet of data that does not meet the standards set in the firewall's filter configuration is not permitted to pass through to the LAN.

Without a firewall in place, all of the nodes on a home LAN are directly accessible to anyone on the Internet. A skilled **hacker** can probe these computers, make FTP or TELNET connections to them, or alter data in any number of ways. If only one user on a LAN leaves a computer unprotected, **hackers** can gain access to that machine, exploit its lack of security, and perhaps attack the entire network.

Threats to Home LAN Security

Hackers and outright criminals use many creative methods to access or abuse unprotected networks and individual computers. Examples of these methods include:

- A **virus** is probably the most common threat to a computer or LAN. It is a small program that can copy itself to other computers, and by doing so, can spread quickly from one system to the next. The actions of a virus range from harmless messages to erasing a computer's hard drive or disabling its operating system.

- E-mail bombs happen when someone sends the same e-mail to a recipient hundreds or thousands of times until the receiving e-mail system fills up with messages and cannot accept any more.

- Many applications allow the creation of a sequence of commands that the application can run as a set. These sequences are called **macros.** Hackers can sometimes break into a program and create macros that, depending on the application, destroy data or crash the computer.

- Application **backdoors** are used in some commercial programs. These backdoors contain code that allows for remote access by the developer of the program. Others contain bugs that inadvertently provide a backdoor or hidden access, enabling an outside person to take control of the program.

- A **denial of service** occurs when a hacker sends a request to a server to connect to it. When the server responds with an acknowledgment and tries to establish communication, it cannot find the system that made the request. By flooding a server with these unanswerable communication requests, a hacker overloads the server until it slows down or crashes.

- Like application software, some operating systems have backdoors. Others provide remote access with insufficient security controls or have bugs that an experienced hacker can use to take control of the system and use or damage it.

- Usually harmless, but always annoying, **spam** is the electronic equivalent of junk mail. It may contain Web links or attachments which, if activated, give the sender backdoor access to the receiving computer.

- Hackers change (redirect) the path that information takes by sending it to a different router. These are known as redirect bombs. This is one way that a Denial-of-Service attack can be launched with information requests redirected to a computer that cannot answer them because it cannot identify their origin.

- **Remote login** occurs when an outside person is able to connect to a computer and control it in some form. This can include viewing or accessing files, storing illegal data, or actually running programs on the computer.

- SMTP is the most common method of sending e-mail over the Internet. A hacker with access to a list of e-mail addresses can send thousands of unsolicited junk e-mail (spam) to users. The source of the spam is disguised by redirecting the e-mail through the SMTP server of an unsuspecting host who may then be blamed for the attack. This is known as SMTP session hijacking.

- In most cases, the path a packet travels over the Internet (or any other network) is determined by routers along that path. But the source providing the packet can arbitrarily specify the route the packet travels. Hackers manipulate the source routing to make their disinformation appear to come from a trusted source or even from inside the LAN.

A firewall on the LAN can greatly reduce or entirely eliminate all of these threats. The firewall implements a set of security rules set up by the LAN administrator. These rules closely control what information comes into the LAN from any source outside, and controls what information goes outside to the Internet from any node on the LAN.

A software firewall is a program installed on the computer in the home LAN that has an Internet connection. This computer then functions as a gateway because all access between the home network and the Internet must pass through it and its firewall software.

A hardware firewall itself functions as the stand-alone LAN gateway. It is wired directly to the modem or other Internet connection, and all data coming from the Internet or moving from the LAN to the Internet passes through it. The firewall has a built-in Ethernet card and hub. Nodes on the home network connect to the firewall/router, which provides their only access to the Internet. The firewall is configured via a Web-based interface reached through the browser on one of the computer nodes of the LAN.

Hardware firewalls are extremely secure and not very expensive. Home LAN versions that include a router, firewall, and Ethernet hub for broadband connections can be found for under $100.

Software and hardware firewalls use one or more of three methods to control traffic flowing in and out of the network:

- **Packet filtering** Packets of data are analyzed against a set of filters that define what information is acceptable. Packets that make it through the filters are sent to the requesting system and all others are discarded.

- **Proxy service** Information requests coming from the Internet are received by the firewall, analyzed, and then sent to the addressed node, if it is acceptable. Outgoing information requested by an outside node is similarly sent to the firewall, analyzed, and sent on to the Internet, if its release is permissible. The firewall serves as a proxy to receive and transmit incoming and outgoing information, thus preventing direct data transfer between the Internet and the LAN until the information has been validated.

- **Stateful inspection** This method uses a state table to compare certain key parts of each data packet to a database of trusted information. Information from inside the firewall going to the outside is monitored for specific defining characteristics. Incoming information is then compared to these characteristics. If the comparison shows a reasonable match, the information is allowed through. Otherwise, it is discarded.

Firewalls can be customized for the individual LAN. This means that the user can add or remove filters based on several conditions. Some of these are:

- **IP addresses** Specified IP addresses can be blocked from access by nodes on the LAN, as can a range of IP addresses. Similarly, if certain IP addresses outside the LAN are seeking access to information inside the LAN, the firewall can block all traffic to or from those IP addresses.

- **Domain names** The firewall can block access to certain domain names or domain names containing specific words or letters. Some firewalls can also be set to allow access only to those domain names specified by an administrator.

- **Protocols** The firewall can be set to accept or exclude information sent via a certain protocol. These restrictions can be very specific, allowing some nodes on the LAN access to a protocol while denying it to others. Protocols for which firewall filters can be set include:
  - **Internet Protocol (IP)** This is usually done for a specific address rather than the entire protocol because IP is the main information delivery system for the Internet. Blocking the entire protocol eliminates almost all Internet access.
  - **Hypertext Transfer Protocol (HTTP)** This protocol is used for Web pages, and all or parts of it can be restricted to limit access on the Web.

- **File Transfer Protocol (FTP)** This protocol can be restricted to be used only by a proxy server on the LAN or by one node so that access to LAN files is limited.
- **User Datagram Protocol (UDP)** This protocol is used to send data that requires no response from the receiver, such as streaming audio and video, which some LANs may want to limit.
- **Simple Mail Transport Protocol (SMTP)** A protocol used to send text-based information (e-mail) on which many restrictions may be appropriate.
- **Simple Network Management Protocol (SNMP)** A protocol used to collect system information from a remote computer. It would not be needed by most LAN nodes unless the user is trying to hack another system.
- **TELNET** Since this protocol is used to perform commands on a remote computer, it too is often restricted to prevent hacking into or out of the LAN.

(Note: A home LAN firewall might set up only one machine on the network to handle a specific protocol and ban that protocol on all other machines.)

- **Ports** All servers make their services available to the Internet using numbered ports, one for each service that is available on the server. A firewall can open or close any of these ports to grant or restrict access.
- **Specific words and phrases** The firewall can search through each packet of information for an exact match of any text listed in the filter. It could, for example, block any packet with the word "sex" in it, or any other string of characters. Any number of strings can be set, but the filter only blocks those packets with strings that exactly match the ones specified.

The level of security a firewall establishes determines how many threats can be stopped. The highest level of security is to simply block everything, but that defeats the purpose of having an Internet connection. The most secure method of determining access is to first block everything, then select what types of traffic to allow. Making selections singly is time consuming, however, and it may be more restrictive than intended because the administrator forgets to allow some desired data to be accepted. For most home LANs, it is best to start with the default settings provided by the firewall developer and change them only if there is a specific reason to do so.

Firewalls offer the best available protection against unauthorized entry into a home LAN. No firewall offers perfect security, and some of the items in the preceding list are hard, if not impossible, to filter entirely. While most firewalls offer virus protection, it is worth the investment to also install antivirus software on each LAN computer. Some spam is likely to get through even the best firewall, if it accepts e-mail. Spam can also contain viruses, so the individual machine protection is useful.

PPTP

PPTP is an acronym that stands for Point-to-Point Tunneling Protocol. It is a new technology developed jointly by Microsoft Corporation, U.S. Robotics, and several other companies that are known collectively as the PPTP Forum. PPTP allows network users to create virtual private networks (VPNs). A VPN is a private computer network in which access is limited to a specified group of users, but which uses the public Internet to connect some of its nodes.

Although the Internet is an open network, PPTP allows private users to transmit secure messages from one VPN node to another. No one who is not part of the VPN can access these messages, which are said to "tunnel through" the public traffic on the Internet. PPTP, which is an extension of the Point-to-Point Protocol (PPP) commonly used on the Internet, allows users to dial in to their corporate network via the Internet and still maintain their private network security.

Proxy Server

A function that is often combined with a firewall is a proxy server. A home LAN may have some data that the owner wants to make available to others on the Internet. The most secure way to do this is by using a proxy server. When an outside computer requests data, it is retrieved from the LAN node by the proxy server and then sent to the requesting outside computer. This isolates the home LAN from the Internet while still allowing access in both directions. Only the proxy server can retrieve data from nodes on the home LAN, and requests for data go only to the proxy server, never directly to other nodes on the LAN.

Proxy servers can also make the home LAN's Internet access work more securely and more efficiently. If a page is requested on a Web site by a LAN node, it is cached (stored) on the proxy server, from where it can be accessed by the requesting node. No data entering the home LAN from outside can be stored anywhere but on the proxy server. This prevents any contaminated data from infecting any node on the LAN except the proxy server. It also allows frequently accessed pages to be kept on the proxy server rather than retrieved from a remote Web site for each use.

Summary

- A home LAN is a network design enabling intelligent communication and mutual data transfer between computers and peripherals, digital home appliances, automated utilities, security systems, and other devices, as well as providing a gateway and access to the Internet.

- If a home has installed UTP or STP wiring, that's the best option to use for a network. For homes without installed wiring, installation of wiring is more difficult and more expensive. For many of these, other alternatives may be preferable.

- HomePNA employs frequency-division multiplexing (FDM) to put computer data on a voice phone line using separate frequencies from the voice signals being carried by the line.

- Power-line networking technology uses a form of orthogonal frequency-division multiplexing (OFDM), similar to the technology found in DSL modems, to put computer data on AC power lines at the same time as the high-voltage power, but at different frequencies.

- Wi-Fi (Wireless Fidelity) is an IEEE 802.11b wireless technology standard that can transmit data at a rate of 11 Mbps. It is presently the most popular wireless standard.

- WiFi5 is a version of the IEEE 802 standard that allows for fast 54-Mbps data transfer speeds through 802.11a-capable hardware, but is not compatible with WiFi (IEEE 802.1b) technologies.

- Home Radio Frequency (HomeRF) is a wireless technology that uses Shared Wireless Access Protocol (SWAP) and operates in the 2.4GHz band at data transmission rates up to 10 Mbps.

- Bluetooth is a short-range (100 meters) wireless technology that uses high-speed frequency hopping to transmit data at speeds up to 742 Kbps.

- Installing and configuring a NIC in a computer using Windows 2000 or Windows XP is an easy process. After a NIC is physically installed, the operating system automatically detects the card and guides the user through the process.
- TCP/IP is the primary protocol used on the Internet. Any PC must have TCP/IP loaded in order to access the Internet, and even most networks that do not connect to the Internet still use TCP/IP.
- The most important TCP/IP services and functions are Hypertext Transfer Protocol (HTTP), TELNET, PING, IPCONFIG, and TRACERT.
- Domain names were developed as a means of making IP addresses more user-friendly. They are always linked to an IP address.

- Dynamic Host Configuration Protocol (DHCP) enables a computer to create a pool of IP addresses that are given to other nodes on a LAN when they need them, and then taken back into the pool when no longer required.
- NAT is a temporary solution to the IP address shortage and allows a single device, such as a router, to act as an agent between the Internet (or "public network") and a LAN so the LAN uses only a single IP address.
- A firewall is a program or hardware device that filters the information coming from the Internet connection into a private network or computer system.

Key Terms

Backdoor A means of accessing a software program other than the normal one (front door).

Bluetooth A short-range (100 meters) wireless connection technology now being used for networking.

Category 5 (Cat5) cable A type of network cable that permits high-speed data transmission; an advanced version is Category 6 (Cat6) cable.

Denial of service A method of attacking a LAN or server by sending thousands of unanswerable requests to it.

Direct Sequence Spread Spectrum (DSSS) A method of signal hopping or rapidly changing frequencies in a specified sequence to transfer data at high speed.

DNS server Special computers on the Internet that keep databases of IP addresses and their corresponding domain names.

Domain name A unique name assigned to a network and registered with ICANN.

Domain Name Service (DNS) A part of the TCP/IP protocol that translates domain names into their corresponding IP addresses.

Dynamic Host Configuration Protocol (DHCP) A method of automatically assigning IP addresses to nodes on a LAN.

Filtering A method of analyzing data by comparing it to preset parameters; data meeting the parameters can then be blocked, approved, or otherwise directed.

Frequency Hopping Spread Spectrum (FHSS) A method of signal hopping or rapidly changing frequencies in a random sequence to transfer data at high speed.

Firewall A program or a hardware device that filters the information coming from the Internet connection into a private network or computer system.

Firewire IEEE 1394 data transmission protocol.

Hacker A person who attempts to gain unauthorized access to a computer or network.

HomePNA Original standard of HomePNA technology currently in use for networks using telephone lines for connectivity.

HomeRF (Home Radio Frequency) A wireless network technology for home LANs.

HPNA 2.0 Network standard currently in use for HomePNA networks.

IEEE 1394 A fast serial protocol that transmits data at speeds from 100 to 400 Mbps.

IEEE 802.11b Wi-Fi protocol based on the IEEE 802.11 wireless network standard.

IEEE 802.11g Advanced fast protocol based on the IEEE 802.11 standard.

Interference Noise and conflicting signals that can occur in transmissions in unlicensed radio bands such as the ISM band.

Internet Corporation for Assigned Names and Numbers (ICANN) The group that assigns and regulates domain names and IP addresses through accredited registrars.

IP address Internet Protocol address, a 32-bit address consisting of four numbers separated by periods, used to uniquely identify a device on a network.

IPCONFIG A TCP/IP utility that displays the computer's adapter address, IP address, subnet mask, and default gateway, and allows the DHCP to be renewed or released by the user.

Network Address Translation (NAT) A method of assigning temporary IP addresses to overcome the shortage of available unique IP addresses.

Macro A programmed sequence of commands that direct a computer to perform certain functions; sometimes used by hackers to gain control of a computer or LAN.

Passport Obsolete discontinued type of power-line technology for networking.

PATHPING An improved version of PING.

PING A TCP/IP utility that enables a user at one computer to determine if that node can communicate with another computer connected to a network.

Power-line network Network technology that transmits data over a home's AC power lines while high-voltage power runs on the lines at a different frequency.

PowerPacket A high-speed power-line technology for networking.

Public Switched Telephone Network (PSTN) The Commercial network of telephone lines and transmission facilities over which most telephone calls are made.

Proxy server A LAN node isolated from the rest of the LAN that services all outside requests for data from the LAN.

Remote login Connecting to and taking control of a computer from a remote location.

Spam The e-mail equivalent of junk mail; unwanted and unsolicited e-mail.

Spread spectrum Spread spectrum signals constantly change frequency, a process known as hopping, to reduce the power requirements for transmission.

TELNET A TCP/IP utility that allows a user in one location to access a computer in a remote location as if the user were physically sitting in front of the remote machine.

TRACERT A TCP/IP utility that shows the complete path that data packets are taking from the computer to reach any given destination.

Universal Serial Bus (USB) A bidirectional, isochronous, dynamically attachable serial interface for adding devices on a single bus.

Virus A small software program that spreads through networks by copying itself onto other computers; it may then do damage to stored data and application programs, or allow hackers to access the system.

Wi-Fi (Wireless Fidelity) IEEE 802.11-based wireless transmission protocol with an 11-Mbps transmission rate. It is presently the most popular wireless standard. Also known as WiFi/b.

WiFi5 IEEE 802.11-based wireless transmission protocol, a very high-speed technology.

WiFi/g Higher-speed version of WiFi/b that is backward compatible with it.

WINIPCFG Windows 9x version of IPCONFIG, a utility for displaying a computer's adapter address, IP address, subnet mask, and default gateway and renewing or releasing its DHCP.

X10 A wired technology that transmits data on existing high-voltage AC power lines in the home, and thus requires no new wires for installation.

Review Questions

1. Which of the following is not a wireless protocol for home LANs derived from the IEEE 802.11 standard?
   a. Wi-Fi
   b. HomePNA
   c. HomeRF
   d. Bluetooth

2. How does X10 technology work?

3. How does home power-line technology differ from X10?

4. Which of the following is not a service offered by TCP/IP?

   a. HTTP

   b. PING

   c. TELNET

   d. ICANN

5. The two main types of cable used by Ethernet today are _____ and _____.

6. What is a network firewall and what is its function?

7. Wi-Fi stands for _____ and is a technology that uses signal _____ to achieve high-speed data transmission.

8. HomePNA technology uses _____ lines to transmit data and _____ information.

9. The long range of Bluetooth technology makes it ideal for linking wireless laptops. True or false?

10. Which of the following is probably not a valid domain name?

    a. yahoo.com

    b. wireless.org

    c. home.lan

    d. Bluetooth.net

11. An IP address always consists of _____ eight-bit numbers separated by _____.

12. Network Address Translation (NAT) is a method of temporarily assigning a registered IP address to a(n) _____.

13. The acronym WAN stands for _____ and this system forms part of the _____.

14. TELNET is a utility that allows a user in one location to access a computer in a remote location as if the user were physically sitting in front of the remote machine. True or false?

15. FHSS stands for File Head Sharing System and is used to store information on a disk. True or false?

16. What is a Denial-of-Service attack?

17. FTP stands for _____ and is a protocol used to transmit _____.

18. If a domain name ends in .edu, the kind of organization it belongs to is an _____; if it ends in .org, it belongs to an _____.

19. X10 technology is less expensive than a Wi-Fi system, but has equal capacity because X10 is wireless and Wi-Fi is wired. True or false?

20. Power-line technologies connect network devices by using _____ wiring.

21. The ISM band is a shared-use unlicensed radio frequency. Explain what this means.

22. Interference in wireless data transmission is usually dealt with through use of _____.

23. Which wireless technologies can be combined with Ethernet in a single LAN?

    a. Wi-Fi

    b. HomeRF

    c. Bluetooth

    d. All of the above

24. What is a virus?

25. If using X10 technology when nearby homes are also using it, a(n) _____ must be used with X10 commands to prevent commands from other systems from entering the home.

Hands-On Projects

Project 3-1: Build a Simple Network

In this project, you will build a simple network using NICs and a Microsoft network protocol. For this project, you will need access to two computers. Both should have Windows 2000 or Windows XP operating systems. If the computers do not have network cards installed, you will need to acquire two NICs that can be installed in the computers. You will also need a crossover cable.

1. Check the two computers to see if they have NICs installed. If they do, go to Step 2. If they do not, follow these instructions to physically install the card:

   a. Unplug the computer from its power source.

b. Remove the computer's casing or a side panel to gain assess to the interior of the machine.

c. Remove one of the small cover plates that cover the card slot openings on the back of the computer so the NIC can fit into the opening.

d. Install the NIC in the slot designed for it on the motherboard.

e. Attach the NIC to the back of the computer frame with the screws or other fasteners that formerly held the plate.

f. Replace the casing or side panel.

g. Plug the computer into its power source.

h. Repeat these steps to install the NIC in the second computer if necessary.

2. Connect both computers using the crossover cable. Insert a cable end into the jack on each computer's NIC.

3. Start both computers.

4. On Windows 2000 or Windows XP computers, the system detects the NIC automatically, and an installation window appears with instructions to guide you through the process. Follow the instructions in the window.

5. When installation is completed, you can verify that the card is correctly installed by clicking Start, and then clicking Control Panel. Double-click Network and Dial-Up Connections.

6. When the dialog box opens, right-click Local Area Connections, and then click Properties to view the card's properties.

7. From the Properties window, install the NetBEUI protocol.

8. On the Windows desktop, right-click My Computer, and then select Properties from the shortcut menu. Click the Computer Name tab, type a name for the computer, and click OK.

9. Close the System Properties window and My Computer window.

10. Repeat Steps 4 through 9 to complete the installation of the NIC in the other computer if necessary.

11. Click the My Network Places icon on one of the computers, and then double-click Computers Near Me. You will be able to view both networked computers.

Project 3-2: Wire an X10 Control Network and Troubleshoot It

You can follow these instructions to install and troubleshoot a simple two-node network using X10 devices.

1. Obtain an X10 lamp module and a control unit.

2. Plug the lamp module into a wall outlet in a classroom or home. Plug a lamp into the lamp module.

3. Plug the control module into a wall outlet in another part of the classroom or home.

4. Turn the lamp on and off with the control module.

5. Relocate your X10 lamp module and control module so they are separated as far as possible in the classroom or home. Plug both units into wall outlets.

6. Test the network to see if it still works correctly.

7. If the lamp does not respond properly, troubleshoot the network to see if you can get it working again at the greatest possible distance.

   a. Check for interference, mismatched power legs, and attenuation.

   b. If necessary, install a noise filter, bridge, or repeater.

8. If you have a friendly neighbor, place the lamp module in another classroom or home and see if you can control it with the control module in yours.

Project 3-3: Trace Data from a Government Web Site

In this project, you will use the TRACERT utility to trace data packets from your computer to a Web site located outside your local area. For this project, you will need access to a computer running Windows 2000 or Windows XP.

1. Log on to the Internet.

2. To open the command prompt, click Start at the lower-left corner of your computer screen.

3. Point to Programs, and then click the MS-DOS prompt.

4. In the command prompt window, type tracert www.nasa.gov. This command directs TRACERT to show the path to the National Aeronautics and Space Administration Web site.

5. In the command prompt window, type tracert www.royal.gov.uk. This command directs TRACERT to show the path to the Queen of England's Web site.

6. Try to determine from the display what other countries the data may have passed through on its way to its destination. At the end of the trace, the words "trace complete" should appear if the complete route was traced. If the words "timed out" appear, the trace took too long and was canceled before completion.

Project 3-4: Configure a NIC or Review the Configuration of a NIC

In this project, you will configure an installed NIC or review the configuration of a NIC using Windows 2000. Configuring a NIC in a computer under Windows 2000 or Windows XP is an automated process.

1. After a NIC is physically installed, turn the computer on and wait while Windows 2000 or Windows XP loads and detects the NIC.

2. Follow the instructions that appear on the screen in sequence to install the necessary drivers.

3. After the installation, verify that the card is installed with no errors by using Device Manager. To activate Device Manager, click Start, point to Settings, and then click Control Panel. Double-click System. On the Hardware tab, click Device Manager.

4. In Device Manager, the network card should be listed under Network Adapters. Right-click the card and select Properties to view the card's properties.

5. Another way to access the NIC Properties window in Windows 2000 or Windows XP is to use the Network and Dial-Up Connections applet in the Control Panel. Open the Control Panel and double-click Network and Dial-Up Connections.

6. When the dialog box opens, right-click Local Area Connection.

7. From the shortcut menu, select Properties to view the Local Area Connection Properties window.

Case Projects

Case Project 3-1: Connect to a Public Wireless Network

For this project, you need to obtain the use of a wireless laptop computer for a short period. Find an area in your school or at another location that is equipped for wireless data transmission. Your task is to determine how well and at what range the laptop functions as a wireless node in the area set up for it. To do this, you must first connect to the network through a wireless access point and then log on to the Internet. Once you've brought up an Internet site, move to a different location within the wireless LAN area and log on to a different Web site. Move to some other locations within the wireless access point's range and see if your laptop remains connected. If it loses its connection, make note of where you were when the connection was lost. If the laptop does not lose its connection in the access area, move outside that area and see how far you can get from the access point before the connection fails.

Case Project 3-2: Research a Home LAN

A client has asked you to research a LAN design for her 1200-square-foot three-bedroom apartment. The apartment is only a year or two old and is wired for a telephone outlet in every room. It has standard electrical wiring, and the service panel for each apartment is located within the unit. The apartment owners will not permit new wires to be pulled within the apartment's walls, but they will allow wiring to be installed in raceways on the walls surfaces. The client has to pay for the installation and for its removal when she moves from the apartment. The client wants to network a computer in each bedroom and another in the kitchen. She also needs to connect a color printer and a black-and-white printer so that all the computers can use them. Finally, she wants a lighting control system for the apartment and a security alarm on its two entrances and three windows that she can monitor from her bedroom or from the kitchen. Based on this information, recommend the network technology or technologies you would use for this installation and the basic design for the system.

Case Project 3-3: Research a Home Wireless Personal Area Network (WPAN)

A client operates his business from his home. In his home office, he has two workstation computers and a server. He also has two printers and a laptop computer that he uses when making calls out of his office. All of his computer equipment is in one room and he wants a secure network linking it and allowing him occasional access to the Internet. Cost is not as important a factor to this client as is a secure reliable network. Write a short report recommending what technology he should use and how his network should be configured. Note how you would connect his laptop to the network, and how you would interface this WPAN with the Internet. Give reasons why you chose the technology you did and why you rejected others.

Case Project 3-4: Link Two Computers with a Home Network Technology

Place two computer workstations in a room. Set up both computers so they are operational, but not connected to any network. Choose one of the home network technologies discussed in this chapter (HomePNA, Home PowerPlug, Wi-Fi, HomeRF, or Bluetooth) and connect the two computers using one of these technologies. Obtain and install the NICs, wiring, and other devices and connectors you need to complete the connection between just the two computers. Keep the cost of networking the two stations as low as you can and still get a data transfer rate of at least 500 Kbps over the network. When you have the computers networked, time the transfer of a 500KB file across the network and compare this time to what is required to transfer the file manually using disks or whatever output devices the computers have.

Installing Network Components and Low-Voltage Wiring

OBJECTIVES   *After studying this chapter, you should be able to:*

- Plan a home LAN.

- Understand the basic central LAN components.

- Install concealed and **surface wiring** for a LAN.

- Configure hardware and cables of a LAN into a working system.

OUTLINE   Planning a Home Network

Basic Central LAN Components

Wiring a Home LAN

Configuring the Hardware and Cables

Introduction

Now that you are familiar with the various network designs and the components that make up a network, you are ready to plan and physically construct a home LAN. In this chapter, you will learn how to evaluate the location where a LAN is to be built, determine how many nodes the LAN should have and where each should be positioned, map a wiring or wireless connection path to each node, and develop a written diagram of the complete LAN on a floor plan of the home. You will also learn how to pull cables in both new construction and existing buildings, how to position central components and nodes, and how to connect the cables, connectors, and components and integrate them through a **patch panel** into a working LAN. Finally, you will learn how to install a **home run cable** to connect the LAN more directly with its ISP.

Planning a Home Network

As you prepare to install a home network, some planning at the beginning means less work in completing the project and better performance of the network. To design an efficient home network, you need to ask six basic questions, and then use the knowledge you have acquired about the different types of networks, network infrastructure, and hardware devices to answer them for the network installation project on which you are working. Each home network is different, both in its setting and in what it is designed to accomplish. For this reason, you need to individually plan each networking project to provide the most effective solution for the client.

The questions for planning a network installation are as follows:

1. Where will the network be built?
2. What nodes will be in the network?
3. Where will each node be located?
4. How will each node be connected?
5. Do I have a proposed diagram of the home LAN?
6. Is my proposed diagram correctly labeled?

At first the answers to these questions seem obvious, but as with so many apparently easy questions, the problem is in the details. Examine each question closely to find the answers that can make any networking project run more smoothly and turn out better.

Note that planning a home network is as much an art as a science because the answers to each of the six preceding questions influence or even determine the answer to some of the others. The type of construction of the home where the network is to be installed, for example, may determine how some or all of the nodes are connected. (Old heavy brick construction, twice remodeled, suggests a wireless LAN for most nodes.) The data requirements of some nodes may determine how they must be connected in order to get the required transmission speeds. (Multiple high-definition television nodes and extensive audio nodes in the network may require a wired Fast Ethernet.)

The following sections discuss some details of each of these questions. Keep in mind that a good network plan requires that you evaluate each factor according to its importance in the particular situation in which you are working. Your job is to envision the home network as a whole entity, then create an integrated plan that fits the home LAN to the needs and wishes of the client, the house in which it is placed, and the technology available to do the job based on the available budget. That's what home technology integration really is.

Where Will the Network Be Built?

The "where" for a network is not so much the geographical location as the physical construction of the building where it will be installed. The type of network chosen and the technology used to connect the nodes is heavily influenced by whether the network is installed in new construction where wiring is easy, or in an existing structure which must be retrofitted with greater difficulty.

Prewired New Construction

Examine the building site in detail. If it is new construction, is network wiring being installed? If it is not in the plans, can it be added so the contractor can include it with the other wiring already planned? If that cannot be done, can you do the network wiring while the house is under construction? Almost any way to install the network wiring while the home is under construction is easier than installing it after the house is finished.

This also applies to remodeling. If a basement is being finished or an addition put on the house, and a network is planned as part of the project, get the wiring in place before any of the finish work is done. Use the area being added or finished to run as much of the network wiring as possible, even if doing so means longer wire runs to some of the network nodes. Plan each wire run so it extends from the node's connection point toward the patch panel or central hub, but in a manner that gets it quickly from the finished area where it must be concealed to unfinished areas where it can be run exposed. The exposed wiring is later covered by the finished surfaces installed as part of the home's construction or remodeling.

If the new or remodeled home has any network wiring installed, be sure you get a schematic drawing of the existing wiring before creating the plan for the LAN. When you have the schematic, examine the home itself and locate all the connecting points that are supposed to be there. Some may have been inadvertently covered over by plaster or paint and you may need to locate them from the drawing. Sometimes the wiring technician moved a connection because he or she encountered a problem placing it where the plans specified. It is far better to note these changes or possible omissions before creating the network plan than to discover later that a needed connection is not there.

You want to make use of as much of the existing wiring as possible, even if it is not all placed in exactly the locations you would have chosen. Good planning can make the most of a general wiring installation done before the design of the LAN was completed. For example, if the LAN has several nodes (computers, printers, file server) located in one room, but only a single wiring connection is available in that room, you may still be able to use the existing wiring by segmenting the network nodes in the room and connecting them all as a single unit via a switch or router to the rest of the network.

Retrofitting Existing Structures

If the network is going into a home that's already built, what wiring (that is already installed) can be used as part of the LAN? The newer a home is, the more likely it is to have some wiring installed beyond the minimum required for its original utilities. There may be additional telephone **jacks** installed, cable television connections, or even a basic network cable system. None of these is likely to be found in a home more than 15 years old, and the closer a home is to that age, the less future-oriented wiring it will have.

The same general rule applies to wiring an existing home as to one being remodeled: Install the wiring so it runs from each node's connection point in a finished area to an unfinished area in as short a distance as possible. In an existing home, the unfinished areas are usually the basement or the attic. In these situations, cables should be fished inside the finished wall of the room to the basement below or the attic above and then run in the unfinished space to the central hub or patch panel.

Carefully plan the wiring run for each node in an existing home. You cannot run wiring inside brick walls. If the home is brick construction, all the outside walls (and maybe some interior walls) are not available for wiring. This may not apply if the construction is brick veneer (a facing of brick over wood frame outer walls), but all outer walls are more difficult to run wiring inside than interior walls. Exterior walls are nearly always filled with insulation and have obstructions at the top and bottom that are hard to drill through for wiring. Pulling wire up or down an exterior wall is at best a frustrating and time-consuming experience.

Plan the network so you can use interior walls for wiring wherever possible. In a multistory home, do not plan on running any wires from one floor to another inside a wall. Homes are constructed so the floor extends across the entire area inside the foundation or walls and is covered with one or more layers of subflooring. The walls of each story are then built on the floor and attached to it. Holes are drilled through the floor for any pipes or wiring that extend between floors or into the basement. A wall may appear to be continuous through two or more floors, but the floor will usually interrupt the open space inside the wall, preventing wiring from being fished between floors after the wall is finished.

What Nodes Will Be on the Network?

What does the client want to include in the home network? What is he or she likely to want added to it later? Even if the home's builder did not plan for the future, you should do so when designing the home network. Plan for every network node you can think of and how you will wire it into the system. You may not actually install everything you plan for, but the network will have a better design, and the installation of future nodes will be easier if you plan for them in the beginning.

More than 35 million U.S. homes currently have two or more PCs. The vast majority of these households will install a home network to share Internet access among the PCs and new Internet appliances, and avoid paying for two or more monthly Internet services. The number of homes equipped with multiple PCs will probably double over the next 10 years with an even larger number acquiring other networking devices and automation systems.

At a minimum, a home network should connect all the computers in the home and give each of them access to the Internet. Start your network plan with that core. Decide if the existing computers should be connected through a central computer or through a hub. How will you set up a firewall to protect the home network from abuse? How will additional computers be added, and where will they be located? Where are the printers, and how are they connected? Scanners? File servers? Other input or output devices? Is wireless access needed in some or all parts of the home? In the yard? Do not forget the garage.

Beyond the basic computer network, does the client want television and audio entertainment systems connected? What about a security system? A lighting control system? A water control system? A heating and air-conditioning control system?

Plan for how each of these systems will be connected to the network, even if they will not be installed immediately. Build in enough excess capacity to the core network nodes and the wiring or wireless connections you install so the network can be expanded later without the need to completely rewire it. Try to think beyond what you see in current home networks to what might become common in the near future.

It is possible, even probable, that not all of the home network will be a wired Ethernet. Specialized network functions operating on different standards can be designed into the LAN and connected to it as network segments through a compatible device. In this manner, any number of segments using standards such as HomePNA, HomePlug, Bluetooth, X10, and others can be incorporated into the total LAN design.

Wireless hubs can be used to give laptop access throughout the home and in the yard. Each access point can be set up as a node on the Ethernet or all of them can be wired as a separate segment and then connected to the network through a bridge. The latter method reduces wireless traffic on the main network, but this is only an important consideration if wireless traffic is likely to be heavy, such as multiple laptop users playing video games or streaming music and video.

A lighting control system can be installed using HomePlug or X10 technology to set up all the controls. The system can then be connected to the network through an Ethernet-compatible device that translates commands into the auxiliary technology. This setup allows overall control of the lighting system from the network's central computer while still permitting remote lighting control from other locations set up with HomePlug or X10 devices. Further information on how these technologies can interface with an Ethernet LAN is available at http://www.homeplug.org and http://www.x10.org.

Utility management systems such as sprinkler controls, heating and air-conditioning controls, de-icing systems, and timed electrical functions such as outdoor lighting, pumps, ventilators, and other devices can all be set up as network segments. All can use the most efficient technology available for their individual control modules, and connect to the network through an Ethernet-compatible hub or other device. Again, this segmentation minimizes traffic on the main network while allowing monitoring and overall control of each system to be exercised from the central computer.

The home security system can be set up as a network segment, connected to the main network for monitoring and central control, but isolated from other network segments and from the Internet to prevent any tampering with, or disabling of, its functions. Good security practice requires that monitoring of sensors and video surveillance nodes be separated from control of security devices such as door locks and alarms. This can be accomplished by setting up each part of the security system as a separate segment, both linked to the main network, but neither accessible to the other or to anyone except an authorized user.

Where Will Each Node Be Located?

Which rooms in a home will have nodes located in them is generally apparent from the needs of the client, but the specific location of each node within the room can often be adjusted to make installation easier and minimize wiring requirements. If nodes are to be set up in adjacent rooms, for example, a double wire can often be fished in the wall between them, and outlet jacks can be installed on either side of the wall to accommodate both nodes.

If the node is a computer, printer, or other device that must be easily accessible to a user for long periods of time, then it needs to be positioned conveniently at a desk or table with its connection wiring out of the way of rolling chairs or people's feet. Nodes where devices require **AC power** connections should have their network connections near enough to power outlets so extension cords do not have to be strung long distances to reach them.

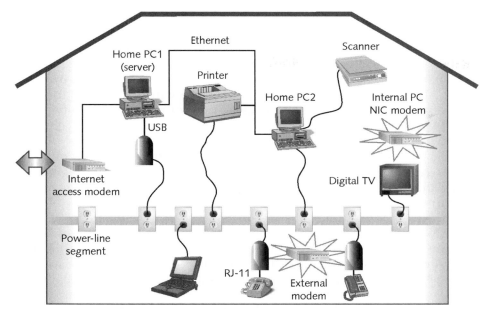

FIGURE 4-1   Conceptualized Multiple Technology Network

If other than Ethernet technologies are being used for segments of the network, then an access hub, bridge, or other device should be planned where each of these segments connects to the main network. These access devices for network segments all need to be connected to the LAN with a wired connection to the patch panel. The connecting devices should be located where most convenient for running a cable from them to the patch panel and also connecting the network segments running on alternate technology to them. For HomePlug or X10 hubs, this can be wherever there is a power outlet, often right next to the patch panel and allowing a direct connection to it. For HomePNA hubs, it can be any convenient telephone jack near which an Ethernet jack can also be installed. Figure 4-1 shows a conceptualized diagram of a multiple technology LAN in which the computers, printer, and modem are connected in an Ethernet topology, and other home appliances are connected in a power-line network such as HomePlug. The network segments are interfaced through the central computer, which is a node on both segments.

Wireless hubs should be sited in locations where they offer the most effective coverage of the desired range, but that criterion can usually be met while still placing them in a convenient spot for wiring into the network. In some cases, a wireless hub may need to connect to another wireless hub or bridge to get its wireless data transmission close enough to the LAN's wired segment so the bridge can be connected as a node to the patch panel.

How Will Each Node Be Connected?

Most electrical contractors now wiring new homes are now specifying **Category 5 (Cat5) cable,** an enhanced form of **UTP telephone wire** that can simultaneously handle up to 100 Mbps of high-speed data. Many are using Cat5e or **Category 6 (Cat6) cable,** both of which are even faster than standard Cat5 cable.

Cat5 is more expensive than other categories of UTP or **STP cable,** but is still a bargain considering the transmission speeds it offers and the fact that installing it is only slightly more difficult than installing the cheaper wire. When using Cat5 cable (and Cat5e or Cat6), extra care is needed to avoid pulling the wire too hard, bending it too sharply (kinking), or forcing it through too tight an opening. All of these errors can alter the way Cat5 wires are twisted inside their plastic jacket and thereby reduce the cable's performance. Each network node jack also needs its own separate **cable run** from the patch panel or central gateway. Cat5 cable cannot be spliced along the way, nor can it run closely alongside AC electrical wires because the **electromagnetic fields** that high-voltage wires generate when in use can interfere with the relatively weak digital signals in the cable. Keeping the two types of lines at least a foot apart protects against such interference.

Because it is the labor, not the cost of cable, that is the main expense in any home LAN wiring job, it makes sense to install the highest quality and capacity cable, even if it will not be immediately used to full measure, to ensure against future obsolescence. To be doubly sure of meeting future needs, some wiring specialists use **Siamese cable** consisting of bundled Cat5 cable, plus coaxial or HDMI cable for video transmission, plus two 18-gauge copper wires for audio transmission. All the voltages on these bundled wires are low and they can run alongside one another without any of the lines producing interference in the others.

Cat5 cable is the minimum standard for wiring home LANs. For the best wiring value and expansion capability, use composite cable, which combines two Cat5 cables, or Siamese cable that combines all the types of cable your network will need (now and in the future) into a single bundle. The top-of-the-line version of this type of cable contains Cat5 and RG-6 wires, as well as a fiber-optic line, the fastest available transmission medium. Figure 4-2 shows some of the available cable types.

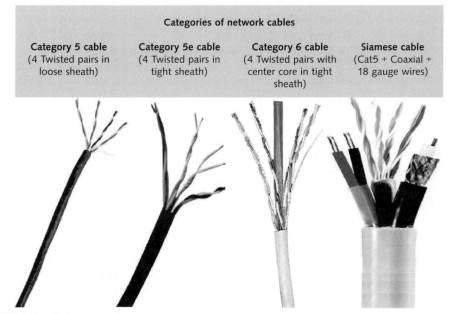

FIGURE 4-2   Network Cable Types

Fiber-optic lines, which carry light-based data through strands of glass no thicker than a human hair, are the fastest and most expensive transmission media yet devised. Few home LAN users currently want to purchase the costly conversion equipment required to change electrical impulses into light photons, but the future price of these devices is virtually certain to decline and their use increase.

At present, home use of fiber-optic cable is mainly for video data distribution. LANs that include a home theater or whole house video distribution, can include a fiber-optic network segment for this purpose, while the rest of the LAN functions through wired or wireless technology. Cost is still much higher than for other types of cables: a 30-foot fiber-optic cable costs about $400. Fiber-optic cable is also more expensive to install in a network because it requires wider radius curves than metal cables. For best performance, fiber-optic cable is usually installed inside a 0.5-inch or larger **PVC conduit.** Rigid conduit is very difficult to install in existing homes, except on the surface of finished walls, which gives an unsightly appearance. Consequently, the use of fiber-optic cable for home LANs is mainly employed in new homes where it has been installed during construction.

Do I Have a Diagram of the Home LAN?

Before starting installation, it is a good idea to diagram the home LAN noting the location of all node connecting points, wiring runs, wiring closets and pipe chases, the range areas of wireless hubs, and all the specific components included in the LAN. The diagram can be best drawn as an overlay on a schematic diagram of the home and surrounding yard that shows the location of existing electrical wiring and other circuitry as well as the basic construction of the home. A typical diagram of a home LAN is shown in Figure 4-3.

As shown in Figure 4-3, note the location of all the network nodes and, very specifically, the location of each of the wall jacks through which they connect

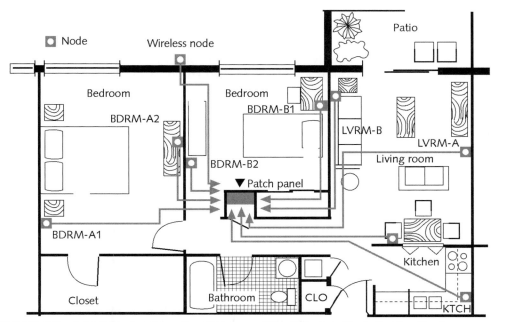

FIGURE 4-3   A LAN Design for a Small Home with a Wireless Node for the Yard

to the network. If you plan to have part of the network running on technologies other than Ethernet, also note where the devices for these technologies are located and, if they require AC power, how they connect to it. If you use wireless, HomePNA, HomePlug, or X10 segments in the network, you must connect them at some point to the Ethernet. The most convenient place to do this is usually near the patch panel and gateway setup, so plan to place those devices where you also have enough room for additional connecting devices linking non-Ethernet segments into the main LAN.

The first priority is to select a location for the patch panel and other connecting equipment. Place these in an unfinished area of the home, if at all possible. All the wiring for the network comes into the patch panel and fishing all of them through finished walls enormously increases the labor of installing the network. It is far better to work in an unfinished area (which can be finished after the network is completed). If such an area is not available, use a closet or utility room where exposed wiring is acceptable.

The location of the patch panel should also be a spot that is conveniently situated for pulling wire from all the other areas of the home where nodes will be placed. This likely will not be a distant corner of the building, but someplace centrally located. Under stairs, in a utility (furnace) room, or in a centrally positioned closet are all good possibilities. All of these are likely to have good access to the basement or attic, both places through which much of the wiring can easily run. In a two-story home, these are also the areas most likely to provide access from the lower floor to the upper. If these areas do not have access already, holes to provide it can be cut or drilled in the unfinished floor beneath and ceiling above. These access points can later be covered with finished wallboard or cabinetry when the wiring infrastructure is in place.

The LAN's Internet connection must also be brought to the patch panel from its location outside the home. This will be a new line (home run cable) to the **Network Interface Device (NID),** the data line connection point on an outside wall of the home; a cable line to a modem; or a cable link to a satellite dish antenna on the roof of the home. You must determine how these cables will run from the patch panel to their exterior terminals. If the home has an attic or basement, these lines can probably run in one or the other to a point near the outside terminus, then pass through a wall or roof to the outside. If the LAN is being installed in an apartment, locating the patch panel near an outside wall may be necessary because cables cannot be run through the walls to reach an outside connection point. In this case, the connecting cable may have to pass through a window fitting.

If you choose a utility room for the patch panel, be sure to position it as far as possible from electric motors and other devices that could interfere with the network. If you do find interference from other electrical appliances in the network, you may need to replace or filter the offending devices.

Once you have the patch panel location set, note it on the diagram, and then plan where the cables will run from each jack to the patch panel. The diagram is two-dimensional, but you have to think in three. Each cable run has to reach its destination within the length and breadth of the home, but also run up and down in the vertical dimension so it remains concealed either within the building's structure or surface mounted on the walls. **Retrofit** wiring requires more cable than wiring in new construction. It also creates more waste because cable segments often have to be cut long in order to be pulled into place and then trimmed off later.

Is My Diagram Correctly Labeled?

When your diagram for the network is complete, label everything on it with a short unique designation. Start with the rooms: Living room equals LVRM, master bedroom equals BRM1, and so on. The jacks in each room can take their names from the room: LVRM-A, LVRM-B, and so on. If each jack has only one cable running to it, then the cable ends can be labeled with the name of the jack. If the jack has multiple ports for data, telephone, or video, then each port must have a separate name: LVRM-A2, LVRM-A3, and so on, and the cable running to each must be correspondingly identified.

Labels are among the most critical items in the network installation process. Nothing saves more labor than accurate and complete labeling of every port and cable end. Nothing causes more frustration and makes troubleshooting more difficult than cables you cannot identify running you know not where. Label everything, and use labels that stick fast and will not come off when the cable is handled and pulled through narrow openings.

Basic Central LAN Components

Some components of a network serve the entire system rather than only a single node. These components are those that connect the network nodes into a linked whole and those that connect it to the Internet. Carefully setting up the gateway, patch panel, cables, and home run line results in a smoothly functioning network that can be easily upgraded or expanded over time.

Patch Panel

A patch panel is essentially a group of **RJ-45 jacks** mounted in two rows. The jacks mounted above one another are wired together in a straight-through manner. This means that a device plugged into the top jack in any pair will be connected straight through to a device plugged into the bottom jack in the same pair. The panel is installed in a convenient location, and each cable line in the home LAN is run from its outlet RJ-45 jack to terminate (with an RJ-45 plug) at one of the RJ-45 jacks on the patch panel. When all LAN cables are connected to the patch panel, the whole LAN can be connected together with patch cables running to appropriate jacks on the patch panel.

Patch panels add some cost to a home network, but they also add a lot of flexibility and allow the network to be more easily upgraded and expanded. They are well worth their cost and the amount of extra effort needed to wire cable lines into them. Most patch panels you buy off the shelf are designed to fit in 19-inch wide equipment racks. These can be bought in many sizes from mini versions that are 10 inches high (a suitable scale for most home LANs) to 6-foot-high stand-alone units. The patch panels have holes on the ends for screws to mount them in the rack, but they can also easily mount in a wall panel box, in a cabinet, or between wall **studs.**

An Internet gateway or a hub is typically located close to the patch panel, and patch cables are used to connect each cable line in the LAN from its termination point at the patch panel to a port on the gateway. The gateway is connected back through the patch panel to an ISP line coming from the outside. Figure 4-4 gives an idea of how these elements are connected. It shows a patch panel, switch, and firewall installed in a small rack.

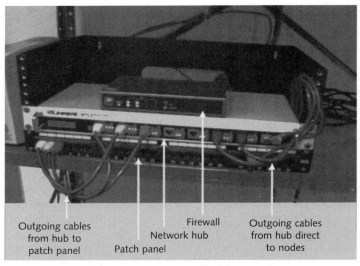

Outgoing cables
from hub to
patch panel

Patch panel

Firewall
Network hub

Outgoing cables
from hub direct
to nodes

FIGURE 4-4   Central Network Components Installed in a Rack

A patch panel is important for several reasons. The jacks to which nodes connect on the network are fixed in place permanently. They are part of the home's infrastructure, as are the cables that run through a building to connect them to a router, switch, or hub. The devices that connect to the jacks are not permanent. They are equipment that can, and will be, replaced or upgraded periodically. The same is true of the router, switch, or hub at the other end of the cable line. Connecting the LAN through a patch panel allows it to be reconfigured easily by simply changing the patch cables on the panel. It also allows any node or piece of equipment to be easily detached for repair or replacement.

By running cable between fixed jacks and a permanently placed patch panel, you avoid any strain on the cable or its connections that might impair the network's function. You also make all network devices easily removable from the network if the need arises. The patch panel allows for network connections to be changed and updated quickly, making the network more flexible and open to expansion as new additions are made.

Straight-Through and Crossover Cables

The cables that connect the network nodes to the wall RJ-45 jacks, and the patch cables that connect hubs, gateways, and other devices to the RJ-45 jacks in the patch panel, are **straight-through cables,** which means that pin 1 of the **plug** on one end of the cable is connected to pin 1 of the plug on the other end and all other pin numbers are correspondingly connected to their matching numbers.

A **patch cable** should not be confused with a **crossover cable,** another type of cable in which some of the wire connections are crossed end to end. Figure 4-5 shows the wiring at both ends of a straight-through or patch cable on the left, a crossover cable in the center, and a third type of cable called a rollover cable on the right. A rollover cable is used on certain routers, but rarely has an application in a home LAN. Neither a crossover cable nor a rollover cable is the same as a straight-through cable and neither can be used in place of one.

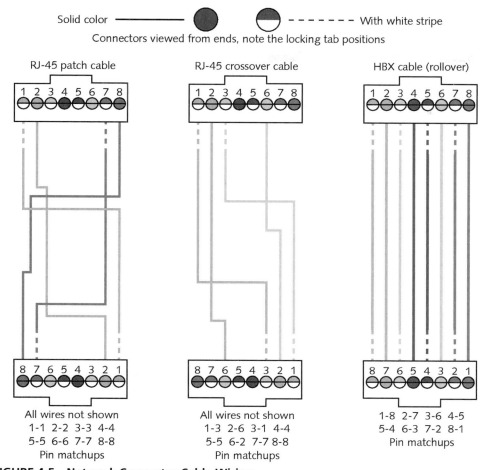

FIGURE 4-5 Network Connector Cable Wiring

The only time connections are crossed over is when two Ethernet devices are directly connected together without a hub. This connection can be between two computers connected without a hub, or between two hubs connected via standard Ethernet ports in the hubs. In either case, a crossover cable, which crosses the transmit and receive pairs (the orange and green pairs in normal wiring), is required. In a crossover cable, one end is normal, and the other end has the crossover configuration. If a crossover cable is used, it should be labeled so it will not be used as a "normal" patch cable in the future.

Note: Remember that only two computers can be networked together using a crossover cable. To add extra PCs to the network requires a bus topology or a hub.

Residential Gateway

A **residential gateway** connects the home's local area network (LAN) to the Internet. A hardware device similar in function to a router, the residential gateway provides a unique combination of features. By definition, a gateway joins two networks together. In the home network setting, this means joining the home LAN with a WAN that forms part of the Internet.

If the LAN has only one or two computers, one of them can be wired as the network's gateway to the Internet and the other can receive its Internet data through the gateway computer. If the LAN includes multiple computers, or might in the future, a separate residential gateway (also called a home portal) is a good idea. It adds $100 or more to the network's cost, but like the patch panel, it also allows for more flexibility, easier expansion, and greater capacity for the network.

As home networking continues to grow in popularity and capability, vendors are working to make home networks easier to build and use. Many elements of a home network, such as IP addressing, present unnecessary complexity to the user and are being simplified and automated in emerging gateway products. Gateway hardware exists in multiple forms, including general-purpose servers with multiple network adapters (also known as multi-homed computers) and routers.

Home or residential gateways vary significantly in their capabilities. As a result, no one typical home gateway exists. However, most residential gateways have some basic features:

- POTS and DSL cable service connectivity
- Multiple ports for Internet connection sharing
- Firewall security

The popularity of broadband network access has also stimulated the improvement and flexibility of home gateways. Broadband brings Web-based applications that require large capacity and fast data transfer into the home, including real-time high-quality audio and video streaming and online gaming. A central device such as a residential gateway is best able to support this new generation of Web technologies.

New residential gateway products with improved capability and specifications are continually being introduced. Home networks may include segments using several technologies. Any or all of these systems, each using a different networking technology, might be included in a single LAN:

- Ethernet home network
- Wireless home network
- HomePNA network
- PowerPlug Network
- Bluetooth Network

Residential gateways are now available that provide connectivity for all of these network technologies as well as **Home Audio Video interoperability (HAVi)** and **Open Services Gateway initiative (OSGi),** two additional connection standards designed to allow user-friendly connectivity of security, video, audio, lighting, and other technologies to the home LAN. If the home LAN you are installing uses more than one technology, a residential gateway with multiple capabilities is the ideal solution for connecting the various network segments.

The ZyWALL 2 Plus Internet Security Appliance is a residential gateway with four switched Fast Ethernet ports. It provides for two separate Internet connections through dual ISP ports and provides highly encrypted data security for all outgoing traffic. It also provides security as a firewall for incoming and outgoing data between the local network and the Internet. The ZyWALL 2 Plus permits two separate ISP connections to be maintained simultaneously, allowing maximum bandwidth connection to the Internet and providing backup by automatically

FIGURE 4-6   The ZyWALL 2 Plus Internet Security Appliance

switching all traffic to one provider if the other goes down. The firewall protection uses Stateful Packet Inspection, DoS and DDoS protection, and provides real-time attack alerts and logs. Its content filtering includes Java/ActiveX/Cookie/Proxy blocking, URL and Keyword Blocking, and Web content filtering. This unit connects to any Ethernet device and multiple types of machines and operating systems and sells for about $185. Figure 4-6 shows a picture of the device.

D-Link's Xtreme N Gigabit Router (DIR-655) is a wireless gateway and router that delivers faster speeds and farther range than other 802.11g-compliant routers. It adheres to the proposed 802.11n standard while staying backward-compatible with 802.11g and 802.11b devices. This router connects to a cable or DSL modem and provides high-speed Internet access to multiple computers, game consoles, and media players.

Wiring a Home LAN

Wiring is the backbone of a network. It is also the part most vulnerable to performance problems caused by poor installation practices. Wiring in new construction is generally a straightforward process, but it needs to be carefully and precisely done if the wiring is to perform at capacity and endure for years. Wiring in existing structures, whether done within the walls or on the surface, can be a frustrating experience, but this type of wiring needs to be as professionally installed as that in new construction. Forcing cables around corners and through openings too small to accept them may enable an installer to get a connection made, but it rarely performs up to standard. No network is better than the quality of the wiring on which it runs.

Wiring in New Construction

If you are lucky enough to get the network installation job while the house is under construction, take a few minutes to celebrate your good fortune. You should wait until the **rough-in** for the electrical, plumbing, and HVAC systems are installed in the home before doing the rough-in for the network wiring. Letting these other pros do their work first means you will not have to worry about them cutting through any of the cables you install. You will also be able to keep the network cable runs well away from any infrastructure that might cause interference in the network.

Install the boxes for the jacks in the walls first. You can use plastic or metal **outlet boxes** commonly used for AC wiring, and these are easily obtained at any building supply. Place the boxes as near as possible to the locations specified in your wiring diagram. Move them only to avoid obstructions already in place or that are likely to block the run of the cable to the box. Also be sure that neither the cables nor the jacks are too close to other electrical lines or equipment in the locations you chose. Put the boxes at the same height from the floor as the AC outlet boxes installed by the electrician (a stick cut to the correct height for the bottom of the box saves measuring each one) so that the **face plates** match when installed.

When all the boxes are in place, drill the holes in the wall studs, floor **joists,** and ceiling **trusses** that are necessary to make the cable runs on your diagram. Cat5 and Cat6 cable is just shy of a quarter inch in diameter. For single cable runs, a 3/8-inch hole is plenty large. Two Cat5 cables together, however, will bind when being pulled through a half-inch hole. For these double cable runs, use a 5/8-inch hole. This size can also accommodate three Cat5 cables in a bundle. If you are pulling any other wires that are larger in diameter (RG-6 video wire or Siamese cable, for example), you should try a set of all the cables that are in a run to see what size hole they can easily pull through without binding.

Drill the holes for cable runs as aligned with one another as you can to make pulling the cable easier. Wherever possible, drill the holes in the same direction you will pull the cable. Clear splinters from the backsides of the holes. Adjust hole locations to make turns in the cable as gradual as possible to avoid altering the twist structure inside the cable and reducing its performance. Keep the cable holes centered in the middle of the wall studs and far enough from the edges of joists and trusses that the cables will not be in danger from wallboard fasteners or molding nails driven into the wall later.

Run Cat5 or Cat6 cables as required from each box to the patch panel location. Leave at least a foot of cable protruding from each box and a full loop (two or more feet) at the panel end for each run. Run the cable loosely and avoid bending it sharply around corners. Be gentle with the cable to avoid damaging it as you pull.

Pulling cable is a lot easier and faster with two people working, one to pull the cable and the other to feed it in from the starting point. If a second person is not available, he or she can be partially replaced by a cable reel (to pay out the cable as it is pulled) and some pull cords and rods.

Installing Concealed Wiring in Existing Construction

Caution: When drilling or cutting into any existing wall, the AC power should always be shut off for the area in which you are working. This cannot be stressed enough. In addition, before drilling or cutting, carefully check that you will not be hitting an AC power line with your tool(s).

Working with **concealed wiring** has its tricks. For example, for cable runs that are concealed inside an existing wall, start with a remodel style outlet box that is designed to be installed into the finished wallboard. To place the box, cut a rectangle into the **drywall** in the correct location, slide the box in, and tighten up the binding screws. Tightening the screws pulls some plastic tabs tight against the backside of the drywall to secure the box in place.

Cut the opening for the box, and then pull the cable for it before placing the box. If the opening is located near the floor, you may be able to slip a small power drill inside the wall and drill a **pilot hole** for the cable down through the floor plate and subflooring into the basement or crawl space. This pilot hole guides you in drilling a larger hole with a bigger drill up from below.

The pilot hole should be no more than 1/8 of an inch in diameter, just big enough to shine a penlight through and locate the position from beneath. The final hole should be large enough for the cable to pass through easily without binding: 5/8 or even 3/4 of an inch.

If you cannot drill a pilot hole to locate the correct position of a cable wire hole from above, measure the position from a heat vent, pipe, or other construction feature that passes through the floor. Measure the exact distance from the outlet box to the feature (in two directions, making sure both are at right angles to one another and are aligned with the direction of the wall itself). Then, measure from the same feature in the basement to locate the hole position, making sure the directions of each measurement are the same as above.

Try to place outlet boxes midway between wall studs. Doing so helps reduce the risk of later drilling wiring holes up from below inside the wall that are on the wrong side of a stud. Also remember that most interior walls have **fire-stops** (cross pieces of wood that block off the interior space between studs) about halfway up the wall. If you wire from the attic down to an outlet near floor level, you will probably have to go through a fire-stop in the wall. There's almost never an easy way to do this. It generally requires cutting an access hole in the wall, drilling through the fire-stop and then patching the hole after the wire is in place. Accessing the outlet box from below is always the better option, if available. Figure 4-7 shows typical stud wall construction in older homes and where the fire-stops are placed.

Pulling or fishing cable is as much art as science. If you have not done it before, experiment a little on an easy cable run before trying the longer and/or harder ones. Cable must be either pushed or pulled through the inside of a wall. Experience and practice is more helpful than instruction for doing either one.

Cat5 and similar cable can only be pushed a short distance (up to 2 feet) before the cable will bend under its own weight and strike an obstruction. You can extend the distance a cable can be pushed by taping the end of it to a 1/4- to 1/2-inch dowel and then pushing both dowel and cable through the wall. Distances up to

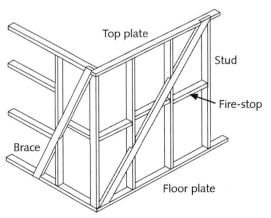

FIGURE 4-7   Wood-Frame Construction Showing Fire-Stop Locations

8 feet can be pushed this way. A dowel can also be helpful in pushing the cable through a small opening you cannot reach directly.

Pulling a cable into place is the preferable method, but it takes a little skill and sometimes a lot of patience to do it well. Pulling a cable involves feeding a string through the route the cable is to go, then attaching the cable to the string and pulling the cable through the route with the string. You can sometimes pass a string through an inaccessible area by dropping a weight tied to the string through from above, then attaching the string to the cable end and pulling it back. String can also be passed while taped to a dowel. It can even be shot across a long open space tied to a spring-fired toy dart. Whether passing string or pulling cable, whenever possible let gravity work for you. Pass the string or cable downward from above rather than up from below.

Use strong string for cable pulling, and before attempting to pull a cable back with it, be sure you attach the cable end to the string very securely. Use only strong non-stretching tape and knot the string on the outside of the tape so it cannot pull away under the tape. Join the string and cable securely enough that you will have to cut the cable end to separate them after the cable is pulled through.

When pulling cable in existing construction, especially Cat5 or other multiple-wire cable, leave plenty of slack in the line. A full loop of extra wire inside the wall at an outlet box allows the box to be easily connected before it is placed in the wall and secured. This also allows you to easily pull the outlet box out should you ever need to work on it in the future. If the jack should ever be damaged later, the extra cable length allows the repairer to cut off the damaged end when installing a new jack.

When cutting openings for outlet boxes, fit one box in place with the jack inside and put the face plate over it to see if the face plate lays flush against the wall surface. Some boxes require that their lip be mortised (fitted into a notch cut around the edge of the opening) into the wall surface before the face plate will lie flush on the wall. If this is necessary, it should carefully be done for each outlet box after the cable is pulled to avoid breaking away the edges of the opening too far while getting the cable in place.

When securing cable in unfinished areas (to walls, floor joists, etc.), do not use staples or nails that must be hammered in. You are almost sure to hit the cable or drive a staple too deep and damage the cable. Use small clamps on the cable and secure them with small Phillips-head screws. The clamps can be removed if changes in the cabling are needed, and a cordless screwdriver is far less likely to damage the cable.

Tips for Surface Wiring

When there is no way to get behind the finished walls in a room where a node outlet is required, the alternative is to run the cable on the surface in a **raceway** or under molding. Using surface wiring takes some planning. When using a raceway, a little advance planning makes for an easier installation and a better finished appearance.

- Mark out the exact route of the raceway and be sure you get fittings that can adapt to all of it, leaving no uncovered parts.
- Be sure the raceway you select is deep enough to accommodate all the cable(s) you are using. Test a piece with the cable inside and the cap in place before you purchase the raceway. If the raceway mounts with screws, make your test with the screws also in place.

- Buy all the raceway fittings and the outlet boxes that attach to them from one manufacturer. Do not try to fit one company's boxes to another's fittings. They usually will not mate perfectly and will not look good when forced together.

- If the walls (especially the corners) of a room are not flat and true, do not use raceway that attaches with an adhesive backing. If there is tension on the raceway after it is fitted in place, the adhesive may stick for a time, but will eventually come off. For these situations, use raceway that attaches with screws or add screws to adhesive-backed raceway wherever it cannot adhere properly.

- Raceway looks best if run near (but not at) the bottom of the wall, just above the baseboard molding. When painted the same color as the molding, it blends in and is hardly noticeable. If you want the surface-mounted outlet boxes a little further up the wall, use a T-junction or elbow in the raceway, and run a short vertical piece up to the level where you want the box.

- Try to avoid running raceway around door casings and window openings. It is almost impossible to fit it perfectly, and anything less looks tacky. Wherever possible, bring the cable into the room at a point from which it can reach the outlet box without passing a doorway. If an opening cannot be avoided, run the raceway to the casing, then remove the casing and cut a channel in the back of it in which to run the cable around the door. The raceway can be continued on the opposite side of the door.

- Leave some slack in the cable as it enters the raceway. This is useful if adjustments are necessary later and it also avoids unnecessary pull strain on the cable as the temperature changes.

If the cable needs to run near the ceiling, it is best to run it a few inches below the top of the room. Adding a geometric-patterned wallpaper border under the raceway may help camouflage its appearance. Alternatively, a ceiling-level cable can be run inside **crown molding,** which has adequate space behind it to easily hold a cable. From the crown molding, a cable can be dropped vertically to the outlet box in the room either inside the wall or in a surface raceway. Crown molding is more expensive than raceway, both for the materials and installation, but it results in a fully concealed installation and a much cleaner finished appearance. Figure 4-8 shows a section of crown molding and a diagram of the space behind it in which cable can run.

Cable can also be run inside baseboard moldings and chair rails, but because both of these are installed flat against the wall, they present an additional challenge. A groove or channel must be cut into the molding or the wallboard under it to accommodate the cable. This makes installation more time consuming and expensive. Neither base molding nor chair rail can pass cable around a doorway, so they are only suitable covers for short cable runs within a room.

Plenum Wiring

A **plenum** is an enclosure in a building that is used to move air for heating, cooling, or humidity control. It may be created by a false ceiling, a false floor, metal duct work, or a variety of other construction methods, but its main purpose is to move air that is environmentally controlled in some manner. A secondary purpose of a plenum may be to contain high- or low-voltage wiring. Because plenums often connect rooms in a building, they provide convenient paths through which to run wiring.

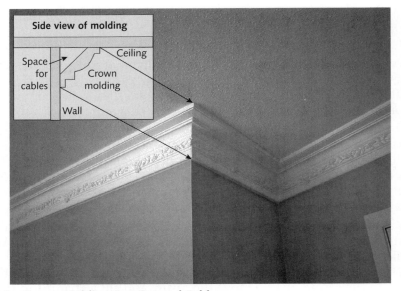

FIGURE 4-8   Crown Molding Can Conceal Cables

Cables that run in plenums must meet applicable fire resistance and environmental requirements. These are important because the plenum-run cables may be subjected to temperature and humidity extremes not encountered in normal wiring paths. Plenum wiring also poses a greater hazard than wiring run inside walls because, if a fire occurs in the plenum-run wiring, smoke and heat will be carried by the moving air in the plenum to other parts of the building, thus spreading the fire faster than it would otherwise move.

Protection for plenum-run cables may mean enclosing them in conduit (inside the plenum) or using cables having jackets and other components made of materials that are resistant to open flame and are non-toxic at high temperatures. Plenum cabling is often covered with Teflon and is more expensive than ordinary cabling. In the event of fire, its outer material is more resistant to flames and, when burning, produces less smoke than ordinary cabling. Twisted pair and coaxial cable are both made in plenum cable versions.

Wiring Connector Plugs and Terminator Jacks

Once cables are installed for the network, the next important step is terminating the cables with jacks at the node end and terminating them with connector plugs at the patch panel end. Stripping the cable ends and installing connectors requires great care. Stripping dimensions must be observed and the wires and insulation must not be damaged, particularly in high-speed Ethernet connections in which the connectors must have the same transmission characteristics as the cable.

Cable Connector Standards

RJ-45 connectors look a lot like the regular RJ-11 snap-in telephone connectors except they are a little larger. The RJ-45 connects eight wires as opposed to six in the RJ-11. The jacks with which the two types of connectors mate have corresponding conductor counts and different sizes. An RJ-45 connector will not fit into an RJ-11 jack.

TABLE 4-1
Wiring Standards for RJ-45 Connectors

Pin Number	Wire Color T568A Standard	Wire Color T568B Standard
1	Green/White	Orange/White
2	Green	Orange
3	Orange/White	Green/White
4	Blue	Blue
5	Blue/White	Blue/White
6	Orange	Green
7	Brown/White	Brown/White
8	Brown	Brown

The "RJ" in the jack's designation simply means "Registered Jack," and the "45" refers to the specific wiring pattern used for the jacks and connectors. The EIA/TIA-568-A standard defines two wiring patterns for Ethernet CatX cabling: T568A and T568B. These standards specify the pattern in which the color-coded wires in the cable are connected to the pins of the RJ-45 connector or the jack.

If you hold an RJ-45 connector in your hand with the tab side down and the cable opening toward you, the pins are numbered from left to right: 1 through 8. The pin numbers connect to the following colored wires in the cable for T568A and T568B. Table 4-1 shows how the colors are usually designated in wiring diagrams and instructions.

T568A standard is preferred for residential applications and T568B for commercial applications. Both, however, are electrically identical as long as you use the same color pattern to connect both ends of a given cable. If you are consistent in this, pin 1 at one end of a cable is always connected to pin 1 at the other end, and pin 2 on one end is connected to pin 2 on the other end, and so on, regardless of which of the two color patterns you use.

For the network you are working on, pick one standard and use it for all the wiring. It does not matter which you choose. If you were to buy a pre-made Cat5 patch cable that has been made to the other standard, it would still work on the network because both ends of the cable are wired to the same standard.

Terminating Cable to an RJ-45 Plug

To terminate an RJ-45 plug on the end of a Cat5 or Cat6 cable, use a **CatX cable stripper** tool to remove about a half inch of the cable jacket. Next, untwist the **twisted-pair wires** and arrange them in the order (left to right as you hold the cable pointing up in your hand) of the color pattern you have selected to use. Press the eight wires flat on a hard surface or between your thumb and index finger so they can slide into the RJ-45 connector. When you have the wires in the right order and flat, cut the ends so the flattened portions are all half an inch long, if necessary.

Insert the flattened wires into the RJ-45 connector and use an **RJ-45 crimping tool** to crimp the connector in place on the cable. You must push the cable into the connector until the jacketed portion of the cable goes all the way up into the plug because the crimping tool presses down on a hinged tab that grips the cable's outer jacket to hold the connector in place on the cable without pulling on the wire connections. The crimping tool also presses the connector's electrical contacts down until they pierce the insulation of the wires and make contact with the copper conductors.

Follow the directions for using the crimping tool carefully. Practice two or three connecting operations on a scrap piece of cable until you feel comfortable doing the job correctly. Then you can start on the network cables. Figure 4-9 shows the combination CategoryX cable stripper/crimper and the steps described in this section for terminating cable to an RJ-45 plug using the tool.

To terminate Cat5 or similar cable to an RJ-45 jack you must use an **RJ-45 punchdown tool,** which can be purchased in a wide range of qualities and prices from $10 to $90. The more expensive versions use a compression spring hammering action to drive the cable into position. The spring-loaded impact is like that of a staple gun and strikes with a consistent force to produce a uniform connection every time. These tools also have replaceable bits to extend their service life. The cheaper models use only hand pressure to set the cable and are not serviceable for a large volume of wiring.

Both RJ-45 and RJ-11 jacks use type 110 IDC terminals, and to terminate them requires the use of a type 110 bit in the punchdown tool to both cut and terminate the wires. To begin the termination process, remove about 1.5 inches of jacket from the cable. Untwist the full length of the exposed wire pairs. Place the cable end onto the jack and arrange the wires into the IDC slots on the jack using the color code printed on the side of the jack. As with attaching an RJ-45 connector, the jacketed portion of the cable must go all the way up into the jack. Do not leave any of the twisted wires in the cable exposed without a jacket covering them.

Using the punchdown tool, press the tool's bit down onto the terminal with the side of the bit that cuts the wire pointed to the outside of the jack. Press down on the tool to compress the spring until the tool hammers down the wire into the terminals slot. At the same time, the hammer action drives down the cutting edge of the bit to terminate and cut the wires. Snap on the protective cover that is included with the RJ-45 and RJ-11 jack to cover the area where the IDC contacts are, and the job is complete.

Installing a Home Run Cable

If the home LAN's connection to the Internet is a DSL, a telephone line that transmits both high-speed data packets and voice signals, a home run cable running directly from the LAN's patch panel or from a new inside jack to the DSL service hookup is a good investment. This can make the LAN's operation as trouble-free as possible. The home run cable, which bypasses all the home's internal telephone wiring, and is isolated from it by a **DSL splitter** at the service connection point, carries data directly to the LAN without any possibility of interference or noise from other parts of the home's telephone system.

Planning and Preparing a Home Run Cable Route

A home run cable installed in combination with a DSL splitter ensures a clean path for the DSL signal from the splitter to the patch panel and then to the DSL modem.

CategoryX stripper/crimper

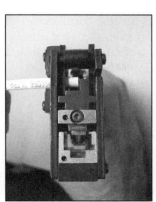

Inserting cable to strip

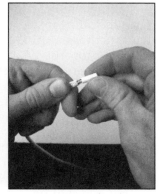

Removing cut sheathing

Flattening the cable wires

Inserting wires in plug

Checking wires placement

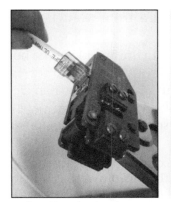

Inserting plug in crimper

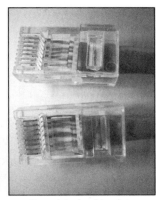

Crimping plug to cable

Completed cable plugs

FIGURE 4-9   Terminating Cable to an RJ-45 Plug

The splitter, as its name suggests, splits the DSL signal into voice and data segments and isolates all of the existing telephone wiring in the home and any equipment operating on it from the data signal. The data signal (and the separated voice signal if desired) is sent directly to the patch panel via the home run cable, and the splitter blocks any interference to these signals caused by problems with old wiring or malfunctioning equipment on the existing voice side of the line inside the home.

Installing a home run and splitter combination is a somewhat more involved job than installing inside wiring because it requires a cable run through an exterior wall and perhaps around the outside of the home, but it is worth the effort and expense to know the LAN has a clean data signal.

For a home run cable installation, you need a DSL splitter, a roll of exterior-use rated Cat5 or Cat6 cable long enough to do the job, and some mounting hardware. You also need a jack and mounting box if that is what the home run cable connects to inside. If it is going to the patch panel, then an RJ-45 connector for the inside end is required.

Before starting installation, plan where the home run will go. The inside end of the cable connects either to the LAN patch panel or, if there is no patch panel, to a wall-mounted jack to which the modem is connected. First determine where the inside end will be located, and then map a route to the Network Interface Device (NID) on the outside of the home.

The NID is a device installed by the telephone company that connects the home's inside telephone wiring to the outside telephone network. It is a gray (usually) box mounted on the outside of the house, often near the electrical meter. In addition to connecting the home's phone lines to the outside, the NID contains a modular plug that allows the homeowner to test whether the local telephone exchange network is working. The NID has two sides, one accessed by the telephone company and one accessible by the homeowner. All of the wiring for a home run is connected on the customer side of the NID. Some older homes may not have an NID, but have a "protector block," a small box where the line coming out of the house connects to the telephone company's incoming cable.

As with other retrofit wiring, the route that the home run takes is based on the most reasonable path for installing it between the NID and the patch panel or jack. Many factors may influence that path, including overall distance, obstructions along the route, and appearance requirements for the finished line.

Three main ways can be used to get a home run from the NID to the patch panel:

- Run the cable through an exterior wall, and attach it to the outside of the home from the exit point to the NID.
- Run the cable through the attic to an inside point near the NID, then run it through the wall and down the outer wall to connect to the NID.
- Run the cable through the basement or in a crawl space under the home, then out through a wall and to the NID.

Before starting the work, survey the interior and exterior walls of the home to determine exactly how and where you are going to run the cable from the NID to the desired jack or patch panel location. If aesthetics are of concern, try to run as much of the cable as possible inside the house. Run the outside portion along the sides and back of the home, away from the street. Once you have decided on the route the home run will take inside the home, the point where it exits the house through a wall, and the path from there to the NID, you can begin the installation.

Running the line through the attic or basement is similar to pulling the other patch panel-to-jack cables for the network. Getting the cable through an outside wall and connected to the NID requires a little more care and a few specialized tools.

Mark the point at which to drill a hole through the exterior wall for the cable. If the cable connects to a DSL jack on the inside of the wall, the hole should be in the center of the square where the jack mounting box is located on the wall.

Mark the box position on the wall and the center point in the square. If the cable is coming from a basement or attic run, mark the most convenient point to drill a hole through the wall to the outside.

Use a 3/8-inch drill bit (masonry or wood, as appropriate) to drill a hole from where the wall is marked to the outside. Drill from inside the house outward, but be careful to avoid hazards such as water, sewer and gas pipes, electrical wires and conduit, and cable TV lines when drilling. The hole should be drilled at a 15-degree tilt downward toward the outside of the wall. This prevents water from running inside the home along the bottom of the cable.

On the outside of the house, the cable mounts in **drive rings** for a cable run to the NID. These can be installed under the **eaves** on the building's wall or under the eaves next to the fascia board. Installing the drive rings under the eaves next to the fascia board hides the cable better and allows it to go around corners in two 45-degree bends.

Starting 6 inches to a foot from the wall hole (in the direction of the NID), drive rings should be installed every 3 feet on straight runs and no more than a foot on each side of a corner. They should be nailed through the eave into the roof trusses, which can be seen through the attic vents or located by the nails in the eave. If the drive rings are installed against the fascia board, use two rings, spaced 6 to 12 inches from each side of a corner to go around it. Do not install a ring at the corner. The two rings allow the cable to make two 45-degree bends instead of one 90-degree bend. If the drive rings are installed on the wall, apply two layers of friction tape or six layers of electrical tape to the cable where it contacts the corner of the building. A short section of plastic tubing can also be used to ease the bend and provide the same protection.

Running the Cable

If the cable terminates at a jack just inside the hole in the outside wall, it is most efficient to pull in the cable working from the NID back to the jack location as you install the drive rings. If most of the cable run is inside the house going through the attic or basement, work the other way from the jack or patch panel toward the NID. Either way, run the cable to the hole, and fish the end through the wall. Pull sufficient cable through to complete the run outside to the NID or inside to the jack.

When running it outside on the building's wall, keep the cable behind other cables and pipes, if possible, but do not pinch it. Inside and outside, do not run it parallel to PVC (plastic) electrical conduit or AC wiring. To avoid possible interference, maintain at least 12 inches of clearance from other wiring, and do not fasten the cable to any type of electrical wiring or fixture.

Allow plenty of slack, at least 2 feet at either end. The excess can be cut off at the connected ends, but the cable cannot be spliced if it is short. If the cable runs for some distance inside the house, leave at least a full loop of slack at the patch panel or the jack located on an interior wall. If the cable is going to the patch panel, secure it to the inside of the wall above the hole so it cannot pull outward. If the cable is going to a jack that is on the exterior wall and goes directly outside from the back of the mounting box, follow the instructions in the next paragraph to wire the jack before completing the cable run on the outside of the house.

For a jack connection, fasten the back box or mounting ring in place on the inside of the wall. The mounting procedure varies depending on whether the wall is brick, block, brick veneer, siding, or stucco. Strip off 8 inches of the cable's jacket. Separate the four pairs, but leave the individual wires that make up the pairs twisted. Follow the instructions supplied with the jack to connect the jack to the cable. You only use one or two pairs from the cable, so wrap the other pairs back around the jacket to prevent any shorts. When all wires are connected, carefully push the excess cable into the outer areas of the back box or into the hollow space between the drywall and the outside wall of the home. Place the jack on the mounting ring and fasten it in place.

For a patch cord connection, attach an RJ-45 connector to the cable, following the instructions in the "Terminating Cable to an RJ-45 Plug" section of this chapter. This completes the inside part of the cable run.

For the outside run, start from where the cable exits the wall and extend it up to the top of the wall where it can run horizontally to the NID. The vertical part of the cable run is fastened to the wall with galvanized straps. Place the first strap an inch to the side and 2 inches below the hole where the cable exits the wall to form a drip loop. Place the next strap level with the hole and 2 inches to the side of it. This strap is the bottom one in the vertical run. Keep the cable vertical and put another strap at the top of the wall about 6 inches below the level at which the cable turns for the horizontal run. With the cable secured at the top and bottom, install additional straps every 3 feet to hold it in position. Seal the hole around the cable with caulking to keep out water and insects. Loop the cable into the drive rings, working from the vertical run back to a point above the NID.

Install the Splitter

Install the DSL splitter adjacent to the NID, but not so close that the cover of either one blocks the other when they are open. Use screws to secure the splitter to the wall. Run the home run cable to the splitter using straps as you did going up the wall from the exit hole.

Wire the Splitters

Open the NID and locate the wiring for the DSL line. After you have identified the correct wiring, unplug all of the modular connectors from the Entrance Bridge Network (EBN) demarcation point in the NID. This disconnects all the incoming lines and prevents shocks.

Make notes of where each wire is connected to the EBN so you can reconnect them exactly the same way. Note the colors, but also keep those on the right and left or top and bottom together. When you are sure you have notes of how the wiring is connected, disconnect all the wires from the EBN of the ADSL line. Figure 4-10 shows the NID as it appears: closed on the left, then opened in the center, and a close view of the wiring terminals on the right.

Run a length of cable between the NID and the splitter. Secure the cable where it enters the NID, leaving a foot or so inside the NID. Strip off about 9 inches of jacket and connect the white/blue and blue wires to the binding posts upon which the existing inside telephone cable was. If the existing cable was connected to more than two binding posts, tighten down the unused posts to prevent static.

Using connectors, connect the white/orange and orange wires of the cable from the splitter to the existing inside telephone cable matching the wires color for color or using the following color code: white/orange = green or black, and orange = red or yellow. Wrap any remaining conductors of the cable running from the NID to the splitter around the jacket.

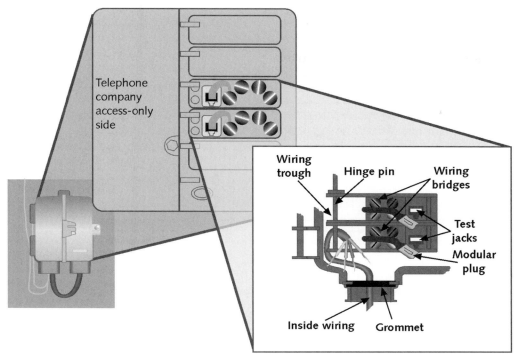

FIGURE 4-10   Network Interface Device Wiring

Run the other end of the cable from the NID and the home run cable into the splitter and secure them. Cut the cables, leaving about 9 inches inside the splitter. Strip off the insulation on both cables as before. Working with the cable to the NID, connect the white/blue and blue wires to the network binding posts in the splitter, and the white/orange and orange to the voice binding posts. If you are installing a two-port jack on the home run, connect the white/orange and orange wires of the home run to these terminals, also matching the colors of wire on each terminal. Connect the white/blue and blue wires of the home run cable to the data terminals, again matching the colors, to complete the splitter wiring. Close the splitter. Reinsert the modular plugs into the EBN(s) according to your notes, and close the NID. The home run cable is complete.

Configuring the Hardware and Cables

With all the cables in place and connected to the jacks, and the central hardware components waiting with the patch panel, the network is ready to be configured. How everything connects depends on the size and complexity of the network you have created. Some network setups are illustrated in the figures that follow. Figure 4-11 shows a basic network with a DSL connection, a gateway, and one hub or switch.

In this simple network configuration, all components connect through the patch panel. The DSL line comes into the panel and connects through it to the modem. The modem connects to the hub. Each of the hub's ports connect back to the patch panel and through it out to the nodes. For most home LANs, this setup is sufficient, at least at the beginning. It has plenty of excess capacity and can readily

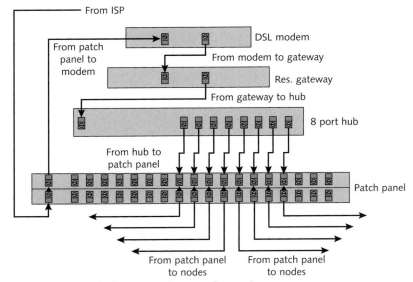

FIGURE 4-11   Basic Network Connection Configuration

be expanded later with additional Ethernet components or network segments using other technologies.

Figure 4-12 shows a larger network with a DSL, gateway, and two hubs or switches. This is a larger but still basic configuration. The second hub connects to one of the ports on the first hub. Its outgoing ports are again connected through the patch panel and through it out to the nodes of the network. The second hub is

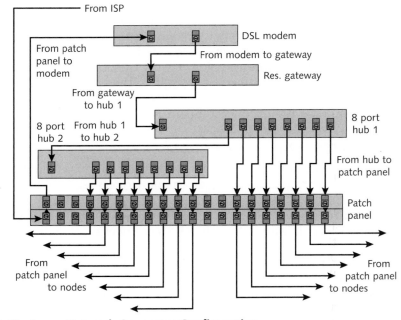

FIGURE 4-12   Large Network Connector Configuration

actually a segment of the network and could use a wireless or power-line technology. This architecture gives the LAN more nodes, but the rest of the configuration is the same.

Wiring Standards and Quality

Wiring and installation standards for both low- and high-voltage wiring have been developed by the National Electrical Contractors Association (NECA), which was founded in 1901. It is the country's largest association of electrical contracting firms, with over 70,000 member companies. These contractors and others not associated with NECA employ over 650,000 electrical workers who install the vast majority of high-voltage and low-voltage wiring in new construction and existing buildings throughout the United States. NECA includes 120 U.S. chapters and has additional affiliates in other countries around the world.

NECA is the organized voice of the electrical contracting industry. Its objectives include the promotion of high standards of electrical installation work through quality workmanship and the training of its members' workforce. NECA provides detailed publications and workshops on virtually every aspect of the electrical construction industry. It develops printed manuals and on-line resources to train workers in the latest technical advances and industry trends for installation of electrical and electronic wiring and devices.

NECA is the main source of training and technical information about electrical contracting. Among its educational activities are publication of Electrical Contractor Magazine and many other information manuals, and sponsorship of the annual NECA Show, the premiere event for showcasing electrical contractor equipment and techniques, new technologies, and changing trends in the installation of electrical wiring and equipment.

NECA also provides courses to help contractors and their employees broaden their knowledge and skill. These courses are offered to all member and non-member contractors at the NECA Show and also through the National Joint Apprenticeship and Training Committee (NJATC) for the electrical industry, which NECA co-sponsors. This program trains thousands of electrical apprentices and provides in-service training for journeyman electricians.

Summary

- Planning for a home network should answer questions about where the network will be built, what nodes it will include, where each will be located, and how each will be connected. The plan should be diagrammed and labeled in detail.

- The plan for a home LAN should be diagrammed on a floor plan of the house and should show the location of the central patch panel and other main components, the location of each node connection point, and the cable runs that will connect them.

- Network cable runs and connection points should be kept clear of other electrical circuits

and equipment that could introduce interference into the network.

- Installing network wiring while the home is under construction is always easier than installation after it is finished.

- If a new or remodeled home has any wiring installed, you should make use of as much of the existing wiring as possible, even if it is not all well placed. Good planning can make effective use of a general wiring installation done before the design of the LAN was completed.

- Most home LANs are wired as Ethernet topologies, but specialized functions operating on different standards, such as HomePNA, HomePlug, Bluetooth, and others, can be designed into the LAN and connected to it as network segments through bridges or other Ethernet-compatible devices.

- When diagramming and installing a network, label everything, including each room, each node access point, each cable run, and all components with a short unique designation.

- Rough-in network wiring in a house under construction should be done after rough-in work for the electrical, plumbing, and HVAC systems is installed in order to avoid cable damage from the other systems and to enable the network wiring to be kept away from any wires or devices that might cause interference in the network.

- In existing homes, network wiring should be run wherever possible inside finished walls to unfinished areas and then to the central patch panel. Where inside wiring is not possible, wires can be run on wall surfaces in raceways, or wireless access points can be set up that can provide coverage in the inaccessible areas.

- Cat5 of Cat6 cable is connected by means of RJ-45 connector plugs, which connect eight

wires in a standard pattern and plug into RJ-45 wall-mounted jacks, which are also connected to Cat5 or Cat6 cable using the same standard color pattern.

- A patch panel is a group of RJ-45 jacks mounted in rows or a block and installed in a convenient location, to which each cable line in the home network connects from its outlet box.

- Cables that connect the network and patch cables that connect nodes to jacks and the ports on the patch panel to one another are straight-through cables wired to identical pins at each end. Crossover cables are crossed end to end so two wire pairs are reversed on opposite ends of the cable.

- A residential gateway is a hardware device similar to a router, but also including a firewall feature, that connects a home LAN to the Internet by joining the home LAN with a WAN that forms part of the Internet.

- A home run cable installed in combination with a DSL splitter ensures a clean path for the DSL signal from the splitter to the patch panel and then to the DSL modem, and blocks any possible interference that might originate from other existing wiring in the home.

Key Terms

AC power Alternating current electric power; the standard electric service in homes.

Cable run A cable installed between two connecting points such as a patch panel and jack.

Category 5 (Cat5) cable A high-speed eight wire UTP data cable for Ethernet.

Category 5 (Cat5) cable An enhanced (higher-speed) version of Cat5 cable.

CatX cable stripper A tool for cutting and removing the outer jacket from Category 5 and other similar types of cable.

Concealed wiring Retrofitted wiring installed within the finished walls of a structure.

Crossover cable A connector cable in which two pairs of wires are connected to different pins (crossed over) at each end of the cable.

Crown molding Angled decorative slats installed at the junction of interior walls and ceiling. It can conceal cable.

Drive ring A metal ring that can be hammered into walls to carry cable on the exterior of buildings.

Drywall The paper-covered gypsum board that is fastened to studs to finish the walls in most homes.

DSL splitter A device attached to the NID, which splits an incoming DSL data line, isolates other telephone lines and equipment, and allows a home run to be installed to the modem.

Eave The underside of a roof that overlaps the outer walls of a building.

Electromagnetic field Force field of electrons generated by high-voltage equipment and wires.

Face plate The decorative cover on wall boxes that contain data jacks or AC power outlets.

Fire-stop   A wooden crosspiece set between studs in interior walls to retard fire.

Home Audio Video interoperability (HAVi)   A standard for audio and video connectors and cables designed to allow multiple technologies to interface with one another.

Home run cable   A data cable running direct from a DSL splitter to the network modem assuring a clean incoming signal.

Jack   A connecting device terminating a cable into which a plug is mated to connect a node.

Joist   A horizontal support beam in a floor.

Network Interface Device (NID)   A device that connects the ISP service line to a home's inside telephone and data wiring.

Open Services Gateway initiative (OSGi)   A standard for audio and video connectors and cables designed to allow multiple technologies to interface with one another.

Outlet box   A wall fixture containing a data jack or AC power receptacle.

Patch cable   An eight-wire straight-through cable in which all wires are connected to the same numbered pins at each end.

Patch panel   A device consisting of a row or block of jacks, used for connecting all components of a network.

Pilot hole   A small hole drilled to locate a position or to guide a larger drill bit.

Plenum   An enclosed passage for moving air in a building. Also sometimes used as a channel for electronic wiring.

Plug   A terminator on the end of a cable that mates with a jack to make a connection.

Pulling cable   The process of drawing a cable through an existing structure from one connector to another.

PVC conduit   Plastic pipe commonly used to transport water, but also used as a tube through which electronic wires and cables can be run.

Raceway   An enclosed track in which to run cable; the track is attached to the surface of walls.

Residential gateway   A device that connects a LAN with a WAN that is part of the Internet; the gateway controls the data coming into, or going out of, the LAN.

Retrofit   To add new wiring or other infrastructure to an existing building.

RJ-45 crimping tool   A tool used to attach an RJ-45 plug as a terminator on the end of a cable.

RJ-45 jack   A terminator device on a cable into which a plug is mated to make a network connection.

RJ-45 punchdown tool   A tool used to attach an RJ-45 jack as a terminator on the end of a cable.

Rough-in   To install the outlet boxes and cable runs for a network.

Siamese cable   Any cable that consists of two or more types of cable (Cat5, coaxial, fiber-optic, etc.) bundled together in a single sleeve.

STP cable   Shielded Twisted-Pair wire. Consists of up to four pairs of wires twisted in pairs and wrapped in a metal shield with an insulated outer sleeve.

Straight-through cables   Cable wiring that is connected to the same terminator pins at both ends of the cable.

Studs   The 2×4 or 2×6 wood uprights in the walls of homes.

Surface wiring   Wiring run along the outside of a wall, usually in a raceway, although it can be bare.

Truss   The triangular structures that support a roof.

Twisted-pair wires   A set of two wires twisted around one another in a specific manner to improve data transmission in a high-speed cable.

UTP Telephone Wire   Unshielded Twisted-Pair wire. Consists of up to four pairs of wires twisted in pairs and wrapped in an insulated sleeve. Typically used to wire telephones.

Review Questions

1. What questions should be answered when planning a home LAN?

2. When diagramming a network, three dimensions must be considered to allow for the network connections extending up and down in the home as well as over its floor plan. True or false?

3. Network cables and connecting points should be kept away from electrical equipment because it can cause _____ in the network.

4. If network wiring is being installed in a home under construction, the rough-in wiring should be done before the electrical and HVAC rough-in is done. True or false?

5. In network wiring, "fishing" a cable means _____.

    a. wetting it before using it

    b. pulling it through an inside wall

    c. lubricating it with fish oil

    d. bending it into a fishhook shape

6. Why can't cables be run as easily in exterior walls as interior walls?

7. Why is it important to label each part of a network design and each physical component with a unique label?

8. The currently preferred cable for wiring an Ethernet network is _____.

9. Why should Cat5 cable be installed gently without sharp angles or deforming of the wire?

10. What is a home run cable?

11. A residential gateway controls a home LAN's access to the _____.

12. A patch panel contains a number of _____ into which RJ-45 plugs are mated to connect the network.

13. A combination stripping/crimping tool is used for attaching RJ-45 plugs to cable. True or false?

14. Cat5 cable contains how many pairs of wires?

    a. One

    b. Two

    c. Three

    d. Four

15. In a crossover cable, which two pairs of wires are reversed when connecting the two ends to RJ-45 plugs?

    a. 2 and 6, 1 and 3

    b. 2 and 4, 1 and 5

    c. 4 and 6, 5 and 8

    d. 5 and 6, 7 and 8

16. The device that terminates a cable run from the patch panel to a node on an Ethernet network is a _____.

17. A crossover cable is used to connect two _____ _____ in an Ethernet without a hub.

18. When running concealed cable in an existing home, why is it better to run the cable to the basement rather than up to the attic?

19. What is a raceway and what is its function in network wiring?

20. The diameter of Cat5 cable is _____ and the diameter hole that should be drilled to accommodate two Cat5 cables going through it is _____.

21. The most convenient place for a HomePlug segment to be connected to the network is at or near the _____.

22. What are two configurations for setting up three wireless access points in a LAN?

23. Why should staples, nails, and similar fasteners not be used to secure Cat5 cable?

24. A punchdown tool is used to connect RJ-45 plugs to cables. True or false?

25. The term NID stands for _____ _____, and it is the connecting point from the network to the _____.

Hands-On Projects

Project 4-1: Diagram a Network of the Floor Plan of a Building

In this project, you will draw a diagram for a network using a scale floor plan of a building as the basis for your design. In the network diagram, include a node in each room, a central patch panel, and a home run to the NID.

1. Obtain a floor plan from a home design magazine or an architectural floor plan drawing.

2. Assume the building is under construction and the network wiring can be installed while building is in progress.

3. Diagram the network on the floor plan as closely to scale as you can.

4. Estimate the amount of cable that is required to complete the network. Use the scale of the floor plan (usually ¼ inch per foot) to determine the approximate length of the network cable runs and add 30% to the total to allow for vertical segments and slack.

5. Count the number of jacks, RJ-45 plugs, outlet boxes, and face plates the network requires. Add 10% to each total to allow for breakage and spoiled installation. Round each count up to the nearest whole item.

6. If cable is $.38 per foot, outlet boxes are $1.65, jacks are $.79, face plates are $.49, and plugs are $.28, how much will the materials for the network cost?

Project 4-2: Terminate a Cat5 Cable with an RJ-45 Jack

In this project, you will terminate one end of a 4-foot cable to a jack. The cable length will also be used in Project 4-3, so it should not be shorter than 4 feet.

1. Use a CatX cable stripper to strip the cable jacket 1.5 inches.

2. Place the stripped end of the cable in the jack and arrange the wires in the correct pattern. Be sure the stripped wires are pulled as far into the jack as the outer sleeve allows so the excess stripped wire is cut off when the wires are set.

3. Use an RJ-45 punchdown tool to set the cable, connect it, and cut the wires.

4. Cut off the jack from the cable end, leaving a 3-foot piece of cable that can be used for Project 4-3.

Project 4-3: Make a Patch Cable

In this project, you will terminate both ends of a 3-foot Cat5 cable with RJ-45 plugs.

1. Use a CatX cable stripper to strip the cable jacket a half an inch.

2. Untwist and flatten the wire pairs in the correct pattern for a straight-through cable.

3. Cut the stripped wires to a maximum half-inch length, if necessary.

4. Place the plug on the cable end with the flattened wires in the correct position. Be sure the wires are fully inserted into the plug.

5. Use an RJ-45 crimping tool to crimp the plug in place.

6. Repeat the preceding steps to terminate the other end of the cable with an RJ-45 plug. Use the same wiring pattern so the patch cable is a straight-through cable.

Project 4-4: Rough-In an Outlet Box and Cable Run in an Open Stud Wall

In this project, you will rough-in an outlet box in a wall under construction. The wall can be a practice wall set up in a classroom or an actual wall in a residence.

1. On an open stud wall, attach an outlet box one foot above the floor at the far-right end of the wall.

2. Attach a second outlet box 1 foot from the floor at the far-left end of the wall.

3. Drill half-inch holes in the studs for the cable run between the outlet boxes.

4. Run Cat5 cable between the boxes.

5. Tie back the cable ends in the boxes so they will not be damaged by the drywall installation.

Project 4-5: Install an Outlet Box and Cable Run in an Existing Wall

In this project, you will install an outlet box in a finished wall. The wall can be a practice unit set up in a classroom or an actual wall in a residence.

Caution: When drilling or cutting into any existing wall, the AC power should always be shut off for the area in which you are working. This cannot be stressed enough. In addition, before drilling or cutting, carefully check that you will not be hitting an AC power line with your tool(s).

1. In an existing interior finished wall, cut a hole for an outlet box 1 foot above the floor.
2. Create a path for a cable run from the opening down to the basement or up to the attic. Depending on the wall's construction, this may require you to drill up from beneath the wall or down from above it. If the wall is a classroom practice unit, assume an attic exists above it and run the cable up into it.
3. Fish a cable from the basement or attic to the outlet box inside the wall.
4. Secure the outlet box to the wall face and bring the cable through an opening into it.

Project 4-6: Run a Cable in a Raceway

In this project, you will run a cable inside a raceway on the surface of a wall. If possible, use the same outlet box for this project that you used for Project 4-5. The wall can be a practice unit set up in a classroom or an actual wall in a residence.

1. Install an outlet box in an existing wall 1 foot above the floor.
2. Attach raceway to the wall 1 inch above the baseboard or 6 inches above the floor if there is no baseboard.
3. Using a corner angle, run the raceway around at least one corner of the room and over to the box. Run an elbow extension up to the outlet box.
4. Run cable in the raceway around the corner and up into the box.
5. Put the raceway covers in place so the installation has a finished appearance. Finish the installation so the cable is completely concealed from the point where it enters the raceway to the box.

Case Projects

Case Project 4-1: Select Technology and Design for a LAN

A client for whom you are designing a home LAN lives in an apartment constructed in the 1950s of brick exterior walls with wood and plasterboard interior walls. The apartment has multiple-line telephone service, with a jack in each room of the apartment, and regular electrical service, but the owners will not permit any cables to be installed through the outside walls or on the inside of any interior walls. The home LAN needs an Internet connection and at least one node in each room of the apartment. In this situation, what networking technology do you recommend to the client in order to install the LAN without damage to the apartment? How would you connect the network to the Internet, and where would you try to place the gateway?

Case Project 4-2: Troubleshoot a Network Connections

You are helping a friend determine the cause of a failure in one of his network nodes. Everything is working on his newly installed LAN except the computer in his daughter's bedroom, which the network does not recognize as a node. The LAN's owner says that the cable from the patch panel to the daughter's bedroom has been tested and is working correctly. You check the connecting cable from the jack to the computer and the cable from the hub to the patch panel plug connected to the daughter's bedroom. Both appear to be wired correctly, although their outer sheathings are different colors. Suggest some possible reasons why the bedroom computer is not receiving data from the network. How would you determine if your diagnosis is right? How would you correct the problem?

Case Project 4-3: Find a Solution to a Node Placement Problem

A LAN you are designing needs a network node located near the center of a 20-foot-long wall in the family room of a residence. The home is a large and expensive modern design and the owner wants all the network wiring concealed so as not to detract from its interior decoration. This presents a problem for the node because the 20-foot wall is made of solid brick. The home has cathedral ceilings so there is no attic space in which to run cables. It has wood floors with a small crawl space underneath. The network patch panel is located in a closet on the opposite side of the family room and at least 20 feet from the brick wall. Suggest two ways in which you could connect a network node located along the brick wall without using a raceway or having any exposed wiring.

Case Project 4-4: Recommend a Trouble-Free Network Technology

The person who has hired you to install a home LAN lives in a condominium development in which each building contains 12 residential units. She tells you she wants a wireless network so the installation will not damage her condominium. She says that 8 other condominiums in her building already have wireless networks and she wants the same kind. She also says she wants her network to be as fast as possible with as much power in the equipment as you can buy because some of her neighbors' networks seem to operate very slowly. Would you recommend a wireless network for this installation? Could another wireless network operate faster than those already installed in the building? If not wireless, what other technology could you recommend that might meet the client's needs? Why might it be better than wireless?

High-Voltage Wiring

OBJECTIVES *After studying this chapter, you should be able to:*

- Define electrical safety hazards and describe good practices in performing high-voltage wiring.

- Describe how to determine load requirements, install new circuit breakers in a service panel, and connect them to the electric service.

- Pull high-voltage AC cable in existing walls and install and wire electrical outlet boxes and devices in new circuits.

- Describe how to install ground wires, surge and spike suppressors, and interference filters in AC circuits.

OUTLINE Introduction

High-Voltage Electrical Safety Standards

Calculating AC Load Requirements

Planning and Installing New AC Circuits

Circuit Protection

Introduction

In this chapter, you will learn how to identify high-voltage electrical deficiencies and safety hazards and how to eliminate them by upgrading and improving home wiring according to current code standards. You will learn how to evaluate a home's AC wiring infrastructure and determine whether it will adversely affect a home LAN, either by failing to provide adequate, clean, and consistent electric power for the network's nodes or by introducing noise and **interference,** which reduce the network's speed and efficiency. You will also learn how to add new AC **circuits** to a home wiring system in order to supplement the existing circuits or to provide a more interference-free source of power for the network. Finally, you will learn how to minimize or eliminate the high-voltage phenomena that can affect the performance of low-voltage networks: static electricity, **electromagnetic induction,** and interference.

High-Voltage Electrical Safety Standards

Before working with high-voltage electrical wiring, you must be certain you meet the required knowledge and licensing requirements for the state and local area in which you are working. You should also know what inspection procedure, if any, your work must pass before going into service. Specific state requirements are found in the *State Electrical Regulations: Guide to Electrical Codes, Enforcement and Licensing* (2002), which is available without charge on-line at: http://www. necanet.org/store/index.cfm?fuseaction=catlist&category=1&cfid=1510705&cft oken=71054237.

Only qualified electricians should perform high-voltage wiring and electrical circuit installation. Many states have laws requiring that commercial and residential wiring be performed only by licensed electricians. Some states' laws allow others to do electrical work, but require that it be approved by a qualified inspector before it is put into service.

Caution: The information in this chapter is provided only to enhance your knowledge. Do not attempt to install or work on high-voltage wiring unless you are a licensed electrician or are working directly with, or under the supervision of, someone who is.

Organizations that Develop Electrical Standards

Electrical wiring and installation requirements also vary from state to state. Some state laws subscribe entirely to the **National Electrical Code (NEC)(r)** which is a safety standard developed by the **National Fire Protection Association (NFPA).** Other states subscribe in part to the NEC, but amend or add to it as their individual state legislatures decide.

The NEC standard is revised every few years, and it is called NFPA Code #70. The current edition is from 2005. It defines minimum requirements for the wiring and installation of "electric conductors and equipment installed within or on public and private buildings or other structures, including mobile homes and recreational vehicles, floating buildings; and other premises such as yards, carnivals, parking and other lots and industrial substations; conductors that connect the installations to a supply of electricity; and other outside conductors and equipment on the premises." Note that the code covers all electrical installations in buildings of any type, as well as the electric power grid that supplies electricity to those end users. Any wiring you do in an individual home becomes an integrated part of a much larger electrical system. Your work must meet the same installation and safety standards as the rest of the system or it puts not only the home where it is installed at risk, but also other parts of the system.

The primary intent of the NEC is to promote safety in the installation and use of electrical wiring and equipment. The Code has been developed and refined over many years and represents the combined experience and judgment of literally thousands of electricians and other experts in the field. Its standards and requirements are taken very seriously by state building inspection and permit departments nationwide.

Other organizations also develop electricity-related standards. The Electronic Industries Alliance (EIA) and the Telecommunication Industry Association (TIA) provide forums for industries working in the electrical field to develop standards and publications in several major technical areas: electronic components, consumer electronics, electronic information, telecommunications, and Internet security.

TIA oversees the telecommunications sector, and EIA oversees the others. The electrical standards developed by these organizations primarily deal with low-voltage products and processes. They have not produced any standards for high-voltage wiring, either in homes or in commercial buildings, but they do have standards for many electronic products and components that are powered by high-voltage AC circuits. These products include many of those used in computer networks, audio systems, video systems, and home automation systems. You can learn more about the various EIA and TIA standards at their respective Web sites: http://www.eia.org and http://www.tiaonline.org.

The Institute of Electrical and Electronics Engineers (IEEE) is a professional association of individuals who work in the electrical and electronics areas. IEEE (the acronym is pronounced eye-triple-e) has over 375,000 members in 150 countries and is a nonprofit organization that publishes many professional papers and standards. Its standards in areas such as telecommunications, electric power, consumer electronics, and computer engineering do not have the force of government regulation, but are consensus-based and accepted in the various fields because of the technical authority that IEEE represents. You can learn more about IEEE standards (established and under development) at http://standards.ieee.org.

Underwriters Laboratories, Inc. (UL) was founded in 1894 as a product testing facility. It is an independent nonprofit organization not affiliated with the government or any manufacturer. Its technicians perform product safety testing and certification of products that meet its standards for many companies worldwide. Companies submit their products to UL for safety testing, and if they are judged to be safe for consumer use, the manufacturer of the product can attach a UL-approved sticker to each unit as an indication to the public that it has met UL safety standards. Very few electronic or electrical products sold in the U.S., whether manufactured in this country or abroad, do not have a UL sticker attached. The UL listing has come to be the minimum accepted safety standard for all consumer products. You can learn more about UL and its safety testing procedures at http://www.ul.com.

Avoiding Electrical Safety Hazards

Electrical safety hazards in a home environment consist mainly of two types: fire and **shock**. Both can be avoided by correct wiring and installation practices; both are virtually certain to occur if good practices are ignored. Many potential errors may put you at personal risk; many others may leave behind an electrical problem that can cause serious harm years later. As a participant in the DHTI field, your career depends on performing safe reliable electrical installations that stand up to immediate inspection as well as the test of time.

Electricity always flows through a complete circuit, and all electrical circuits produce heat. The amount of electric **current** that flows in a circuit is determined by the voltage (electrical **potential** or pressure) coming from the power source and the **resistance** to the flow in the wires of the circuit and in the electrical equipment connected to it. The higher the voltage (potential) and the lower the resistance within the circuit, the larger the electric current flow. The more current that flows, the more heat is generated. A short circuit, one with little or no resistance to current flow, produces enough heat to cause a fire or melt metal. If a short circuit passes current through a person's body, it can kill.

The NEC specifies wire sizes and other standards for electrical installation in order to prevent heat from building up in the circuit to the ignition point. A large number of home fires are caused each year by overheated electrical circuits: too much current flow forced through wires too small to carry it adequately. When this happens, the wiring accumulates heat over time until the insulation on the wire ignites or melts away, allowing the hot wires to touch something else flammable and causing a fire. Most home electrical fires are not caused by a short circuit; they are heat-produced fires caused by overloaded wiring.

A short circuit is caused when the circuit does not have enough resistance to limit the flow of current passing through it. Touching a **"hot" wire** to a **"ground"** wire produces a short circuit because there is nothing in the circuit to resist the flow of current and too much electricity flows instantly. The amount of current flowing produces a lot of heat, which can quickly cause a fire. The current can also cause sparks at any point along the circuit where some of the current can arc or jump to another conductor and bleed off from the overloaded circuit. These sparks are like miniature lightning bolts, and they are almost as hot. An electrical spark can ignite wood instantly or fuse metals like a welding torch, creating the potential for even more current flow and a worse fire danger.

Short circuits can also cause severe shock to anyone touching the circuit or near enough for the current to jump to his or her body. Depending on how much current then flows through the person's body and the path it takes, an electrical shock can result in a stab of pain, a serious burn, or death.

Whether you are working on a client's home or your own, if you perform any high-voltage wiring work, always be sure you follow the requirements of the electrical code applicable in the area where the home is located. Adhering to the code ensures that your wiring functions properly and will not turn into a liability in the future.

Caution: The information in this chapter is provided only to enhance your knowledge. Do not attempt to install or work on high-voltage wiring unless you are a licensed electrician or are working directly with, or under the supervision of, someone who is.

Calculating AC Load Requirements

Alternating current (AC) is the type of current that almost all homes receive from the electric utility company. AC is so named because the electric flow reverses direction in the wires several times each second. 60-cycle (60Hz) AC, the U.S. standard, reverses 60 times each second. The reversing nature of AC enables it to be sent longer distances over wires than the other type of current, which is called DC, or direct current. The reversing nature of AC is also used to control the speed of motors and for other purposes. DC, the type of current that comes from batteries or AC to DC adaptors and is used for low-voltage circuits such as networks and electronic devices, functions much better in transistors and other electronic parts than AC.

Load requirements refers to the amount of electric power a home, or a circuit within the home, must have to meet the maximum needs of all the electrical devices in the home or wired to the circuit. To effectively participate in the DHTI industry, you need to understand how load requirements can affect the function of home networks. This section begins by discussing the factors involved in bringing electricity to the home, and then explores how to calculate loads for modern dwellings.

Electrical Use in the Home

Residential **electric service** is delivered to a house in two "legs" of electricity with a total potential of 240 **volts.** Some appliances in a home, such as the stove, oven, electric clothes dryer, and air conditioner, operate on 240-volt power. These appliances are connected to both legs of the residential service. The balance of the home's electrical needs are usually 110–120 volts, and all of these circuits are connected to only one leg of the 240-volt service, thus cutting the voltage by half. You will recall from your knowledge of basic electricity how **Ohm's Law** describes the relationship among **amperes (amps),** or the amount of current flowing; volts, which is the electrical potential between the two ends of a circuit; and ohms, the resistance within the circuit to the flow of current:

$$\text{Ohm's Law: } V = IR \text{ or Volts} = \text{Amperes} \times \text{Ohms}$$

The amount of electric power used by any device is usually expressed in **watts,** which are commonly calculated in a circuit by multiplying the volts by the amps. Using this formula, a light bulb in a 110-volt circuit through which 0.9 amps of current flow can be said to consume about 100 watts of power. This simplified calculation does not allow for resistance in the rest of the circuit (the wires outside the light bulb), but it gives results accurate enough to calculate load requirements in a home system.

Most average-size homes constructed before 1950 were typically wired for electric service of 60 to 100 amps, usually divided into five or six circuits of 15 amps (at 110 volts) or 30 amps (at 240-volts) each. Such a home could have an electric stove (the 30 amp, 240-volt circuit), 20 or 30 electric lights (up to 100 watts each), and still have plenty of capacity for electric radios, toasters, coffee makers, furnace motors, phonographs, and the new black-and-white television sets that were then sweeping the country. Electric clothes dryers and air conditioners were uncommon.

In the third quarter of the twentieth century, typical home electric service began to rise to around 150 amps as the use of electrical appliances, lights, and some electronic devices expanded. By the end of the century, most average homes were being wired for 200 amps, and larger ones for more. A power company research group calculated in the 1990s that the average home had 86 electric motors running at various times within it. Electronic devices, particularly computers and peripherals, became fixtures in most homes, and video systems (television sets and video recorders) and high-powered audio systems increased exponentially in both size and number.

One result of the dramatically increased use of electric power has been that the power use of electrical devices in many homes exceeds the capacity of the home's wiring. Many older homes have been rewired to increase their service ceiling, but many others have not. Some of those that have been rewired were done in a manner not meeting code requirements, and still others had one or two circuits added, but still come up short of what is needed to meet the demand for electricity in the post-2000 home.

Calculating the Load

The average power consumption of a home can be calculated from the electric bill or as a percentage of the available maximum service, but **average use** is not a good method of determining whether a home needs more electrical service capacity and additional wiring. Average home power use has been found by numerous tests to be only about 5% of **peak use.**

Peak use is when everything electric in the home is operating. Think of a hot summer evening when the air conditioning is running, the lights are on, the entertainment center is in use, dinner is cooking in the oven, someone is doing laundry, kids are playing games on every available network node, and the refrigerator and freezer have to work hard to overcome the heat. That is peak use, and if the wiring and electric service in the home cannot meet that demand for a few hours each day, an upgrade may be needed.

A home LAN does not require a great deal of electric power. A computer will not often use more than 200 watts. Printers that have heating elements may use more, but only for short periods. Network hardware such as gateways, routers, switches, and similar devices, mostly measure their power consumption under 100 watts. Consequently, if a home is currently powering one computer and a printer, adding a couple more and wiring them into a network probably will not overload the electrical service.

Nevertheless, calculating the current and likely future load requirements for a home in which you plan to install a network is a good idea for several reasons, among them:

- Computers, peripherals, and network hardware all work best on **clean power** that is at, or very close to, the recommended voltage. The closer a home's power consumption approaches to its capacity, the more vulnerable the power supply becomes to voltage drops, **spikes,** power **surges,** and outright blackouts.

- Even though the total power consumption in the home is well within normal limits, individual circuits may be at or above their rated capacity. If any of these circuits are running electric motors or electronic equipment, their performance may be affected and they may generate interference, which can find its way into the network.

- In AC circuits operating at a constant frequency (60 cycles per second in the United States), the speed of most electric motors (furnace blowers, air conditioners, etc.) is determined by the current frequency and is nearly always an even multiple of it. Most AC motors run at 3600 rpm or at 7200 or 1800. If the frequency remains constant, so does the motor's speed. If the voltage drops, however, the motor consumes more power in order to maintain its speed. As it consumes more power, it runs hotter and the wires supplying it get hotter, too. A serious voltage drop can burn out the motor, cause a fire in the wiring, or blow the circuit **breaker** or fuse. It is not good to have computer equipment running on the same circuit (or in the same house) if this happens.

- A network that includes wireless segments or any of the technologies that rely on data transmitted over existing wires (HomePlug, X10, or HomePNA) is much less likely to have radio frequency (RF) interference if the home's existing wiring is not straining (and overheating) to meet demand. This type of interference is much easier to eliminate by replacing or supplementing old or inadequate wiring than by filtering it out after it has entered the network.

For all these reasons, you should check the electric load requirements in the home before you install a home network. If all the circuits are operating well under capacity, then you know you can proceed with confidence in the existing wiring. If additional circuits are needed, they can be installed before you begin the network installation, and outlets for the new circuit can usually be placed so that the network's equipment runs only on them. This setup can help isolate the network from any interference generated in older wiring or equipment.

Checks for Inadequate Home Wiring

A couple of easy checks can help determine if the home in which you are planning to install a network has adequate wiring to provide the new equipment with consistent clean power, while still meeting the needs of the rest of the home's electrical system.

Start your check by turning on all the lights in the house and all the electric appliances that normally stay in operation for considerable periods of time (radios, TVs, etc.). If lights flicker or dim before you finish activating everything that could normally be running, the home definitely needs additional circuits. If a breaker trips or a fuse blows during this part of the check, the house wiring is probably unsafe and needs to be upgraded to reduce the risk of fire as well as to provide adequate electrical service.

If all the lights and regular appliances can operate at the same time, leave them on for a few minutes and try these additional checks.

- Turn on two burners on the electric stove, and watch for lights dimming or breakers tripping. If either happens, upgrading is needed. An electric stove is connected on a separate circuit in a home. If its operation causes problems in the rest of the circuits, all are probably inadequate, not just the stove circuit.

- Start each of the large appliances that run periodically in various rooms of the house: the refrigerator, the clothes dryer, the microwave, the dishwasher, the furnace blower, and the air-conditioning compressor. As each starts up, watch for lights dimming or breakers tripping. Electric motors draw more current when starting than they do when operating at normal speed. For this reason, dimming or circuit overload is more likely to be observed when a motor in an appliance is starting than when it is running at regular operating speed.

- Take a vacuum cleaner into each room of the home, plug it into a wall outlet, and turn it on. Watch for lights dimming or circuit breakers tripping. Again, this is most likely to happen as the vacuum cleaner starts up.

- In the kitchen, turn the electric stove on and start both the dishwasher and the microwave. (Do not run the microwave empty; put some water in it to heat.) Both these devices have heating elements that draw significant power. If both run without tripping a breaker, while other appliances and electric lights are also operating, the home's wiring is probably adequate.

If any of these tests causes observable dimming, blown fuses, or tripped breakers, then some upgrade to the home wiring is called for. If only one or two circuits continually shows symptoms of overload, then the solution may be as simple as moving some of the devices on those circuits to another existing circuit or to a new one you install. If the symptoms show up randomly throughout the home, then additional circuits need to be installed and the overall electrical load in the home redistributed.

Installing new circuits in a home does not mean the whole house has to be rewired. That can be done if there is reason to believe that the original wiring, even if used at its intended load level, is a fire or safety danger, but in most cases it is not. Adding some new circuits that take a portion of the load off the old ones almost always solves an **overload** problem. You may also want to replace some switches and other fixtures, either for improved performance or an updated appearance, but the wiring will generally perform acceptably, if it is not pushed beyond capacity.

You will want to install new circuits where they can take over as much of the home's electrical load as possible, but require the least amount of new wiring. Before planning or installing any new electrical circuits, find out what is operating on the ones already in place. To do this, turn on all the lights and appliances again and trip one circuit breaker at a time to find out what goes off. As each breaker is turned off, check the unused outlets in the area to see if they are connected to the disconnected circuit. Circuits are not usually wired to outlets on more than one floor, but you should check to be sure, especially if the basement or a top floor was finished some time after the original house.

Planning and Installing New AC Circuits

Once you have determined that some new circuits are needed for the home, the next step is determining exactly where they are needed and how near to these areas you can actually install them. Before cutting any holes in walls for new outlets, you want to make sure you can connect an outlet at that location to the **service panel** on a new circuit. Not all areas in a finished home are accessible for installing new wires and outlets easily. The location of new circuits in most cases is a compromise of where you would like them and where you can place them.

Planning the Electrical System Upgrade

When you have mapped the circuits in the home, decide where you can install new ones to take over some of the heavy electrical load. If the laundry room is not already on a separate circuit, it is often a good place to start because the washer and dryer can be put on a circuit with one duplex outlet. Even if the dryer is gas heated, it has an electric motor to spin the load, and a new circuit can take the load of both washer and dryer off the old wiring. The existing circuit can then easily handle the remaining load still wired to it.

Other places for new circuits are the kitchen, where several appliances can be conveniently grouped on a new breaker, and the furnace or utility room. The latter is often a good location for the network's central components and an easy place to wire into because it is generally unfinished (all or part of the walls aren't covered with wallboard, the ceiling is completely or partially open to the joists or rafters above, and the floor may have open communication with the basement below). If you are lucky, the service panel for the home's electrical system will also be there, making the installation of additional circuits even easier.

A new circuit installed exclusively for the home network gives plenty of capacity for hardware components and the probability of cleaner power for the system. If all the main devices of the network are on one circuit, they won't be affected by other tripped breakers, and you can put an uninterruptible power supply (UPS) for all the network components on the network-only breaker.

If you are installing additional circuits to increase the home's total electrical service, in addition to wiring one for the network, keep a general rule in mind. A 15-amp circuit should have a maximum of five outlet receptacles on it, and a 20-amp circuit not more than seven outlets. This allows an average of 300 watts usage for each outlet if all are in use simultaneously. That probably won't happen, but some outlets often have loads greater than the average, so the five or seven outlet limits are reasonable. If lights are included on a circuit, a two- or three-light fixture is equivalent to one outlet receptacle in calculating the circuit load.

In an existing home, you need to plan not only where the new circuits will be, but how to run **AC cable** to them. Take some time to calculate the best routes for running the wiring. Plot the route of each wire by observation to be sure the paths you select can be followed all the way from the service panel to the outlets with no insurmountable obstacles in between. Careful route planning can save you the lost time and frustration of literally running into a brick wall or other obstacle that you cannot get a cable through.

Installing Outlet and Switch Boxes

Mount **receptacles** at the same height as the existing wall outlets in the home, which is usually about 12 inches above the floor. In a kitchen, utility room, or closet, you may want to mount the outlet boxes at about a 3-foot height for easier access, but before doing so be certain you can get wiring to them in that location. Where an outlet is placed above a counter top, the cabinet underneath can often be removed and a hole punched in the wall behind it in order to install the outlet wiring.

In most rooms, a receptacle should be placed on each wall and a second one on walls with 12 or more feet of usable space (open areas without windows or doors). Walls with less than 3 feet of space do not need an outlet. These are guidelines, but there is no maximum number of outlets, and it is better to have too many than too few.

If **switches** are to be added on some circuits, boxes for them should be mounted on the inside of rooms near the doorway and opposite to the hinge side of the door. Place switches close to the door opening and at the same height as existing switches in the home. In old homes, the switches may be mounted lower than seems preferable now, but it is better to maintain a consistent appearance throughout.

Outlet boxes mounted in finished walls must be the type that can be secured in the wall by means of flanges that tighten against both sides of the wallboard. These are called remodel boxes and are nearly always made of hard plastic. Do not try to use the stud-mounted boxes intended for new construction. You will not often find a stud located near the hole you cut in the wall for the box and even if you do, it is almost impossible to attach the box to the stud with only the small wall opening to work through. Wallboard-mounted remodel outlet boxes work well, but should be permanently installed only after the cable going to them has been pulled into place. Remember also, boxes that mount flush in the wall and accept a hidden cable from behind are different from boxes that mount on the surface of the wall or protrude from it and accept cable from a surface raceway. Figure 5-1 shows three types of boxes: on the left is a standard stud-mounted metal box for use in unfinished walls, in the center is a wallboard-mounted remodel box (with grip tab extended) for use in finished walls, and at the right is a surface-mounted box with mounting plate for use on wall surfaces with a raceway.

In unfinished areas, stud-mounted boxes or boxes with mounting brackets attached can be used and secured in place with nails or screws. Mount the outlet box on a stud, flush to the stud's outer edge if the wall will not be finished later, or protruding from the stud edge the thickness of the wallboard used in the home (3/8 inch to 5/8 inch) if it will.

Drill cable holes in the studs from the unfinished area outlet boxes in as direct a path as possible to the service panel. Also drill holes for cable going to finished

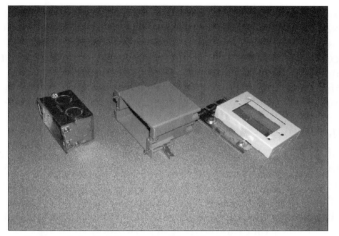

FIGURE 5-1   Three Types of Outlet Boxes for Electrical Outlets

areas through unfinished routes. Then, after the openings for outlet receptacles and switches are cut and holes for cables are drilled, it is time to pull the cable for the new circuits.

Installing AC Wiring

Circuits in a home are wired with cable made up of three or more copper wires in a casing. Each copper wire consists of a single copper strand in the smaller sizes or multiple copper strands in the larger sizes. **2-conductor wire** actually contains three wires (one black, which is the hot wire, and one white, which is neutral). These are insulated with plastic and are encased together with an uninsulated third wire in a plastic sheathing to form the cable.

The standard by which wire size is measured is the **American Wire Gage (AWG).** Small wire sizes have higher numbers such as #20, which is a small wire indeed. As the number decreases, the wire size increases, which means that #14 wire is smaller than #12, and #4 wire is larger than both. The size of wire determines how much current it can safely carry. Table 5-1 shows wire sizes, current ratings, and breaker sizes for the most common sizes of wire.

TABLE 5-1
Wire Sizes and Their Current Capacities

Wire Size	Current Capacity (Amps)	Breaker Size (Amps)
14	15	15
12	20	20
10	30	30
8	45	40
6	65	60
3	105	100

The most commonly used wire size is 2-conductor wire (plus ground) #14, which is used in most home circuits to wire outlet receptacles, switches, and lights. Use 3-conductor #14 wire for 2-way switches and split outlet receptacles. Use 3-conductor #12 wire for equipment that may require AC power up to 20 amps, such as heaters and air-conditioning compressors. Larger wires are used for 240-volt circuits that can carry up to 40 amps of current, such as for electric stoves and dryers.

Nonmetallic Sheathed (NM) Cable

Nonmetallic sheathed **(NM) cable** (often referred to as Romex) consists of several insulated wires and a ground wire encased in a plastic or composite sheath. The cable can have two, three, or four insulated wires, along with the obligatory ground wire. NM cable is the most widely used type of cable for AC wiring in homes. The NEC allows this type of cable to be installed exposed in the building or contained in a sleeve or raceway for physical protection. If it is installed exposed, the cable must not be in an area where it may be damaged or subjected to any physical abuse.

The cable can be installed exposed in a residential garage or in an attic or basement. In an unfinished basement, an NM cable must be installed through bored holes in the framing members. In attics, it can be installed in holes through the frames or laid on top of them, if doing so will not subject the cable to damage. In walls, the cable can be dropped from overhead or brought up from the basement without any sleeve requirements. It can also be run through holes in wall studs as long as it is protected inside the finished wall. Cable run on the outside surface of finished walls, or in other exposed situations where it could be damaged must be run in a raceway that covers and protects it. Some state and local electrical codes require that interior exposed wiring be contained in conduit. You should check requirements in your area before installing any new wiring.

In addition to inside wiring, NM cable is often used outside the home to provide power for devices on the exterior of the building. Where the NM cable may be exposed outside to physical damage, the NEC requires that it be protected by installing it in conduit, electrical metallic tubing (EMT), polyvinyl chloride (PVC) pipe, or in surface raceways. Floodlights on exterior eaves, for example, are often powered using NM cable that has been run in conduit to protect it from physical damage and from deterioration from the outside elements, such as dampness or sunlight.

Either metal or plastic outlet boxes can be used when wiring circuits with NM cable. If metal boxes are used, the cable should be secured in them with snap-in plastic connectors. If plastic boxes are used, the connectors aren't required as long as the cable is secured with staples close to the box. When this isn't possible in an existing wall installation, connectors should be used.

Metal Clad (MC) Cable and Armored Cable (AC)

Some local electrical codes require the use of metal clad (MC) cable or armored cable (AC) for interior wiring in homes. AC cable was historically referred to as BX. Both of these types of cable contain insulated wires covered with a flexible metal outer casing that protects the interior wires from damage. Both MC and AC cable can be obtained with different numbers of insulated wires. The significant difference between them is in their grounding technique. **MC cable** has a green insulated ground wire inside the cable, while AC cable has no grounding

2-conductor NM cable
with copper ground wire

MC cable with
ground wire

AC cable with
aluminum bond wire

FIGURE 5-2   NM, MC, and AC Cables Can Have from Two to Four Connector Wires

wire, but uses the outside metal casing as the ground. AC cable now also has a bonding strip on the inside of the cable armor, which improves its ground continuity.

Both MC and AC cable require the use of metal boxes and connectors. With MC cable, the boxes must be insulated from the metal cladding of the cable by using a plastic anti-short bushing at every point where a box connects to the cable. Both AC and MC cost more than NM cable and both are more time-consuming to install because of the increased connector and grounding requirements. MC and AC cable can be purchased with a plastic or rubber outer covering for use in damp areas, but without such protection, neither of these cables should be used for wiring in wet conditions.

Figure 5-2 shows an example of NM cable at the top, MC cables at the bottom left, and AC cables at the bottom right.

Wiring New Circuits

Each new circuit should be wired using the correct size two- or three-wire insulated cable (#12 or #14, depending on code requirements in the area). The cable must be pulled from the service panel box to the new outlet boxes in much the same manner as Cat5 cable is pulled for the network (see Chapter 4). NM (Romex) cable need not be handled with as much care as Cat5 because it is not carrying data, and AC electric current isn't affected by sharp turns or twists in the cable.

Run each cable through unfinished areas as near as possible to the outlet box, then fish the cable inside the wall or run it in a raceway on the wall surface to the outlet box. After the cable is in place, run it through an appropriate knockout opening in the box, and mount the box in the wall. Check wall-mounted boxes with an outlet and faceplate in place to see if the faceplate rests flush against the wall. If it doesn't, you have to mortise the box into the wall slightly to get a flush fit. This requires cutting a shallow notch around the face edge of the outlet opening so the flange on the front of the outlet box recesses into the wall far enough to let the face plate mount flush with the wall surface.

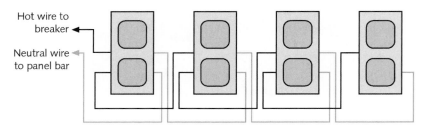

Hot wire to breaker

Neutral wire to panel bar

FIGURE 5-3   A Parallel-Wired Circuit Connects All the Outlets Directly to the Service Panel

For boxes mounted in unfinished areas, run the cable into the box and secure it in place with the holding plate in the box. Leave half a foot or so of slack outside the box and at least 8 inches inside the box.

Follow the instructions that come with the outlet receptacles to wire them in the boxes. Buy good-quality outlets for your wiring. The main difference between low-quality outlets and good ones is that the cheaper ones lose their ability over time to grip a plug firmly. After a period of use, they become loose and allow the plug to fall out. Even when the plug stays in place in a loose outlet, it often makes a poor connection to the current and generates interference in the wiring. Spend a few cents more and purchase outlets that last.

Be sure each outlet is correctly connected to all three wires in the cable: black to the hot side (usually brass-colored), white to the neutral side (usually silver), and bare wire to ground. Always connect the ground wire first. Even though it isn't grounded at the service panel yet, it is a good practice to follow this rule so if you are ever working with hot wires, the habit is established. Mount the outlet in the box, and attach a faceplate to cover it.

Multiple outlets wired to one circuit must always be wired in parallel, never in series. This means that all the wires from all the outlets on a circuit must connect directly to the circuit breaker. Wires from multiple outlets are either brought together in a junction box or connected in a line from one outlet to the next. The outlet at one end of the line or the cable in the junction box is then connected to a cable running to the breaker for the circuit. An outlet box can also serve as a junction box for cables from only one other outlet on a circuit, but cables cannot be spliced outside a junction box inside a wall. Figure 5-3 shows a parallel-wired circuit.

High-Voltage Electrical Components

A new circuit wired in a home may contain different kinds of electrical devices. The most common are electrical outlets into which network devices and other appliances can be plugged, but several others may be included, either to balance the electrical load on the home's circuits more evenly or to provide additional electrical service and convenience along with power for the network.

The following sections discuss in detail the different kinds of electrical devices.

AC Outlets (Controlled and Uncontrolled)

AC power outlets are the standard source of AC power for appliances, electronic equipment, and other devices in the home. Up to five outlets can be wired in a single circuit, but the number and placement of outlets should be planned so their combined average usage does not exceed the 15- or 20-amp capacity of the circuit.

Outlets can be wired in an "always on" configuration directly to the service panel, as shown in Figure 5-4a. This is the most common arrangement, but at least

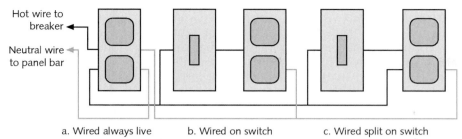

Hot wire to breaker

Neutral wire to panel bar

a. Wired always live   b. Wired on switch   c. Wired split on switch

FIGURE 5-4   Alternative Methods of Wiring an Outlet

two others are possible. In Figure 5-4b, the outlet is wired through a switch so the switch controls power to it. This allows a lamp plugged into the outlet to be controlled by the wall switch, but it also makes the outlet unsuitable for any device that needs continuous power, such as a clock or refrigerator. Figure 5-4c shows a third alternative called a "split" outlet in which the top side of the duplex outlet is wired to the switch, while the bottom side is wired directly. This allows the outlet to be used as a continuous power source or as a controlled source governed by the switch.

Switches

Switches are installed in circuits to control lights or outlets. They can be wired in several configurations, but the most common is as a single switch controlling a light or outlet, as shown in Figure 5-5a. Another wiring configuration is shown in Figure 5-5b. It shows how two switches can control a single device such as a light fixture or outlet. This allows convenient access from opposite sides of a room, the top or bottom of a stairway, or the inside or outside of the home. This arrangement uses double-pole double-throw switches and is known as a **3-way switch.** The light can be turned on or off by either switch, regardless of the position the other (unused) switch is in.

A third configuration, shown in Figure 5-5c, also provides for a 3-way switch, but gives one of them ultimate control (either on or off). With this arrangement, switch 1 can turn the light on and it remains on regardless of the position of switch 2. If switch 1 is in the off position, switch 2 can turn the light on or off, but only so long as switch 1 remains in the off position. This last configuration is more useful in an automated system than with a strictly manual one, as you will see later in this book.

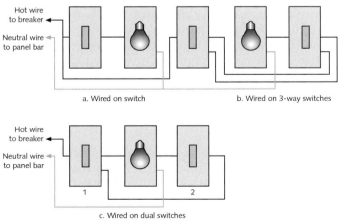

Hot wire to breaker

Neutral wire to panel bar

a. Wired on switch   b. Wired on 3-way switches

Hot wire to breaker

Neutral wire to panel bar

1   2

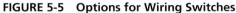

c. Wired on dual switches

FIGURE 5-5   Options for Wiring Switches

Dimming Modules

Dimming modules that can dim a single room light or several lights wired together are now common in automated homes. Manual dimming modules can be incorporated into any wall switch and used to control the lights connected to the switch. These modules can now be controlled with either X10 technology or RadioRA wireless technology in order to automate the lighting system completely. Both X10 and wireless lighting controls can be connected to the home LAN through an Ethernet-compatible device. We will discuss automated home lighting controls later in this book, but any needed additional or upgraded AC wiring should be in place before the automation is installed. X10 and RadioRA controlled dimmer modules can be manually operated from the switch. They can be installed and used before the automated control system for them is activated.

Dimmer modules are wired in the same manner as regular switches and can be configured as single-pole switches, single-pole double-throw switches, or double-pole double-throw switches without affecting the dimming component. In a dual switch configuration, the dimmer module must be wired between the regular switch and the light in order for the dimmer to function when either switch is turned on.

Fixtures

Light fixtures are wired in the same manner as outlet receptacles, either directly to the service panel so that power to them is always on, or through a switch so that power can be controlled from the room in which the fixture is located. As with switch wired outlets, the switch for a light fixture should always be wired on the black (hot) wire of the circuit and not on the white (neutral) wire.

Service Panels

New circuits in the home will probably mean that a new service panel is required to hold the additional breakers. This must be done because electricians rarely leave excess capacity in the original box. In some areas, a second service panel can be installed and wired from the first, but in many areas of the country, a second service box isn't permitted, and a larger one must be put in to contain the original and the additional breakers.

Additional circuits that increase the home's total electrical load capacity may also require new wiring to connect to the power grid outside. The utility company can bring new wires to the original connection point. You have to connect them from that point to the new service panel.

Adding a new 100-amp service to a home requires #4 stranded copper three-wire cable from the outside connection point to the main breaker in the service panel. But this is only if the new line runs parallel to the original and is installed in addition to it, not in place of it. If the original wiring to the service panel is replaced, the wire size must be sufficient to carry the total number of amps now coming into the home. A 150-amp service requires #1 stranded copper wire and a 200-amp service requires #2/0 stranded wire.

Figure 5-6 shows a new oversize service panel installed in a home to accommodate increased power requirements. Main **service lines** (two legs of 100 amps, 120 volts each) are connected at the top right and circuit breakers are wired down both sides of the central panel. At the lower left, a new outlet has been installed at table height to connect the LAN's central components on a clean circuit. The service panel has room for further expansion, if needed.

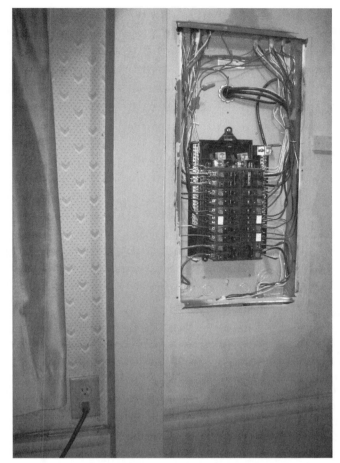

FIGURE 5-6   Service Panel with Cover Removed to Show Wiring in Place

The wiring from the outside connection point to the service panel usually runs in a conduit. If the wiring must be replaced, the old wire can usually be used as a pull cable to get the new wiring through the conduit. This can only be done when the conduit is large enough to accommodate the new wire and it may require that part of the conduit be cut or disassembled to eliminate curves and angles that heavy wire cannot be pulled through.

The incoming service line exits the conduit (which is attached to the service panel box) inside the box. It is then connected to the two main circuit breakers (one for each leg of the 240-volt service or a two-leg breaker double the size), which may also have to be replaced with units of larger capacity. A 150-amp service needs 75-amp main breakers on each leg. A 200-amp service needs 100-amp breakers on each leg.

The main breakers in the service panel connect to the branch circuit breakers, as shown in Figure 5-6. Each circuit is wired through the breaker that controls it. Only the hot side of the circuit goes through the breaker. The neutral side is always connected to main neutral (which is a grounded bar in the service panel), and no breaker can interrupt it. From the breakers, the home wiring extends out to the individual devices and fixtures in the home.

Caution: If you are not licensed as required and absolutely certain that you can correctly install the service panel wiring, get someone who is fully qualified to help. There is no room for error in wiring a service panel and the circuit breakers that go in it.

Circuit Protection

Circuit protection is a form of insurance for a network and the data on it. Protecting the AC circuits that power the network's equipment, as well as the other electric lights and appliances in the home, helps prevent problems from developing in these high-voltage circuits and in the low-voltage data circuits of the LAN. This section discusses the factors involved.

Ground Connections

The first level of protection for a network is to carefully ground all the devices on it. Grounding is a safety precaution in case a short circuit should develop in a device, but its more important function for network devices is to eliminate static charges and other noise-producing electric impulses before they reach the network and interfere with data transmission.

All the AC circuits in a home must be grounded either through a grounding rod or water pipe passing deep into the earth. The rod or pipe is connected to a ground bar or plate in the service panel box, and the ground wire of each circuit in the home is connected to that ground bar.

Because grounding is a part of the NEC and the wiring requirements of every state code, all circuits in a home are grounded at each outlet. When a network is connected to an outlet by a three-wire power cord, the device is grounded through the power cord. No further action is needed.

If the device connects to the outlet with a two-wire power cord, it should be grounded by connecting a ground wire from the metal frame of the device to a ground connector in a wall outlet or to a metal pipe. Ground wires should not be connected to other metal objects in the home (metal studs, rails, etc.) as these frequently do not connect to ground in a complete circuit.

Surges and Spikes

An electrical surge is a sudden momentary increase in the rate of current flow or the voltage in a circuit. Surges can be produced by power equipment starting up or shutting down either inside the home or at a more distant location. They can also be caused by lightning strikes or other atmospheric electrical disturbances. Another name for a power surge is a spike, although some people define a spike as having a shorter duration and a higher voltage than a surge.

Both surges and spikes mean that for a short period of time there is too much electricity in a circuit. These electrical excesses may last only a few millionths of a second or they may persist for many times that long. In either case, they can do serious damage to electronic equipment.

The electric potential forced into the circuit by a surge may discharge by arcing from wiring to a ground connection. The heat of such an arc can fuse components or burn them out just as a light bulb filament shatters when subjected to excess power. Even relatively small surges, repeated over time, reduce the life of low-voltage electronic parts by subjecting them to momentary current flow and resulting heat far above their designed maximums.

Surge protection on the main AC power lines coming into the home usually takes the form of a circuit interrupter that cuts off the electric current when the voltage rises above a preset maximum or drops below a preset minimum. Either condition can damage electric equipment, although electronics are more vulnerable to high voltage than low. Electric motors, such as those on furnace fans, air-conditioning compressors, refrigerators, and freezers, can overheat and burn out if the voltage drops too low over a period of time.

Main line interrupters are expensive and are probably only warranted in areas where the danger of surges originating outside the home is fairly high. A little research into the weather history of a location and a check of whether any large-capacity commercial or industrial electrical equipment is operating nearby can help determine whether a power line interrupter should be installed in a home system.

Because surges can originate from within a home or from outside, protection from both sources is necessary to prevent damage to home LAN equipment. This dual protection is particularly important when part or all or the network's data transmission wiring is shared with or connected to television cables, telephone lines, or outside antennae.

A middle course of protection may be to install small interrupters on individual circuits that power expensive equipment. Individual plug-in devices are available to protect against surges and spikes in AC circuits. These can be installed at the point where network devices connect to the AC power. For other circuits that are potential carriers of surges, such as telephone lines, and television cables, surge protectors should be placed at or near the point where these circuits enter the home.

Surge **suppressors** that can be installed at electric outlets vary widely in price and quality. The primary consideration for a surge protector is how fast it acts when a surge occurs. If suppression starts only after .1 seconds of increased voltage, for example, most surges will have come and gone, doing whatever damage they are capable of, before the suppressor even begins to respond. Such devices are useless for protecting against surge damage.

Devices that can suppress only a few thousand surge amps are also useless. The most damaging surges often attain up to 50,000 surge amps for a few millionths of a second, during which time they can burn through low-capacity suppressors and wreak havoc among low-voltage electronic circuits.

Home LAN equipment should be protected by grounded surge suppressors that are rated to act within a few picoseconds and suppress at least 50,000 surge amps. These industrial-grade suppressors are available in the same one-, two-, four-, and six-socket configurations as the less effective "power strip" models, and they look much alike. The high-quality units can be distinguished by two factors: They have their response time and suppression strength ratings printed on them (low-rated suppressors rarely publish their numbers), and they cost more than unrated units. Expect to pay between $50 and $100 each for these suppressors, depending on their configuration and amperage.

Combination surge suppressors and noise filters are available for installation on telephone lines, cable television lines, and satellite antenna lines. In all cases, these suppressors should be placed in the lines before they connect to any network equipment. The vast majority of their effect on the home network is improved performance from noise and interference reduction, but the few times they intercept a surge or spike prevent far more damage than their cost, which is about the same per unit as a good AC suppressor.

Causes and Cures of High-Voltage Interference

High-voltage interference in a home LAN can originate from a number of sources and affect several parts of the network. The best defense against this interference is to know what conditions cause it and then prevent those conditions as much as possible. The first part of this section discusses the causes of interference and suggests ways to minimize them. The second part discusses methods of filtering or suppressing interference that cannot be prevented.

Causes of Interference

Outside interference can be caused by lightning and the atmospheric conditions that produce it. Wind can also generate static electricity, either by the friction of air moving over stationary objects or by the motion that wind pressure produces in everything from tree leaves and flags to tumbleweeds and windmill blades. Other sources of interference from outside the home include radio stations, citizens band, and police radios.

Inside the home, interference sources include household appliances and fluorescent lights. Inside interference is mainly produced by differences in electrical potential between different parts of the home or between objects in the home. These differences cause electric currents to flow for very short periods of time between some of the points with differing potential. The brief electrical flows tend to neutralize the potential between the different points, but the current flow often passes through wires, metal component parts, and other conductors that form part of the network structure. When this happens, the high voltage of the flow disrupts the low-voltage flow of data in the network, or may even entirely obliterate it for a time.

Interference can also be caused by operating pieces of electrical equipment that create electromagnetic fields around themselves. Imperfectly wired electrical connections can produce tiny current arcs. These arcs result in interference being generated around the circuit.

Static Electric Discharge

Almost any movement of an object through the air or while in contact with another object can generate a **static electric charge** on the object, which when discharged produces interference. This can mean anything from feet walking on a nylon carpet (and discharging with a painful spark when the walker touches a metal object) to clothes tumbling in the drum of an automatic dryer.

Static electric discharge, correctly termed electrostatic discharge (ESD) occurs whenever the static charge on two objects is dissimilar. If the two objects touch, electricity from the one with the higher voltage charge flows to the one with a lower voltage charge until the two charges are equalized. Static discharges can attain very high voltages. If you touch a metal object and feel an electric discharge, the static charge (voltage difference between you and the object) was 3,000 volts or more. If you saw a spark when the discharge occurred, the voltage difference was at least 8,000 volts.

Such voltage discharges can produce high levels of interference and also damage or destroy low-voltage electrical circuits and parts that normally function in a voltage range of 5 volts or less. Even a mild static discharge can wipe out a data packet running in a low-voltage wire, or completely fry a millivolt-rated capacitor or other electronic part.

Static charges can also be produced by high-voltage devices. All cathode ray tubes (computer monitors, television screens) contain high-voltage electron beams and create static on the face of their screen and also on surrounding objects. Laser printers, copiers, and power supplies in computers and many types of electronic equipment all produce static charges as part of their normal functions. These static charges can remain for long periods on the equipment even after power to it is shut off. The charge can also transfer to other objects or people, creating interference as it does so and forming a new voltage difference on the object or person that can then create more interference when it discharges again.

Finally, fluorescent lights produce interference through a phenomenon called electromagnetic induction (EMI). The high-voltage transformer, called a ballast, in a fluorescent light causes an electromagnetic field to be generated around the transformer by the current passing through it. This field is what causes a fluorescent light to glow. It induces a current to flow through the fluorescent tube, causing the phosphorus inside the tube to emit light. The electromagnetic field can also induce a similar current in other nearby objects, including network cables and electronic equipment. This is why network cables (and other low-voltage wiring) must never pass close to fluorescent lights or other high-voltage devices. Induction currents in the network wiring can destroy data and can damage equipment if the discharge reaches it.

How much the interference generated in and around a home affects a home LAN depends on climatic conditions in the area (dry windy conditions produce more static electricity and hence more interference), its location (proximity to radio stations, power lines, industrial plants), and how well the home is constructed to deal with the causes of interference. Most interference cannot be prevented, but it can be minimized and its danger to the functioning of the home LAN almost entirely eliminated by careful wiring, good grounding, and perhaps a few filters.

Preventing Interference

The first defense against interference is to create as little of it as possible. Clean securely connected wires and adequate separation of low-voltage lines from high-voltage equipment and circuits reduce interference. So does good maintenance of electrical equipment. Devices with electrically connected moving parts such as motors, relays, switches, solenoids, and sensors all produce interference if their parts are worn or dirty. Keeping all the home's electrical systems, not just the network-connected devices, in top working order minimizes the interference that must be dealt with after the fact.

The next step to zero interference is to ground everything electrical. Again, this applies to all the electrical devices in the home, not just those connected to the network. It may even apply to a few non-electrical objects, if you suspect that static charges are being generated on their surfaces. A carefully installed ground wire will not harm any object and may reduce interference on the network by eliminating a source that cannot easily be filtered because it is not part of a circuit.

Floor surfaces, furniture, and glass do not ground well, but they all take a static charge quite readily. If these surfaces are near network wiring or devices, they can bleed interference into the network. Antistatic sprays, grounding mats, and removal, where possible, are all methods of eliminating this type of interference. You must determine on a case-by-case basis whether it is easier to eliminate a source of interference or **filter** it out after it is created.

For the interference that remains in a home after as many sources as possible have been removed, two other defenses remain: **shielding** and filtering.

Shielding

Shielding applies primarily to the network's cables and is actually a refined form of grounding. The data-carrying wires in a shielded cable are surrounded for the full length of the cable by a webbing of metal wires. Interference entering the cable through its insulation is intercepted by the web of shielding wires and grounded before it can reach the data-carrying wires in the cable. If shielded cable is used in the network, it is important that the connectors are all properly attached so the shielding is grounded and can discharge any electrical interference it intercepts. If the shielding is not grounded, it can accumulate an electrical charge and eventually discharge part of it into the data line.

Twisted-pair cable also helps eliminate electromagnetic interference induced in the cable by proximity to AC power lines or equipment. This is not as big a problem for most home LANs as it is in commercial networks, but wherever network wiring comes near AC wires or devices, shielded twisted-pair (STP) cable is a must.

Filtering

Filters are electronic devices designed to permit the normal function of a device, but block or suppress any other signal coming from it. Filters can be placed on either the source of interference (the preferable location, if it can be found) or on the recipient of the interference. The latter is the usual practice because the sources of interference are often impossible to locate.

AC power-line filters are often built into high-quality surge and spike suppressors. They allow the AC current powering a device to pass, but block any other frequency of signal. They are designed to be placed on equipment that might produce power-line interference or network devices that may be the recipient of the interference. One multiple outlet filter/suppressor can protect up to half a dozen devices for a reasonable cost.

Radio frequency interference (RFI) generally originates outside the home, and enters the LAN through a telephone or cable modem. An RFI and electrical noise filter placed on the incoming connection cable in front of the modem (so the incoming signal passes through the filter before reaching the modem) can eliminate this type of interference. These filters cost around $100, but can greatly speed up a modem connection with serious interference by eliminating the need to resend many data packets corrupted by interference. The filter can also reduce lost connections to the ISP caused by interference.

An interference filter can also be wired into the network itself. These filters (which also often function as surge and spike suppressors) operate by eliminating high voltage from the network lines. Since network data is transmitted at plus or minus 5 volts, the filter simply suppresses any voltage significantly above that level and thereby eliminates interference. The key to good suppression is speed, and a quality network filter should act within a couple of picoseconds in order to block interference effectively. Network filters cost about $50. For additional information on the causes and cures of power disturbances and interference, see Chapter 13.

Summary

- Electrical wiring and installation requirements are mainly based on the National Electrical Code (NEC), which is a safety standard developed by the National Fire Protection Association (NFPA). Most state codes incorporate all or part of the NEC.

- The main electrical safety hazards in a home environment are fire and shock. Both can be avoided by correct wiring and installation practices.

- A short circuit occurs when too much current flows because the circuit does not have enough resistance to limit the amount of electricity passing through it.

- Ohm's Law states that Volts = Amperes × Resistance and describes the relationship among amperes (amps), the amount of current flowing; volts, the electrical potential between the two ends of a circuit; and resistance within the circuit to the flow of current.

- Most homes need to be wired for 200 amps of electrical service, but many older homes are insufficiently wired and do not meet current electrical needs.

- If additional circuits are to be installed in a home, it is a good practice to put most or all of the home LAN equipment on a new circuit in order to isolate the network from any interference generated in older wiring or equipment.

- A home in which a LAN is to be installed should be checked for inadequate wiring and electrical service, both in the home as a whole and in its individual circuits.

- New AC circuits added in an existing home should be placed to take some load off existing circuits and allow sensitive electronic equipment to run on the new circuits as much as possible.

- Wiring for new circuits should be installed using NM cable of an appropriate size. MC and AC cable should only be used if local codes require them because they are more expensive and more difficult to install than NM.

- Wire sizes are measured by AWG numbers, a system in which larger wire sizes are given smaller numbers. Circuits with 15-amp breakers should be wired with a minimum of #14 wire. Circuits with 20-amp breakers should be wired with a minimum of #12 wire.

- All outlet boxes, switches, and fixtures in a circuit must be grounded by a separate wire from the conductor wires.

- Manual dimming modules can be incorporated into any wall switch and used to control the lights connected to the switch. These modules can also be controlled with either X10 or wireless technology in order to automate the lighting system.

- An electrical surge is a sudden momentary increase in the rate of current flow or the voltage in a circuit caused by power equipment starting up or shutting down, or by lightning strikes and other atmospheric disturbances.

- Interference is any signal that blocks or corrupts data flow on a network. It is caused by lighting, radio transmitters, operating electrical equipment, and differences in electrical potential between different parts of the home or between objects in the home.

- Networks and the electrical circuits to which they are connected must be protected from surges, spikes, and interference by grounding the equipment, reducing interference sources, and installing filters and suppressors.

Key Terms

2-conductor wire A high-voltage cable containing a hot wire, a neutral wire, and a ground, which is not counted as a conductor.

3-way switch An arrangement by which either of two switches can control power to a device.

AC cable Armored cable; type of cable sheathed in metal. Also, any cable used for AC wiring.

Alternating current (AC) The type of power that almost all homes receive from the electric utility company.

American Wire Gage (AWG) The standard for electrical wire sizes.

Ampere The unit used to measure electric current flow.

Average use The amount of electricity used in a home over a period of time, usually a day.

Breaker A safety device that is wired into a circuit to cut off current flow if the circuit becomes overloaded.

Circuit A conducting "circle" in which electricity flows from a source to a device, through the device, and then back to the source.

Clean power AC current that does not contain noise, interference, surges, or spikes.

Current The flow of electricity, measured in amps.

Electric service Electric power purchased from a utility; also, the cable bringing the electricity to a home from the utility.

Electromagnetic induction The process whereby a high-voltage current flowing in a circuit produces (induces) a current in an adjacent circuit even though the circuits are not connected to one another.

Filter A device to remove interference from an electric circuit.

Ground A wire that connects an electric device to the earth so excess current can flow to the ground rather than elsewhere.

Hot wire The wire in a cable that is connected to the source of electric potential, usually colored black.

Interference Any signal that corrupts or blocks a data signal.

Load The amount of current flowing in a circuit, measured in amps or watts.

MC cable Metal clad cable used for AC wiring.

National Electric Code (NEC)(r) A safety standard for electrical wiring and installation developed by the NFPA.

National Fire Protection Association (NFPA) Publisher of the NEC.

NM cable Nonmetallic cable used for interior AC wiring.

Ohm's Law Describes the relationship among amperes (amps), the amount of current flowing; volts, the electrical potential between the two ends of a circuit; and resistance within the circuit to the flow of current.

Outlet box A metal or plastic wall box into which an outlet receptacle is wired.

Overload Current flow greater than a circuit can carry without danger of burning out.

Peak use The maximum load of electrical consumption in a home.

Potential The flow force of electric current, measured in volts.

Receptacle A device into which AC-powered appliances can be plugged to obtain power.

Resistance The force inhibiting the flow of electricity in a circuit, measured in ohms.

Service line The cable that brings electric power into a home from the utility.

Service panel The wall box containing a home's circuit breakers to which the service line is connected.

Shielding Metal webbing around a data line that grounds noise and interference before it can reach the data line.

Shock Electric current flowing through a person.

Spike A large but very brief increase in voltage or current flow in a circuit.

Suppressor A device for blocking surges and spikes in a circuit.

Surge A brief increase in voltage or current flow in a circuit.

Static electric charge Electric potential (voltage) that develops on the surface of an object and can discharge if the object touches any conductor with a lesser charge.

Static electric discharge Also called electrostatic discharge. A current flow caused when two objects that have different static charges touch or come near enough for the discharge current to arc between them.

Switch A device for controlling the flow of electricity in a circuit.

Volt The unit used to measure electric potential or flow pressure.

Watt The unit of electric power, often calculated for a circuit by multiplying the number of volts by the number of amps.

Review Questions

1. NEC stands for _____ _____ _____ and was developed by the _____.

2. What is the primary objective of the NEC?

3. The two main hazards of AC electrical circuits in a home are fire and shock. True or false?

4. What is Ohm's Law and how is it used in designing electrical circuits?

5. What is the voltage of one leg of a 240-volt electric service in a home?

   a. 60 volts

   b. 120 volts

   c. 240 volts

   d. 480 volts

6. Can 12 light bulbs, each drawing 100 watts of power, be safely wired in a single 15-amp circuit?

   a. Yes

   b. Yes, but only 10 can be on at one time

   c. Yes, but only 6 can be on at one time

   d. No

7. If the average power use in a home is 1,200 watts and this is 5% of the peak usage in the home, how many amps of electric service should the home have?

   a. 100 amps

   b. 150 amps

   c. 200 amps

   d. 250 amps

8. What can cause lights to dim momentarily in a home when an air conditioner or furnace blower turns on? Why don't the lights remain dim?

9. The most common speeds of electric motors running on AC power in the United States are 7,200 rpm, 3,600 rpm, and 1,750 rpm. True or false?

10. If the voltage going to an AC electric motor drops, what happens to the motor?

    a. It stops.

    b. It runs faster and uses less power.

    c. It uses more power and runs hotter.

    d. It does not change at all.

11. Give some reasons why network equipment may function better on a new circuit in a home rather than an old one.

12. If a breaker frequently trips when most of the equipment powered through it is running, what does this indicate?

13. What is the difference between NM cable and MC cable, both of which are used to wire home AC circuits?

14. Which of the following have to be grounded: outlet boxes, outlet receptacles, switches, or light fixtures?

    a. Outlet boxes and switches

    b. Outlet boxes and receptacles

    c. None of them

    d. All of them

15. In a 2-conductor NM cable, the hot wire is _____ _____, the neutral wire is _____, and the ground wire is _____.

16. Three-conductor #14 NM cable can carry more current than 2-conductor #12 MC cable. True or false?

17. If NM cable is likely to be damaged after it is installed, how must the installation be changed?

18. On an outlet receptacle, the hot connector is colored _____, while the neutral connector is colored _____.

19. What is a split outlet receptacle, and how is it wired?

20. Why should ground wires not be connected to metal wall studs or fasteners?

    a. They may not be grounded.

    b. They do not conduct electricity.

    c. They do not have water in them.

    d. They may be connected to electrical fixtures.

21. What is a surge in an electric circuit and what can cause it?

22. When buying a surge or spike suppressor, the two most important specifications to look for are _____ and _____.

23. What is interference and what are some of its causes?

24. How does a filter on a low-voltage data line act to eliminate interference?

25. How does shielding in a cable act to suppress interference?

Hands-On Projects

Project 5-1: Licensing and Electric Code Standards in Your State

You can download the State Electrical Regulations file of the National Electrical Installation Standards (NEIS) by following these steps. Note that you will need to provide a credit card number to complete the download. Although the credit card number is needed to complete the transaction, there is ultimately no charge for the download.

1. Get on the Internet, and go to the Web site http://www.neca-neis.org/catalog.
2. Click Other Code and Technical Publications.
3. Click the State Electrical Regulations price box (it is free), and then click the Add To Cart button.
4. Complete the required name and address information, and then follow the instructions you receive for a free download of the booklet.
5. Look up the state where you will be working to find out what the licensing and electrical code standards are. If the text notes there are local government regulations in addition to the state's, find out what these are for the area where you will be working.

Project 5-2: Map the Circuits in a Home

Use your own home or a comparable building where you can spend some time evaluating the electrical system, and follow these steps to map the electrical circuits.

1. Draw a floor plan of the home as accurately as you can. Scale the drawing at 1/4 inch to the foot so it is large enough to note the information you need.
2. Mark all electrical fixtures, outlets, switches, and appliances in their correct locations in each room of the home.
3. Turn on all the lights, electrical fixtures, and appliances in the home.
4. Turn off the first breaker in the home's service panel and observe which electrical devices stop running. Test unused outlets with a tester or portable device to see if they are working.
5. Note each electrical device that is off by marking a number 1 near it to indicate that it is connected to the first circuit.
6. Repeat Steps 4 and 5, turning off each circuit in the home service panel in succession and noting what devices go off with each. Complete your home electrical map.

Project 5-3: Install an Outlet Box and Two Switch Boxes and Rough Wire Them

Complete this project on an unfinished wall section in a home or on a classroom test wall where a new electrical circuit can be installed.

Caution: High-voltage wiring should be installed only under the supervision of a licensed electrician. Be certain all electric power is turned off in the area in which electrical work will be done. Test all installations thoroughly using low-voltage test current before connecting to full electrical power.

1. On the lower-right end of the wall, use screws to install an outlet box on a stud 12 inches above the floor. (You could use nails to attach the box, but the screws allow it to be removed so other students can repeat this exercise.)

2. Install another box above the first one and 44 inches above the floor. Install a third box at the same height on the left end of the wall.

3. Drill half-inch holes in the studs in a line from the high box on one end of the wall to the high box at the other end.

4. Rough-wire a cable between the two high boxes and another from the right high box to the outlet box below it. Run the cables out of a knockout hole in one box, through the holes between the boxes, and into a knockout hole in the other box. Leave the cable slack between the boxes with at least 6 inches of cable protruding from each box.

5. Run another short length of cable out of the left switch box to represent the cable to the service panel.

6. Clamp the cables in the boxes with 6 inches protruding from each box.

Project 5-4: Connect a Switch to an Outlet or Light

Use the rough-wired wall section and boxes you completed in Project 5-3 to complete this project. Use a standard outlet receptacle and a single-pole single-throw switch.

Caution: High-voltage wiring should be installed only under the supervision of a licensed electrician. Be certain all electric power is turned off in the area in which electrical work will be done. Test all installations thoroughly using low-voltage test current before connecting to full electrical power.

1. Connect the cable in the outlet box to an outlet receptacle, starting with the ground wire (green or copper), then the neutral wire (white), and then the hot wire (black).

2. Install the receptacle in the outlet box with the two mounting screws in its face tabs.

3. In the switch box above the receptacle box, connect the ground wires to ground. Connect the white wire from the outlet box to the white wire of the cable going to the other switch box. Twist the wires together clockwise, and use a screw-on connector to secure the pigtail splice.

4. Connect the black wire from the outlet box to one side of the switch. Connect the black wire in the cable from the other switch box to the opposite side of the switch.

5. Mount the switch in the box using the two screws in its face tabs.

6. If you have a low-voltage source of current available, connect it to the cable in the second switch box, and test your wired circuit by plugging an electrical device in the receptacle and turning the switch on and off.

7. Disassemble the switch and outlet so you can use the same wall section for Project 5-5.

Project 5-5: Connect a 3-Way Switch Circuit to an Outlet or Light

Use the rough-wired wall section and boxes you completed in Project 5-3 to complete this project. Use a standard outlet receptacle and two double-pole double-throw switches for this project. Replace the 2-connector wire between the switch boxes with a 3-connector cable.

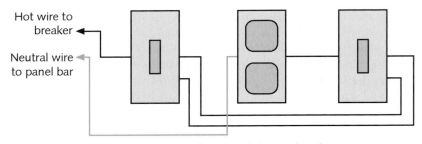

Hot wire to breaker

Neutral wire to panel bar

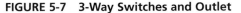

3-way switches and outlet

FIGURE 5-7   3-Way Switches and Outlet

Caution: High-voltage wiring should be installed only under the supervision of a licensed electrician. Be certain that all electric power is turned off in the area in which electrical work will be done. Test all installations thoroughly using low-voltage test current before connecting to full electrical power.

1. Connect the cable in the outlet box to an outlet receptacle, starting with the ground wire, then the neutral wire, and then the hot wire.

2. Install the receptacle in the outlet box.

3. Follow the wiring diagram in Figure 5-7 in this step and the following steps. In the switch box above the receptacle box, connect the ground wires to ground. Connect the white wire from the outlet box to the white wire of the cable going to the other switch box. Twist the wires together clockwise, and use a screw-on connector to secure the splice.

4. Connect the black wire from the outlet box to the side of the switch that has only one connecting screw.

5. Connect the black wire and the other colored wire from the 3-connector cable to the two screws on the opposite side of the switch. These are called traveler wires.

6. Mount the switch in the box.

7. In the left switch box, connect the ground wires to ground. Connect the white wires from the two cables in the box together with a screw-on connector.

8. Connect the two traveler wires from the right switch to the two-screw side of the left-side switch. Connect the black wire from the short cable (to the service panel) to the single screw side of the left-side switch.

9. Mount the switch in the box.

10. If you have a low-voltage source of current available, connect it to the short cable from the second switch box, and test your wired circuit by plugging an electrical device in the receptacle and turning the switches on and off.

11. Disassemble the switches and outlet.

Project 5-6: Identify a Surge Suppressor and Filter Suitable for Use on a Home LAN

In this project, you search on the Internet for an AC power line surge suppressor and filter to use for a home LAN. Try to find a suppressor that can act in 2 picoseconds and can suppress at least 50,000 surge amps, and a filter that suppresses as much signal noise and interference as possible in AC wiring.

1. Get on the Internet and go to a search engine such as Google or Yahoo. You can use any search engine with which you feel comfortable.

2. Search for the phrase "AC surge suppressor" and examine the responses you receive to find a suitable suppressor.

3. You may want to refine your search by adding additional qualifying words, such as "surge amps" or "filter" to find the exact type of product you want. (Hint: If your search engine has an image search mode, using it lets you see pictures of the suppressors and more easily identify them.)

4. Find at least three suppressors, and compare their features and price. Choose the one you would use for a home LAN.

Case Projects

Case Project 5-1: Recommend a Wiring Solution

A client has asked you to recommend a solution to an electrical problem and a related network problem. He wants to install a home LAN, but his house is more than 100 years old, constructed of solid brick, and has been remodeled many times. The home is inadequately wired, but installing new circuits would be almost impossible without doing great damage to the historic building. The only unfinished room in the house is the main floor furnace room in the center of the house, which also contains the electric service panel. Write a short proposal telling how you would get at least one new circuit into the house, what you would use it for, and what technology you recommend for a home network in this house with old and inadequate wiring.

Project 5-2: Wire a Room for Switched Power

You are installing new wiring in an older home. You have the design complete except for the two children's bedrooms. In these rooms, the client wants an outlet on each wall of the room, except the doorway wall. On that wall he wants a switch that turns off all power in the room, except to the electric clock plugged into an outlet on one wall, which must remain on. Draw a diagram of how you could wire the room so every outlet and light is controlled by a switch except one half of one duplex outlet.

Case Project 5-3: Find and Install a Wallboard-Mounted Outlet Box

For a wiring job you are completing, you need to install a new outlet box in the center of a finished plasterboard wall. You have not done this installation before and want to practice a bit before cutting a hole in the client's wall. Mount a piece of plasterboard between two studs or other uprights at least a foot apart. You can mount the board with nails, screws, or clamps. Purchase a wall-mounted box at a building supply store (Hint: they are often called "remodel boxes" and they cost less than a dollar.) Follow the instructions that come with the box to cut a correctly sized hole in the mounted wallboard, and mount the box securely in the hole. The instructions will vary depending on the type of box you get. After mounting the box, decide if the box you selected was a good choice for ease of mounting. Would you buy the same type again, or do you want to find another type that might be easier to install?

Case Project 5-4: Find Sources of Static Electricity and Interference

You have installed a network for a client, and it works well, except there seems to be a lot of interference that occasionally slows down data flow. You think it is caused by something electric operating at intervals and producing the interference. Obtain a portable battery-operated AM/FM radio you can use for a short while. Take it outside and tune to a place on the FM dial where you can hear no station at all. Find the quietest place on the dial when the volume is turned up high. Now walk into a building with the radio still tuned to the quiet frequency. Hold the radio very close to each piece of operating electrical equipment you can find. Can you hear static on the radio when it is near some equipment? If so, what you are hearing is radio frequency interference (RFI) produced by the equipment. Note which pieces of equipment produce the most RFI and which produce the least. What do you think makes the difference? How would you silence the static producers and stop the network interference?

Video and Audio Fundamentals

After studying this chapter, you should be able to:

- Describe the difference between **analog** signals and **digital** signals and how each is created.
- Describe how analog radio broadcasting and recording is accomplished and how analog audio signals are transformed to digital files of various formats and **compressions.**
- Describe how analog television functions and how television images are converted to digital form in a variety of formats and compressions.
- Identify the sources of audio and television signals and describe how each reaches the home user.
- Define the various types of digital video available on the Internet and the features and limitations of each.
- Describe the system design issues surrounding the connection of audio and video.

OBJECTIVES

Introduction

In this chapter, you will learn how sounds and visual images are changed into analog signals that can be broadcast on radio and television or recorded. You will discover how analog audio signals are converted into digital data that provides a higher quality for the broadcasting and recording of sound. You will also find out how television began as an analog picture that was converted into electronic signals for broadcasting and recording, and how it has now been transformed into digital formats that are higher in picture and sound quality and easier to transmit over broadcast systems, cable, the Internet, and satellite systems. You will discover how digital audio and video files are now created in a multitude of sizes, compression formats, **aspect ratios,** and file types so they can be transmitted, stored, and used in a variety of ways. Finally, you will learn how digital versions of both audio and video signals are transmitted and played on the Internet and over home networks.

From Analog Recording to Digital Transmission

When Guglielmo Marconi developed the first method of wireless or radio communication (patented 1900) and Philo T. Farnsworth developed the first television transmitter (1927), both used analog technology. Some years previously (1876), a gentleman by the name of Alexander Graham Bell had also used analog technology when he invented a device by which one person could speak to another over a wired connection called a telephone.

The term "analog" is derived from the word "analogous." When something is analogous, it has a similar or related pattern to something else. An analog device converts a pattern such as light from an image or a sequence of sounds into an analogous pattern of electromagnetic waves. An example of an analog device is a video recorder, which converts the light and sound patterns of a scene into electrical signals with similar patterns. Analog is simply a form of data transmission that is created using a continuously varying electromagnetic signal, which can then be stored (recorded) or sent to a distant destination where it can be converted back into the original sounds or pictures.

Analog Electronic Signals

Sound is created by the rapid vibration of objects, which produces waves in the air surrounding the objects. The vibrating object can be a guitar string, human vocal cords, a bell, or anything else that can move in two opposite directions rapidly to create a wave pattern in the air. Just how rapidly the object vibrates determines the pitch of the sound it makes. The sound of middle C on a piano keyboard is made by wires in the piano vibrating at 262 cycles per second. Thus, the frequency of middle C is 262 cycles. Higher-pitched sounds have higher frequencies; lower-pitched sounds have lower frequencies. Sound waves can be transformed into analog electromagnetic waves of the same frequency or a multiple of the frequency. These electromagnetic waves can then be transmitted to a distant location and changed back into sound waves of the original frequency. This is the process that radio uses to broadcast audio programs.

Visual images are also created by waves, light waves that are reflected from objects and that strike the retinas of our eyes producing a visual image. Each color of light is a different frequency and the human eye can distinguish about a million different colors. The frequencies of the visible light spectrum are hundreds of millions of times higher than the frequencies of sounds, but these frequencies can still be converted into **radio frequency (RF)** analog electromagnetic signals and transmitted to a distant location. The signals are then converted back into light patterns on the screen of a television or other visual display on which we can see a reproduced image of the original scene. The process of converting light waves into analog signals is more complex than the process for converting sound, but the result is still an analog: a continuously varying signal that reproduces a visual image on a screen.

Digital Electronic Signals

Digital technology, which was developed after analog, is not based on a continuously varying signal, but on discrete bits of data that are precisely defined. These bits consist of various combinations of only two numbers (digits), 0 and 1. Using this binary numerical system, analog data can be encoded into digital form.

Analog signals such as sound waves or visual images are converted to digital form by taking samples of the characteristics of the sound waves and light images at regular intervals. The more often the samples are taken (up to thousands or even millions of times per second), the more digital data can be recorded and the more closely the digital representation resembles the original analog signals.

Figure 6-1 shows an analog sound wave and how a digital representation can be made of the same sound wave by using the **sampling** technique. The sound wave's form is checked frequently and its height and horizontal position noted at each sample point. This waveform position data is then converted to binary numbers—that is, digital sound. The more frequent the samples taken, the more accurate the digital representation of the analog wave becomes.

Both sounds (voices, music, etc.) and visual images (motion pictures and television) were transmitted for years as various types of analog signals. Today, audio and video data is mainly recorded, stored, and transmitted in digital form.

Digital data offers several advantages over analog, making it the preferred choice of most listeners and viewers, as well as most professionals in the business of creating and distributing audio and video data.

- Copies of an original digital audio or video recording can be made without any loss of quality. Unlike copies of analog recordings, which lose some quality as they become further removed from the original, a copy of a copy of a copy of a digital recording contains exactly the same numerical data as the original; the numerical data is, of course, the binary number combinations that were created by sampling the analog signals made from the original sounds of voices or music.

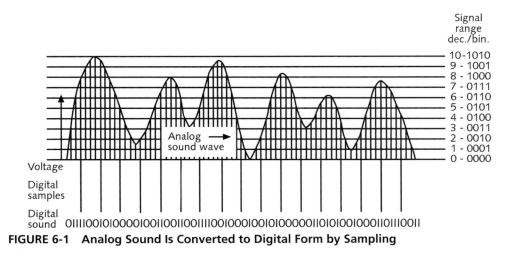

FIGURE 6-1 Analog Sound Is Converted to Digital Form by Sampling

- Digital data recordings are virtually unaffected by noise or interference because digital sampling techniques record data only about the sound or visual image, not the background noise or interference around it.

- Transmitting digital data does not degrade its quality as it does an analog signal, which can pick up noise and interference on the way to its destination or even after arrival. The numerical data of digital signals arrives unchanged or not at all.

- An error correction algorithm is usually included in digital recording systems and reconstructs any bits of lost signal using adjacent data. A "hole" in a digital image can be filled in with numbers derived from areas around it. This type of correction is not possible with analog signals.

- Computers can only record, interpret, and output digital sounds and images. They cannot handle analog signals at all because computers are completely digital devices. The ability of computers to manipulate digital data enables them to enhance, modify, edit, copy, store, and otherwise use digital audio files and video images in many ways that are impossible with analog signals.

The first method of recording and transmitting both sound waves and visual images was to create analog signals from the original sound and light patterns. These analog signals were then stored on media or broadcast as RF signals to receivers in homes where they reproduced the original sounds and images. Digital sound and visual recording was developed after the analog processes. Digital recording uses sampling techniques to encode continuous wave analog signals into binary numerical data that can also be recorded and transmitted using RF signals.

Although analog video signals have some limitations, they have been the predominant form of television broadcasting in the U.S. for more than 60 years. In early 2009, however, American television stations will cease broadcasting analog signals and will convert entirely to digital format. Thereafter, all television sets must have digital capability in order to receive programming.

Digital audio signals can be transmitted and stored more easily and in more ways than analog signals. Like digital video signals, they also reproduce higher-quality audio than is possible with analog signals. Digital radio broadcasts are now becoming dominant in the U.S. and most stations broadcast a digital signal in addition to analog. There is no cutoff date for converting audio broadcasts to digital format, so some analog radio will probably be available for years to come.

Audio Recording and Broadcasting

We perceive sound by means of continuous sound waves in the air striking our eardrums. These waves cause the eardrums and the mechanisms behind them to vibrate and send messages to our brains that we hear as sound. The highest point of a sound wave is called the peak and the lowest point the trough. The wave is composed of a continuous sequence of peaks followed by troughs. The shape of sound waves (as well as all electromagnetic waves including radio waves, light waves, and even x-ray waves) is a sine wave. This shape is set by the trigonometric ratio of the same name and is shown in Figure 6-1.

The frequency of a sound is defined as the number of wave peaks that pass a given point in a second. The loudness (amplitude) of a sound is determined by the size of the wave—that is, by the height of its peaks and the depth of its troughs. The higher the peaks and deeper the troughs, the louder the sound is. But it will have the same pitch as long as its frequency remains unchanged. In sound, frequency equals pitch, while amplitude equals loudness.

The pioneers of audio technology sought some means of preserving sound waves so they could be reproduced artificially and then transmitted over distance. Once they learned enough about sound to record it, a method of transmitting sound—first over a wire and then by wireless broadcasting—soon followed. Improved methods of storing and transmitting sound are still being developed today, even though many of the old ones are still in use. Home technology uses several methods of audio and video recording and transmission and will probably continue to do so for years. Technicians need to be familiar with each of these technologies and how to compatibly fit them into an integrated system.

Audio Recording

Thomas Edison first recorded the human voice in 1877, and by 1900 at least three competing styles of phonograph cylinders and discs were being manufactured and sold. Discs or records, as they came to be called, eventually won the competitive battle for recording sound, but proved to be a somewhat limited medium because of the short length of their recordings. Making the grooves in the record smaller and closer together and reducing the speed at which records turned from 78 rpm to 45 rpm and then 33-1/3 rpm increased the playing length to a maximum of about 30 minutes, but the development of **magnetic tape** recording began to overshadow records after World War II.

Magnetic recording on steel wire was patented by Valdemar Poulsen in Denmark in 1898 and in the U.S. in 1903. Magnetic recording (still using wire) was developed in Europe and somewhat in America during the first half of the twentieth century, but never seriously competed with records because it offered no extended length of recording. This was not true of magnetic tape, however, which was developed in the late 1940s and had the ability to record 40 minutes of material on a 1,000-meter reel of tape. Later improvements in tape and recording machines allowed the recording speed to be reduced until a reel of tape could record several hours of voice or music at high quality. The same intense competition that characterized the first century of audio recording and broadcasting continues today. Reel tape recording was challenged by cassette tapes, which have now been mostly replaced by CDs. All three systems are sometimes still included in new audio systems such as those DHTI technicians install. Some users may even want a phonograph (turntable) for records wired into an audio system.

Magnetic recording of analog sound is accomplished by recording an "image" of the sound waves on tape, which is coated with metal particles that can be arranged in patterns by passing the tape over a magnetic recording head. Sound waves from a microphone are fed into the recorder head, which then magnetizes an analog pattern of the sound waves on the magnetic tape as it passes over the recording head. As with records, the analog magnetic recording does not reproduce the sound perfectly. Noise and interference can occur in magnetic recording and, in addition, magnetic tape has a unique type of noise called hiss. This is a high-frequency narrow-band noise that occurs in all tape recording. It can be reduced but never entirely eliminated in analog recordings.

Figure 6-2 shows how a sound wave signal is recorded in analog form on magnetic tape. The figure shows the tape divided into two tracks, or channels, which can each record separate sound information. The separate tracks allow sound to be recorded in both directions on the tape or stereo sound to be recorded using one track for each side.

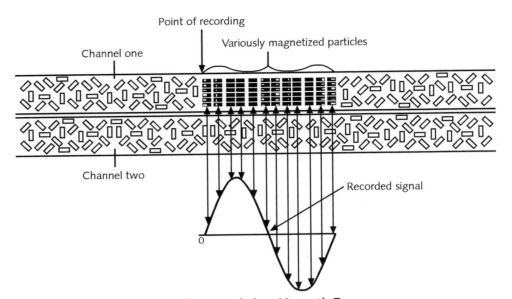

FIGURE 6-2   How Analog Sound Is Recorded on Magnetic Tape

Monaural, Stereo, and Surround Sound Audio

All audio recording was originally monaural or one track: a single microphone recorded the sound and a single speaker played it back. Later additional microphones were added at the recording end to better capture the audio of an orchestra or choral group, or the multiple voices of a play or panel discussion, but the playback remained monaural.

Stereo sound, recorded on two separate tracks (recording channels) using microphones placed on the right and left sides of the sound source, and played through left and right speakers gave the listener the audio impression of being present in front of the sound source. Two-track, and later four-track, stereo became the standard for high-quality music recording. But where stereo provided directional sound across a large arc in front of the listener and gave a good "being there" experience for any musical or even stage performance, it could not replicate the illusion of being "in the action" of a movie or television program with sound coming from all directions, behind the listener as well as in front, and sounds moving as their sources moved in the action of the program.

The answer to this higher audio need is multichannel surround sound, a system of five or more sound channels, including a left, right, and center in front of the listener, and two connected to speakers behind the listener. These five channels permit full 360-degree directional sound and a complete range of sound source movement in any direction.

Most surround sound systems also include a subwoofer speaker located in front of the listener that is specifically designed to enhance (magnify) low frequency sounds that the system's other speakers may not be able to fully reproduce. This means that when you see a movie of the space shuttle blasting into orbit, the ground beneath your feet will shake with the thunder of the launch just as the ground beneath the rocket's engines did when its liftoff was recorded.

The first surround sound was an analog system based on analog stereo recording, but present systems are fully digital. The dominant name in digital surround sound is Dolby Laboratories, the company that developed the original analog system in the 1980s. Dolby's only serious competitor is Digital Theater Systems (DTS), which has invented another surround sound encoding format that is equal in quality to Dolby's, though the latter firm still has a large majority of the surround sound market.

The **Dolby Digital 5.1** surround sound system encodes six separate channels of sound, one for each of the main front and rear speakers, plus one for the subwoofer. This requires a data stream of up to 448,000 bits per second. Dolby's competitor, DTS, requires up to 1.5 million bits per second to achieve the same result. Obviously, both systems are data intensive and require a high data throughput capability if either is included in a home LAN.

If your home technology integration includes a home theater, digital surround sound is essential to get the full benefit of major film productions and gaming devices. The accompanying video signal for a digital home theater is also data intensive, so your network must be wired with high capacity to accommodate these signals.

AM Radio Broadcasting

The first radio broadcasts sent a simple sine wave from the transmitter to the receivers. This worked well, but was very limited in the amount of information that could be sent on the wave. Since the wave was a signal of constant strength and frequency, no information of a varying nature could be sent on it. The only way to transmit data was by patterns of starts and stops in the signal transmission. This was why early radio used the same Morse code as the telegraph (sets of short signal pulses [dots] and long pulses [dashes]) to send messages. Information had to be sent one letter at a time in this "manual digital" format by a radio operator who could never exceed a speed of 50 words per minute. The message had to be decoded at the receiving end before it could be read, so early radio could not have a wide audience. Only those who could decipher Morse code could understand a transmitted message.

To transmit voice and music by radio, a way had to be found to attach this information to the sine wave of a given frequency that a radio station transmitted. The method used is called **amplitude modulation (AM)** and it was an adaptation of the original Morse code radio. Instead of simply turning the transmitted signal on and off to get the dots and dashes of code, radio engineers developed a means of varying the strength of the wave from low to high in a pattern that matched the sound waves of voices or music. The frequency of the radio wave (called the carrier wave) remained constant; only its strength or amplitude changed. The result was that the carrier wave could now carry analog voice and music signals. Figure 6-3 shows a music or voice sound wave at the top, a constant frequency, an unmodulated carrier wave in the center, and an analog AM radio wave (carrier wave modulated by the sound wave) at the bottom.

All the early radio stations used AM transmissions for their broadcasts, and its success made radio the first truly mass method of communication around the world. AM was not, however, the perfect form of radio transmission. Its main drawback was the fact that most natural and man-made radio noise (interference) occurs in the frequency range used by AM. This noise interferes with AM signals and reduces the quality of their voice and music transmissions. AM is also restricted in the distances it can travel by the strength of its signals. Since amplitude modulation means that, by definition, a low-frequency voice or music sound

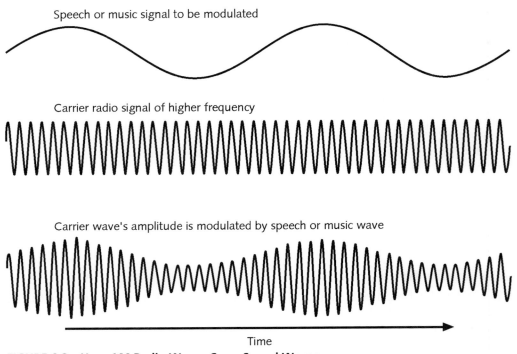

Speech or music signal to be modulated

Carrier radio signal of higher frequency

Carrier wave's amplitude is modulated by speech or music wave

Time

FIGURE 6-3 How AM Radio Waves Carry Sound Waves

has a lower-power (weaker) signal than a high-frequency sound, AM signals tend to break down before they have traveled many miles from their source. Increasing the power of the transmitter and building receivers that filtered some noise and compensated for signal strength variations helped AM radio increase its popularity around the world, but radio engineers were still looking for a better form of radio transmission.

FM Radio

The man who found a better way was Edwin H. Armstrong. Instead of modulating the amplitude of a radio carrier wave (AM), Armstrong modulated the frequency of the carrier wave. The result was **frequency modulation (FM)** radio, which had a signal of constant strength (amplitude) but could still carry voice and music data. The peaks of a sound wave were transmitted as higher frequencies, and the valleys as lower frequencies. By this method, the frequency of the sound wave could be represented almost perfectly in the analog FM radio signal. Figure 6-4 shows how this representation was created. An unmodulated carrier wave is shown at the top, a sound wave in the center, and a frequency modulated (FM) wave at the bottom in which the sound wave is incorporated in the varying frequency of the carrier wave.

FM also operates at different, more noise-free frequencies than AM. It has become the standard of radio transmission today, although many AM stations still operate efficiently in local markets and a few as long-range clear-channel stations (those that have exclusive rights to a frequency nationwide) with up to 50,000 watts of transmitting power.

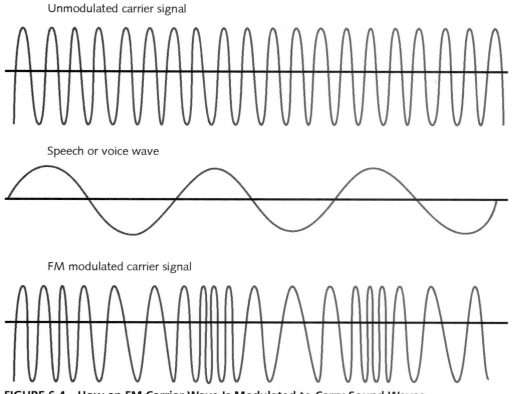

Unmodulated carrier signal

Speech or voice wave

FM modulated carrier signal

FIGURE 6-4   How an FM Carrier Wave Is Modulated to Carry Sound Waves

Digital Radio

Both AM and FM commercial stations in the U.S. are now converting to digital radio broadcasting, known as HD Digital Radio. They use a system called In-Band On-Channel (IBOC) digital radio broadcasting that received the endorsement of the National Radio Systems Committee for full-time use on FM radio stations and daytime use on AM radio stations. The AM restriction was imposed because of nighttime skywave interference issues. Station WOR in New York City became a test station for IBOC digital AM radio in 2003.

The IBOC digital system allows stations (AM and FM) to broadcast a digital signal and an analog signal at the same time. The digital signal (which can carry other information in addition to the audio radio program) is compressed and bundled together with the analog signal for broadcast. Analog radios receive and play the analog portion of the signal. Digital radios receive, decode, and play the digital signal. The dual capability of the IBOC system allows for the gradual transition from analog to digital radio.

Most U.S. radio stations have converted to digital or are planning to do so soon, but no time limit has been set for when stations must go digital. Eventually, analog radio broadcasts will cease and only digital radio will be available, but before this can happen, consumers must participate in the conversion process by purchasing HD radio receivers. Although sales of these are accelerating rapidly, it will likely be several years before the transition to digital is complete.

HD radio enables analog stations to broadcast CD quality audio and, like analog radio, broadcasts, the service is free. For a home LAN, digital radio is a "must have" option. Incorporated as part of the LAN's audio visual segment, clear noise-free radio can be distributed on the LAN throughout the home. No subscription fees are required and only a single digital receiver is needed.

Another way of incorporating digital radio into a home LAN is through the Internet. While U.S. stations have been converting to digital technology for their commercial broadcasts, most of them have also launched Web sites for streaming their digital radio signals over the Internet. Thousands of schools, colleges, hobbyists, and others have also started Web-based radio streaming sites that offer a tremendous variety of listening options. A new law is intended to allow these digital stations to expand their online audiences while paying modest royalties for the music they broadcast. The development of relatively inexpensive streaming equipment and software has further encouraged Internet radio and now the home LAN user has an almost infinite choice of online listening options available through a Web connection.

Another form of digital radio broadcasting is satellite-based Digital Audio Radio Service (DARS), a subscription radio service offered by two companies in America and a third in most of the rest of the world. For a monthly fee of about $15, XM Satellite Radio and Sirius Satellite Radio offer hundreds of radio channels beamed from satellites in orbit to the user's ground receiver. WorldSpace, the third satellite radio company, offers similar service throughout Europe, Asia, Africa, and Latin America.

The U.S. satellite radio companies use either two satellites placed in geostationary orbits, or three satellites orbiting in a pattern where one or two are always over North America. Ground stations transmit signals to these satellites, which then send the signals back down to radio receivers on the ground. Only one uplink is required (from the ground station to the satellite) and the downlink area of the satellites covers all of North America. This means that, unlike ground-based digital radio stations, satellite channels never go out of range. This feature has made them ideal for use in cars and other vehicles. Mobile radios formed the original user base for satellite radio, which was initially supported and partially funded by auto manufacturers.

But satellite radio is just as good for stationary listeners as those on the road, and many home LANs now include it as part of the network's audiovisual segment. As with HD radio, a digital receiver is required, but to receive satellite radio, the receiver must be equipped with the proprietary decoding chip of the broadcasting company to which you subscribe. Only a Sirius-equipped receiver will play Sirius Radio; it is the same for XM and WorldSpace.

The radio receivers are programmed to receive and unscramble the digital data signal, which contains up to 100 channels of digital audio. In addition to the encoded sound, the signal contains additional information, which is displayed on the radio's LCD display: the song title, artist, and music genre. This supplementary service is known as datacasting. Content other than music, such as news and weather broadcasts, is also offered on satellite radio, but no local radio stations are available through this service.

Digital Audio Recording

Audio sounds can be digitally recorded on magnetic tape using the same sampling technique described at the beginning of this chapter. The sound wave is sampled at a high rate of speed and its amplitude in digital form is recorded on the tape.

To get a high-quality recording, the rate of sampling must be at least twice the frequency of the highest sound the human ear can detect. The range of sound that most people can hear consists of frequencies from a low of 20 cycles per second to a high of 20,000 cycles per second. Consequently, digital sound recording techniques sample the sound waves being recorded at a rate of at least 40,000 times per second (40 KHz). The sampling process produces a steady stream of signal–waveform values 1/40,000 of a second (25 nanoseconds) apart. The actual standard sampling rates for digital recording are 44.1 and 48 KHz, both somewhat higher than the 40 KHz example given. Some formats of digital audio recording use sampling rates that are much higher in order to further increase the quality of the sound.

An analog-to-digital **converter** (ADC) performs the sampling process and expresses the values as binary numbers (0's and 1's only). It is this digital information, rather than the analog waveform itself, that is recorded. Because there is a lot of digital information to record, and it must be recorded rapidly in order to keep up with the playing music or speaking voices, **digital audio tape (DAT)** is not recorded in straight tracks running down the length of the tape, as shown with analog recording in Figure 6-2.

Digital sound is recorded by a rotating head that very rapidly "swipes" short tracks of recorded data across the tape at an angle as the tape moves forward on the reels. The tape head is rotating at 2,000 rpm and so can record the large amount of data required for high-fidelity digital sound. Digital tapes cannot be cut or spliced, however, because the dense, angular recording method is disrupted by any break in the tape. Figure 6-5 shows how digital audio tape (DAT) is recorded with angled tracks. As the measurements on the figure indicate, the recording tracks must be very precisely placed and can be read only by a digital tape player configured in exactly the same manner as the recording machine.

Digital sound can also be recorded on other **magnetic media** such as computer disks and hard drives, although usually not in the same format as on

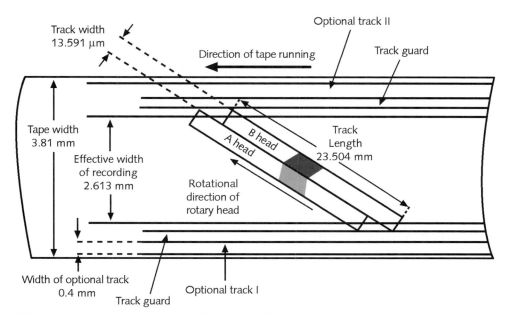

FIGURE 6-5 Digital Sound Is Recorded on Angled Tracks across the Tape

magnetic tape. Digital audio intended for storage on disks or hard drives must also be organized in named files so the computer can locate a desired music track or voice clip among many that may be stored randomly on the disk.

CD Audio Recording

Compact discs (CDs) are the most common form of recorded music today. They also offer the highest quality of music or voice recording. CDs use a recording system called **pulse code modulation (PCM)** to convert analog sounds into digital form and store them on a CD. The PCM process is a **lossless** technique for digital recording, meaning that none of the audio information is lost in the digitizing process due to compression of the data. A few other lossless formats for audio recording exist, mainly some used for computer audio files, but many other formats are **lossy codecs,** which lose some of the audio sampling information through compression or other factors.

CD audio format samples the analog sound 44,100 times per second, which enables it to record sounds up to 20,000 cycles per second, the limit of human hearing. As it records, the system must deal with the errors between the sampled discrete values and the actual continuous sound. This is referred to as the **signal-to-noise (S/N) ratio,** and in recording it is the ratio between the difference of the highest and lowest frequencies and the average background noise level. The higher the S/N ratio, the better the sound. S/N ratios must be greater than 70 **decibels (dB)** or the background noise becomes audible when the sound file is played. PCM audio format achieves S/N ratios in the low 90 dB range, which gives it a very low noise level while maintaining a high quality of sound reproduction.

PCM produces very high-fidelity sound quality, but also requires a large amount of storage space. There is a direct relationship between the rate of sampling the analog sounds and the amount of storage space required to store the digital data derived from the sampling. A single CD can hold 600 million characters of text, but only 74 minutes of uncompressed music recorded in CD audio format. One minute of stereo audio recorded in PCM format requires 10MB of storage space. Five seconds of recorded music require as much storage space as 135,000 words of text, enough to write a complete novel. Only the large storage capacity of a CD makes the CD audio format practical for high-fidelity music recording.

CD and DVD drives in computers can all play PCM audio format music recorded on a CD, but computers rarely store music in this format on any other type of storage device because doing so requires too much space. For any type of magnetic disk or hard drive storage, a magnetic media format, such as those described in the next section of this chapter, is generally used because they compress the data into smaller file sizes. While this causes some loss of quality compared to PCM audio, the loss is not detectable by most listeners in sound recordings. These formats are referred to collectively as lossy formats, because they do lose some audio information, but the savings in storage space justify their use for any but the most high-fidelity recordings heard by discriminating listeners.

Computer Audio Recording

Computer audio files must also be recorded in a compatible format, which the computer can read using software that is contained in its memory or stored on its disks. While original digital audio is usually recorded on tape in a continuous stream using a lossless format, such recordings are usually too large to be conveniently stored on a disk or hard drive.

Only Waveform Audio Format, which is a part of the later versions of the Windows Media Player software supplied with most PC computers, and the Apple Lossless Encoder, which is part of the iTunes software on Apple computers, are commonly used computer formats for lossless recording of audio.

Most computer-based audio storage formats compress the full-size audio files into smaller size files that the computer can more easily store and handle. Among the formats used for digital sound recordings, which can be played back on a computer, are the following:

- **MP3 files** are the most popular audio standard for computer files. The German company Fraunhofer patented this format, and users must pay royalties for compressing files with it. With MP3 compression, computers can compress the content of a music recording to less than one tenth of its original size. This allows up to 20 hours of music to be stored in 1 gigabyte of storage space. MP3 files are also streamable. This means that a computer can begin playing an MP3 file when only the first part of it is loaded in the computer's memory. It also allows an MP3 file to be sent over the Internet and played on the receiver's computer as soon as the head of the file arrives, while the rest of the file is still being transmitted.

- **Windows Media Audio (WMA)** is part of the Windows Media software package that is patented by Microsoft. It is an integrated part of the Microsoft Windows Media Player, which can also play video files. WMA files are about a third the size of CD files, but are still larger than MP3 files. WMA files can also be streamed over the Internet and, unlike MP3 files, they contain a built-in protection against the copying of copyrighted music. This is the primary reason why many music and entertainment companies publish only in WMA format and not in MP3.

- **Ogg Vorbis** (OGG) audio format is also known as "Squish." OGG is an open source format and hence is free of any patents. It was designed as a substitute for MP3 and WMA and is almost as popular and well known as MP3. The algorithm is still being developed, but the files are backward-compatible so that older recordings can be played with newer versions of the software. OGG also supports multi-channel compression (stereo or surround files) and is streamable.

- **MP3PRO** is a newer version of MP3 and has only been available for a short time. MP3PRO offers the same quality as MP3, but requires files only half as large. MP3PRO will likely replace MP3 over the next few years. Like WMA and OGG, MP3PRO is backward-compatible, so MP3PRO files can be played with older MP3 players. It also offers copyright protection for published music and streaming capability for the Internet.

Television Recording and Broadcasting

The most common analog television format in the United States is the **National Television Standards Committee (NTSC)** standard, which has been in place for more than 50 years and is now badly outdated. The improved standard, **digital television (DTV)** with high definition, is now being widely adopted. On February 17, 2009, all analog television broadcasting will cease and all television stations in the U.S. will broadcast digital signals only.

Most consumers will probably purchase a digital television by that date. Those who do not must purchase some variety of digital **decoder** (housed in a set-top box) and connect it between their television antenna and their television set. The decoder will enable an analog television to display the digital signal, although only in NTSC form, not at the higher digital television picture quality.

Digital television equipment, though still more expensive than analog sets, is declining rapidly in price as millions of units are sold. The vast majority of DHTI audio visual systems installed during the next few years will be digital and most will include HD television. The NTSC standard is now a poor investment for new buyers, although the last analog video products, such as **VHS** cassette players and TV sets, may continue in use for many years. The future of television, like that of audio, is digital.

Most analog television broadcast stations transmit a picture that contains 525 horizontal **interlaced scan lines** or rasters with the equivalent of approximately 720 pixels (picture elements) per line at 30 interlaced scans per second. A **raster** is one scan line of a television image. The NTSC **resolution** for each raster is expressed as a **pixel** equivalent because analog television does not have pixels. The rasters that make up an analog TV picture are continuous horizontal lines with varying colors along their lengths.

Europe and some other parts of the world adopted the **PAL** or **SECAM** formats for television, which both have an equivalent resolution of about 720 pixels (picture elements) wide by 625 lines high. These formats operate at 25 interlaced **frames** per second to more easily function on the European standard power grid, which is 50 cycle AC current at 240 volts. Even though the PAL/SECAM formats have higher resolution than NTSC, their frame rate is slower and so their picture quality is comparable.

Because analog television is still a large (though declining) part of the installed television base, it is useful to know how it is created, how it is recorded, and how it is broadcast. It is also important to know how digital television differs from analog television, which it is rapidly replacing, and how elements of both systems can be integrated into a home technology system.

Analog Television

To broadcast analog television as a radio frequency (RF) signal, the television image seen by an analog camera, and its accompanying sound, must be broken down into elements that can be transmitted on radio waves. When the signal is received by a television receiver, it must be interpreted and reassembled into picture elements that the **cathode ray tube (CRT)** display in the television can scan onto its screen. This process has to be repeated 30 times each second for picture and sound. A standard NTSC screen includes 525 raster lines of picture data, and stations broadcast at 15 frames per second (30 interlaced scans). That means that a tremendous amount of data must be squeezed into the 6 MHz bandwidth of a television channel. The crowded bandwidth of analog television channels means that any improvement in analog transmission (more raster lines or faster scanning) is difficult or impossible. This is one of many reasons why the switch to digital television is essential.

Not all of the NTSC standard screen's 525 lines by 720 (equivalent) pixels wide can be seen on the screen. Some scan lines at the top and bottom are used to convey other information. In addition, the NTSC standard calls for some "over-scan" on all sides of the video picture to allow for TV sets in which the picture is misaligned. The lines used for other data and the overscan reduce the visible lines to about 480 from top to bottom of the picture and the equivalent of about

640 pixels from one side to the other. Despite the fact that part of the lines are not visible onscreen, they are still part of the NTSC standard and must be included in all NTSC broadcasts and any digital conversions of that standard.

Analog video is turned into a broadcast signal by first breaking each picture frame down into its component rasters. Each raster is recorded along its full width by the camera in terms of its color and luminosity (brightness). To the raster content signal is added two additional signal elements, the horizontal sync and vertical sync, which together specify precisely where each raster was scanned by the camera and, when received, where it is to be scanned onto the CRT screen display. All this information (263 rasters of color and luminosity, with horizontal and vertical sync signals for each raster) is broadcast 30 times per second on an analog television signal. This information is received and interpreted by the receiver to produce a color television picture. In addition, on the same frequency band, the sound portion of the signal is broadcast as an FM radio signal.

Because analog television signals are large and complex, they are increasingly subject to noise, interference, and attenuation with distance. Nearly all analog television sets display some of these signal problems in such effects as snow, static lines, color shifts, ghosts, dropout lines, or just overall poor picture quality. Sound is usually not badly affected because it is an FM signal and stays strong through the full distance that the picture signal can be received. Analog television arrives at the receiver through an antenna as RF (radio frequency) signals, which also include stereo sound. The system has worked well for more than 50 years, but it has long since reached its maximum potential for quality transmission. The demand for better television resulted first in recorded video programs that did not suffer from all of the same defects as live broadcast television, then in the digital television revolution.

Analog Video Recording

When television broadcasting began, the only way to record a program was to photograph a television set that was receiving the show using a synchronized motion picture camera. This process was known as kinescope recording and a number of early television shows were recorded in this manner, even though the quality of the image was poor and the filmed recording of a television program could not be used to transmit the program electronically. If the program was to be rebroadcast, the kinescope film had to be projected into a television camera using a device called a film chain to synchronize the film projector and the TV camera. The quality of the second (and all subsequent) broadcasts was even lower than the original kinescope recording.

Magnetic recording of analog video programs was much more difficult than recording sound because of the large amount of information contained in each video image. The changing data in each of 525 scan lines in each frame of an NTSC video picture had to be recorded in real time. The first attempt to do this with a linear recording system was made in 1955 and required a tape recorder that operated at a speed of 200 inches of tape per second. It was not practical for recording programs more than a few minutes in length and an improved system was soon developed.

Angling the recorder's recording head relative to the tape's direction of travel proved to be the solution for video recording, as it later would for digital audio. The angled heads rotated rapidly and recorded the video information in a series of short tracks running at an angle across the tape. Even with four recording

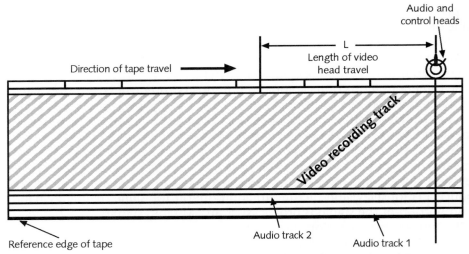

FIGURE 6-6   Analog Video Is Recorded with a Slanted Rotating Four-Head Drum

heads on the drum recording information simultaneously on magnetic tape that was 2 inches wide, the early video recorders still had to run at fairly high speed to record the amount of information required for video. Later improvements allowed reductions in the speed of the tape and its width, first to the one-inch-wide C type, then to 3/4-inch, and finally to the half-inch VHS standard used in consumer cassettes today. The smaller tape size caused some quality loss, but was far more economical for home use. Figure 6-6 shows how analog video is recorded on tape. Note that the tape is wider than that used for audio recording, which allows more information to be placed on it, and the recorder has four recording heads mounted on the rotating drum, each of which records a portion of video data each time the drum revolves.

Digital Video

Storing and transmitting a video picture in digital form is actually a more direct process than storing sound in digital form. Rather than sampling an analog picture and then re-creating it based on the samples, video images can be converted to digital form direct from the original scene using the picture data from a digital video camera that photographs the scene perfectly. This first generation conversion is possible because video images are all made up of the small individual rectangles called pixels (picture elements).

Digital video is created by assigning a numerical value to every pixel that makes up a video image. This is done either by a digital video camera, which converts the image seen through its lens directly into digital data, or by a scanning process in which a digital video scanner converts each frame (individual picture) of an analog video program or a motion picture film into digital data. The numerical value assigned to each pixel describes its position on the screen, its color, and its luminance (brightness). Since there are more than 375,000 pixels on an NTSC resolution screen, and millions of possible colors for each, the amount of data to be stored for each video image is many times greater than a sound file, which samples a sound wave a mere 44,000 times per second. Digital television formats of higher resolution and wider aspect ratio than NTSC standard require even more data per frame.

For each second of digital video, 30 frames of picture data must be recorded, but in some digital formats this actually requires a **refresh rate** of only 15 complete images because the pictures are interlaced in the same manner as NTSC analog television. This means that each time the digital image is refreshed, only half of the screen is actually scanned with new digital information. This half-screen scanning is not detectable to the viewer because it is accomplished by scanning every other line of pixels on the screen. In one scan, all the even-numbered lines (one, three, five, etc.) are refreshed, and in the next scan, all the odd-numbered lines (two, four, six, etc.). Only half as much data is thus required for each interlaced scan as would be necessary for a complete frame refresh. Consequently, an interlaced program requires much less storage space than a non-interlaced or progressive-scanned program in which every line in every frame is completely refreshed.

Figure 6-7 shows how one half of an interlaced screen is refreshed each time the screen is scanned and how two interlaced scans are combined to create a complete picture frame on the screen. The illustration shows only the parts of the screen refreshed in each interlaced scan. In an actual television picture, the white lines of each interlaced **field** are filled by picture data from the preceding scan. A complete interlaced frame is shown below the two scans.

Interlaced video requires less data storage than **progressive**, but progressive scanning produces a better picture and is the preferred technology for **high-definition television (HDTV).** As we will discuss later in this chapter, the amount of digital data required for peak resolution HDTV (1920 pixels wide by

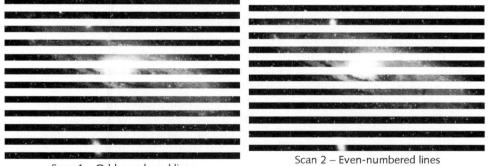

Scan 1 – Odd-numbered lines   Scan 2 – Even-numbered lines

Interlaced scans appear as complete picture

FIGURE 6-7   Two Successive Scans Make Up a Complete Analog or Digital Interlaced Television Frame

1080 high) is so large that storage and broadcast technologies are only now reaching the level where this standard can be consistently used.

All digital television is a significant improvement over analog, even when the NTSC-based resolution standard remains unchanged. As with digital sound recording, digital television removes nearly all of the noise and interference from the recorded images and defines the picture information much more precisely. When digital television is displayed, the picture and sound are almost flawless. Reception does not deteriorate with distance as analog signals do. If the digital signal is received, the picture display will be near perfect.

The "if" regarding digital reception results from a characteristic of digital transmission called the **cliff effect.** Unlike analog signal reception, which gets progressively weaker and declines in quality as the receiver's distance from the transmitter increases, digital signals retain their quality to the limit of their transmission range, but then just suddenly stop, as though they have fallen off a cliff—hence, the name cliff effect. This effect can mean that a house in one block will get excellent digital reception, while a house only a block further from the transmitter will get no reception at all.

The cliff effect is caused by weakening of the digital signal with distance. This eventually causes transmission errors in some of the digital data. Digital equipment has built-in error-correction technology to correct or "fill in" some lost digital data in a signal, but when the amount of lost or corrupted data reaches a certain threshold, the receiver can no longer make up the lost data and therefore ceases to display the entire signal.

Digital Television Compression

Because digital television is more data intensive than analog television and would require much more bandwidth for broadcasting, its use would be impractical except for the fact that digital data can be compressed. Nearly all digital video is compressed either as it is recorded or before it is broadcast. As with digital audio compression, there are two types of digital video compression: lossless and lossy.

Lossless compression, as the name implies, means that after compressing the video, and then decompressing it, the restored video has all the exact same data as the original. Lossless has the advantage that no matter how many times it is compressed and decoded, none of the original data is lost. The decompressed video has all the quality of the original. The limitation of lossless compression is that it cannot save nearly as much space as lossy compression algorithms. Consequently, lossless video codecs (codec is an acronym for compression/decompression) such as **Huffyuv** and Lossless MJPEG are not used nearly as much as lossy codecs which compress further, and thus save more storage space.

Lossy compression is the form of video compression that 95% of all video codecs use. This means that when they compress video and then decompress it, the decompressed video does not still have all the data of the original. Lossy codecs include **MPEG-1, MPEG-2,** MPEG-4, Quicktime, DV, Digital-S, RealVideo, Sorenson, Indeo, and Cinepak. Each of these loses some data when it compresses the original video. The objective of lossy compression is not to lose any data that will be observable on the screen as a deteriorated image. None of the codecs achieves this standard completely, but some come close to it.

The two main methods of lossy compression are called intra-frame and inter-frame compression. Intra-frame means that each frame of video data is compressed

Encoder/transmitter

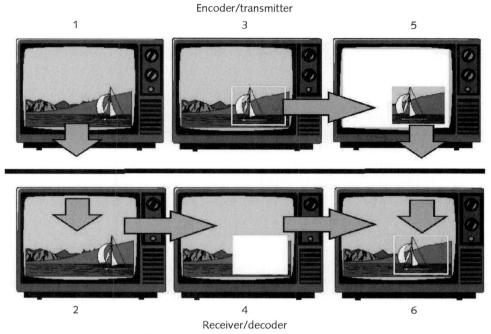

Receiver/decoder

FIGURE 6-8   Inter-Frame Video Compression

independently without regard to what is contained in any other frame. MPEG is an intra-frame compression method that uses the familiar JPEG graphic compression found in computer graphics and on the Internet to compress each frame. This type of compression results in very little loss of quality in the picture, but is limited in the amount of compression possible.

Inter-frame compression exploits the fact that most of a television image is not moving all the time. Inter-frame codecs look at the preceding frame (and sometimes the following frame) to see what has changed in the frame they are compressing. They then compress only the changed parts of the new frame and repeat the unchanged parts of the preceding frame to get a complete picture for the next frame. Figure 6-8 shows a simplified example of how this is done. The numbered list that follows explains what is happening in each frame of the figure.

1. Frame 1 of a new scene is compressed by the video encoder in digital format and transmitted. The encoder retains the data from Frame 1 for comparison with later frames.
2. Frame 1 is received by the receiver, which displays it on screen and retains the data from the frame for use in other frames, if needed.
3. The video encoder compares Frame 1 of the scene with Frame 2 and compresses only the part of Frame 2 that is different from Frame 1.
4. The receiver is instructed to repeat the unchanged part of Frame 1 to use as part of Frame 2.
5. The transmitter sends the changed part of Frame 2 to the receiver.
6. The receiver combines the unchanged part of Frame 2 it retained from Frame 1 with the changed part it receives from the transmitter to create a complete Frame 2. Frame 2 requires only one quarter as much data transmission as Frame 1.

Digital compression technology is continually improving. Most compression systems use both intra-frame and inter-frame algorithms. Some of the newest compression techniques have achieved almost 100-to-1 compression ratios; that is, they are able to reduce the data in a digital broadcast signal to one one-hundredth of its original size, without significant loss in the quality of the picture or sound.

All digital video broadcasting is done in compressed format in order to squeeze the required quantity of information into the available bandwidth. Using digital compression enables a broadcaster to send five channels of standard-definition television (digital television with a **4:3** aspect ratio) in the same bandwidth as a single channel of analog television. Compression also enables the much more data intensive HDTV (digital television with a **16:9** aspect ratio) to be squeezed into the same bandwidth as an analog channel.

Changing Video Formats

The NTSC standard of 525 lines of resolution and a 4-to-3 width-to-height screen ratio has been the format for television in the U.S. since commercial broadcasting began in the middle of the last century, but new and more advanced digital television formats are now rapidly replacing it. Computer monitors display video images in several formats, which are different from the NTSC standard and from the new formats used for television. Video streaming on the Internet has also produced new formats. Finally, cable and satellite television transmission have brought many international television broadcasts within range of American receivers, and many of these programs arrive in formats different from NTSC.

Currently, American citizens own nearly 250 million television sets. As digital television (DTV) replaces analog broadcasting on the NTSC standard, the change will make all of these sets obsolete, because digital television signals cannot be displayed by sets with only analog tuners. At a minimum, all analog TVs will require a converter box that will convert digital programming into analog signals for display. Owners of analog TVs will be then able to receive digital programs by connecting the TV's antenna to the set-top box that converts the digital signal for analog display. Since the TV was designed for the NTSC quality standard, it will not display the increased sharpness and resolution of the digital transmission. The reception will, however, be free of snow and other distortions of analog transmissions, which will be a great improvement for many viewers.

Standard-Definition Television (SDTV)

When the U.S. decided to make the transition from analog television to digital television (DTV), the Federal Communications Commission agreed to let broadcasters decide whether to broadcast **standard-definition television (SDTV)** or high-definition television (HDTV) programs. Most have decided to broadcast SDTV programs in the daytime and to broadcast HDTV programs during prime time in the evening. Both SDTV and HDTV are supported by the Digital Video Broadcasting (DVB) and **Advanced Television Systems Committee (ATSC)** set of standards.

SDTV is a digital television (DTV) format that provides a picture quality similar to that recorded on standard (not HD) **digital video disc (DVD).** SDTV is the lower-quality standard of DTV (HDTV is the higher-quality standard) that has been adopted for U.S. broadcasters. SDTV can be broadcast in either a 4:3 aspect ratio (720 pixels wide by 480 pixels high) or in a 16:9 aspect ratio (850 or more pixels wide by 480 pixels high). It can also be broadcast as either an interlaced or progressive scan signal, but must have resolution of at least 480 lines interlaced. The SDTV minimum resolution (4:3 aspect ratio, **480i**) provides the same picture format as NTSC analog TV, but much better resolution because of the digital precision of its picture.

Another standard that some manufacturers have adopted for digital television sets is **enhanced-definition television (EDTV),** which includes the same screen aspect ratios as SDTV but requires that the screen resolution be a minimum of **480p** lines rather than the interlaced 480i of SDTV. EDTV thus includes any resolution of 480p and above, up to the **720p** lines that are the minimum for HDTV.

Digital-receiving television sets can be either SDTV-capable, HDTV-capable, or both with receivers that can convert the digital signal to their native display format. SDTV and HDTV both use the MPEG-2 file compression method. SDTV with a 4:3 aspect ratio will be able to display the minimum SDTV standard broadcast format (4:3 aspect ratio, 480i) as a full-screen picture. If an SDTV set receives a 16:9 aspect ratio signal, it will display it as a **"letterbox"** picture in which the top and bottom parts of the screen are blacked out to adapt the wider picture to the narrower TV set. Conversely, if an HDTV (with a 16:9 aspect ratio screen) receives an SDTV format signal with a 4:3 aspect ratio, it will display the picture with **"barn doors,"** blacked-out parts of the screen on either side of the screen to adapt the narrower picture to the wider format. Figure 6-9 shows the relative difference in screen size between a 4:3 aspect ratio screen and a 16:9 aspect ratio screen. It also shows the "letterbox" and "barn doors" effects of displaying wrong-format images on different screens.

"Barn doors"
4:3 picture on 16:9 screen

"Letterbox"
16:9 picture on 4:3 screen

FIGURE 6-9   The 16:9 Aspect Ratio Shows a Much Wider Picture than the 4:3 Aspect Ratio

Another factor that affects the resolution of all digital (as well as NTSC) television displays is the fact that in television technology the pixels are not square but rectangular in shape with a height slightly greater than their width. This means that an array 480 pixels high and 720 wide has an aspect ratio (width to height) of 3 to 2. When the pixels are displayed, however, their rectangular shape makes the actual screen size show a narrower aspect ratio of 4 to 3. This 12.5% squeezing of the pixels horizontally is the same for all television screens regardless of their aspect ratio. The rectangular pixel format simply means that more of them must be stored for each video image in order to maintain the aspect ratio of the displayed picture.

High-Definition Television (HDTV)

High-definition television (HDTV) provides a higher-quality display than SDTV. It has a minimum vertical resolution of 720 lines progressive (720p). Most broadcasters are using a higher standard of 1080 lines interlaced **(1080i)** and some are using a 1080 progressive standard **(1080p).** 720p HDTV has 1280 horizontal pixels. 1080i and 1080p HDTV both have 1920 horizontal pixels per line. All formats of HDTV are thus much higher in picture quality (resolution and sharpness) than SDTV.

Regardless of the resolution used, the HDTV aspect ratio (width to height ratio of the screen) is 16 to 9, about the same as a widescreen movie. HDTV, in common with SDTV, uses the MPEG-2 file compression method for broadcast programming.

HDTV also includes 5.1 digital surround sound as part of the broadcast signal. This provides six separate sound channels to be connected to speakers in a surround sound configuration. HDTV receivers must be able to receive and distribute the surround sound signal, but not all sets are equipped with the full range of speakers required. A separate speaker array may need to be purchased with an HDTV to achieve the full potential of surround sound.

HDTV is the preferred standard for new home television purchases, particularly if the set is to be part of a home theater installation. Although HDTV sets cost more than SDTV sets (and both cost more than the old analog sets), the increased picture and audio quality of HDTV makes it well worth the extra cost. Since some form of conversion to digital television is required for all users by early 2009, the cost of this conversion will be of most benefit when invested in a quality HDTV set.

Table 6-1 shows a comparison of several television formats with popular computer screen resolutions, photo print sizes, and digital camera formats. It also shows the number of pixels required to create one frame of data in each of these formats.

Digital Television Receivers

Analog television sets can display digital television programs, but only with the help of a digital decoder. These are available as set-top boxes that receive the incoming digital signal from an antenna, a cable feed, or a satellite dish. The box decodes the digital signal, transforms it into analog form, and sends it to the television for display. A digital-to-analog decoder is the minimum digital equipment that will be required to receive television broadcast signals (which will all be digital) after February 17, 2009.

While a converter box allows viewers to see a program that has been broadcast in digital format, it does not display the visual clarity and CD-quality sound that a digital set provides. The reception may be a considerable improvement over ordinary analog, but its quality is defined by the limits of the analog set that is displaying it. Only with a digital television set can the full capability of digital television be displayed. Figure 6-10 shows some of the differences between an NTSC image and an SDTV image.

TABLE 6-1
Digital Image Resolution and Data Sizes

Format and Other Info	Aspect Ratio (H:V)	Horizontal Pixels	Vertical Pixels	Total Pixels
3×5-Inch Photo Scanned at 100 dpi	5:3	500	300	150,000
SVGA Computer Screen	4:3	640	480	307,200
Standard MPEG-2 Screen	16:9	724	408	295,392
NTSC Television Standard	4:3	720	525	378,000
SDTV (minimum standard)	4:3	720	480	345,600
Anamorphic (HD) DVD	16:9	960	540	518,400
8×10-Inch Photo Scanned at 100 dpi	5:4	1,000	800	800,000
XGA Computer Screen	4:3	1,024	768	786,432
HDTV 720p	16:9	1,280	720	921,600
HDTV 1080i	16:9	1,920	1080	2,073,600
2.35:1 HDTV	2,35:1	2,538	1080	2,741,040
3.3 Megapixel Digital Camera	4:3	2,048	1,536	3,145,728

FIGURE 6-10   An SDTV Picture (right) Is Sharper and Clearer than NTSC Analog (left)

Some "digital" televisions come with a built-in **digital tuner,** while others have only an analog tuner but are "digital-ready," which means they can display a digital picture, but only if they receive a decoded digital signal from a decoder. There is no way to tell whether a television has a built-in digital decoder by looking at it. When buying a new set, check the product specification sheet to see if it has a decoder. Television sets with digital decoders may also have analog tuners so they can receive the analog broadcasts that will continue until digital conversion is complete.

Television sets with built-in digital decoders for cable and broadcast reception are more expensive than those without, but the price difference is not large. The FCC has adopted labeling rules for digital sets specifying three levels of **digital cable ready** television receivers:

- Digital Cable Ready 1: A consumer electronics TV receiving device capable of receiving analog basic, digital basic, and digital premium cable television programming by direct connection to a cable system providing digital programming. There is no 1394 digital connector or other digital interface. This device does not have two-way capability using cable facilities.

- Digital Cable Ready 2: A consumer electronics TV receiving device that in addition to the features of the Digital Cable Ready 1 sets also includes the 1394 digital interface connector that may be used for attaching the receiving device to various other consumer appliances.

- Digital Cable Ready 3: A consumer electronics TV receiving device that in addition to the features of the Digital Cable Ready 1 sets is capable of receiving advanced and interactive digital services by direct connection to a cable system providing such service.

—FCC Media Release, September 14, 2000

Television sets in Category 2 can be attached through the 1394 connector to a video recorder or to a computer so the digital video signal can be transmitted over a LAN (with sufficient bandwidth capacity). It is important to remember, however, that the decoding device is only capable of decoding one channel of digital television at a time, and the channel to be decoded must be selected on the decoding device. This means that if the decoded signal is transmitted on the network, every display that receives it shows the same channel. A change of channel can only be made selecting it on the decoding device, not the individual displays, and any change made affects all displays.

Sources of Audio and Video Services

A home network can receive audio and video data from several sources. Which combination of these sources is chosen will, to some extent, determine how the LAN will be set up to distribute the audio and video data throughout the home.

Broadcast Digital Radio and Television

The most widespread source of audio and video programs, and the least costly, is commercial broadcast stations. Every large American city has television and radio stations, as do most mid-size and many smaller ones. Rural areas are served by repeaters and translators, which extend the range of commercial stations into areas beyond the reach of the primary transmitters. Commercial television and radio are free and require only an antenna (usually built in to radio receivers) to receive the broadcast signal.

By 2006, more than 1,500 television stations were broadcasting in digital format and nearly all the rest had applied for licenses to begin digital transmission. Over 99% of U.S. households are now served by a digital station and 75% have eight or more stations available.

The government has set a deadline of 2009 for the end of analog broadcasting. The deadline has encouraged nearly every television station to commence digital broadcasting. Over 97 percent of all stations are now licensed for digital. Many are still broadcasting analog signals along with their digital output because millions of analog receivers that cannot receive digital signals are still in use, but these must be converted to digital by the deadline when all analog television broadcasting will cease.

Analog broadcast television service usually is not linked to a home LAN because it is easier to simply receive the signal through the antenna to each television in the home. The same is true for analog AM and FM radio broadcasts. Digital television and radio broadcasts can also be received through an antenna, but must be decoded before they can be played.

Digital broadcast radio and television are the only free source of digital programming. Either one can be received by a decoder that is connected to the home LAN. The decoded audio and video signals can then be distributed throughout the LAN to as many players (radios and TVs) as desired. The limitation of this arrangement is that only one television channel and one radio station can be received at a time. Since the system has only one decoder for audio and one for video, the decoders must be set to one channel or station and only the decoded signal for that station or channel can be distributed.

The alternative for broadcast digital audio and television is to distribute the broadcast signal on the LAN to decoders in each television and radio receiver. Each node can then select its own channel or station for decoding and switch channels independently of other nodes. As with distribution of decoded signals, a signal booster may be needed in the LAN if many nodes are connected to the audio visual signal.

Cable Television Broadcasting

Much of the cable television reception in the U.S. today is still analog and not digital, although digital is gaining rapidly. Consequently, even though cable companies send their television signals to homes in digital format, the set-top boxes in most homes not only determine which cable channels are received, they also convert the digital cable signal to analog so the consumer's television can display it. This will probably continue to be true for many users after the 2009 analog deadline, because cable companies are not subject to the deadline and do not have to end their analog broadcasts on that date. Still, most cable television transmitters will probably upgrade to digital and most consumers will want to purchase digital sets to get the full benefit of digital cable television.

Cable companies offer SDTV to their customers and HDTV in the 720i format. Recent improvements in television compression codecs have permitted cable companies to broadcast some programming in 1080i format, but existing cable infrastructure still does not have the bandwidth to permit total cable conversion to HDTV. Only by replacing the cable network with high-speed fiber-optic cables would enough bandwidth be available for full HDTV via cable. No plans exist at present to make this conversion, but further improvements in digital television compression may make it unnecessary. If better compression techniques do not solve the problem, increased numbers of digital cable customers may warrant the expense of the cable upgrade.

Digital Satellite Television Broadcasting

All satellite television broadcasting is digital. This technology was developed after digital television was already beginning to overtake analog in ground-based broadcasting and so was designed from its beginning to take advantage of digital technology. Satellite television broadcasting differs from ground-based digital television broadcasting primarily in the location of the transmitter and the power of its transmission.

Satellite television signals are first transmitted from a facility on the ground to a satellite orbiting the earth. Television transmission satellites are placed in geosynchronous orbits. This means that the satellite stays locked in a stationary specific location above the earth. In effect, it is "parked" 22,300 miles above the earth. It orbits in sync with the earth's rotation so its position overhead does not change. Unlike most satellites, which rotate in orbit, television satellites do not. They hold a constant steady orientation in space with their solar-power panels facing toward the sun 24 hours a day and their signal receiving and broadcast antennas pointing steadily toward earth at all times. Because the satellite is stationary in space relative to the ground, transmission signals can be aimed at it very precisely. Receiving antennas can also be aligned accurately to receive the strongest signal from it.

Once the signal reaches the satellite, it is rebroadcast to customers on the ground. Satellite transmitters do not have the power of ground-based television transmitters, but they do not need to because their signals are received by **dish antennas** that are precisely aimed at the satellite. These aimed antennas concentrate the received signal at a point in front of the dish and send it to the decoder. The decoder converts the signal into a digital display and sends it to the television.

Figure 6-11 illustrates how local and national television programming sources send their signals by satellite or cable to a digital television broadcast center. The center then broadcasts the programming signals to a satellite that relays the signals to a home dish antenna. From the dish antenna the signals pass through a digital decoder and are displayed on the home's television screens.

Satellite systems can broadcast 150 digital television channels from a single satellite, but all satellite companies have more than one satellite in orbit in order to have backup capability and multiple coverage over all areas of the country.

The satellites broadcast all 150 channels to all customers. The decoder selects which channel it decodes for display based on instructions it receives from the satellite and selections made by the customer. The decoder can only process one channel at a time, so if the decoded signal is sent to more than one television, all must show the same decoded channel. If more than one channel at a time is to be available for separate television sets, a splitter is required on the dish antenna. It sends the signal to independent decoders attached to each set. Up to four separate decoders can operate off of a single dish antenna equipped with splitters.

As with cable television systems, most of the television receivers linked to satellite antennas and decoders now display analog pictures, superior to broadcast

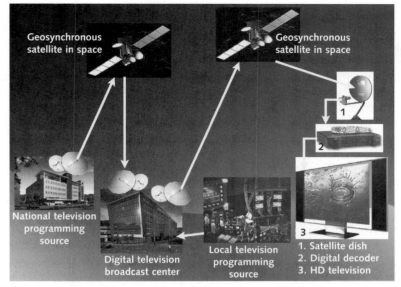

FIGURE 6-11 Satellite Television Broadcasts Nationwide from One or Two Geosynchronous Satellites

analog reception, but still not equal in quality to SDTV, let alone HDTV. The decoders in these systems convert the digital transmission to analog so that a standard NTSC television can display it. Satellite decoders will continue to function with analog TV sets after the 2009 deadline ending analog broadcasting. Since these decoders are already performing the function of digital-to-analog converters, satellite subscribers who have them can continue to use them until they upgrade to digital television sets.

A satellite system can display full digital pictures (SDTV or HDTV) on digital sets that have sufficient capability. Satellite signals have the bandwidth to carry full HDTV (1080i by 1920). Most of the HDTV being offered today is transmitted over satellite systems. Satellite companies broadcast mainly SDTV programs during the day and HDTV programs at night and as pay-per-view specials. This mix is likely to continue as the cost of SDTV sets is still about one-third the cost of a comparable size HDTV set. Digital converters in set-top boxes for analog sets cost about $100, while SDTV sets with digital converters built in to the television start at around $250. For cost reasons, most digital television upgrades are likely to be digital to analog converters or SDTV rather than the more expensive HDTV.

Digital Video Storage Files and Formats

Analog television programs are stored only on video tape. Analog video programs cannot be compressed. Digital video files are almost always compressed to save storage space.

Digital video is stored on media (magnetic tape, magnetic disks, DVDs) in many different formats. Because computers are digital devices and computer networks handle only digital data, all video formats used on the Internet or by computers to store and transmit video data are digital formats.

Internet Video

Four prominent formats of video files are available on the Internet. Each is compressed using a lossy algorithm and consequently exhibits some loss of quality from the original (uncompressed) video format. Internet video files are typically short, a few minutes in length rather than a full-length movie or even a half-hour program, although many of the latter shows are available for those who have the patience to download them over a home connection.

MPEG-1 files offer very good quality in an aspect ratio of 4 to 3 and an image size 320 pixels wide by 240 pixels high. This size is one quarter of a 640 × 480-pixel SVGA resolution computer screen and is large enough to be viewed easily on most monitors. MPEG-1 files cannot be streamed over the Internet, which means that the entire file of a video clip must be downloaded before the file can be played by the receiving computer. This lack of ability to play while completing a download makes MPEG-1 less popular than the other formats that do not have this limitation.

QuickTime files are played using the QuickTime player and are compressed using any of the supported codecs. They have better quality than MPEG-1 video files and they are streamable. This means the file can begin playing as soon as the first part of it has been downloaded into a buffer file on the receiving computer's hard drive or memory. The video program plays smoothly as long as the speed of downloading new data in the file stays ahead of the player. If the player overtakes the download, the video display stops until additional data has been loaded. QuickTime is made by Apple Computers Inc. and was originally a Macintosh-only video format. Later versions can also be used on PCs. QuickTime is a very well-supported format and is often used in video editing, as well as for distribution over the Internet.

Windows Media Player uses the MPEG-4 compression system and can stream video files. Windows Media is a relatively new format, but its popularity is growing quickly. It provides very good quality video, and given it is manufactured by Microsoft, it is included as a standard feature in all new Microsoft operating systems (Windows 2000, Windows XP, and Vista).

Real Player offers somewhat lower quality in video, but the Real media format was the first to deal with streaming video and has the widest user base. Its maker provides authoring tools for creating streaming video files. The current version of the software, RealOne Player, also supports full-screen (640 by 480 pixels) video playback over a connection fast enough to maintain the download while playing.

Most video files available on the Internet can be downloaded in at least two of the common formats. Generally, one of the available formats is of lower quality, but is also smaller in file size so it can be conveniently downloaded over a 56K modem connection. The other format is often higher in quality with larger files and is intended for use by those who have a high-speed connection.

Digital Video Storage on a Computer

Most current model computers have internal DVD recorder/players installed for video storage and playback. These devices offer the largest video storage in the smallest space currently available and are the most durable video storage medium as well.

CDs can also be used for storage of short video clips, but lack the capacity for more than a few minutes of video in any of the formats noted previously. They are not suitable for MPEG-2 recording of SDTV or HDTV, both of which have much larger data streams. Like DVDs, CDs are durable and convenient to use, but should be limited as storage devices to the smaller video formats and short programs.

Computers also store video on hard drives and these have the capacity to record both SDTV and HDTV programs of full motion picture length. 250GB (gigabyte) hard drives are now available for around $200. These will store several hours of video in lossless compressed formats. They can be used to store and play back complete movies or digital television programs and are particularly useful for storing raw video scenes from a video camera for later editing into a finished program. All digital video editing software works with video files stored on hard drives.

There is no limit to the amount of video that can be stored on hard drives for use in a home LAN, but hard drives take up more space than DVDs and are not as durable. DVDs are separate pieces of media that are inserted in a player/recorder. If the player fails, the DVD is not affected. Hard drives use magnetic media that is integrated into the mechanics of the recorder/player. If the hard drive fails, the media usually is destroyed with it. For both space and security reasons, DVDs are a better choice for large scale video storage than hard drives.

Analog and Digital Video Storage Media

Analog Video Home Standard (VHS) videotape cassettes have been until recently the most common means of duplicating and distributing video movies and other programs. Although analog video will soon be obsolete in television broadcasting, millions of these cassettes are still being manufactured and more than 80% of American homes have a videocassette recorder/player (VCR).

VHS cassettes are analog only, so neither SDTV nor HDTV programming can be recorded on them or played from them. Their popularity is declining and has been overtaken by digital video discs (DVDs), which are digital recording media only and which provide higher-quality video.

Some VHS cassettes, and the VCRs to play them, will probably continue in use even after the digital conversion is complete. Some movies and older television programs, especially those originally recorded by kinescope, would not improve in quality by being upgraded to digital format and do not justify the cost of converting them. Some demand for these programs will continue to exist, however, and VHS cassettes will likely continue to be the storage medium for them.

DVD technology offers crystal-clear pictures, superb multidirectional sound, massive data storage for computers, and a host of interactive features such as selectable camera angles and a choice of movie endings. A majority of U.S. homes have DVD players that convert the digital data recorded on a DVD to analog video that an NTSC standard television can display. These DVD player converters are now available for less than $100, a price that assures their popularity will continue to grow and that DVDs will rapidly replace VHS cassettes as the core of most home video libraries.

DVDs are recorded in the digital MPEG-2 compressed format, which allows up to 4 hours of digital video to be placed on a single DVD. Quality is excellent and DVDs have now become the standard against which any other video compression format is measured for both picture and sound quality. Most DVDs at present are recorded in an SDTV standard of 4:3 aspect ratio. These play full screen on an SDTV set and will also play to an HDTV set, but with black "barn doors" filling either side of the wider HDTV screen.

Anamorphically encoded DVDs are used to record movies filmed in a widescreen format. They have full HDTV format (16:9 aspect ration) and fill an HDTV set screen at full resolution (720p or 1080i). If they are displayed on an SDTV, black bars appear at the top and bottom of the SDTV screen, giving the picture a "letterbox" appearance.

This ability of DVD technology to be forward-compatible so it can provide improved viewing on existing analog and SDTV television equipment and still be compatible with more advanced digital technology added later by the consumer is one of the most appealing features of DVDs. It is made still more attractive by the fact that a DVD movie never wears out like VHS cassette tapes do after frequent viewing.

Connected Audio/Video System Design

The ideal networked video system might be one that allows users to simultaneously view HDTV programs or movies on any of several screens located throughout the home, or to play graphics-intensive three-dimensional action games across the LAN and with other players on the Internet. In addition, the ideal system would allow the user to video conference with others around the world while still monitoring the home security systems, children's play areas, summer vacation home, and sailboat. Unfortunately, while such a system is possible, it is not easy to accomplish technologically, and it is not cheap.

Transmission of entertainment video over a network puts a heavy load on the LAN because quality video, even when compressed by the most sophisticated codecs (coder/decoder algorithms), still requires the fast transmission of huge amounts of data and the fast processing of that data in order to render it viewable on screen in real time. Both the bandwidth of the LAN and the processing power of its nodes must be not just momentarily rapid, but consistently so for as long as the video stream is running. Any significant drop in transmission rates or processing speed can result in dropouts in the video display, jumps in the action, unintended freeze frames, and even audio distortion. None of these are acceptable when watching a movie or playing an action game, and they are equally annoying when participating in a video conference, watching a downloaded video clip, or scanning the yard for any sign of intruders. Some unpleasant truths about networks make these problems more likely to occur than most system vendors want to acknowledge.

The first of these truths is that most network technologies do not operate consistently at their maximum rated speeds. Tests have suggested that some, like HomePlug, and Wi-Fi, average only about half their rated maximums. Since these are both rated at speeds above 10 Mbps, even an average data transmission speed of 5 Mbps appears more than adequate for all network functions, including video, but this may not always be true.

The reason becomes evident when we consider how the average network speed is attained. An easy analogy illustrates the problem: If a person sits down with one half of his backside resting on a hot stove and the other half on a block of ice, on average, he's comfortable. But not really, because the average "comfortable" temperature does not exist anywhere on his seat and the deviations from that average are too extreme to tolerate.

The same thing happens on many networks: Sometimes they operate at maximum speed, and sometimes a momentary run of interference, noise, or heavy traffic will slow transmission to a tenth of maximum or even stop it altogether. When the slow periods and fast ones are averaged, there is plenty of bandwidth for most transmissions because the data is simply being stored at the receiving end until the file is complete and can be displayed or otherwise used.

Compressed video, however, cannot easily tolerate the slowdowns or the stops because it is a data stream in which each frame is interrelated. A slowdown or momentary break in the data stream leaves the processor unable to decode a frame on time. When the data starts up or speeds up again, the processor has to relocate where it was in decoding the stream, perhaps decode a couple of previous frames again because parts of succeeding frames depend on the data content of earlier ones, and then proceed forward. Buffering the video stream at the receiving end can mitigate some of the effects of slowdowns in data transmission, but to transmit a smooth uninterrupted video stream, a LAN must be able to function consistently for long periods at high speed with very few slowdowns in its transmission rate.

The most popular home LAN technologies can usually meet this performance standard, but not always, and hence have not yet proved popular for networking SDTV or HDTV real-time video. All of them handle full-screen (640 by 480 pixels) Internet-based video in all four of the most popular formats without difficulty. This includes HomePlug, Wi-Fi, HomeRF, and HomePNA. The manufacturers of each of these technologies have developed upgraded (faster) versions that should answer the future need for entertainment-level video data transmission.

10-Mbps Ethernet and 100-Mbps Ethernet can both network entertainment video with consistently high quality, provided that the cabling and connectors of the system are professionally installed so as not to degrade the performance of the LAN. A network's speed is only as fast as the slowest link in its infrastructure, so the quality of work done by the installing technician is critical if video is to be carried on the LAN. Ethernet networks can also be slowed by heavy traffic, which makes smooth video transmission impossible.

If video programs of any length are to be downloaded from the Internet, or games played with players on other networks, a high-speed Internet connection is a must. Video programs can be downloaded with a 56K modem, but the user needs to start the download in the evening and plan to play the program the following morning. Modems do not always connect with the ISP at their full-rated speeds. Large providers often limit their low-speed connections to 20K or 25K, especially during peak hours of usage, in order to still have adequate bandwidth available for high-speed customers who are paying for that service specifically. A recent test of a Microsoft video tutorial download using a 56Kbps modem revealed a speed of only 4.5Kbps in the connection. A 9MB video file required 23 minutes to download. Real-time video games are virtually impossible unless the Internet connection speed is at least 1 Mbps and performance improves proportionally as the ISP connection is raised above that level.

Entertainment video, whether it comes into the home as a digital signal from a broadcast antenna, a cable connection, or a satellite dish, should be distributed on a specialized segment of the LAN separate from other nodes that do not require the large data capacity. The incoming signal is distributed to all the nodes (receivers) connected to the video LAN segment, each of which has a converter to decode the signal for display. A switch connects the video segment to the rest of the LAN so Internet video and video clips from other nodes can be downloaded to display receivers.

With this arrangement, a DVD played on any computer connected to the LAN can also be distributed across the network, either to another computer node, or to a display television connected to the video segment.

Video programs that display in formats less than a full screen (MPEG-1, for example, at 320 pixels by 240) can be networked easily by any of the home LAN technologies, as well as Ethernet. This type of video includes most Internet downloads and video conference programs. Likewise, most security system videos can be networked, even if they display full-screen, because they typically transmit no more that two or three frames per second, often in black and white, and thus are not nearly as data-intensive as entertainment video.

Audio files can also be networked on any reasonably fast system. Audio formats that download completely before playing generally present no difficulty with dropouts or interference and have excellent quality. Those that are streamable, such as MP3, are subject to the same potential problems as afflict streamed video programs. The digital audio decoder often has difficulty recovering bits of data lost due to interference or noise on the network, and the lost bits translate into dropouts and sound distortions in the playing file.

Summary

- Analog audio and video signals are radio frequency (RF) transmissions, which have patterns that are related to the sounds or images they represent and can be interpreted and displayed by radio and television receivers.

- Digital audio is a conversion of the analog sound signals into numerical form by rapidly sampling the analog data and converting each sample into numerical values. Digital audio can be recorded or transmitted and received as RF signals over wires and on radio.

- Compact discs (CDs) are the most common form of digital audio recording. A CD can contain over an hour of audio and has a response range from 20 to 20,000 cycles per second, equal to the total range of human hearing.

- Audio files can be recorded on computers using any of several formats such as WAV (Windows Audio), MP3 (MPEG Audio Layer 3), WMA (Windows Media Audio), OGG (Ogg Vorbis), and MP3PRO (Motion Picture Professional).

- Analog television is transmitted over RF frequencies and received by an antenna, which is connected through a tuner to the television display. Cable transmission of analog television is also common, though declining.

- Digital video is created directly as digital data by a video camera either from the original scene or by converting an analog television program or motion picture film into digital data frame by frame.

- Analog television is still the most common type in the U.S. and is always in NTSC 4:3 format. Digital television is rapidly replacing analog and can be in SDTV 4:3 format or in HDTV 16:9 format.

- Digital television broadcasts are converted to analog for display on analog sets, but digital sets that display a true digital image have higher resolution and sharper images. Digital television will be the only U.S. television format after February 17, 2009.

- Digital video discs (DVDs) are the most common form of digital video-recording media. They are recorded using the MPEG-2 compression format and can each contain up to four hours of compressed video.

- Digital video is always transmitted in compressed form because of the large amount of data involved. Lossless compression (Huffyuv) is the best type because it loses no data, but 95% of compression is done in lossy formats (MPEG-2, QuickTime) that lose some data, but can compress the video files to much smaller sizes.

- Digital television is transmitted by broadcast to antenna-equipped receivers, by cable networks, and by satellite broadcasts to aimed receiver dishes. Digital signals are not affected by noise or interference as analog signals are, but they do terminate their range suddenly in the cliff effect.

- Digital video is transmitted over the Internet and on LANs using any of four formats:

MPEG-1, QuickTime, MPEG-4 (Windows Media Player), and RealOne Player. All except the last two display pictures smaller than full-screen size (640 by 480 pixels) and all use lossy compression technology to reduce the size of transmitted files.

- The data transmission requirements for entertainment video (speed and quantity) require a LAN technology with a consistently high speed. DVDs can be played over home networks, but must be controlled at the node where the DVD is physically located.

Key Terms

16:9 Wide-screen aspect ratio for HDTV and SDTV screens; 16 units wide by 9 units high.

4:3 Standard aspect ratio for SDTV, EDTV, and NTSC analog television; 4 units wide by 3 units high.

480i Screen resolution of 480 interlaced lines; minimum resolution standard for SDTV screens.

480p Screen resolution of 480 progressive scanned-lines; minimum resolution standard for EDTV screens.

720p Screen resolution of 720 progressive-scanned lines; minimum resolution standard for HDTV screens.

1080i Screen resolution of 1,080 interlaced lines; resolution standard for HDTV screens.

1080p Screen resolution of 1,080 progressive-scanned lines; top resolution standard for HDTV screens. Not yet in common use, but products that can deliver it are in development.

Advanced Television Systems Committee (ATSC) The committee that developed the U.S. digital television standard (DTV), which is also known by the acronym ATSC.

Amplitude modulation (AM) A method of converting sound waves to radio signals by varying the amplitude (strength) of the signal, but not its frequency.

Analog Sound or video signals that are analogous to (have patterns similar to) the actual sounds or images.

Aspect ratio The ratio of a video screen's width to its height. NTSC screens have a 4:3 ratio; HDTV screens a 16:9 ratio.

Barn doors A term that describes the effect of viewing a 4:3 aspect ratio image on a 16:9 aspect ratio screen; the viewer sees black bars on either side of the screen that are referred to as barn doors.

Cathode ray tube (CRT) The display tube in a television set or monitor on which the picture appears.

Cliff effect A description of the sudden termination of the range of digital video transmissions which can end as if they "fell off a cliff."

Codec Coding/decoding: a software program or hardware device that encodes and/or decodes digital transmissions.

Compact discs (CD) A plastic disc on which audio files are encoded using the pulse code modulation method.

Compression Any technique that uses math algorithms to reduce the size of digital files for storage or transmission.

Converter A device that changes digital signals into analog, or vice versa.

Datacasting Providing additional program material or non-program-related resources with a digital television or audio broadcast signal (video, audio, text, graphics, maps, services, etc.) to specially equipped computers, cache boxes, set-top boxes, or DTV receivers.

Decibel (dB) A unit for measuring sound level. Used in audio engineering.

Decoder A device that changes digital video or audio files to signals that can be displayed or heard on audio visual systems.

Digital Any kind of data that is recorded in numerical (discrete) form rather than analog (continuously varying).

Digital audio tape (DAT) Standard for recording uncompressed digital audio on tape at the same quality level as a CD.

Digital cable ready A government designation for television sets denoting their capability to play digital cable programs.

Digital television (DTV) Television signals that are in numerical format and create a picture in pixels rather than rasters as does analog TV.

Digital tuner (Digital converter, digital encoder/decoder) The electronic device that decodes digital audio or television signals so they can be played on radio or television receivers; a tuner can be an internal part of the receiver or contained in a separate set-top box.

Digital video discs (DVDs) Media for recording digital video.

Dish antenna A parabolic-shaped antenna that receives satellite broadcasts.

Dolby Digital 5.1 A digital surround sound technology that is broadcast as the audio part of HDTV signals and used in home theater systems to enhance audio. It includes five separate channels of audio that each connect to a separate speaker plus a low-frequency subwoofer-connected channel that together produce true-to-life audio.

Enhanced-definition television (EDTV) Digital television format with a 4:3 or 16:9 aspect ratio and a minimum resolution of 480p.

Field One scan of an interlaced television frame that refreshes one half the frame.

Frame One complete refresh of a television screen, which can be two scans (fields) if interlaced, one scan if progressive.

Frequency modulation (FM) A method of broadcasting sound by varying the frequency of the carrier wave, but not its strength.

High-definition television (HDTV) Digital television format with a 16:9 aspect ratio and a minimum resolution of 720p or 1080i.

Huffyuv A lossless codec for digital video compression.

Interlaced Method of refreshing a video screen in which only odd-numbered lines are scanned on one pass and even-numbered lines on the next.

Letterbox A term that describes the effect of viewing a 16:9 aspect ratio image on a 4:3 aspect ratio screen; the viewer sees black bars above and below the image that are referred to as the letterbox effect.

Lossless Type of video compression in which no data is lost. Huffyuv is an example.

Lossy Type of video compression in which some data is lost, but files can be made smaller than with lossless compression.

Magnetic media Any media with a ferrous coating capable of storing analog or digital data recorded on it by a magnetic head.

Magnetic recording Any analog or digital data recorded on magnetic media.

Magnetic tape Plastic tape coated with ferrous material for recording data.

MP3 file A file created with an audio compression algorithm with the same name.

MP3PRO A file created with an audio compression algorithm with the same name; higher compression than MP3, but equal in quality.

MPEG-1 A file created with a video compression algorithm with the same name.

MPEG-2 A video compression algorithm used for DVD video recording.

National Television Standards Committee The government and industry group that established the NTSC analog television standard in 1953.

NTSC format Standard analog U.S. TV format with 525 scan lines and a 4:3 aspect ratio.

Ogg Vorbis (OGG) An open source video compression format that is, hence, free of any patents.

PAL format European analog TV standard equivalent to U.S. NTSC standard.

Phonograph A machine that plays analog recordings from a plastic disc embossed with grooves bearing the sound wave impressions.

Pixel A picture element in digital television; the unit of color and brightness that forms the picture in digital television.

Progressive Method of refreshing a video screen in which all lines are scanned in sequence.

Pulse code modulation (PCM) The method used to record compact discs using MP-2 compression. Not a magnetic process. Uses light diffraction to record data.

QuickTime An Internet and computer video file format that employs compression and is widely used by PC and Apple computers.

Radio frequency (RF) Any electronic wave with a frequency in the radio band of the electromagnetic spectrum. Includes all radio and TV frequencies.

Raster One scan line on a television screen.

RealPlayer A computer and Internet video compression algorithm.

Refresh rate The number of times per second a television picture is "painted" on the screen; usually 15 times per second for interlaced broadcasts or 30 times per second for progressive-scanned broadcasts.

Resolution The amount of clarity, detail, and sharpness in a television picture. Usually measured in scan lines (interlaced or progressive) or pixels (e.g., 480 × 720).

Scan line One horizontal line of picture data on a television screen.

Sampling A technique for converting analog signals into digital form by taking quantified samples of the analog data.

SECAM A French-created European equivalent of NTSC analog standard.

Signal-to-noise ratio (S/N ratio) The difference in sound level between the recorded audio and the background noise on any type of audio recording.

Standard-definition television (SDTV) The approved minimum format for U.S. digital television with a 4:3 aspect ratio and an interlaced 480 × 720-pixel screen.

Video Home Standard (VHS) The U.S. standard analog video recording format used in videocassettes.

Video image Any image, analog or digital, displayed on a CRT or other type of screen.

WAV file A compression algorithm for sound files.

Windows Media Audio (WMA) An MPEG-4 audio compressed file.

Review Questions

1. Analog sound signals can be converted to digital form using a _____ technique.

2. Lossy is a form of _____.
   a. Video compression
   b. Audio compression
   c. Analog-to-digital conversion
   d. Recording

3. NTSC is a digital television standard with an aspect ratio of 4:3 and 480 scan lines. True or false?

4. Most radio stations in the United States have been broadcasting in digital format since 1992. True or false?

5. Frequency modulation is used to broadcast high-quality _____.

6. HDTV has a 16:9 aspect ratio and 1,080 interlaced scan lines. True or false?

7. Hertz (Hz) is a unit of measurement for the frequency of all _____.

8. DVD stands for _____.
   a. Direct video display
   b. Directional vane dominance
   c. Digital video display
   d. Digital video disc

9. What is the difference between an interlaced screen display and a progressive one?

10. DVDs are recorded using what form of compression?
    a. OGG
    b. WAV
    c. MPEG-2
    d. Huffyuv

11. For an analog television to display digital television images, it must have a digital-to-analog converter. True or false?

12. What is the difference between inter-frame and intra-frame video compression?

13. Satellite TV broadcasts originate on the ground and are broadcast to the satellite, which then rebroadcasts the signal to customers on the ground. True or false?

14. Radio Stations in the U.S. are now broadcasting both analog and digital radio signals at the same time. True or False?

15. A dish antenna is used to _____.
    a. Receive analog audio broadcasts
    b. Receive analog TV broadcasts
    c. Receive digital TV broadcasts
    d. Download cable signals

16. SDTV has a minimum of 480 vertical pixels and 720 horizontal pixels, giving it an aspect ratio of _____.
    a. 4:3
    b. 3:4
    c. 9:16
    d. 5:3

17. The "cliff effect" applies to _____.
    a. Analog TV signals
    b. All TV and radio signals

    c. Digital TV signals

    d. Audio recording

18. Why are digital television images always compressed before being broadcast?

19. If television satellites orbit the earth, how do customers receive their signals when the satellite is on the other side of the world?

20. A raster is _____.

    a. A device for mounting a dish antenna

    b. A single scan line on an analog TV screen

    c. A security program for video

    d. A connector cable

21. The most significant difference between pixels on a computer screen and those on a television screen is _____.

    a. The TV pixels have more colors

    b. The computer has more pixels

    c. The TV pixels are not square but rectangular

    d. The TV pixels invert the colors

22. Why is analog television not broadcast from satellites that could allow one station to cover the entire country?

23. MPEG and QuickTime video files can be quickly transmitted on the Internet because they compress the video data and because they display a screen that is _____ than a digital television broadcast.

24. The LAN technology best suited to transmit video at present is _____.

25. To stream video on the Internet or download streamed video, a LAN should have at least a ___ _____ ISP connection.

Hands-On Projects

Project 6-1: Download the QuickTime Player

In this project, you will download a video player program from the Internet. You will need access to a computer that is connected to the Internet. If the computer you are using already has the QuickTime player loaded on it, download another video player such as **RealPlayer** or Windows Media Player.

1. Log on to the Internet and go to http://www.apple.com/quicktime/download.

2. Select the operating system the computer is using and register your name if you wish. If not, click off the news group subscription, and click Download. The QuickTime file downloads automatically.

3. When the security warning displays, click Yes and proceed with the download of the installer program.

4. When the installer program has downloaded, it launches the download for QuickTime. Follow the screen instructions to start the download of the program, which is about 5MB in size.

5. After the program has downloaded, it installs automatically. Select the default choices at the various screen prompts unless you have specific reasons for changing them.

6. When the program completes installation, close the installer. Return to the Internet and check the installation page to be sure the program installed correctly. If any problems are noted, select the appropriate prompts to correct them.

7. You can test your download by playing one or more of the movie trailers listed on this site. Compare the time it takes to download and play a trailer at low resolution and high resolution.

Project 6-2: Download an Animation Video from the Internet

In this project, you will download and play an animation video from the Internet. You will need a computer that has access to the Internet.

1. Log on to the Internet and go to http://mars.jpl.nasa.gov/gallery/video/animation.html.

2. Click the QuickTime video captioned "Rover Launch and Cruise."

3. Select Save from the window that appears and then choose a directory on the computer in which to save the file. (If your computer has a QuickTime browser installed, the file will play immediately, and you can save it afterward.)

4. Download the file. It is about 5MB. See how long it takes the computer to download it. Do you have a slow or fast Internet connection?

5. When the file has downloaded, close the window and play the file using the player you downloaded in Project 1.

Project 6-3: Download an Audio File and Play It from the Computer

For this project, you will need access to a computer that has a CD-ROM drive. You will also need access to a music CD and the Internet.

1. Insert a music CD into the CD-ROM drive and close it. Windows 2000 or Windows XP will open the CD player. If the player does not open automatically, open it in Windows 2000 by clicking Start, Programs, Accessories, Entertainment, and Windows Media Player. In Windows XP, you can open it by clicking Start, pointing to All Programs, and then clicking Windows Media Player.

2. Select a music track to play and then listen to it as played from the CD.

3. When you have listened to the selection, click Stop, and then click the Copy From CD selection on the left side of the Media Player screen.

4. When the selections on the CD appear, check only the one you want to copy to the left of its title. All the other check boxes should be blank. If they are not, click them off.

5. Click the Copy Music button at the top of the Media Player screen.

6. Click OK to leave the copy protection on if the work is copyrighted (as almost all are). If it is not, click the Add Copy Protection button and follow the screen instructions to continue making the copy.

7. The copy is recorded on the computer in the default directory listed at the bottom of the Media Player screen.

8. When the file has completed loading, close the Media Player and remove the CD from the player.

9. Click the Start button, and then click My Music or go to the alternate directory where you saved the music file.

10. Right-click the file and choose to play the music.

11. Listen to the music again. Can you hear any difference in the quality of the music?

Project 6-4: Explore Available Media on the Internet

In this project, you will learn media types.

1. If you are using Windows XP, click the Start button, point to All Programs, and then click Windows Media Player. For Windows 2000, click Start, Programs, Accessories, Entertainment, and then Windows Media Player.

2. When the player opens, click Guide on the top bar of the window.

3. Explore the video downloads available at this site. Select and download one of the video clips and play it. How does its quality compare with the 5MB file you downloaded in Project 6-2?

4. Select one of the movie Web sites listed in the media guide and visit it. Download the highest-quality movie trailer you can find. If you are using a computer with a 56K modem, limit your search to files that are 5MB or less. If you have a high-speed connection, you can download larger files.

5. Play the large high-quality video on your computer. Does it play in Windows Media Player or QuickTime? If not, what player is required? How can you get it? What is the cost?

6. How does this video compare to the video you downloaded in Project 6-1?

Project 6-5: Record Digital Television from a Broadcast Source

For this project, you will need access to a television set capable of receiving digital television signals, either through an antenna or via satellite or cable connection, and a DVD recorder/player that can be connected to the television. You will also need a blank recordable DVD.

1. Connect the DVD to the television so it can record a program playing on the television. Plug one end of two RCA cables into the TV's video out and audio out jacks. (These are usually found on the back of the TV, but may be on the front in some models.)

2. Plug the video out cable into the DVD recorder's video in jack. Plug the audio out cable into the DVD recorder's audio in jack.

3. Tune the television set to a channel that is broadcasting in your area.

4. Place a blank recordable videocassette in the VCR and begin recording the program that is playing on the television. Press the record button on the VCR to begin recording. (Some VCRs require you to press the Record and the Play button together.)

5. After you have recorded at least 5 minutes of programming (you can record a full program if you want to), stop the DVD recorder.

6. Change the cables connecting the TV and DVD recorder. Connect one cable to the DVD recorder's video out jack and the TV's video in Jack. Connect the other cable to the DVD recorder's audio out jack and the TV's audio in jack.

7. Set the TV's tuner to channel 3 or the input channel, if it has one. Play the recorded material on the DVD through the TV. Observe how it appears. Is the recording as clear and well focused as the broadcast program?

8. Switch back and forth between the recorded program and a broadcast program. See if you can detect differences in quality between them. DVDs record

digital video at very high quality so there should be little or no difference between the broadcast television picture and that produced from the recorded DVD.

9. Play a commercially recorded DVD in the DVD player and compare its quality to the recording you made. Do you see a difference? The commercial DVD was recorded from a hardwired connection to the original camera image or video recording, so it does not have any of the defects and may appear slightly higher in quality than your recording. If it is significantly better, you should check the DVD recorder for problems.

Project 6-6: Compare the Monaural and Stereo Playing of a Music Selection

For this project, you will need access to a stereo audio system set up to play with left and right speakers. The audio receiver must have a monaural select button on it so it can be set to play in single track format. You will also need a stereo CD recording that can be played on the audio system.

1. Place the audio recording into the player and turn the system on.

2. Start the recording and play a selection through in stereo format. Listen for the balance of the instruments or voices on the recording. You should be able to discern them separately as if each were standing in a different location in front of you and playing at that spot. The more widely spaced the speakers of the audio system are, the more widely spaced the players of the music should sound.

3. Switch the stereo audio system to monaural. This causes the system to play both tracks of music through both speakers, effectively combining the two tracks into only one.

4. Play the same selection of music again. Does it sound the same? To which version would you prefer to listen? Can you describe why the stereo version sounds better?

5. Switch the audio system back to stereo and play the music selection a third time. For this playing, keep your hand tightly pressed over one of your ears. Does the stereo playing sound better than the monaural with only one ear to listen? Why?

Case Projects

Project 6-1: Research a Service Provider for HDTV

A client has asked you to recommend a service provider for HDTV service as well as SDTV with as wide a selection of channels as can be obtained. He lives in your area. Find out what television services are available in your area and how the client can best obtain the level of service he wants. Be sure to check which channels are actually broadcasting HDTV programming in the area rather than just which services offer HDTV. Do not overlook pay-per-view channels, but be sure to note the costs of these services. Write a short summary of what is available in the local area and how much it will cost monthly for the service.

Case Project 6-2: Determine When Local Television Stations Will Go Digital

Visit the Web sites of the television stations in your area. Find out which of them have commenced digital broadcasting and which are planning to do so. Note what the startup dates are for each station that has not yet begun digital broadcasting and what programming hours they plan to broadcast in SDTV format and HDTV format. If the Web sites do not have this information, contact the stations directly and see if they have made any announcements or have a plan in place. Write down your findings and give your opinion about whether you feel the purchase of an SDTV set or an HDTV set is justified by the amount of digital programming in each format that is (or will be) available in your area on broadcast stations.

Case Project 6-3: Test a Network for Streaming Video Files

Visit a local supplier of home LAN hardware in one or more technologies. Ask if the vendor can demonstrate the ability of the technology being offered to stream video from one node of a LAN to another. If the vendor has no demonstration network set up, ask if he knows of anyone who has a network using the technology that is streaming video files over it. Try to find at least one technology that can successfully demonstrate video streaming in a working LAN using a home LAN technology.

Case Project 6-4: Compare a Large-Screen CRT SDTV Set to an LCD SDTV of Similar Size

Visit an electronics dealer who handles large-screen entertainment centers. Get information on SDTV sets, both CRT type and LCD flat screens, including picture quality, viewing angles, length of service, and cost. Write a short report comparing the prices and features of each type, their sizes, and the quality of images they provide. Note any special requirements they may have that might influence a buyer, such as large space requirements, room darkening, and so on. Give an opinion on which system you would recommend to a client if price was not a consideration, and tell why you chose it.

Audio and Video Installation and Setup

After studying this chapter, you should be able to: **OBJECTIVES**

- Identify the major components of audio and video systems and describe the function of each.

- Install and set up an audio or video system as part of a home network.

- Configure an audio or video system for receiving and displaying external programming.

- Configure an audio or video system so it can receive and display internal and streamed programming.

- Perform preventive maintenance for audio and video systems.

Introduction

In Chapter 6, you learned how analog and digital audio and video programs are created, broadcast, received, and displayed. In this chapter, you will discover the specific components that make up audio and video systems. You will also learn how to install these components into working systems, and how to configure the systems for peak performance, both for displaying external programs received from commercial providers and for receiving streamed audio and video programs from the Internet or from the home LAN. Finally, you will find out how to maintain audio and video systems to prevent high-repair costs and preserve performance standards.

Audio and Video System Components

The audio and video systems we enjoy today are the culmination of many years of development. They are still evolving as new and more advanced technologies provide a means of still higher-quality audio and video programming than what has been available previously. As new technologies are adopted, older ones fall into disuse and are gradually abandoned, but this is a slow process. The investment people have made in their existing entertainment systems, plus the often higher cost of purchasing a newer technology, makes many content to stay with their tried-and-true systems, as long as they continue to function normally. Hence, millions of people are still listening to audio cassette tapes and analog radio broadcasts many years after CDs and digital broadcasting have raised quality standards tremendously. Millions more are still watching analog NTSC television and VHS cassettes when HDTV and **DVD** movies could greatly enhance their viewing experience.

This tendency to retain old technology is increased by the fact that neither the equipment nor the storage **media** of obsolete systems are compatible with newer ones. A record or tape collection cannot move up to CD audio quality, even if the music recorded on them is rerecorded to the newer media. The same limitation applies to VHS cassettes: They cannot be turned into DVDs with anywhere near the level of quality of the newer medium. Each time a new technology is adopted, a new media collection must begin for the consumer. Figure 7-1 shows some of the development of video and audio media from early forms (at the bottom part of the picture) to current forms (at the top part of the picture). None of these audio and video media are forward-compatible.

FIGURE 7-1 Some Outdated and Current Audio and Video Media

This part of the chapter discusses components of audio and video systems, some of which are older than the current digital technology, because many home LAN users want to include some of these devices in their home entertainment systems. Most home users probably also want to continue using their existing entertainment collections until they acquire new versions in the current media. Retaining older equipment as a subset of a new audio or video system costs very little and allows the transition to CDs and DVDs to be more gradual and easier on the budget.

Audio System Components

Nearly all audio broadcasting and recording, except AM radio, is done in **stereo** or multiple sound tracks. Stereo recording creates two separate soundtracks, one recorded from the left side of the performer and the other from the right side. Stereo recording is accomplished with multiple microphones placed around the performers being recorded so as to record sound from multiple angles as it would actually be heard in the room. The sound is recorded on multiple tracks that are combined and broadcast as a stereo signal. The stereo tracks are **synchronized** together when recorded and played back. All current audio equipment plays in stereo and many systems are capable of playing four (quadraphonic) or more tracks.

All the components described next function in stereo audio systems, although some are **monaural** (single track) devices. To maintain high quality in an audio system, the stereo source signal is divided into its separate tracks when it is output from the **tuner** or player. From that point on, the two (or more) tracks of sound pass through separate **pre-amplifiers, amplifiers,** and speakers, all of which are monaural devices. For the highest quality sound reproduction, it is important that the two separate tracks be configured and wired identically so synchronization of the separate stereo tracks is not lost.

Surround sound (Dolby Digital 5.1 or higher, DTS) is recorded and transmitted as six individual tracks (five regular speakers plus a subwoofer) and requires six speakers to be fully functional. Surround sound is mainly used for HDTV, either in an entertainment center or home theater, and can play stereo or even monaural audio recordings or broadcasts with equal or better quality than a stereo system. A surround sound system can take the place of a stereo audio system and uses the same basic components as a stereo system, although in expanded form. If a surround sound system is in the home, a separate stereo system is not needed unless the user wants to listen to the stereo while a television program or movie is playing in the home theater.

Audio programs are brought into the home through commercial digital and analog radio stations, satellite broadcasts, cable channels, and the Internet. All of these audio signals can be converted back into audible sound by a set of the same basic components (vastly improved) that were used to receive early radio broadcasts. By adding player components to these, music and voice programs stored on CDs, cassette tapes, and records can also be played over the same audio system.

An analog radio receiver requires only four basic components to function: an antenna, a tuner, a speaker, and a ground. A digital receiver requires these plus a fifth component, a **converter** (or **decoder**), which renders digital data back into audio waves. Everything else in the modern radio set are refinements. When you think of a radio receiver, you usually think of an integrated device that contains all of these components. That is an accurate view of most radios, but in high-quality

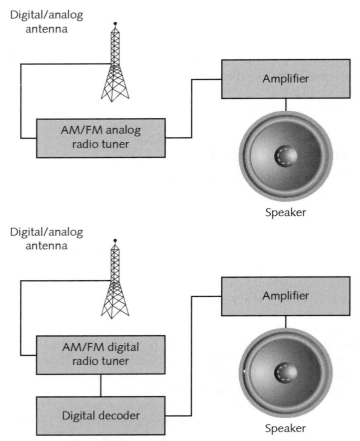

FIGURE 7-2   Basic Digital and Analog Radio Components

sound reproduction systems, the major components are almost always separated into individual units, often produced by different, highly specialized manufacturers. These systems usually include additional components that supplement the basic ones found in the radio receiver, or provide additional sound quality, volume, or distribution.

Figure 7-2 shows the basic configuration of an analog radio receiver at the top and a digital radio receiver at the bottom. The main difference between them is the decoder that the digital radio requires to convert its numerical signal into an analog format that can be amplified and sent to the speaker for playing. These components are essential for a radio, whether they are housed in a single box or widely spaced from one another.

Antennas

Most AM and FM radios, digital and analog, have a built-in **antenna** that is part of the radio itself. They do not need an outside antenna connected to the radio by a wire. In rural areas distant from broadcast transmitters, an external antenna, sometimes called an aerial, can improve radio reception by bringing in a stronger signal.

For digital radios, an outside antenna may make reception possible at distances greater than would otherwise be possible (due to the cliff effect of digital signals). If received, the quality of the digital signal will be fully equal to that received by any other user. The digital signal has to be received perfectly or within the error correction capability of the receiver. If it is not within that tolerance it will not be received and decoded at all by the digital decoder. Because an outside antenna brings in a stronger signal, a digital decoder may be able to read the enhanced signal and play the radio station even though it would be unable to do so using only its built-in antenna.

For analog radio, a stronger signal from an external antenna requires less amplification and therefore contains less noise and interference. This means better quality sound and a better signal-to-noise ratio for the system.

When an antenna is used for an audio system, it should be mounted on a roof or other structure as high as possible in order to have the best chance of picking up **attenuated** signals. It should be connected to the receiver by a shielded wire (in order to prevent noise and interference from corrupting the signal on its way from the antenna to the receiver) and it must be grounded, both for safety from lightning and other electrical phenomena and to complete the signal-carrying circuit to the receiver, which is also grounded.

Tuners

A radio tuner is a device that enables the radio to receive only the selected frequency of the station that the user wants to hear and ignore all others. The tuner is variable so that all stations can be received, one at a time, by simply tuning to the correct frequency for each.

The quality of a tuner is measured by how precisely it can tune to a selected frequency and how completely it can block all other frequencies, thereby reducing noise and interference. Two criteria are used to determine how good a tuner is: its sensitivity and its selectivity.

Sensitivity refers to how well the tuner can pick up radio signals and is measured in decibel femtowatts (dBf). dBf is a measure of the strength of an incoming signal. The fewer dBf the signal has, the weaker it is, and the better quality a tuner must be to receive it. Hence, the lower the dBf rating of a tuner (good ones are rated 11 dBf or less), the higher its quality.

Selectivity is a measure of how clearly and precisely a tuner can receive one frequency while avoiding interference from others. In other words, how wide or narrow is the selectivity of the tuner? In fact, a good tuner will be both wide and narrow with a switch to let you choose which of these features you want. Wide selectivity lets the user tune in a close or powerful station clearly without **distortion** of the signal. Narrow selectivity lets the user tune in a weak or distant station precisely without interference from other nearby frequencies. It is a good idea to look for adjustable selectivity when buying a tuner.

Analog tuners process the continuous signal they receive into a continuous output wave that is sent to the amplifier to be strengthened, and then to the speakers. Digital tuners receive the frequency they are tuned to and send the received signal to a decoder (which may be built in to the tuner) that converts the digital sample data of an audio program (as discussed in Chapter 6) into a very close approximation of the original continuous wave signal and then sends that audio signal to the amplifier. If the digital audio signal is in the format used for CDs or another very rapid sampling technique that uses very little compression, the sound quality is actually better than a comparable analog signal and has less noise. If the digital signal uses a less frequently sampled format, or is more compressed so that some of the data is lost, the reconstituted audio sound is lower in quality, but still equal to almost any analog signal.

Tuners are often combined with an amplifier in a radio, but stereo audio systems usually have separate amplifiers (although the tuner in a stereo system may have a preamplifier as well). The amplifier increases the power output of the signal received from the broadcast station or other audio source so it can drive speakers at a higher volume. Strictly speaking, an amplifier is not a required component of a radio, but without it, the output volume of the speaker is barely audible to the listener.

Amplifiers

Amplifiers come in many sizes and qualities, but only in two basic categories: pre-amplifiers (pre-amps) and power amplifiers (amps). Do not confuse the shortened name *amp* with the shortened term for amperes. They are not the same. In fact, they have nothing to do with one another except a common name.

Pre-amplifiers amplify the source signal coming from one of the audio system's tuners or players. A pre-amplifier typically receives a source signal of perhaps 5 millivolts (5/1000 of a volt) from the tuner or player. It boosts (increases) this signal strength several thousand times to a strength of up to 5 volts and sends the strengthened stereo signal on to the power amplifier.

In digital audio systems, a pre-amplifier, if one is used, is placed between the tuner and the decoder. Unlike an analog pre-amp, which simply raises the power level of the signal it receives without altering the signal's characteristics at all, a digital preamplifier may actually be a repeater similar to those discussed in Chapter 2. The repeater pre-amp receives the digital signal from the tuner and regenerates it at a higher power level. Because it is generating a new signal that exactly duplicates the one it receives, the pre-amp can produce a completely clean digital audio output to send to the decoder. This process eliminates any possible loss of data or corruption in the signal and allows the decoder to generate a perfect stereo analog output to send to the power amplifier.

The power amplifier increases the output strength of the analog signal several times in order to drive the speakers that actually produce the sound in an audio system. The size of power amplifiers is measured by the output wattage of the signal coming from the amplifier. This power is what drives the speaker or speakers and it can range from only 20 or 30 watts of output to several hundred watts or even more in amplifiers designed to power huge speaker systems. More power is required to produce low notes on the sound scale than high ones, so much of any amplifier's output power is only used when it is producing low-frequency sounds. These low-frequency notes are what give reproduced music from a radio or recorded source the full tonal qualities that otherwise can be heard only in a live performance. High-quality sound reproduction requires these sounds and, hence, powerful amplifiers are usually a part of any good sound system.

The quality of both pre-amplifiers and power amplifiers is measured by how cleanly they amplify sound frequencies without adding any noise or interference and without distortion in their output signal so it sounds different from the original input signal. Signal-to-noise ratio is as important for an amplifier as for a tuner. Both must produce clean output signals in order to generate good sound quality.

Speakers

Speakers are the audio system components that render electronic signals audible to the ear. They are electromagnetic devices through which a varying frequency current is passed. Speakers convert this electrical energy into sound waves. Figure 7-3 shows a diagram of a speaker and its internal working parts.

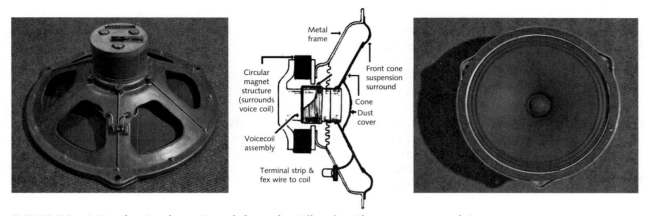

Metal frame
Circular magnet structure (surrounds voice coil)
Front cone suspension surround
Cone
Dust cover
Voicecoil assembly
Terminal strip & fex wire to coil

FIGURE 7-3   A Speaker Produces Sound through a Vibrating Electromagnet and Cone

Electrical current from an amplifier travels through the speaker's flex wire to the voice coil. Current flowing through the voice coil turns it into an **electromagnet.** The magnetic field created by the voice coil causes the coil to be attracted to, or repelled from, one pole of the magnet structure, depending on whether the current in the voice coil is positive or negative.

As the current in the coil varies in both positive and negative directions in a pattern analogous to the original sound wave signal, the voice coil acts like a piston, moving rapidly back and forth as it is attracted to, or repelled from, the magnet structure. Since the voice coil is connected to the **cone,** they both move as a single unit. The cone vibrates in the air, alternately compressing and evacuating the air immediately around it, thus producing sound. The vibrating cone produces sound waves that match those originally converted into an electric signal when the sound was recorded.

As the preceding description suggests, speakers are extremely inefficient. They convert only 1 or 2% of the power fed into them into sound. All the rest of the energy is lost as heat. This heat radiates from the voice coil and circuits inside the speaker. If a speaker's voice coil overheats, the speaker is destroyed by the distortion caused from intense vibration combined with the heat. The speaker is then said to be "blown," a condition often produced by driving it at too high a volume.

The output sound of good-quality speakers closely matches the frequencies of the current fed into them without distortion or addition of noise or hum (a low-frequency sound similar to tape hiss, but caused by something different). Although the human ear can perceive sounds at frequencies ranging from 20 to 20,000 cycles per second (cps) or Hertz (Hz), no single speaker is capable of responding accurately to that broad a range. For this reason, several speakers (each responsive to a range of frequencies less than the range of hearing) are nearly always connected to an amplifier through a **crossover network** of wires and electronic circuitry in high-fidelity sound systems.

The crossover network divides the sound signal coming from the amplifier so that the high-frequency portion is sent to speakers designed for high-frequency response (these speakers are called "tweeters" for obvious reasons), the middle frequencies are sent to mid-range speakers that are responsive to them, and the lower frequencies are sent to low-frequency responsive speakers (called "woofers," again for obvious reasons). Stereo sound systems require two crossover networks and two sets of speakers. Surround sound systems require three connected to speakers in front of the listener, plus two separate connections to the speakers behind the listener and one to the sub-woofer.

The power of the sound signal sent to a speaker or group of speakers must not exceed the capacity of the magnetic circuit of the speaker to carry it. If it does, the speaker overheats and eventually burns out. When a speaker is overdriven (too much power is sent to it), its sound output also distorts because the speaker cone cannot vibrate far enough and fast enough to utilize all the power available to it. When this happens, the cone simply stops each vibration at the limit of its movement capability and the resulting sound has its peaks suppressed.

Small single-speaker audio systems have electronic circuits built into them that limit the power output and the frequency response of the tuner and amplifier. The single speaker may, for example, have response limits of 60 to 10,000Hz and a maximum of 10 watts input power. This means it cannot move fast enough or slow enough to reproduce frequencies above or below its limits, and it overheats if more than 10 watts of power is applied to it. The output signal to the speaker is limited by the system's electronics so its maximums do not exceed the capacity of the speaker. The frequencies of sound below 60 and above 10,000 are simply cut off and not reproduced by the speaker at all. At average voice volume, less than one person in every hundred can hear any difference between a sound track that includes only the mid-range (60 to 10,000Hz) frequencies and one that covers the whole hearing range.

Equalizers

One other nonessential, but desirable, component that is often found in audio systems is an **equalizer.** This device allows the user to increase or decrease the volume of various bands of sound frequencies in the output of the system. This is valuable in adjusting sound for maximum listening pleasure according to the acoustics (sound reflection and absorption qualities) of the particular room where the sound system is located.

An equalizer divides the range of audible sounds into three or more bands of frequencies and allows the user to adjust the volume level of each band independent of all the others. If high frequencies tend to be absorbed by the furniture and carpet in the room, the level can be increased to compensate for the loss. If the soprano's high notes threaten to break the windows, they can be reduced to a more pleasant level.

Many types of equalizer circuits exist, and digital equalizers operate differently than analog, but all are intended to allow the listener to adjust the sound output to his or her preferred taste. Equalizers are used extensively when recording voice and music media in order to produce a recorded sound that is as close as possible to what the performer intended. For most listeners, the artist's rendition may be perfect, but if it is not, the listener also has the option to change it.

Audio Players

Audio that does not come into the home through broadcast signals, cable channels, or the Internet almost always arrives stored on some form of media: CD, cassette tape, or record, with perhaps an occasional flash drive or zip disk on which is recorded a rare piece that a friend has, or a personal performance. All of these media can be played on the same audio system that plays radio broadcasts, provided a player unit is substituted for the antenna and tuner that receive the broadcast signals.

CD Players

CD players are the main player component of home audio systems, and a CD-ROM drive, which can also play CDs, is part of nearly all new computers. Many CD-ROM drives are also **CD burners**—that is, they can record CDs as well as play them. CD drives that can record CDs (designated CDR-R for recordable format or CD-RW for rewritable format) enable users to record voice or music from the Internet, from tapes or cassettes or from records. These recorded CDs can be played over a home LAN from the computer's CD drive or played on the audio system's CD player.

When choosing a CD player, the main consideration should be its versatility in input and output. If the CD player will output to a stereo amplifier, it needs only analog output connections. If it will output to a home theater A/V receiver, digital output may be preferable and the CD player should have digital output (either optical or coaxial cable).

The CD should be able to play not only commercial CDs, but also CD-Rs and CD-RWs, as well as WMA and MP3 files recorded on a CD. If you are into very high-quality music and surround sound for audio other than movies, you may want a CD player that can handle the two new formats for CDs: Super Audio Compact Disc (SACD) and DVD-Audio Disc. Both of these have higher sampling rates for digital sound than the standard CD format and both can record 5.1 channel surround sound as well as traditional stereo sound on the same disc. CD players that can handle all these formats are known as universal players. They can connect digitally or analog to an A/V receiver, but may cause problems connecting other devices to the receiver because they require so many input ports.

If you want to record CDs from other sources in your stereo system, the CD player should have this capability. Other features that are useful on a CD player include multidisc capability (players are available that can hold from 5 to 100 CDs or more) and remote control.

Computers provide the most convenient platform for creating CDs, not only to record **digital audio files** from the Internet and other sources, but also to encode analog music selections and standard **MIDI (Musical Instrument Digital Interface)** files. Any computer with a sound card is capable of converting analog sound to digital using a microphone or tape recorder as the analog input device. The frequency response of most computer sound systems is not very high, but they can record both voice and music adequately for many uses.

CD player/recorders with high-quality **encoders** are available for converting analog music to digital format for storage on CD. An encoder allows the user to convert both records and audio tapes to digital format. CD player/recorders with digital audio encoders are more expensive than simple players, but for serious music enthusiasts or people with large record or cassette tape collections that they want to convert to digital format, they are a good investment. CDs are more durable than tapes and records, and also produce higher-quality sound than either of the older storage mediums. But remember, CDs recorded from tape or record sources cannot improve the quality of the original sound source. The frequency response limits and the noise level of the original will still be present in the digital recording made from it.

Many DVD players, those in computers and those that are part of a home entertainment system, are capable of playing CDs as well, but only one or the other at a time. If you want music played in parts of the home while the DVD player is playing a movie on the television, a separate CD player is a necessity. A CD drive in a home network computer can play or send audio files to any other node. It can also play to an audio system, provided that system is also connected to the network.

Audio Cassette Players

Very few new music recordings are being made on audio cassettes, but many voice recordings (books, educational programs) still use this format. Millions, if not billions of audio cassettes, which could be converted to digital format and played on a digital audio system, still languish in private collections around the country. Either of these continuing uses provides a strong enough reason for anyone who already owns a cassette player to keep it connected to their audio system. New cassette players can be bought starting at $50 and that price may justify adding one to a system, if the owner wants to convert any significant quantity of cassettes.

A digital encoder is required to convert cassettes to digital format. This can be either in an CD player/recorder that the cassette player is connected to as an input device, or in a computer. If the encoder is in a computer node on the network, the cassette player can also be connected to it as an input device, but it cannot be connected through the network as a node on the LAN. Analog sound from the cassette player cannot travel on the network, which carries digital data only, but must be fed directly into the computer's sound card input and converted to digital form. It can then be recorded on a hard drive or on a CD, or transmitted over the network to other nodes.

Phonograph Record Players

A phonograph (also called a record player or turntable) is now an antique. The main reason to have one in a current audio system is to play records featuring music that cannot be obtained on other media, or to play records so their content can be converted to digital format and stored on new media. If conversion is the objective, locate the record player near the encoding computer so its output can feed through an amplifier into the computer sound card input. If playing the records is still the principal pleasure, locate the player with the audio system, and connect it to the audio amplifier. If a digital receiver with analog-to-digital encoding capability is part of the audio system, the record player's output can be digitized like any other analog signal and played or recorded on digital equipment.

Audio/Visual Receivers (A/V Receivers)

A/V receivers are really more video than audio devices—they are the main component of a home theater system, but they also have audio functions. If you have an A/V receiver, you may not need some of the other audio devices noted earlier.

A/V receivers can perform all of the following audio system functions:

- Receive digital and analog radio broadcasts. Some A/V receivers can also receive satellite radio, but the same paid subscription is required as for a stand-alone satellite receiver.
- Decode surround sound audio on six channels, either from an HDTV source or from a DVD.

- Amplify audio signal power to drive surround sound speaker systems.
- Connect audio (and video) input and output hardware.

An A/V receiver also has additional video functions that we will discuss later in this chapter, but its audio capabilities are considerable in and of itself. With an A/V receiver, you will not need a radio receiver (analog or digital with decoder), a pre-amplifier, or a separate amplifier for your audio system. You also will not need a separate decoder for digital audio input for HDTV, CD player, or DVD player. The A/V receiver has a decoder for these.

An A/V receiver is the heart of a home theater system and more than half of a good stereo audio system at the same time. It is the most cost-efficient way of getting a good audio system in your home, but as with all combination function equipment (such as the combo DVD and CD player) its limitation is that it can perform only one of its functions at a time. If you are watching a movie in the theater, you cannot listen to the radio in another room or play a CD in a third. For this reason, you may want to keep your stereo system separate from your home theater by getting independent components for it.

Television System Components

Like audio systems, television systems require only a few basic components to function. Some of these components are very similar to those used in audio systems. Others are much more complex because they must process and display more complex signals than those required for sound reproduction.

The sound portion of television signals, whether analog or digital, are broadcast as radio signals, and television sets have a separate radio tuner, amplifier, and speaker to reproduce sound along with the picture.

Antennas

Analog television sets usually have antennas built into them, but these antennas are not as efficient in picking up a television signal as set-top models or roof antennas. Because the quality of analog television reception is directly related to the strength of the incoming signal received, an efficient antenna, correctly adjusted for maximum signal strength, can make a very significant difference in the clarity and interference level of the picture.

Digital television sets, the only kind useable for broadcast reception after February 2009, also need efficient antennas to function well. Because of the cliff effect of digital television broadcasts, a good antenna may make the difference between receiving a near perfect digital signal, and receiving none at all.

Television antenna reception can be affected by the electrical fields that exist in city areas where consumption of power is high. It can also be affected by weakening of the broadcast signal due to obstructions such as densely concentrated buildings, especially those with large amounts of concrete or metal in their construction, land masses, and foliage. RF signals and other television signals reflected from the ground or from layers in the atmosphere may also affect broadcast television signals.

Interference in an analog television signal produces all kinds of picture defects in the displayed image, including "ghosting" (multiple shadows of the image), "snow" (moving white flecks in the image), and fine lines running through the screen horizontally, vertically, or at an angle. The image may also be distorted or "roll" because the television set cannot interpret the stabilizing portions of the signal.

None of the analog interference problems appear in digital television, even if interference is present in the digital signal. That is one of the great advantages of switching to digital television reception. The other is the much higher resolution and picture clarity of digital compared to analog television. The picture quality of digital television always appears nearly perfect (that is, an exact reproduction of the image that was broadcast). Interference in digital signals is corrected by algorithms built into the digital decoder so the effect of missing or corrupted data is not evident on the screen. Only when the interference is so strong that it overwhelms the decoder's ability to compensate for it does the effect become evident on screen. The effect is simply no picture on the screen. Until the decoder can receive enough data to produce a complete picture, it displays none at all. As soon as the interference is corrected sufficiently, the picture reappears at full quality.

This "all or nothing" characteristic of digital television happens because digital television signals are highly compressed as they are encoded. Nearly all of the digital data in the signal must be received in order for the signal to be decoded and displayed. If enough of the data is lost in transmission, the loss does not degrade the picture quality—it simply prevents the decoder from converting the signal into an actual picture. This is the "cliff effect" of digital transmission, which was discussed in Chapter 6. The abrupt cutoff of digital transmission can be caused by interference, as well as distance from the transmitter. Adjusting a digital antenna, or moving it to a better location may be all that is needed to obtain reception. Antennas should be kept away from tall buildings, metal utility towers, high-voltage power lines, and water towers. Smaller versions of these objects, such as wooden utility poles, other antennas, and local power lines rarely cause significant interference.

Satellite or dish antennas can suffer from a weak signal and interference, but not the same type as broadcast antennas. Because these antennas are aimed directly at the single transmission source, the geosynchronous-orbiting satellite, the only interference that affects their reception is that which comes from the same direction as the satellite to which they are pointed. Interference from the sky is almost entirely caused by electrical storms in the earth's atmosphere and electromagnetic storms on the sun's surface, which produce corresponding disturbances in the Earth's atmosphere that can disrupt satellite broadcasts. In addition, the relatively weak signal strength of satellite broadcasts can be attenuated by heavy rain or snow falling between the satellite and the dish antenna. A blizzard or a summer downpour can temporarily block satellite TV reception.

Television Tuners

We noted in Chapter 6 that the millions of television tuners now in use are analog receivers that cannot process digital signals. Although the ratio of analog to digital sets is changing rapidly as more digital broadcasting becomes available and DVDs take the place of analog VHS cassettes, the government estimates that about 40 million analog sets will still be in use without digital tuners (also called decoders or converters) on February 17, 2009, when all analog television broadcasts cease in the U.S. All of these sets will need to have digital tuners attached to them if they are to receive digital television broadcasts. It will be up to each set's owner to decide whether to get a tuner for an existing set (and continue receiving analog quality television on the set) or buy a new digital television that can receive digital signals and display their higher-quality image on the screen.

The descriptions for digital television formats put out by the **Advanced Television Systems Committee (ATSC),** the television industry and government people who determined the standards for the new digital system, now clearly differentiate between a digital television set and a digital tuner or receiver.

SDTV is the minimum government-approved standard for digital broadcasting and HDTV is the premium digital television standard. In between the high and low standards are **enhanced-definition television (EDTV)** and several other versions of SDTV that are above the minimum, plus one version of HDTV that is below the premium standard.

An SDTV television is a fully integrated television receiver that receives all ATSC terrestrial digital transmissions and decodes all ATSC Table 3 video formats to produce a viewable picture on its screen. No aspect ratio is specified for SDTV, but manufacturers have adopted mainly a 4:3 or 16:9 aspect ratio. An SDTV set must also receive a useable audio signal, but it does not have to be stereo.

SDTV television sets cannot display HDTV images at full quality. As we noted in Chapter 6, the minimum standard picture standard for SDTV is a 4:3 aspect ratio with 480i lines height and 720 pixels width. The least expensive SDTV sets may have this minimum picture quality with monaural digital sound and no more, and still be rated as meeting the SDTV standard. Other (more expensive) SDTV sets may have higher picture quality but not as high as the HDTV standard. They can have an aspect ration of 4:3 or 16:9 and a height of up to 720i with a width of up to 1280 pixels. Any set that falls within these parameters is within the government standard for SDTV.

An **SDTV tuner** (as opposed to a complete SDTV television set) is defined as an RF receiver that receives ATSC terrestrial (transmitted from the ground) digital transmissions and decodes all ATSC Table 3 video formats. Its output can be in the form of an NTSC signal so that it can be played on an analog TV, but all such tuners also output an SDTV signal in a 704 by 525 pixel format for display on a digital SDTV.

The SDTV tuner is the standard digital decoder that can be attached to an analog NTSC television set so it can display digital television. If the tuner is limited to the minimum SDTV standard, then the consumer cannot upgrade to a higher-quality digital television by purchasing a higher-resolution monitor. Even if the monitor is capable of HDTV picture resolution, its display will be limited to the quality of the data output by the SDTV tuner.

SDTV tuners will likely be the standard hardware that analog television owners who do not upgrade to a digital television set will purchase. These tuners will bridge users from analog to the first digital television generation because they can play to NTSC televisions or monitors, but also play to digital monitors that have a 4:3 aspect ratio. A digital television system based on this standard should also be able to play downloaded video from computers and the Internet with a suitable decoder program.

SDTV sets that have capabilities above the SDTV minimum discussed earlier will usually fall into the Enhanced Digital Television (EDTV) category. To meet this standard, a digital television must display a picture in 4:3 format with 480p height and 720 pixels width. The progressive picture display gives EDTV a higher picture quality than the minimum SDTV standard, but the two formats are otherwise very similar. Like SDTV, EDTV can also have a 16:9 aspect ration and picture resolution up to 720i by 1280 pixels wide.

HDTV tuners and televisions, with their 16:9 aspect ratio and high resolution, are more difficult to connect to a home network, either to send video to a computer or to receive a video program from the network. HDTV tuners receive a

TABLE 7-1
Television Formats

Standard	Aspect Ratio	Width (pixels)	Height (scan lines)	Scan Sequence
NTSC	4:3	720	525	Interlaced
SDTV (minimum)	4:3	704	480	Interlaced
SDTV (maximum)	16:9	1280	720	Interlaced
EDTV (minimum)	4:3	704	480	Progressive
EDTV (maximum)	16:9	1280	720	Interlaced
HDTV	16:9	1280	720	Progressive
HDTV	16:9	1920	1080	Interlaced
HDTV	16:9	1920	1080	Progressive

signal that produces a picture with resolution of at least 720p by 1280 pixels wide. This is just above the resolution of the best SDTVs. Other HDTV formats are 1080i by 1920 pixels wide and 1080p by 1920 pixels wide. There is, as yet, little television and few DVDs available in the latter format, but it will become much more popular, if not the dominant standard, in the near future.

When selecting a digital tuner, SDTV or HDTV, it is important to differentiate between the type of signal a tuner can accept for decoding and the quality of the signal it sends after decoding it to the display screen. Some SDTV tuners can accept all SDTV signal formats (up to 720i by 1280 pixels wide) as an input signal, but their output is limited to NTSC analog, or a 480i by 720 pixels wide 4:3 digital display. Similarly, some HDTV tuners can accept 1080i or even 1080p input, but their output is only 720p.

Television sets with built-in tuners usually match the tuner's output to the highest resolution the screen can display. If the set has a 1080 screen display, the tuner will be able to make full use of that resolution with 1080i output (though perhaps not 1080p). If you buy a television display and tuner as separate units, be sure you get the highest resolution the display is capable of in the tuner's output signal as well. Table 7-1 shows television formats and their characteristics.

Television Displays

Television displays come in the same three types used for computer screen displays: cathode ray tubes (CRTs), flat panel displays, and LCD projection systems. The technology of each of these was discussed in Chapter 2. One additional technology for television displays is now competing with these three: plasma screens.

Plasma screens are flat and thin like LCD screens. They are similar to LCD screens in that their surface consists of an array of pixels set in rows and columns that can be activated individually. In a plasma screen, the pixels are made of gas-filled cells sandwiched between two panes of glass. An electrical grid behind the cells activates the cells with an electric current that ionizes the gas in the cells into a plasma form (hence the name of the technology). The ionized gas causes a phosphorous layer on the outer glass of the sandwich to light up, in much the same way as a fluorescent tube lights up when the gas inside it is ionized by an electric current.

Variations in the current to each cell determine the color of the phosphor in front of the cells and the result is a smooth color picture with very even color distribution, no picture distortion, high picture brightness, and good resolution. Unlike LCD flat screens that often have a limited angle for viewing (you must be almost directly in front of the screen to see it clearly), plasma screens have a wide viewing angle much like a CRT display.

Most of the disadvantages of plasma screens are being overcome as new display models are introduced. Plasma screens are not as high in resolution (usually 720p) as LCDs and they have somewhat poorer color reproduction, especially for reds (which look more orange) and black (which looks dark gray). They also have a shorter lifespan than LCDs and are susceptible to burn-in on the screen if static scenes remain on the screen for long periods of time. The resolution and color problems are declining, although high-resolution plasma screens are more expensive, and the lifespan is being improved, but the burn-in feature is part of the basic technology (as it is in CRTs) and cannot be overcome.

CRT television displays differ from LCD and plasma television and computer screens in two important particulars: They create their images in **rasters** (scan lines) rather than pixels (square or rectangular picture elements) and they have a somewhat lower resolution than the digital displays. Because the signal that paints the picture in a CRT is continuous, the raster lines of a CRT TV screen are also continuous, rather than being divided into individual pixels set in rows across the screen, as LCD and plasma screens are. The color and brightness of each raster varies continuously along its full length, in contrast to pixels, each of which has only one color and one brightness level set by its digital designation.

Until recently, CRT monitors and televisions were only available in 3:4 aspect ratio, but some 16:9 models have recently been introduced. The wider screen of the 16:9 format requires a greater angle of the electron beam inside the CRT to reach the outer edges of the screen. This factor makes it more difficult for a CRT to display an HDTV picture without distortion. LCD and plasma screens, in which each pixel is set in color and brightness by digital data, do not suffer from this problem.

The pixels of a television screen are rectangular (slightly higher than they are wide) where those of a computer screen are square. This difference makes computer images difficult to display without distortion on a television screen and vice versa. The fact that television screens (analog and digital) also come in many different sizes, aspect ratios, and resolutions, as opposed to relatively few for computers, makes the conversion of images from one format to the other still more difficult.

Computer-created graphic images are difficult to display on television screens without loss or quality or distortion due to the different aspect ratios and display formats. And television images do not play much better on computer screens. Both transitions can be accomplished, but both take a lot of compromise and usually result in some loss of picture quality in order to force a video program created in one format to play in a different technology.

Because both digital television tuners and digital television displays do not adapt well to a networked environment, the question naturally arises: How can a television system be connected as part of a home LAN so television reception and computer/Internet images can be played over the network to computers and television sets distributed throughout the home? The answer is that it cannot, at least not with presently available equipment. If a home LAN owner really wants high-quality audio and HDTV distribution throughout the home, the audio and video systems must be wired separately from the LAN using wiring and connectors suited to this purpose. The audio and video system can be connected to the rest of the network through an SDTV or HDTV tuner set up as a node on the network. With this connection, compatible video and audio programs can be transmitted to and from the network. Video distribution networks are discussed in the next section.

Video Player/Recorders

The price of a VHS videocassette recorder/player (VCR) has now fallen to around $50 in discount stores across the country. This is a sure indication that DVDs are rapidly taking over as the medium of choice for viewing recorded motion pictures and other video programs. Nevertheless, the multitudes of VHS cassettes sold over the last two decades will not be replaced overnight and so a VCR should continue to be a component of most home video systems for the next few years.

VCRs, which play VHS cassettes, are analog only and so cannot play to an HDTV display, or any digital video signal. They can, however, play on an SDTV display which, though digital, also has the capability to display an NTSC analog image. All SDTV sets will play analog NTSC, but this is not true of sets built to the newer EDTV standard or to HDTV displays. Because television sets and monitors built in either of these formats are only required to display the digital screen formats specified by the ATSC, with nothing mentioned about analog or NTSC images, most will not play analog at all. Some EDTVs may also have analog capability, but this should not be assumed. If you need NTSC capability for continued use of a VCR, buy an SDTV set or check when purchasing an EDTV set to be sure it can play an NTSC-format image.

The variety and range of quality available in VCRs has been enormous, but is now diminishing as their market share is being taken over by DVD players. Price ranges for VCRs still range from the low-end $50 models, which can be expected to last no more than a year, to machines costing $1,000 or more and built to last a lifetime. If a VCR is being purchased today, with its end of useful life clearly visible not more than a few years in the future, it is probably best to opt for a mid-range machine with good playing specifications and a fair degree of durability, rather than shop at either end of the price spectrum.

A good VCR should have at least four playing heads, full tracking adjustment, fast forward and reverse, pause, single-frame advance, blue screen off mode, and remote control. All of these features in a major-brand high-quality player are available. Also available are **multisystem VCRs.** These machines can play videocassettes recorded using any of the broadcast standards (NTSC, PAL, or SECAM) on any analog television set. This flexibility allows the user to play a PAL cassette tape on an NTSC television screen, but with some compromise in quality. As an archival system to play the occasional videocassette that will never be available in DVD format, one of these machines may be a good investment.

DVD Players

DVD players are available in an even more bewildering variety than VCRs were a few years previously. A DVD player at the high end of this price range is the Denon DVD-9000, which offers just about every feature to be found on a DVD player and has heavy-grade construction for durability. Among its standard features are the ability to play CDs, **CD-Rs (CD-Recordable format), CD-RWs**

FIGURE 7-4 The Denon DVD-9000 Player, Front and Back Views

(CD-Rewritable format), DVD-R/RWs (DVD-Recordable and Rewritable formats), and CDs that contain JPEG-format graphic files. It also has a de-interlacer that allows it to convert interlaced format discs to play in progressive format, an MP3 decoder for playing encoded CD-R/RW discs, a 2MB memory buffer to decrease pauses and pickup dropouts, six-channel DVD-A playback, PCM, DD, DTS-compatible optical and coaxial digital outputs, and component, **S-video,** and composite video outputs. Figure 7-4 shows front and back panel views of this unit. A more complete description of it can be found at http://www.usa.denon.com.

DVD players are digital devices, but they do not have the resolution and sound capability of HDTV. Standard DVDs do not have the capacity to record a full-length movie in high-definition format. DVDs were developed before HDTV standards were set and so were directed more to the SDTV standard and being able to play to analog television sets as well.

In the absence of a set standard for high-definition DVD players, two competing formats were developed by major manufacturers. Toshiba introduced the HD DVD standard and Sony produced the Blu-ray format. Both provided excellent picture and sound quality at the highest HD resolution (1080p) and for a time they competed fiercely with one another for dominance in the emerging high-definition DVD market. Blu-ray DVD players are more expensive than HD DVD players, but the Blu-ray format allows more data to be recorded on a disc, thus allowing more video and audio content than is possible on an HD DVD. By the end of 2007, about 330 movie titles had been released in HD DVD format and 360 in Blu-ray.

In February 2008, Toshiba announced it was ceasing production of HD DVD products. This effectively ended the competition and left Sony's Blu-ray the dominant HD format. Consumers should now buy only Blu-ray players and DVDs for HD viewing.

When buying a new DVD player, regardless of price, look for as many of these features as possible.

• High-Definition (Blue-ray) playing capability. Blu-ray is now the undisputed gold standard for HD DVDs. All future HD DVD releases will be in this format and only a Blu-ray player can play Blu-ray DVDs with full video and audio fidelity.

- Progressive scan capability. Most standard DVDs use an interlaced format, but better quality ones are progressive. A progressive scan DVD player can de-interlace an interlaced DVD movie and display it on screen in progressive scan format. This gives an improved picture that is closer to HDTV than an interlaced DVD.

- Upconversion capability. This feature allows a DVD player to increase the resolution of a standard DVD (480i by 720 pixels wide) to 480p (progressive scan) or to 720p or even 1080i. The resulting picture is significantly better than with a standard DVD, although it is still not equal to HDTV.

- Multiple connections on the back. The best video connections are made through HDMI connectors that we will discuss in the next section. In descending order, the next best are **component video** connectors, then S-video, and finally composite video. A DVD player with the high-end connectors (and plenty of them) will transmit the best quality picture signal to the display.

- Support for multiple disc formats. If you want to play computer-made DVDs that contain video clips in formats like DivX and xvid, your DVD player must have the capability to play these formats. A versatile DVD player should also be able to play CDs of several types (commercial, CD-R, CD-RW). In addition, some DVD players can display computer picture files (jpeg or other bitmaps) from your homemade CDs

- Surround sound capability. Some standard DVDs and all Blu-ray discs are recorded with surround sound. You cannot get surround sound out of any DVD unless the player has a decoder to send the six surround sound channels to the appropriate speakers.

- Recording capability. Recording DVDs on a home system is an art as well as a science, but it is growing in popularity and better DVD players should be able to do it.

- Universal remote control. Most video systems have too many remote controllers (tuner, display, VCR, DVD player, etc.). It is nice if the DVD player can be controlled from the same remote unit that handles other components in the system.

Installation and Setup of Audio and Visual Systems

Whether they connect directly to the home network or are wired independently as separate networks, audio and video systems can use the same wiring and wiring techniques that are used for connecting an Ethernet network. Cat5, Cat6, or coaxial cable can all be used for wiring the long cable runs that connect an audio or video system's components and connect it to the network. All of these have the data-transmission capacity to handle audio and video streams without difficulty. Although the long run cables for audio and video may be the same types as are used for the rest of the network, the connectors that attach to them are different. We will discuss these connectors and the specialized cables that go with them next.

There is no need to worry about running audio and video cables close to network transmission cables. Because all the cables are carrying low voltage signals (plus or minus five volts), none of the cables has a sufficient electromagnetic field to induce interference in the others. Like network lines, however, video and audio lines should not be run near high-voltage AC power lines or equipment because these definitely can introduce interference into the data circuits. In new construction installations, this means the audio and video wiring should be installed after the electrical rough-in is completed so the two wiring systems can be kept well separated. In retrofitted installations, it means locating audio and video components away from existing electrical equipment and, as far as possible, running the cables on routes that will not be close to high-voltage wiring. If audio or video cables must cross over or under AC power lines, they should do so at right angles (90 degrees) to the AC lines. As with network wiring, it is also important to handle Cat5 cable gently so as not to disturb the internal twisting of the wires that enable it to carry data at high capacity.

Termination Points

Audio and video systems use different connectors than those typically found in computer networks. Audio and video cables are similar to patch cables in that they have plugs at both ends, but they use different types of cable. When audio and video cables are installed as structural wiring, each cable is terminated in a jack which then becomes the connecting point for one end of a two-plug cable. The other end of the cable is connected to another jack in the audio or video device. Table 7-2 shows the common audio and video **connector** plugs and the main uses for them. Each of these plugs mates to a corresponding jack in the equipment.

The cables to which these plugs are connected are usually made in set lengths with connectors attached at both ends. When specifying cables for an installation, it is necessary to state the length of the cable as well as the connector at each end. Some combinations of connectors will not be available as a pre-made cable, but the necessary connection can nearly always be made by using one or more **adapters.**

Adapters are connector devices that convert one type of connector that is already attached to a cable into another type of connector needed to plug into a jack in a piece of equipment. Figure 7-5 shows six examples of adapters that change connector types, and that sometimes change the gender of the connector as well.

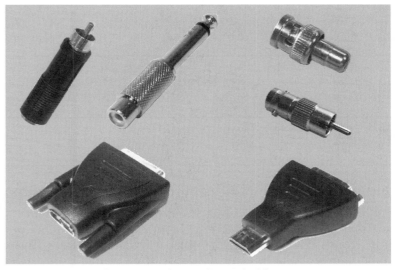

FIGURE 7-5   Six Examples of Adapters for Audio and Video Connectors

TABLE 7-2
Common Audio and Video Connectors

Connector	Connector Name	Common Uses
	¼ inch phono	Speakers and headphones
	3.5 mm sub mini	Computers, speakers, microphones, headphones
	Coaxial cable	Cable TV, satellite TV, broadcast antennas
	USB cable	Computer peripherals, video
	FireWire cable	Computer peripherals, video
	RCA cables	Audio and video equipment
	BNC 75-Ohm cable	Television, VCRs
	S-video and SVHS cables	Television and audio equipment
	DVI cable	HDTV
	HDMI cable	HDTV

As with low-voltage cables run inside walls, the connector cables that attach audio and video units together or link them to wall jacks should be kept away from power cables for the devices to which they attach. The same potential for induced interference exists from high-voltage wires that are free standing as from those in walls. Low-voltage lines need a foot of clearance from the equipment power line to avoid interference. Most equipment is wired with the power cord at one end and low-voltage connectors at the other end to preserve this separation. The installer should be sure that the lines do not converge anywhere along their lengths.

Many connectors and the jacks that they plug into have gold-plated surfaces where the connectors make electrical contact with one another. The purpose of this plating is to help make the connection better and longer lasting, particularly if it is outside. Gold is not only a good electrical conductor, but it is also inert to all forms of corrosion. Plating a connector ensures there will be no oxidizing of metal to diminish the electrical flow. Because the plating is only a few millionths of an inch thick, electricity quickly passes through it to the metal beneath it.

For audio and video installations, new types of cables and connectors have been developed that provide better data transmission and require fewer cables to make the connections. All three of these audio and video cable types are now in wide use. Since they all serve the same need, one of them will probably become dominant and the others decline in use, but that has not happened yet.

FireWire is a cable and connector system that was developed by Apple Computer and is widely used to transfer video on computer-based video editing systems. It is also good for downloading digital video camera tapes (mini-DV tapes, etc.) to digital hard drive storage, and has lately been included as a connector type on video displays, tuners, DVD players, and A/V receivers. Because FireWire came from the computer world to the video world, it is the most common way to connect a video switcher as a node on a LAN: FireWire is at home in both environments.

Digital Visual Interface (DVI), another connector and cable system, also came from the computer world. It was developed to connect LCD screens and projectors to computer CPUs. After a copy protection system called **High-bandwidth Digital Content Protection (HDCP)** was incorporated into it, the HDTV industry pushed hard for its adoption over FireWire because it prevented copying of HDTV content. DVI is widely available on digital television components, but it has one drawback that has limited its use: It transmits video only. A separate connection is required for the audio portion of an HDTV signal, and this can mean an additional six cables (one for each digital sound channel) to connect surround sound audio from a tuner to an amplifier. To correct this deficiency, a third connector and cable system made its appearance and is now gaining acceptance over DVI.

High-Definition Multimedia Interface (HDMI) has it all: the highest data capacity of any cable system (up to 5Gb [gigabit] per second, as opposed to FireWire's 400Mb [megabit]) plus the ability to carry up to eight channels of audio simultaneously. HDMI also incorporates the HDMP copy protection scheme, but for most users this presents no problem. HDMI cables are more expensive than most other cable connector systems, but not so much as to be out of reach for the average user, especially considering that one such cable may replace seven others of a different type.

Receiving Components

Audio and video components that receive outside broadcast signals or cable input must be connected to their antennas or incoming cables in such a manner as to avoid any loss of signal strength. Broadcast signals in particular typically reach the antenna at a strength of a few millivolts (thousandths of a volt) and must be boosted by a factor of thousands before being sent to the output speaker or display. Careful antenna wiring can help keep the signal as strong and as clean (interference-free) as possible so there is less chance of data loss during processing.

Use shielded coaxial cable for antenna-to-receiver connections in both audio and video systems. This recommendation applies whether the system is digital or analog and whether the antenna is for receiving ground-based transmissions or satellite transmissions. Shielded cable protects the incoming signal from interference, which can be severe in outdoor areas. Video antenna connections to a television, tuner, A/V receiver, or VCR use a standard coaxial single wire **BNC connector** that locks or screws into place on the jack. Most new audio tuners use the same connector, but some older ones may have a bare wire connection. In this case, an adapter should be used on the wire screws so the coaxial connector can be plugged into the adapter from the antenna.

Audio Components

Audio and video components connect to one another using one or more of the cable types shown in Table 7-2. Cables and adapters with appropriate ends are available at electronic supply shops. Because the cable ends are already attached, you must be certain to get the correct minimum length needed for the hookup. Cables that are longer than necessary can still be used to link components, but the excess length should be coiled out of the way and kept distant from any source of interference or power induction.

Most audio and video cables are not shielded and do not need to be if they are short (6 feet or less) and are kept clear of induction-producing power cords to the components. The signals coming from player components are stronger than antenna signals, and in most cases, have already been boosted to a level where interference is less of a problem. Still, if a cable run to a component is long, or the home is highly infected with interference, a shielded cable might be warranted. These can be purchased in most end configurations, but at greater cost than the unshielded versions.

Audio connections are all one way—that is, the audio signal is traveling into the device or out of the device, but never both ways in the same wire. A typical route for an audio signal to travel, for example, is out of a CD player (where the signal originated by being played from a CD) and into an amplifier, out of the amplifier, and into speakers. The signal travels into the amplifier in one set of wires and out of the amplifier in a different set. Because nearly all audio devices are stereo or multitrack, two or more wires are required to bring the signal into a device and two or more different wires are needed to bring it out. Most connections between audio devices are made with **RCA connector** cables, usually color-coded to help identify the correct connection. Some audio connections are made with BNC connectors, standard and mini phone plugs, and even bare wires (the last usually only for speaker connections). Figure 7-6 shows the back of a stereo audio amplifier with a number of input jacks available and a number of outputs.

Output devices such as CD players and phonograph turntables have only output jacks because they do not receive input. They create signals and output them to other devices. On the left side of Figure 7-6 are the audio input connections (left stereo side is white at the top; right stereo side is red at the bottom) for these devices. This amplifier has connections for a phonograph, a CD player, a radio tuner, a DVD player, and an auxiliary input. To connect any of these devices

FIGURE 7-6   Audio Amplifier Showing Input and Output Connections

to the amplifier, a pair of RCA connector wires is run from the output jacks on the device to the appropriate input jacks on the amplifier.

In the center part of Figure 7-6 are input and output jacks for devices that can have both input and output. These are devices such as tape recorders that can both play a tape and record one. This amplifier has side-by-side stereo input (left) and output (right) jacks for a tape recorder, a VCR, and a second tape recorder, VCR, or other device. To connect a tape recorder to this amplifier, four RCA connector wires are required. Two of these must run from the left and right output jacks of the tape recorder to the corresponding left (top) and right (bottom) input jacks of the amplifier. The other two must connect the output jacks on the amplifier with the input jacks on the tape recorder.

Any other input and output device requires the same connectors: two from the output of the device to the input of the amplifier, and two from the output of the amplifier to the input of the device. Notice that audio connectors always connect output to input. Input never connects to input and output never connects to output. The audio signal is always traveling from one device (output) to the other device (input).

On the right side of Figure 7-6 are four sets of speaker connections. These are output connections that allow the amplifier to power two pairs of speakers (left and right). This amplifier uses bare wire connections for speakers. They are connected by attaching wires to red and black terminals on the amplifier and then attaching them to corresponding connector terminals on each speaker. It does not matter which way the speaker wires are connected because the current passing through them to drive the speaker is an alternating current. All speakers must be connected with the same polarity, however, so their output will be in phase. There must be two wires connecting each speaker, so four speakers require a total of eight wires leading from the amplifier. Although bare wire connectors are used on this amplifier, most speaker connections are made with mini-phone plugs or RCA plugs.

All audio components should be carefully grounded to reduce the risk of interference. This is especially important if shielded connector cables are used as connectors because the shielding grounds to the frame of the component. If the frame is not grounded, the shielding cannot perform its function correctly. Instead of grounding interference, it simply transfers it into the component frame where it may be picked up by other parts and still do harm to the system. The ground connection for the amplifier in Figure 7-6 is the single screw clamp on the far left side of the unit. A bare wire connected from this clamp to ground grounds the device.

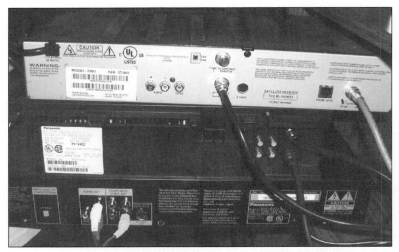

FIGURE 7-7   Video Components Connected by Coaxial and RCA Cables

Video Components

Video signals, like audio, travel only one way: from an output connection to an input connection. Video devices that are strictly output, such as a DVD player, have no input jacks because they do not receive any input. Devices that send output signals and also receive input, such as a VCR, have separate connectors for input and output. The two signals never run in the same wire.

Analog television components are usually connected with RCA connector cables. These connectors are the standard for composite video signals and their accompanying audio signals. Composite video signals consist of three color elements (red, green, and blue) all combined into a single channel signal and delivered on one wire. A second wire carries monaural audio signals, or two additional wires carry stereo sound. Figure 7-7 shows the backs of a set of video components that includes a satellite decoder at the top, a VCR in the center, and a DVD player on the bottom. The satellite decoder and the DVD player are both digital devices, but in this setup they are playing to an analog television set and do not need digital connections to function.

On the right side of the satellite decoder in Figure 7-7, you can see the coaxial cable input that brings the satellite signal into the decoder. In the center of the decoder is another coaxial connection that is the output from the decoder to the television set. Since the television is an analog unit, the output of the decoder is sent as a composite analog signal to the VCR located just below it. The signal goes into the input jack of the VCR, and an output connection (again a coaxial cable) runs from the VCR to the television, which is not visible in the photo.

In this arrangement, the VCR, which is an input and output device, receives its input from the satellite decoder. It can record the input signal on a videocassette (in analog form) or it can output the signal to the television or do both. The VCR also has RCA input and output jacks (monaural audio and composite video) just to the left of the coaxial cable connectors, but these are not used in this setup.

Just to the right of the coaxial connector output on the satellite decoder is an S-video output jack. S-video (separated video) is another type of video connection that does not combine the entire video output into a single signal, but leaves it as separate brightness and color elements, which are sent by separate wires (along with separate audio wires) in the output connection. S-video is also an analog video output, like composite output, but it provides better quality than the composite signal. On the left side of the decoder, the composite video output RCA jacks (white and red stereo audio and yellow video) can also be seen. These are not used in this setup to connect the decoder to the television.

The DVD player in Figure 7-7 (the device at the bottom of the picture) is wired directly to the television. The DVD player is an output-only device and has no need of any input connection. It is wired using three RCA connectors (white and red audio and yellow video), which go to corresponding RCA connectors on the television. This connection provides stereo sound and composite analog video from the DVD player to the television. An S-video output jack is located to the right of the yellow RCA composite video connector, and three component video RCA jacks are located just to the right and above the composite connector, but neither of these are used in this setup.

Component video connectors use three RCA connectors to send a brightness signal and two color signals. The three connectors are labeled red, green, and blue. Together they provide a higher-quality signal to the television than can be obtained with either a composite or S-video signal. Component video connections sometimes use BNC connectors or the VGA-type connector used for component connections on computers, but both of these are unusual on television connections. All component video connections use separate wires for audio in addition to the three video signals.

The separate RCA connection of the DVD player allows it to play on the "input" channel of the television (channel 1). The VCR and TV, as wired in this setup, play on channel 3 of the television, and the individual channels are selected from the decoder using its IR remote. As with all satellite connections, only the channel selected on the decoder can be recorded on the VCR or played on the TV.

All analog video devices connect in much the same manner as the three illustrated in Figure 7-7. The connectors can be BNC type, but these are rare in newer components. Video components can be wired as easily as audio components by remembering the same basic rule. Connectors always go from output jacks to input jacks, never to the same type jacks (output to output or input to input) on different devices.

Digital television devices (SDTV and HDTV) also connect from output to input jacks, but they frequently use different types of connectors. The two reasons for the different connectors are to get a higher-quality signal and to prevent unlicensed copying of HDTV-broadcast and DVD-recorded movies on home devices. HDTV devices with standard RCA connector jacks can easily be wired to DVD burners with similar connections. This enables broadcast HDTV movies or rented DVDs to be illegally copied by anyone with the equipment to do so. To prevent this and provide a better signal for displaying the HDTV programs, new connector types have been introduced for connecting SDTV and HDTV devices. Figure 7-8 shows the rear of a digital A/V receiver with component video and HDMI connector jacks.

FireWire connectors are unique in the HDTV field because they are two-way connections that can send data through the cable in either direction. This connector can also be found on many computers with DVD burners and so would appear to permit an HDTV program to be conveniently recorded. Most FireWire connections on HDTVs and other video devices, however, include **Digital Transmission**

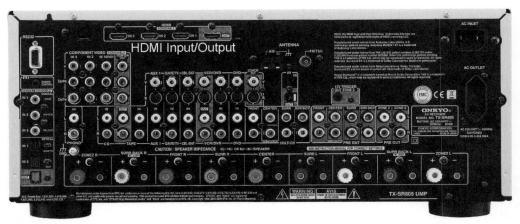

FIGURE 7-8   The Back of an A/V Receiver Showing Various Input/Output Jacks

Content Protection (DTCP), which allows codes to be transmitted with a program that prevent it from being copied.

DVI connectors are 18-pin single cable connectors with D-shaped ends. A DVI connection transmits digital video in a completely uncompressed format, which no consumer device can record, so it also effectively prevents all copying. DVI connections also include the HDMP copy protection scheme that blocks data transmission to any unlicensed device (one without the copy protection).

HDMI connectors are smaller than DVI, but carry more data, including up to eight channels of audio (DVI cables do not carry audio). HDMI connector cables are backward-compatible with DVI connectors. This means you can connect a DVI port to an HDMI cable using an adapter and transmit video data (but not audio) through the cable. Similarly, an adapter can connect a DVI cable to an HDMI port for video transmission, but again the audio capability of the HDMI port is lost and the audio connection must be made separately. Since both DVI and HDMI connector systems incorporate HDMP copy protection, they will function together on licensed devices, but neither will allow copying.

Control Components

Handheld wireless remote controllers for audio and video systems and other electronic equipment are **infrared (IR) devices.** They operate as transmitters, but the signal they send is on a frequency far above that of any radio band. It is just below the frequency of visible light and has nearly all the characteristics of light except that it is not visible to human eyes. Like other light colors, infrared will not pass through most solid objects, and the transmitter that produces it also produces heat. The amount of heat infrared transmissions produce means that the signals have to be kept short to avoid causing damage to the transmitter or the receiver due to heat buildup.

The **carrier frequency** of such infrared signals is around 36KHz. The digital control codes from the controller are sent in serial format by turning the carrier signal on and off (in the same manner as the original Marconi radio, but much faster than a human code clerk). The data transmission rate is actually quite slow, only around 1KHz per second. Many different coding systems are in use, and different manufacturers can use different codes and different data rates for transmission.

Many remote controls are interchangeable, however, because the technology is quite flexible and not very sensitive to either frequency accuracy or data transmission rate. Most receivers respond to data sent at anywhere near their designed parameters.

Television remote controls send commands only one way, to the TV in a low-speed burst for distances of up to 30 feet. They use directed IR produced by **light-emitting diodes (LEDs).** The transmitted infrared signal spreads out from the front of the controller so it does not have to be aimed precisely at the receiver, only pointed in the general direction. Any solid obstruction between transmitter and receiver can block the signal, however.

IR control systems are inexpensive and are generally reliable, although interference from other IR sources can be a minor problem. The problem can originate with other remote controls or other IR sources such as fluorescent lights. One way to limit interference is to use higher IR carrier frequencies for the control transmissions. Some IR systems now use carrier frequencies in the megahertz range. IR controllers do not cause interference in radio or television transmissions because their carrier signals are too low and the infrared light frequency much too high to be a concern to any radio or television band. If, on the other hand, IR interference from multiple remotes or other sources is evident, the solution may be to purchase a universal remote (a unit capable of "learning" the IR signals of all the audio and video equipment in the system) that operates on a high-carrier frequency that is not as susceptible to interference. Such a remote should be tested with all equipment to be sure it can respond to the higher-frequency signals. Some devices, especially older ones, will not accept high-frequency IR signals and their IR receivers are not adjustable.

Configuration and Settings for External Audio and Video

Audio and video systems should be designed as segments of a home network or as independent networks that are internally linked but operate entirely outside the computer-based network. Audio systems can connect fairly easily to an Ethernet topology, a wireless network, or even one of the power-line or phone-line systems. This section describes how the connection can be made using various network technologies.

Television systems will not link well to any of the network technologies because the data flow requirements for television distribution are high and the technology of picture creation, transmission, and storage in television is just different enough from the same technology in computers to make integration of the two systems difficult. Consequently, most television systems do not connect easily to a home network. A **video distribution system** must be wired and connected separately from the home network, but it can connect to the LAN through an Ethernet or FireWire connector linking the video system's A/V receiver or video switcher as a node.

Audio Systems

All the components of an audio system connect to one another through their input and output jacks. The output signal of one device is the input signal for the next device in the configuration. This input/output process continues until the audio signal is transmitted onto the network (and passes out of the audio system entirely) or until it is sent to the speakers. The speakers' output is audible sound, which is input for human ears, but not for any other device in the audio system.

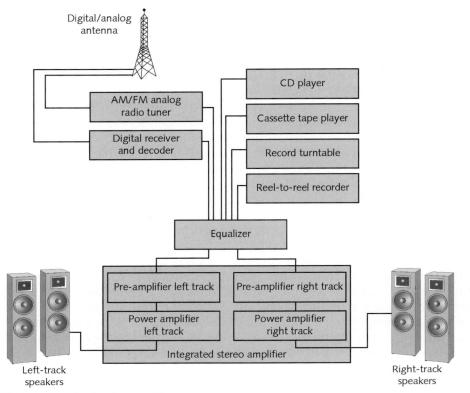

FIGURE 7-9   An Audio System Diagram

Literally hundreds of audio system components can be configured in thousands of combinations. Figure 7-9 shows a diagram of an audio system's major components and how they connect to one another.

The best way to connect an audio system to an Ethernet network is by placing a digital audio receiver (tuner) as a node on the network. The receiver's Ethernet port connects directly to the network, and its audio connections allow a large number of digital and analog players and other input sources to be connected to it. All of these sources can be converted to digital format, if necessary, and transmitted to other network nodes or even out onto the Internet.

The Onkyo TXNR900 7 Channel Receiver with Ethernet Connection, shown in Figure 7-10, is an example of such a receiver. It provides not only an Ethernet connection, but also analog-to-digital and digital-to-analog audio conversion, plus decoders for both THX and Dolby Digital theater five-track sound systems. These can be interconnected to a DVD player, an SDTV receiver, or an HDTV receiver to receive the most complete available multitrack digital sound with video programs and recorded movies. The receiver also has multiple inputs (analog, digital, S-video) and outputs and an audio amplifier output of 110 watts on each of its five tracks. This receiver is not for the price conscious ($2,500), but it does provide state-of-the-art digital audio and a hardwired network connection to facilitate distribution of audio both to and from the network.

Other network-to-audio system connections are available with less features, but considerably lower price tags. **Wireless audio transmitters** can transmit a stereo audio signal up to 300 feet. These devices typically include a transmitter,

FIGURE 7-10   A Digital Seven-Channel Audio Receiver

which is located near and connected to the audio system amplifier, and a receiver that is connected to, or actually installed in, a network computer. The transmitter connects to the audio system with standard RCA input and output jacks. The transmitter sends audio signals to the receiver at the computer, which provides the network connection for the system. Wireless audio devices provide an easier and less expensive (about $150) solution for connecting an audio system to a network, but they also have limitations. The connection is one-way, from the audio system to the network. To send data requires a second transmitter and receiver with their locations reversed. Data traveling in either direction on a wireless connection is much more prone to interference than are other types of connections.

An audio system can also be directly cabled to a computer. Like a wireless transmitter, this connection is technically outside the network, but it has the same result of connecting the audio system through the computer, which is a network node. Direct cable connection requires a separate cable run from the audio system to the computer. The cable type used needs to conform to the type of connection that is made at either end. This connection can be through a USB port, if both the computer and the audio amplifier have one, or through the computer's sound card and the input/output jacks on the audio system. Mini jacks, standard 1/4-inch jacks, and RCA connectors can all be accommodated with appropriate adapters.

Television Systems

At least 11 different companies manufacture video distribution systems that can send video and audio signals throughout a home suitably wired for the system. Some of these systems can interconnect with an Ethernet topology via a FireWire connection that links the system's A/V receiver or **digital video/audio switcher** as a node on the LAN. A/V receivers and switchers can distribute from three to sixteen separate video input sources (cable connection, DVD player, digital broadcast antenna, satellite dish antenna, Internet, etc.) simultaneously to as many as 22 separate locations throughout the house. These systems rely on a wired pathway for distributing video and audio signals to multiple displays. The wiring for these systems is via DVI or HDMI connectors, or by fiber-optic cable with appropriate connectors.

Because the distribution system transmits data in one direction only, from the video sources to the video displays, it is not really a data-sharing network in which each node communicates with all the others. It is a video and audio signal-sharing

FIGURE 7-11   Multiple Displays in a Video Distribution System

system that transmits programming from source to display. Figure 7-11 shows a video distribution system with multiple displays in one room. Additional displays are located throughout the home.

Most video distribution systems are modular so they can be started with display units in two or three rooms and expanded later as the owner can afford to do so. Wiring for the system, however, must be installed during construction of the home or retrofitted in a finished residence, the same as for a home LAN. In either case, the structural wiring is better done as a complete project with connections to every room where they might be wanted, even if some cables have to wait unused for a time.

Connectors for video/audio distribution systems must be compatible on all the display devices used. If DVI or HDMI connectors are used, then the infrastructure cables must be compatible with these technologies and have jacks that will accept the connector cables. Although DVI and HDMI are compatible, the lack of audio transmission in DVI makes mixing the two connector types in a whole-house system impractical. In-house wiring should be installed so one type or the other can be used throughout. HDMI is the preferred choice because it requires only a single cable for both audio and video.

The equipment manufacturer chosen for the video system will determine how the various components are connected. Each of the companies now making video distribution systems has come into this area of home technology integration from another specialization. Some were primarily audio component manufacturers, others control specialists, and still others automation experts. Each has adapted some of the technologies they have previously used to video and audio distribution. The result is working systems, but very little standardization. Figure 7-12 shows a video distribution based on Leviton components.

Most video distribution systems can include **wall-mounted controllers** (keypads) or free-standing infrared remotes that can direct the video and audio components in the room (and sometimes the entire system). The wall-mounted controls are hardwired and may include a video display that allows the user to see the status of every component, source, and receiver connected to the system.

Like audio systems, a video distribution system can be linked to a computer that is also a node on the home network. The link is most easily accomplished by

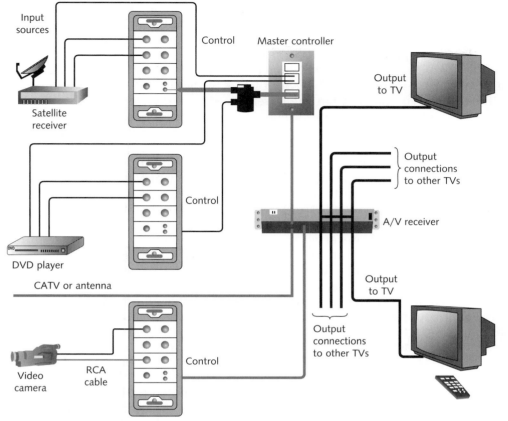

FIGURE 7-12 Video Distribution with Multiple Input Sources

connecting the video system's A/V receiver or video switcher to the computer's video output jack. This setup allows the computer to send any video program playing on its monitor to the video system as a source signal. From the receiver or switcher, the signal can be directed to any video display in the home. This is again a strictly one-way data transfer, from the computer to the video system. Thus, the display seen on the remote screens is subject to the limitations of the computer's video display.

Configuration for Internal Video/Audio

Any audio or video program that can be streamed from the Internet to a computer on the home network can also be distributed to other nodes on the network or to a video distribution system linked to the computer's output. No additional processing of the program is required to distribute it. If the television display is digital, the computer can output a direct digital signal to it, although the television image may have some distortion due to the difference in pixel shapes between the video and computer images. If the television display is analog, the computer output must be processed through a converter in order to be displayed on an NTSC screen.

High-quality low-priced digital video cameras and video editing systems have now made it possible for everyone to create video programs and stream them on their own home network or out onto the Internet. Producing and streaming digital video involves several key steps. Some of these require only software additions to a home computer, while a few need additional hardware components.

The first requirement for creating video is a camera to record original footage. A number of high-quality digital cameras are now available for less than $1,000, and many others are priced above that figure. As with most hardware, you generally get what you pay for, but the lower-priced cameras produce video of a high enough quality to satisfy most viewers. The camera should have a FireWire(r) connection or USB port so its video can be downloaded to the computer without the need for a separate digital playback machine capable of playing the **miniDV cassettes** on which most cameras record. If a user plans to record and download a large quantity of video, a playback/download device may be a good addition to the computer system, but for most home video systems it is not necessary. With the download port, recorded cassettes can be played back in the camera and their content downloaded directly to the computer's hard drive.

Once captured (transferred from the video camera to a computer storage system), the next hardware need of a home network video streaming system quickly becomes evident. About 5 minutes of **uncompressed video footage** occupies a gigabyte of storage space. If the computer has a 60-gigabyte hard drive, it will be completely filled by less than 5 hours of video footage. This heavy storage requirement suggests the need for a **video server** with multiple high-capacity drives that can hold all the user's uncompressed original video footage, as well as edited shows and compressed video that is ready for streaming.

A video server can be set up as a network node so all nodes can request data from it and can store video on it, whether downloaded from a camera, or from the Internet, or from another digital video player connected to the network. A video server should have enough storage capacity to meet the user's likely needs for at least a year or two, and the capability of being upgraded with more storage in the future. It should have a fast processor and fast hard drives so it can respond quickly to video data requests and keep up with the requirements of streaming video data.

Stored video can be edited on any computer node equipped with editing software and sufficient processing capacity to handle the large files. Once a video program is edited, it needs to be compressed for streaming either on the home network or on the Internet. Video should not be compressed prior to editing because all the common compression techniques use lossy compression algorithms causing some of the data in the original video to be lost. If compression is done before editing, the editing process may cause further data loss and a corresponding loss in quality in the edited video. Waiting to compress video until after it is edited requires more storage space for the raw (unedited) footage, but results in a higher-quality finished product.

Video compression is a software function that, like editing, can be accomplished on any computer with sufficient processing power and access to the stored footage. Not all compression systems allow streaming of the compressed program, but if the video is to be downloaded from the Internet, it should be compressed into a format that does permit streaming. Even compressed video files are large and time consuming to download completely before playing. Streaming allows the viewer to watch the front end of the program while the back portion continues to download, and many viewers find this a preferable option.

Compressed video programs, whether produced in the home or downloaded from the outside, can be sent to any network node on request and viewed on the computer's screen. They can also be broadcast on the home's video distribution system, if it is linked to the network. As noted earlier in this chapter, most video system links to the network are through the video output of a computer node rather than a direct connection to the network. The video output link to a video distribution system permits any video playing on the computer node to be distributed through the system. When direct network connections to video distribution systems become common, as they surely will within a few years, the video system will be able to access video programs directly from the video server without needing to pass them through another network node.

Monitoring and Maintaining Video/Audio Systems

Video and audio systems are combinations of electronic and mechanical components, and as such they are subject to the same wear and environmental problems that afflict other similar devices. Because video and audio systems must continually operate at peak performance levels in order to produce quality output, even small maintenance problems can cause a serious degradation in their output. Electronic systems are adversely affected by three main factors: heat, water, and electric power variations. Mechanical systems are also adversely affected by these three, plus a fourth: dirt.

Video and audio systems produce heat that needs to be dissipated or it will build up in components until they burn out. Many components contain **heat sinks** to help prevent heat buildup, and most systems provide for cooling with ventilating fans that carry heat away from the components. These work well if not obstructed by placement of the system in an enclosure where air circulation is restricted. Components should not be stacked on top of one another or enclosed so the cooling fans in them cannot circulate air freely. Likewise, components should not be placed where they are subject to additional heat from outside sources such as heating fixtures, fireplaces, and space heaters. Keep the system in as cool an environment as possible, and its own cooling components will maintain a good working temperature inside.

Water problems for video and audio systems usually occur in the form of condensation of moisture from the air on internal components. This usually happens when the system is turned off and warm parts cool down, picking up condensate from the air as they do so. If the system is again turned on before the condensed moisture has had time to evaporate, short circuits or other damage to electronic parts can result. The key to avoiding condensation in a system is the same as for avoiding heat buildup: Keep good air circulation within the components. The less heat buildup in the system, the less danger there is of condensation when it cools down. Likewise, the more air circulating in the system, the less chance there is for moisture to accumulate. This is another reason to keep systems out of areas or enclosures where slow-moving moisture-laden air can create condensation.

Power fluctuations are the surest way to damage a video or audio system, just as they can damage computer systems. High-quality fast-acting surge suppressors are a must for video and audio systems. Voltage drops should also be avoided. This may mean connecting the system on an independent circuit and making sure the home's power load requirement does not periodically pull down the voltage as equipment starts up.

Finally, dirt, usually in the form of dust particles, is to be avoided in all mechanical systems. It increases wear on moving parts and can sometimes bring them to a halt. Video recorder/player heads rotate at over 1,000 rpm. So do CD players, DVD players, and hard drives. All of these are motor-driven and rotate on shafts set in precision bearings. Even small quantities of dust can shorten the life of these parts. A component's internal parts cannot be cleaned of dirt. The most that can be done is to keep dirt away from the equipment and out of the air that circulates through it. The more dust-free the interior of the home can be kept, the longer its electronic systems (and perhaps its occupants) will last.

Summary

- Video and audio systems in the U.S. are changing from analog to digital formats, but some analog equipment will continue in use after the digital conversion process is completed. Analog television broadcasting in the U.S. will cease on February 17, 2009.

- Digital audio systems consist of five basic components: an antenna to receive broadcast signals; a tuner to select frequencies; an amplifier to strengthen the received signal; a speaker to convert the signal back into audible sound; and a ground connection to complete the broadcast circuit.

- Audio recordings are distributed mostly on CDs (for music) because this medium provides the highest-quality digital reproduction of sound. Cassette tapes are often used for voice recording and millions of recordings on cassettes and plastic records still exist and are being converted to digital format.

- Digital television systems require the same basic components as audio systems in order to receive and display digital television transmissions: an antenna, a tuner, an amplifier, a speaker and display monitor, and a ground.

- SDTV tuners and television sets, HDTV tuners and television sets, and EDTV television sets are all approved digital television standards under the ATSC Table 3 definitions. A total of 18 digital television formats have been approved by the ATSC and American digital components must process and/or display all of them.

- Digital video programs are recorded on DVDs, which are played on DVD players and displayed on digital or analog television sets. Analog VHS cassettes are still widely used for television recording, but are being replaced by DVDs as digital television becomes the standard.

- Audio and digital components always connect from the output jacks of one device to the input jacks of another device. Audio and video signals travel in one direction only. No connection on these systems can be both input and output.

- Audio systems can be connected directly to an Ethernet topology for data transmission on the network. They can also be connected through a computer node's sound inputs.

- Video systems, both analog and digital, are difficult to connect to a network because of the differences in the technologies of both systems. Video can be downloaded from a network to a video distribution system, but the transfer of digital television data from a tuner or DVD to the network requires that the tuner or player be connected as a node on the network as well as to the source of the video signal.

- Both video and audio programs in digital format can be produced by the home network user and streamed on the home network or out onto the Internet. They can be played on an audio/video system connected to the network or on other computer nodes.

- Maintaining audio and video systems requires controlling heat buildup, keeping away moisture and condensation, suppressing electrical current fluctuations, and preventing dirt from entering mechanical assemblies.

Key Terms

Adapters Devices that change audio and video connectors from one form to another, including gender changes.

Advanced Television Systems Committee (ATSC) The government-run and industry-influenced group that sets standards for the television industry.

Amplifier A device that strengthens the power of an RF signal it receives from an antenna or pre-amplifier.

Antenna The component of a radio or television receiver that receives a broadcast signal from a distant transmitter.

Attenuated Weakened by distance from the source; resistance; used to describe radio and television signals.

A/V receiver A digital television system component that serves as the central connecting unit of the system. May include a digital tuner, amplifiers, distribution connections, and other devices.

BNC (bayonet Neill-Concelman) connector One of the standard connector types used for audio and some video system connections.

Carrier frequency In radio transmission, a signal at high frequency on which a signal of lower frequency is carried.

CD burner A device for recording compact discs (CDs). A CD burner can also play a CD, but CD players are not always burners.

CD player A device for playing recorded compact discs.

CD-R (CD-Recordable format) A compact disc that can be recorded once on a CD burner, but is then a permanently fixed recording that cannot be changed.

CD-RW (CD-Rewritable format) A CD that can be recorded multiple times with succeeding recordings added to or replacing older ones on the disc.

Component video A method of transmitting video signals using three separate connector cables, usually terminated with RCA plugs.

Cone The part of a speaker that vibrates in the air producing sound.

Connector Any cable with end plugs designed to transmit electronic signals from one device to another. Includes such systems as RCA, BND, S-video, FireWire, DVI, HDMI, and others.

Converter A device that renders digital data into analog signals in audio and video systems. *See* decoder.

Crossover network A wiring circuit in speaker systems that permits several speakers of different ranges to function seamlessly together.

Decoder A device that renders digital data into analog signals, or vice versa, in audio and video systems. *See* converter.

Digital audio file A recorded music or voice audio segment stored on a computer hard drive or other similar media.

Digital Transmission Content Protection (DTCP) A copy protection system used to prevent unlicensed copying of HDTV programs.

Digital video/audio switcher A device that receives digital signals from multiple sources and routes them as instructed to multiple outputs.

Digital Visual Interface (DVI) A connector system for distribution video (but not audio) HDTV signals. It includes HDCP technology.

Distortion Changes in reproduced sound waves caused by excessive power or other interference in the recording and playback processes.

DVD (Digital Versatile Disc) A small plastic record on which digital video programs are recorded using a laser process.

Electromagnet A magnet created by passing a current through a coil of wire. Used in audio speakers and many other devices.

Encoder A device or software program for changing analog audio or video signals into digital format. Video versions are also called coders.

Enhanced-definition television (EDTV) An ATSC-defined television that can play both analog and DTV digital programs and has a minimum resolution of 480p by 704 pixels wide in 4:3 format.

Equalizer An audio device for balancing the levels at which sound frequencies are recorded or played back.

FireWire(r) A 1394 standard data connection for transmitting digital data at high speed between two devices or nodes on a network.

HDTV tuner A digital television tuner that receives full resolution HDTV signals in 16:9 format.

Heat sink A relatively large piece of metal or other material placed near a heat source to absorb heat from it and dissipate it into the surrounding air.

High-bandwidth Digital Content Protection (HDCP) A copy protection system used in DVI and HDMI connector systems to prevent unlicensed copying of HDTV or DVD programs.

High-Definition Multimedia Interface (HDMI) A connector system for distribution video and audio HDTV signals in a single cable. It includes HDCP technology.

Infrared (IR) devices Devices that function as controllers by sending command signals using infrared light, which is invisible to human eyes.

Light-emitting diode (LED) An electronic device, similar to a vacuum tube, that emits red light or infrared radiation. Arrays of tiny LEDs are used to make television display screens.

Media Any means of storing or recording audio or video information: magnetic tape, CDs, DVDs, and phonograph records are examples.

MIDI (Musical Instrument Digital Interface) file A type of digital music recording in which the file stores an actual musical score. MIDI files can be played on any electronic instrument or software-equipped computer.

MiniDV cassettes A small-format recording medium used in many digital video cameras.

Monaural Having only one track or sequence; not stereo. Used to describe single-track audio recordings.

Multisystem VCR A video player capable of playing NTSC, PAL, or SECAM videocassettes on any analog television.

Pre-amplifier A device that strengthens an original signal from an antenna or other source before it goes to an amplifier.

Rasters The individual horizontal scan lines that make up an analog television picture.

RCA connector One of the standard connector types used for audio and video system connections.

SDTV television A television that receives and displays all ATSC digital formats, but not necessarily at full high resolution.

SDTV tuner An RF receiver that receives ATSC terrestrial digital television signals and decodes all Table 3 video formats. It can play to an analog or digital television set.

Stereo Two soundtracks recorded from the left and right side of a musical performance to give balance and depth to the recording.

S-video connector One of the standard connector types used for video and audio system connections.

Synchronized Operating or playing together at the same rate. Stereo sound tracks are synchronized, as are the picture and sound of a video program.

Tuner The device in a radio or television that sets the one station frequency to be received and excludes all others.

Uncompressed original video footage Digital video recorded without use of any algorithm to reduce file size. Produces large files that contain all digital data in sequential format.

USB port Universal Service Bus port. One of the standard connector types that uses data connections, including video, but rarely audio.

Video distribution system A wired system for distributing video programs from multiple sources to multiple viewing sites in the home.

Video server A computer set up as a storage location for video programming on a network, from which the video footage can be requested by other network nodes.

Wall-mounted controllers In video and audio systems, a device for controlling distributed programming in a room, or for the whole system.

Wireless audio transmitter A wireless transmitter that sends analog or digital audio signals to a computer.

Review Questions

1. A VHS videocassette tape can contain either an analog or a digital television program. True or false?

2. FM radio is broadcast with two audio tracks and is called a _____ transmission. AM radio is broadcast with only one track and is called a _____ transmission.

3. What is the function of a digital radio tuner?

4. A converter renders a digital audio or video signal into analog format. True or false?

5. Signals from an audio or video antenna are usually first sent to _____.

   a. The speaker

   b. The equalizer

   c. The tuner

   d. Ground

6. The function of an amplifier is to _____.

   a. Filter noise from the signal

   b. Increase the strength of the signal it receives

   c. Tune out unwanted frequencies

   d. None of the above

7. Another name for a converter is a _____

8. Distortion of audio output is caused by overdriving the speaker. True or false?

9. The part of a speaker which actually vibrates the air to produce sound is called a _____.

   a. Shield

   b. Voice coil

   c. Frame

   d. Cone

10. The range of human hearing is about 20 cycles to 20,000 cycles per second. True or false?

11. What is the function of a crossover network?

12. An equalizer cannot change the signal coming from the antenna or the pre-amp. True or false?

13. The function of an encoder is to change _____ ____ signals to _____ signals.

14. What happens when interference causes more of a digital signal to be lost than the decoder can correct?

15. Two phenomena that can cause interference with digital satellite transmissions are _____ and _____.

16. Computer graphic images look a little distorted on a digital television screen because _____.

   a. The two systems use different shaped pixels

   b. One is a digital image and the other analog

   c. Signals traveling between them are distorted

   d. The television screen has a different aspect ratio

17. CD-RW stands for CD _____ and means that the CD can be _____ several times.

18. The most common type of connector for analog video and audio devices is _____.

19. DVD stands for _____ _____ _____.

20. An audio system can connect directly to an Ethernet topology, but a television distribution system cannot. True or false?

21. What is an adapter and how is it used?

22. Cables running from an antenna to a tuner or pre-amp should be of what type? _____

23. Remote control devices use _____ signals to send commands to components.

24. A wireless audio transmitter is used to _____.

   a. Send audio signals to a tuner

   b. Send audio signals to a computer

   c. Convert audio signals to analog

   d. Strengthen audio signals

25. When video is "captured" it is transferred from a(n) _____ to storage on a(n) _____.

Hands-On Projects

Project 7-1: Compare Analog and Digital Audio Sources

In this project, you will compare the quality of sound played over an audio system from a CD to that played over the same system from a radio tuned to an FM radio station and a digital radio. To complete this project you need access to an audio system that includes a CD player, an FM radio, and a digital radio.

1. Play a music CD you enjoy on the audio system and listen for the following characteristics:

   a. Signal-to-noise ratio: Can you hear any background noise or hiss at all?

b. Interference: Can you hear any sounds that do not appear to be an intended part of the recording?

c. Sound quality and balance: Does all the music sound clear and precise, as if you were actually in the presence of the performers? Are the high and low notes distorted?

2. Tune to an FM radio station of your choice at a volume that is comfortable for you. Listen to it for a while and evaluate the sound quality on the basis of the same three criteria as the CD.

3. Tune the digital radio to a station of your choice at a volume that is comfortable for you. Listen to it for a while and evaluate the sound quality on the basis of the same three criteria as the CD.

4. Turn the volume of the FM radio station lower and listen to some more music. Then, turn the volume of the digital radio and the CD player lower and listen to more music on each of them. Does the lower volume change the relative quality of the three sources?

5. Turn the volume of the FM radio station higher than it was in Step 2 and listen to more music. Then, turn the volume of the digital radio and the CD player higher than it was in Steps 1 and 3 and listen to more music. Does the higher volume change the relative quality of the three sources?

6. Write a short evaluation of this sound comparison. Which of the three sources do you think is better for music listening? Why? If volume levels made a difference in the comparison, why do you think they did?

Project 7-2: Compare Analog Digital Sources

In this project, you will compare three different sources of analog and digital-to-analog television to see which provides the best picture. For this project you will need access to an analog television set, a VCR, a DVD player, and the cables to connect them.

1. Turn the television on and select a local channel that has good reception in your area. Evaluate the picture you see for the following items:

a. Are there ghosts in the image? Snow? Lines, moving or static?

b. Is the picture clear and sharp, or does it appear fuzzy around edges?

c. Are the colors well defined and clear, or do they appear muddy?

d. Is the sound clear and crisp with no noise and is it synchronized with the picture?

2. Play a videocassette with the output going to the television. Evaluate the picture and sound on the same points as in Step 1. Is the analog videocassette picture better, worse, or the same as the broadcast image?

3. Play a DVD with the output going to the television and evaluate the picture and sound on the same points as in Step 1. Is the digital DVD (playing in analog format) picture better, worse, or the same as the broadcast image?

4. Write a short report noting which picture you considered best of the three you saw, and why you think it was. Give any reasons you can think of for the lack of quality in any of the three pictures.

Project 7-3: Record a Digital Audio Selection

In this project, you will record a voice or music selection on the computer and store it to a hard drive. For this project you will need access to a computer with a sound card and a connected microphone that works on Windows 2000 or Windows XP.

1. To check the microphone and sound recording hardware, click Start, point to Settings, click Control Panel, double-click Sounds and Multimedia, click Audio, and then check to see that the microphone is functioning correctly.

2. To open Sound Recorder, click Start, point to Programs, point to Accessories, point to Entertainment, and then click Sound Recorder.

3. Set the microphone in a convenient position for recording and prepare the material you want to record so you can read it easily.

4. When you are ready to record, click the Circle (Record) button on the Sound Recorder and begin speaking. When you finish speaking, click the Rectangle (Stop) button.

5. Click the Right Arrow (Play) button to play back your recorded voice. If you want to rerecord any portion, stop playing the recording where you want to make the change, and repeat Step 4 to record from that point to the end of the recording again.

6. Click File and then Save to save your recording to the hard drive. The default folder for saving the file is My Documents, but you can change to another folder if you want to. When you have the folder you want open, enter a name for the file and press Enter to save it.

Project 7-4: Record a CD and Play It on a CD Player

In this project, you will record music on a CD and play the CD on a CD player connected to an audio system. For the project you will need access to a computer with an Internet connection and a CD burner/player. You will also need music selections on a CD that you want to copy (record) on your CD.

1. Open the CD drive door, place a CD you want to record from in the drive, and close the door. When the Audio CD window opens and asks what you want to do, click Play Audio CD, and then click OK.

2. When the Windows Media Player opens, click Copy From CD on the left side of the window.

3. Click to check the boxes on the left side of the titles of selections you want to record. When you have all desired selections checked, click Copy Music at the top of the window and wait until the computer finishes copying. The music file is stored in the My Music folder.

4. Repeat Steps 1–3 to record all the music selections you want on your CD. Remember, a typical CD can store up to 650 megabytes of data, so you can put up to fifteen selections or more on a CD.

5. Insert a blank recordable CD into the CD recording drive.

6. Click Start and click the My Music folder. Click the files you want to copy to the CD. After clicking the first file, hold down the Ctrl key while you click each of the others. (Note: You must select at least two files before you can proceed to Step 7.)

7. Under File And Folder Tasks, click Copy The Selected Items.

8. In the Copy Items dialog box, click the CD recording drive, and then click Copy.

9. In My Computer, double-click the CD recording drive. Windows displays a temporary area where the files are held before they are copied to the CD. Verify that the files you want to copy appear under Files Ready To Be Written To The CD.

10. Under CD Writing Tasks, click Write These Files To CD. Windows displays the CD Writing Wizard. Follow the instructions in the wizard.

11. When the process is finished, close the wizard and remove the recorded CD from the drive.

12. Place the CD in the audio system CD player and press the Play button. Listen to your CD on the audio system.

Project 7-5: Connect an Audio System

In this project, you will connect several audio devices into a working audio system. For this project you will need access to a CD player or other audio output device, an amplifier, a cassette recorder or other input device, and two speakers. You will also need 10 RCA connector cables or cables of the appropriate type if the devices use different connectors.

1. Place the equipment on a table with each device close enough to the others that the RCA connectors you are using can easily reach from one device to the others. Do not stack the devices on one another if they have vent louvers or access doors on top. Do not stack the amplifier on any other device.

2. Using two RCA connector cables, one for the left stereo track and one for the right track, connect the output jacks of the CD player to the input CD jacks of the amplifier. If the amplifier does not have a specific set of CD jacks, connect to the Auxiliary jacks.

3. Using two RCA connector cables, one for the left stereo track and one for the right track, connect the output jacks of the cassette recorder to the input Tape 1 jacks of the amplifier. Use two more RCA connectors to connect the Tape 1 output jacks to the cassette recorder's input jacks.

4. Use two RCA connectors to connect the left speaker to the left output jacks of the amplifier. Use two more connectors to connect the right speaker in the same manner. Connect both speakers' wires in the same order so their polarity will match.

5. Connect the AC power plugs of the CD player, amplifier, and cassette recorder to an AC power strip or multiple outlet.

6. Turn the amplifier and CD player on. Switch the amplifier to the CD input (or other input jacks you used for this device).

7. Play a CD to test that the audio system is working. If you do not hear the music, check all the connections to be sure you have output wired to input and each speaker connected by two wires to the speaker output jacks on the amplifier output panel. Test again.

8. Put a blank cassette in the cassette player. Play another music selection on the CD and record it on the cassette player while the music is also playing over

the speakers. Did your system function correctly? Troubleshoot any problem until you can record a cassette from the CD while playing the music at the same time.

9. Play back the recorded sound from the cassette through the amplifier and speakers. Can this sound be recorded on the CD player? Why not?

Project 7-6: Connect a Video System

In this project, you will connect several video devices into a working video system. For this project you will need access to a DVD player or other video output device, a VCR or other video input device, and a television set. You will also need eight RCA connector cables or cables of the appropriate type if the devices use different connectors.

1. Place the equipment on a table with each device close enough to the others that the RCA connectors you are using can easily reach from one device to the others. Do not stack the devices on one another if they have vent louvers or access doors on top.

2. Using two RCA connector cables, one for the left stereo track and one for the right track, connect the audio output jacks of the DVD player to the audio input jacks of the television. Use another connector to connect the DVD player's composite video output jack to the television set's composite video input jack.

3. Plug the television set and DVD player into AC power outlets and turn them on. Put a DVD in the player and play it to the TV.

4. Use three more RCA connectors to connect the VCR's output jacks to the TV. Do not disconnect the DVD player to make this connection. If the TV does not have enough input jacks to allow both devices to connect to it, then connect the output of the DVD player to the input of the VCR and the output of the VCR to the TV.

5. Use three more connectors to connect the output of the TV to the input of the VCR. If you have already connected the DVD to the VCR's input and the VCR does not have enough input jacks to also accept the TV output, skip this step for now.

6. Plug the VCR into an AC outlet. Put a tape in and play it to the TV. Start the DVD player and play it to the TV also. Can you switch between the devices while both are playing, or does one block the other?

7. How can you connect the VCR and DVD player so both can play simultaneously on the TV and you can choose between them by just switching channels from Channel 3 to Input (Channel 1)?

8. Experiment to see if you can get both devices playing at the same time. Remember, output jacks must always connect to input jacks. Can another type of connector such as an S-video cable or a coaxial cable help make this setup possible?

Case Projects

Case Project 7-1: Research a Video System

A client wants to have a new television system installed as part of his home technology. He wants the new system to play DVDs in SDTV format and digital satellite broadcasts in HDTV format, but also still be able to play analog NTSC broadcasts and VHS cassette tapes. Your job is to research on the Internet or at a large electronics store what compatible components are required for this system in order to meet all the requested output formats. Note the components and their prices and compile a list for the client. What does the total system cost, including the required cables to link the components to one another?

Case Project 7-2: Evaluate a Conversion Proposal

A friend tells you he's planning to purchase a DVD burner to add to his computer. He then plans to connect his analog VCR to the computer and copy each of his VHS videocassettes onto a DVD. He can then discard the videocassettes and have digital-quality DVDs of his movie collection. Write a short response to this plan. Can it be done in the manner your friend suggests? Will it work? Will he really be able to get high-quality DVDs in this manner?

Case Project 7-3: Connect Video Components

You are setting up a video system for a home entertainment center. You need to connect a DVD player and a VCR to the television so both devices can play to the TV. The VCR also needs to be connected so it can record from the television or from the DVD player. On checking the equipment, you find that the TV has only one set of video input jacks and one set of video output jacks. The VCR also has one set of video input jacks and one set of video output jacks, and the DVD player has only video output jacks. Using RCA cables and adapters, describe or diagram how you would connect these devices to accomplish the desired setup. (Hint: Check this chapter for an adapter that allows splitting an output or an input.)

Case Project 7-4: Advise a Client about Digital TV

A client asks your advice about installing digital television in her home. The reception of analog TV signals is poor in her area. Several local channels are now broadcasting digital programming, but she is concerned that digital broadcasts seen will not look any better on her analog TV (using a digital converter) than the regular broadcasts. She is also concerned that digital programs seen on a digital TV will be no better. Is she right in both cases? What would you advise her about putting in digital television so far as the quality of the reception she is likely to get?

Security and Access System Fundamentals

After studying this chapter, you should be able to:

- Define the factors that go into designing security systems and integrating them into a home technology system.

- Describe the various types of home security systems and the differences between them.

- Identify the factors that determine the locations of various security equipment.

- Describe the characteristics of the system components used in security systems.

OBJECTIVES

Introduction

Security Design and Installation Factors

Security System Types

Security System Equipment Locations

Security System Components

OUTLINE

Introduction

Security systems form a unique part of home technology. In this chapter, you will learn about the components and structure of a security system and how one is designed to protect the home environment and the people who live in it. You will find out the significant differences between the types of security systems, and when each is best used. You will also discover how to identify effective locations for security equipment and what factors should be considered in setting the system up. Finally, you will learn the function of each of the many system components used in security systems.

Security Design and Installation Factors

Home security has been a concern of homeowners since long before the age of technology. People have tried to ensure their personal safety and protect their property in almost every society since the Egyptian kingdoms of 5,000 years ago. The technology available for home security systems today permits a greater assurance of protection than ever before, but it also requires some security decisions that were not really necessary even a generation ago.

In former times, people often bought all the physical security they could find, or at least all they could afford. Cost is still a consideration in security systems now, but it should not be the only determining factor. Among the questions to ask when designing a security system for a home are the following:

- What or who is the security system protecting?
- What major threats must be warded off?
- What is the level of risk for which protection must be provided?
- Which factors external to the home environment affect the risk assessment?

The three traditional objectives of home security are protecting people, premises, and property. Everyone has a right to be personally safe in their own home and for members of their family and guests visiting their home to also be safe. Consequently, protecting the home's occupants is usually the first priority of a home security system. After personal safety, and often connected to it, comes protecting the premises of the home, first from persons who might try to enter and harm the occupants, and secondly, from fire, flood, or other dangers that could damage or destroy the house and also pose a hazard to those inside. Last, and again often connected to personal safety, is the protection of property inside the home from theft or destruction.

The amount of threat to the people in the home, the home itself, or property within it is partially a constant and partially a variable influenced by individual circumstances. The constant part of the risk is from inanimate disasters: mainly fire, but also flood in some locations, gas or other poisons from the home's systems or the surrounding environment, and weather-related factors such as severe storms and lightning. Wherever any of these exist, and some of them—such as fire danger—always do, they should be guarded against by a home security system.

The variable part of risk is much harder to evaluate. Wherever wealth is stored, theft is a threat. The more wealth, the more threat, especially if the wealth is unprotected. Threats to property often turn into threats to premises as thieves try to break into where the goods are kept. Sometimes threats to property turn into threats against people who may be in the way of a successful robbery, or the homeowners themselves may be a target of kidnapping.

As a general rule, the greater the amount of wealth in a home, the higher the risk to the home, the property inside it, and its occupants. Risk is often enhanced by other factors, including celebrity status, holding a public office or security-related job (bank officer, judge, police officer), or holding controversial views. Every homeowner should evaluate how much his or her personal safety and possessions are likely to be at risk in the area where they live, considering the circumstances of their life. If the risk appears significant, then maximum home security precautions are the surest way to peace of mind for the family.

Outside factors can also influence risk. A home's isolation from help or observation by others may be a reason to increase its fire protection as well as its personal protection security system. It may also warrant perimeter security in the yard to give a more timely warning for people to seek safety in a secure area if danger threatens.

All these factors, as well as cost, weigh in the decision of how much home security to include in a DHTI system. This chapter discusses maximum protection features and systems. Most homeowners will want less than everything, but knowing about the highest levels of home security technology can help you make your own judgment on a factual basis. All of the following factors should be considered in the design of home security and its integration with the rest of the home's technology system.

Existing Home and New Construction Environments

As with other areas of home technology, whether the home security system is installed in a new home under construction or retrofitted into one already completed has a large influence on what type of system is used. Hardwired security systems can be easily installed in new construction, but are much more difficult to retrofit in existing homes. Wiring can also be better concealed in new construction than when retrofitted. Building the wiring into the walls makes it much more difficult to cut or bypass than if it is tucked under a molding or inside a raceway. Like other home technology system wiring, security cables should be roughed in by the DHTI technician on new construction projects after the home's utilities are installed.

The main reason for installing security lines after the utilities are in is to avoid having them cut or damaged by the other utility workers. Being the last to install also allows the security technician to be sure the system's lines are located where they do not provide any easy access points for intruders to cut or bypass them. Alone among home technology systems, security lines and hardware must be considered at risk not only from normal age and use, but from intentional sabotage by intelligent persons. Consequently, they always need to be installed with an eye toward preventing deliberate damage by persons trying to get past the security system's protective screen.

Wireless security systems are usually the preferred choice for retrofitted installations as well as many new homes. Remote sensors with wireless transmitters do not require a lot of electricity to operate and can be battery powered at nominal expense. Wireless sensors can be placed almost anywhere and still send their signals to a central security panel without the need for any physical power or data connection. Many also transmit the state of their batteries so the homeowner can tell from the central control device when the batteries of any **sensor** need replacing.

Concealed wireless sensors may offer better security than wired ones, too. Because they have no wires to bypass, and sensor failure triggers an **alarm,** they are more difficult to evade than some wired units.

Security Zone Layout

Security systems are organized in zones so their protection devices can be easily monitored and the origin of any security breach quickly located. A **security zone** can be any area of a home or yard. It can consist of a single room or multiple rooms and can include outside and inside areas. The limits of a zone are really the number of devices that the security system's central control panel can monitor for a zone. Typically, a zone can include four devices to be monitored from the central control. Some systems allow up to six devices, but more than that calls for the area to be divided into two or more zones, each with their own group of devices to monitor.

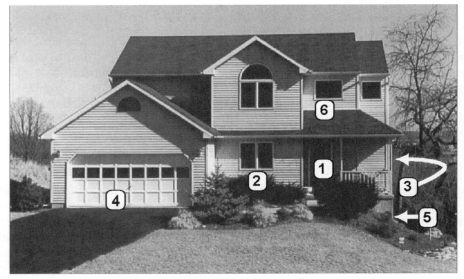

FIGURE 8-1   Common Entry Points for Burglary Attempts

When determining how a home should be segmented into zones, take into account how homes are usually entered by burglars or other unauthorized persons. The strongest protection (and perhaps redundant protection to detect any intruder who manages to disable or evade one device) should be placed in those areas most often targeted for illegal entry.

Using a variety of **detection devices** and sensors in the most vulnerable areas helps assure that any attempt to break into the home immediately sounds an alarm and summons assistance. Figure 8-1 identifies the common entry points in homes where break-in attempts are made. Statistics shown in Table 8-1 are from ADT Security Systems and indicate the percentages of burglary attempts made for each entry point.

TABLE 8-1
Break-In Entry Points and Percentage of Attempts

Rank	% of Break-in Attempts	Location
1	34	Front door
2	23	First-floor window
3	22	Back door
4	9	Garage
5	4	Basement
6	2	Second-floor entry
7	6	Storage area and other (not shown in Figure 8-1)

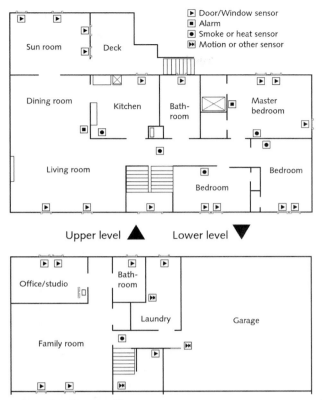

FIGURE 8-2   Home Security System Plan

Standard security systems usually allow for up to eight zones, and large-scale systems allow up to 16. Any system can have a number less than the maximum by simply not connecting devices to some zone monitoring circuits. Figure 8-2 shows a home floor plan divided into zones for a security system and possible locations for sensor devices within each zone.

Home Utility Specifications and Capacity

Securing the home against unauthorized entry is one of the first requirements of a security system. Since over 75% of all break-in attempts are made at first-floor entry points, securing these with door and window movement sensors is essential. Garage and upper floor entry points should also be secured if they offer direct access to the home. A backup detection of illegal entry could be motion detectors installed inside the home. If early detection of a threat is necessary, motion sensors outside the home around the perimeter of the yard are another option.

Fire protection mandates that the security system have a smoke or heat detector in each room. This requirement includes utility rooms and the garage where many fires start. Smoke and heat detectors should sound an audible alarm first to warn the home's occupants to evacuate the building. If the alert is not canceled within a minute or two by the homeowner (due to a false alarm), the security system should be programmed to summon the fire department. It is better to have the summons automatically done by the system than to require a phone call from the homeowner, who might delay evacuating the home or travel some distance to find another phone.

One or more of the zones in a home security system may be devoted to the home's utilities by the use of sensors to monitor whether they are functioning correctly, or at all. Electrical system monitoring is especially important because, if it fails, the home's occupants could be left in a dark building trying to find their way out in the face of an intruder or a smoky fire. A sensor on the electrical service line can let the homeowner know if power is out to the entire house. The security system can also be set up to activate emergency battery-powered lighting in the event of a power failure and to contact the utility with a service failure notification. Individual circuits within the home can also be monitored, but this would only be warranted if a power failure in the circuit might jeopardize the safety of the home in some manner. Circuits where a power sensor might be worthwhile include those to furnace and air-conditioning motors, those to pumps needed to keep the home dry, and those to freezers containing perishable food.

Homes that have natural gas service for heating can monitor that service by means of a pilot light sensor connected to the security system. The pilot light monitor can be either a light sensor or a **thermocouple.** In gas heating systems without continuous pilot lights, a pressure sensor on the gas line can be used to confirm that gas is still available for use. Another useful sensor in the area of a gas furnace or water heater, as well as a gas dryer or stove, is a **gas detector,** which can warn of a leak in the gas system. The vast majority of gas leaks result from an appliance inadvertently left on, or from a crack or burn-through in the gas-fired heating element of the appliance. A gas sensor placed in the room where these appliances are located can quickly alert the security system to a problem.

Water sensors are of two types, both of which can be very helpful in avoiding water damage in the home. The first is a **flow meter** placed in the main water line to the home. This device can monitor the flow of water in the home's pipes and sound a warning if the flow is excessive or if water is flowing at all during hours when it should not be. By checking the flow meter for the early morning hours, say from 2 until 4 am when no water would normally be flowing in most homes, even fairly small leaks in bathroom fixtures can be found and corrected. A flow meter is very helpful in detecting water loss in leaking sprinkling systems as well.

The other type of water system sensor is simply a device that detects the presence of water where there should not be any. Water sensors placed in dry areas or a basement can warn of a broken pipe or a leaking appliance before serious damage is done. If the home is in an area where ground water is a problem, a basement sensor can alert the homeowner to incoming water as soon as it appears. If the problem is pervasive enough that the home has a sump pump, which is a self-starting pump installed in a pit below floor level to keep ground water out, then a water sensor is a necessity as a backup for the pump in case it fails mechanically or from lack of power.

Security System Integration with Home Utility Systems

The utility system sensors that a home security system includes, if any, depend on the degree to which any of them is at risk of failure. Most areas of the country can experience the occasional short duration power failure, caused by lightning storms or equipment failure at power stations. That risk may justify a **power failure warning sensor** in case the problem occurs at night when it would not otherwise be noticed, or in case it occurs through the deliberate effort of someone trying to circumvent the home security system. Other than these two situations, most power systems do not fail often enough to need additional security warnings.

Gas and water are pressurized systems. Leaks in them are rarer than power failures, but they are also less likely to be noticed before doing serious harm. In areas subject to earthquake, catastrophic storm, or heavy flooding, these warnings may be a valuable addition to a home security system. Utility sensors are most easily installed during home construction, but they can be added later, either as hardwired components or as wireless devices.

Safety and Code Regulations

All utility systems are installed according to local codes and regulations. Security systems should supplement the built-in safety devices in utility systems, but never replace or circumvent them. Electrical wiring installed and maintained according to the National Electric Code (NEC) and local regulations ensures home safety by preventing electrical hazards. Security devices can detect a system breakdown only after it has happened. Prevention of a utility problem is preferable to detection and response after the fact.

Gas and water utilities also have installation and maintenance codes that should be diligently followed. The best security for utility systems is careful construction and good maintenance. You should be aware of the requirements for utility installation and maintenance in your area, and adhere to local ordinance requirements when working on them or installing security systems that monitor them.

Many local regulations govern minimum safety standards and security in homes. You should be aware of the requirements for home safety and security in your area so the systems you install conform to local ordinances.

Doors and windows and means of securing them in residences are among the items most frequently addressed by local ordinances. Some communities, for example, prohibit barring lower story windows or permanently locking them closed because doing so presents a hazard to people inside who may be fleeing a fire. Other communities recommend barred windows as a crime prevention measure. Some ordinances require double-keyed locks on exterior doors (locks which require a key to open from the inside as well as the outside). Others ban these locks and require a twist-knob opening on the inside as a safety measure in case of fire. Electric and remotely operated door locks (and in some cases window locks) are regulated by many local ordinances. You should be sure your security system designs conform to the local ordinances of your community.

Security System Types

Security systems, like home networks, transmit data between devices or nodes. Security systems differ from high-speed data networks in that their data is usually low volume and moves relatively slowly in comparison to say, an Ethernet. Most of the data flows only one way: from the security system's input sensors and devices to the central control module that monitors the entire system. The only output from the control module in some systems is the sounding of a siren or alarm bell and the possible automated summoning of outside emergency assistance (police, fire, or medical aid) in certain instances. Other more sophisticated systems may have additional built-in emergency responses such as activating sprinkler systems, shutting off utilities, or locking down entrances to prevent a break-in, but even in these systems, the vast majority of data flows from the sensors and monitors in the home to the central control with only an occasional instruction going the other way to an output device.

Hardwired Security Systems

Hardwired security systems, like hardwired networks, are easiest to install during new construction. Security system wiring is low voltage and so can be retrofitted in existing buildings, but with the same difficulty and additional expense that is encountered when wiring an Ethernet or a video distribution system. The wired security system may actually be a little more difficult to install than other networked systems because most of its sensor nodes are on outside walls where wiring is hardest to pull. Window and door sensors, as well as outside monitors and sensors, are all mounted on exterior walls that, depending on their construction, pose a real challenge for wiring installation.

Figure 8-3 shows the central components of a hardwired security system manufactured by ELK. The basic installation kit includes a wiring enclosure that can accommodate up to 200 security, lighting, and automation connections, a low-voltage **transformer** (power supply), a controller for programming and monitoring the system, a keypad, and a speaker. Sensors, lighting modules, and appliance modules are purchased separately to meet the requirements of different configurations. This system can include wireless elements, and can interface with hardware devices manufactured by many companies. It can also interface with an Ethernet LAN and has available software for PC control of the system through the LAN.

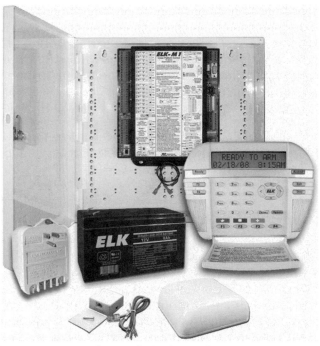

FIGURE 8-3   Main Components of a Hardwired Security System

Hardwired systems, although they have been used longer than the wireless types, offer no inherent advantages over them. Wired systems can be bypassed or disabled by a determined intruder, but this is also true of most wireless systems.

Professionally installed security systems connected to central monitoring stations outside the home were formerly all wired systems, but this is no longer the case. All the major monitoring services now offer wireless installations, and some appear to be moving away from the older wired technology and adopting the less expensive wireless systems. The three largest companies that offer professionally monitored security systems are ADT Home Security Systems, Brinks Security Systems, and GE Home Security Systems. Each of these companies and many other smaller ones install their own brand of security system and provide round-the-clock monitoring of them by telephone line from a remote site.

The most convenient wired security systems are those that use existing home AC or telephone wiring rather than requiring the installation of new cables. HomePlug manufacturers and X10 suppliers both offer extensive lines of home security devices that can be plugged into wall sockets and thereby interconnected. Both technologies can also interface their security systems with the home network so the security system can be PC-monitored and controlled by the system's central computer. X10 technology suppliers have moved mainly to wireless systems, but some X10 power-line components are still available.

HomePNA is the telephone-line network technology discussed in Chapter 3. These manufacturers have not yet targeted home security as a focus of their development efforts, but this technology is adaptable for security system use. Some of the HomePNA controller devices now on the market can easily be set up to control security sensors and to feed security system data into an Ethernet network for PC control.

Wireless Security Systems

Wireless systems are less expensive to install than wired ones and offer the same security features for the most part. Wireless transmissions can be disrupted by interference produced accidentally or deliberately, so the wireless security systems are no more secure from tampering than the wired ones, although they do have more flexibility and economy than their wired counterparts.

Wireless sensors and other components can be installed almost anywhere, without concern for how to get wires to them. This is a big advantage when placement of a sensor or other device may be critical to its function. The fact that most security input and monitoring devices do not need two-way communication with the central control unit of the system means that only a transmitter is required in the sensor or monitor. No receiver is necessary, which reduces costs for these devices.

Figure 8-4 shows the basic components of an X10-based wireless security system. It includes a master control, power-line receiver/transmitter, a keypad controller, and a remote control. Components can be added to the system that expand it up to sixteen zones and permit a variety of sensing devices in each.

At least six other companies manufacture wireless home security systems, ranging from starter kits at $149 up to 16-zone add-on systems that can cost $1,000 or more. Some of these, however, can connect to a PC through proprietary technology, but none operate on the standard wireless networking technologies. Despite the fact that wireless home security systems offer considerable advantages and a large potential market, neither of the major wireless technologies in the home networking market, WiFi and HomeRF, have security systems available at present.

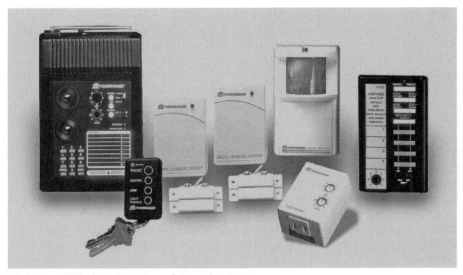

FIGURE 8-4   A Wireless X10-Based Security System

Remote Access Systems

Remote access or monitored security systems have been a standard for businesses and homes for many years. ADT, Brinks, and GE, the three largest firms offering this service, have programs through which the company installs a home security system for a nominal charge (or at times, during promotional specials, without charge) and the customer is charged a monthly service fee for the monitoring of the system by the company's personnel over a telephone line. Monitoring service contracts can range from 2 to 5 years in duration after which they are renewable for as long as the customer desires to maintain the service.

Monitoring companies install either wired or wireless security systems and the extent of coverage within the home and outside is a matter for individual negotiation and pricing. Some monitoring contracts specify that the installed security equipment is leased and will be reclaimed and removed by the company at the expiration or termination of the monitoring contract. Other arrangements allow the customer to either retain the installed equipment at the end of the monitoring contract or buy it at a reduced amount.

Monitoring services offer professional installation of security equipment and 24-hour monitoring. They also maintain the equipment they install. Many professional installers work for firms associated with ADT, Brinks, or one of the other monitoring services.

Security System Equipment Locations

A security system's components may well be the most spread out of any home technology installation. Home security devices are designed to monitor the perimeter of a home environment against intruders as well as the interior against fire and other hazards. Consequently, the system may have sensors and other input devices outside the home directed toward the farthest corners of the yard, and other devices inside the home monitoring conditions there. A control unit, located centrally in the home and probably near the network gateway or central computer, receives the input data from all the sensors. The central location of the control unit allows it direct and easy access to telephone lines and other communications links that may be needed to summon assistance in the event of an emergency. A central location also places the control unit well away from any outside intruder who might attempt to disable it or shut it down.

Lock and Keypad Locations

Keypads should be located near the doors they serve, but also in plain sight so anyone using them will be visible to passersby. Wireless keypads can be mounted directly on the door above the doorknob. Hardwired units need to be placed on a wall with access to power. Most wired keypads and some wireless ones operate on either 9–16 volts or 20–26 volts. The power can usually be either AC or DC, and is stepped down (reduced) from the 120-volt current in the home's electrical system by a transformer.

Some wireless keypads are battery-operated, as are the locks they control. These units require at least an annual battery change, more frequent if use is heavy or if the unit is located where it is subjected to a lot of heating from sunlight. (Heat increases the chemical reaction of batteries, depleting their power more rapidly.)

Battery-operated keypads are usually 6–9 volt units, while the locks often use 12 or more volts. The expense of battery replacement may eventually outweigh the convenience of wireless installation for these keypads and locks. Nonetheless, they offer security protection as good as the hardwired units, for considerably less initial expense. Figure 8-5 shows a battery-powered keypad lock on the left, which is self-contained and functions somewhat like a combination padlock or safe. On the right is a typical front door area keypad location.

Keypads and other access devices such as **swipe slots** for **magnetic keys** and identifier pads for fingerprint impressions are usually mounted on the door lock itself. This placement is low enough so the home's occupants can easily access them for input, but low enough that they may need some protection from sun and weather if the door is in an unsheltered position. This may consist of a protective

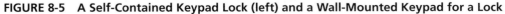

FIGURE 8-5 A Self-Contained Keypad Lock (left) and a Wall-Mounted Keypad for a Lock

cover or similar shield. Such covers should open to the side, not toward the top, so they will remain open without being held, allowing a single hand to access the keypad or identifier unit inside.

Sensor Locations

Sensors can be located throughout a home and yard and connected to one of the zones defined on the security system's main control panel. The type of sensor used determines to a large degree where it can be located to function efficiently, but some general rules apply to placing these devices.

Sensors should be located where they will not be disturbed by curious (or malicious) people. This usually means locating them high enough to be out of reach without the use of a ladder. Most types of sensors can function well when placed at least seven or eight feet above ground level. When a sensor cannot be located high up, either because there is no suitable mounting location or because it will not function in a high position, then the alternative is a low location with good concealment of the device, or strong protection for it against tampering.

Outdoor sensors need some protection from strong sunlight or bad weather. This may mean mounting them under roof eaves or in other sheltered places, or it may require that some type of screen or cover be added to protect the sensor. Constant sunlight and water can eventually damage even most weatherproof hardware, so a protected location or some type of weather cover is always a good investment. Figure 8-6 shows a wireless **motion sensor** located beneath a front porch facing the outside stairs. This unit senses approaching persons and sends a signal to activate lights and a video camera at the front of the house. The sensor is invisible to approaching people (it is below eye level, but easily senses moving feet) and is sheltered from sun and weather by the porch above it. Although it appears nearer in the picture, it is mounted more than a foot back from the stair above it. Its field of view is fairly narrow, but never obstructed. Although the sensor is well protected, it still will not function well if the outside temperature is above 95 degrees.

FIGURE 8-6   A Wireless Motion Sensor beneath a Front Porch

Sensors need to be placed where they cannot be easily blocked from functioning. Motion sensors should not have any large objects in front of them that would mask an intruder's approach. They also should not be placed where a shield can be placed in front of them from the side. Placing such a shield may trigger the sensor once, but if the shielding remains in place, the sensor thereafter regards it as the normal setting of the area and any movement taking place behind it goes undetected.

All wired sensors should be located where their wiring cannot be easily cut or bypassed. Many security systems sound an alarm if a sensor is cut out of the system, but not all do. The installer should make the wiring as difficult to access as possible to prevent tampering.

Sensors also need to be placed where false reports will not be triggered by events or conditions that are not security threats. Motion sensors should not be placed where the normal wanderings of family pets can trigger an alarm. **Heat sensors,** whether inside or outside, should never be placed where direct sunlight falls on them during any part of the day at any time of the year. They also should not be placed where unusual heat from a fireplace or cooking appliance can reach them.

Smoke detectors should not be placed directly above a stove or oven where the occasional smoke or steam from normal cooking may trigger them. Placing such fire-warning devices a little distance away from potential heat and smoke particle sources can prevent constant false alarms, but still give adequate protection from a genuine fire.

Note: NFPA standards call for at least one fire and smoke detector to be located on every floor of a home and outside each sleeping area. Additional units can be placed near hazards such as kitchens, fireplaces, furnaces, and any open flame.

Camera Locations

Video cameras allow the home and its yard to be monitored remotely. A good video surveillance design has all outside areas covered by two cameras, if possible. This requires placing the cameras high up, away from visual obstructions, and using cameras that can pan (move from left to right and back) by remote control. A panning camera under the eaves of a home can often cover 270 degrees of the surrounding yard. With a camera on each corner of the home, all areas of the yard are at least double-covered and some more. Double coverage of yard areas eliminates any blind spots that a single camera might have because of visual obstructions such as trees and other landscaping in the yard.

If cameras are also placed inside the home, double coverage usually is not necessary, but the camera needs to be placed out of the way of visual obstructions. Inside cameras should also be concealed if at all possible. Their obvious presence makes many people nervous and may cause a definite chill in the home's social atmosphere. Placing inside cameras in concealed locations within the home's decor lets them function without the drawback of making guests, or even the home's usual occupants, uncomfortable. Miniature cameras are now available that can be placed in books, ceramic pieces, picture frames, or other home furnishings. These devices can be easily relocated if the occasion requires, and they eliminate the need for any hole in the wall such as was once necessary to conceal a camera behind a one-way mirror.

Cameras, like most electronic equipment, do not like heat or water. This means they should be kept away from direct sunlight, inclement weather, and sprinklers. Cameras can also be blinded by light sources shining directly into the lens. They should be located, as much as possible, so as to avoid this problem. Yard lights can usually be relocated or redirected so they do not shine into the camera lens. The direction of sunlight cannot be changed, however, and may require careful positioning of the camera so sunlight does not hit the camera lens during any part of the day. Remember that the sun's angle in the sky changes with the seasons. A camera safely in the shade of the eaves during July and August may be exposed by the lower angle of light in December and January. This phenomenon is significantly more pronounced in the country's northern latitudes (and is extreme in Alaska where the summer sun circles around the horizon), but it affects all areas to some degree. Reflected light coming off of a nearby window or the surface of a swimming pool can also blind a camera. So can car headlights entering the driveway or passing by on a curved street. All of these possibilities should be checked and, to the extent possible, eliminated when placing video cameras.

Security System Components

A number of individual components make up the typical security system. Not all of these are used in any given system, but you should be aware of their functions so you can select the most effective pieces of hardware for each system you design. Some systems include many of a single type of sensor and none of another. The overall size of the system is determined by the size of the home and yard and the degree of security protection desired. Each of the following devices may find application in a security system you design.

Access Devices

The ideal access formula for any security system is that it allows the home's family and their guests to enter with minimum effort, and prevents everyone else from entering at all, unless specifically cleared to do so by the homeowner. The problem with implementing this formula, of course, is providing a means for the home security system to determine who is welcome and, by elimination, who is not. Access devices are the methods used to do this. They identify those authorized to enter the home and then mechanically or electronically unlock an entrance door. Access to a home through its security system can be governed in different ways. How restrictive these methods are is determined by the homeowner.

Keypads and Identifiers

Almost all home security systems include electrically operated door locks and garage door openers. These can be activated with a remote signal from a key-ring transmitter, but most also include a keypad or another form of access control near the door for use by those who do not have a remote unit.

A keypad is a device with numerical keys in the same pattern as a telephone that allows the user to manually enter the numeric code that opens a **cipher lock.** Keypads can be battery-operated or hardwired to a low-voltage electrical connection. Entering a preset four- to six-digit number on the keypad activates the electric lock mechanism so the door can be opened.

Most doors equipped with keypad-activated locks have no key lock mechanism at all. The door can be opened manually from the inside, but there is no keyhole on the outside for a burglar to attack. Only the correct number combination entered on the keypad activates the lock. Many of these systems also have a **lock-down feature** that allows the outside keypads to be deactivated at night or while the homeowner is away. When deactivated, the keypad does not open the lock even if the right number combination is entered. Only the remote unit can open the door from the outside, although it can still be opened manually from the inside. Even with plenty of time to guess number combinations, a would-be burglar will not get the door open.

Passwords

Passwords are alphanumeric combinations that the security system recognizes. They can be composed of any number or letter combination that can be easily remembered, but not easily guessed by an unauthorized person. The password is entered on a keypad and, when it is recognized by the security system, the door lock is opened to admit the person. If the password is a name or letter combination, it is entered on the keypad using numbers for letters in the same manner as the letters of the alphabet are listed on a telephone keypad. Michael, for example, is 6424235 and Monique is 6664783.

Passwords are good access devices and provide strong security for the home, as long as they are used correctly. To remain effective, passwords cannot be shared with other people (who may share them with others until the wrong person has them), nor written down, used or otherwise divulged in a manner that allows unauthorized persons to find out what they are. Giving a password to a friend seems innocuous, but it is really giving the friend the key to your house, and the right to duplicate it (by passing the password to others) as often as he or she decides. Passwords are only as safe as the authorized users keep them. For this reason, other access restrictions are sometimes added to the home security system.

Keys

Keys are the original method of opening locks and are still effective for access control. They have been stolen and duplicated for hundreds of years and the locks they fit have been picked and forced for an equal length of time, but a well-designed mechanical lock with a closely fitted key offers a good measure of protection for a home even today.

The security potential of keys has also been enhanced by new technology in recent years. Keys can be used alone to open a lock, or used in combination with a password so both are required to gain entrance. In this case, entering the password tells the security system that the person has the right to use the key, so when the key turns the lock, the door opens. Without the password, the door will not open, even with the key. The two devices work independently of one another and neither alone can open the door.

New types of keys and locks have also been developed. Some of these are electronic and connect directly to the home security system. Instead of having the traditional cuts and ridges found on most keys, which fit moving tumblers in standard locks, these new keys are fitted with a numerical combination recorded on the magnetic strip of a plastic card, or imbedded as a code on a metal key.

Figure 8-7 shows a standard double-cut mechanical key on the left, a metal Marlock key with a numeric password encoded on it in the center, and a plastic magnetic-stripe key on the right. Both of the encoded keys work electronically. They have a numeric combination recorded on them that can be unique to every key, and is read by the security system when the key is inserted in the lock or passed through a reader. If the numeric code is recognized, the lock opens.

FIGURE 8-7   A Mechanical Key (Left) and Two Encoded Electronic Keys

Electronic keys have the additional security advantages of being very diffi-cult to duplicate and being recordable every time they are used. Being difficult to duplicate prevents a would-be intruder from copying the key and also prevents the rightful owner from inadvertently or deliberately giving away the password on it. Neither the owner nor anyone else can read the password or see it input, so it remains secure no matter how the key is used. If the key is lost or stolen, the password on it can simply be blocked on the security system. No other keys are affected and no change of locks or other devices is required to maintain security.

Recording the use of electronic keys allows the homeowner to monitor who is using them and when. Each time a key is inserted in a home lock, the security system records its password and the time it was used. All entries into the home are thus recorded and, if any security problem later develops, the users of the keys can be traced through the file in the security system.

Time-Coding

Electronic keys and locks can also be time-coded so the lock either does not open during set times of the day, or automatically sends a notification if it is opened during those times. In home security systems, such times might be set for the late hours of the night and during the day when parents are at work and children at school. Setting the lock to notify a monitor during certain hours still allows an authorized person entry, but also signals that the door has been opened, thus alert-ing the homeowner to check on who has arrived.

Many security systems can also log the date and time when events occur—that is, record each time entry, exit, or other events happen. Some also allow users to set a time schedule on the system so it logs events (either security breaches or normal entry and exit of the premises being secured) only during specified time blocks.

Biometric Identification

Biometric identification systems are presently not used much in home security systems, but may become more common in the future. They simply take the pass-word concept to another level by identifying authorized persons according to something unique in their physical makeup.

Biometric identifiers include a fingerprint (usually a single thumbprint impressed on a sensitive pad and scanned into the security system), a retinal scan, or a photographic image of a person measured for uniquely distinguishing features by the security system. All these identifiers are unique to every individual and so offer an absolutely secure means of identifying the person who is authorized to enter a secured area. Since these identifiers form part of the individual's body, they cannot be loaned out, duplicated, or stolen.

At present, all of the identifying programs used for these body identifiers are so expensive they cannot be justified for most home security systems. These systems are now in common commercial use, however, and their cost is declining to the point where it will soon be economical for many home systems to use them as well.

Cameras and Recorders

Video cameras located at strategic points around the exterior of a home and at the entrances provide a means of visually checking the area in view of the camera. This allows the homeowner to identify visitors personally and also spot any potential intruders. If the cameras are connected to video recorders so that their views are intermittently recorded on tape, then any event that occurs at the home can be reviewed on the videotape later. This record of events, with the time recorded on the tape, is very helpful to police investigating a crime that occurred in the owner's absence and may even be instrumental in solving the crime. If nothing unusual is recorded on a tape, it can simply be reused.

Surveillance video cameras need not be expensive. Good quality units are available for under $100, and the addition of an automatic/remote motor that pans the camera left to right and back again at a set speed will not add more than another $75 to a station. It is more important that the cameras function well in low-light levels than that they view in color. A black-and-white camera with high light-gathering capacity provides better images for identifying people and details than most color cameras, especially at night.

Small digital surveillance cameras are now readily available at costs comparable to analog units. Their price will probably decline further in the future and analog cameras will gradually be phased out. Like their entertainment counterparts, digital surveillance cameras offer better resolution and can function in lower light levels than analog units.

If a video camera has a **panning motor** attached, it needs to be hardwired to a power source since the motor will consume too much power for long-term battery-powered use. A stationary video camera can be battery powered, but will still require frequent battery changes or recharges, especially if the batteries are exposed to high heat in an outdoor location, which tends to discharge them more quickly. Figure 8-8 shows a miniature surveillance video camera on the right and its wireless receiver on the left. This camera is less than 4 inches high. It is intended for interior use and is not weatherproof.

Most surveillance video cameras for home use are still analog with the corresponding limits of resolution that this technology has. Most are also black-and-white units, although color cameras are available at a higher cost. Digital cameras are rapidly overtaking the analog versions, however, and new installations should use the newer technology. Digital cameras are usually color and function better at frame rates of one or two per second than analog. Their output can also be recorded directly on a flash drive or computer hard drive.

FIGURE 8-8   A Wireless Video Surveillance Camera (Right) and Its Receiver

Data from analog cameras must be recorded on tape or encoded to digital form before it can be recorded on digital media. The digital conversion process is more complex and more expensive than analog tape recording. For this reason, a VHS video recorder is sufficient for analog video security systems.

The recorder for a video surveillance system should be a high-quality unit capable of recording all the resolution picked up by the camera. It should also have durable recording heads that are serviced frequently since it will be in constant use recording images, most of which will not be of any value, but a few of which may be very important. These few will need to be as clear as possible in order to reveal valuable details; hence, the need for a recorder that records everything the camera sees.

The best recording unit is the type that records only one frame per second from the camera. This conserves videotape, allows the frames that are recorded to be of the highest quality, and permits up to a continuous week of surveillance to be recorded on a single tape. Alternatively, the recorder can be an extended time type that records 24 frames per second, but runs the recording tape more slowly so that up to 10 hours of surveillance can be recorded on each tape. This type does not record at the same resolution as the single-frame-per-second unit, and its tapes have to be changed daily.

Monitors

A security system with video surveillance cameras does not have to have a monitor if the camera images are being recorded. They can be played back later from the tape if all that is needed is information from a recorded event. A video monitor is a good idea, however, if the cameras are to be used to visually monitor the yard or entrances to the home at any time. Generally, one large-screen monitor that can be switched to display any of the camera images or split to display several smaller images simultaneously is preferable to an array of small monitors constantly displaying each camera's image. The larger display can reveal details better, and there is almost never a need to have more than one camera image on display at a time.

Any good television set can serve as a monitor for a security system. None of the video cameras, analog or digital, used for security has resolution higher than the NTSC standard, and a television set can display that format adequately. Higher-quality monitors may last longer, but they will not produce a better image because the original camera images are not of a high enough definition to begin with.

Sensors

Sensors are the devices that provide input to a security system. They tell whether conditions are normal, usually by reporting nothing except that they are still functioning. If conditions do change, then the sensor reports the altered state of whatever condition it is designed to monitor. At a minimum, it reports only that the condition is not normal. If the device is more sophisticated, it may report the exact degree of change, or how much it has altered from the normal state, but the basic purpose of a sensor is to report that the condition it is monitoring has changed out of the normal range. Many types of sensors exist.

Break Switch and Make Switch Sensors

Break switch and make switch sensors signal when they are either turned on or turned off. One state is defined as the normal condition; the other is the changed condition. As long as the normal condition is maintained, the switch sends no signal except to confirm that it is functioning. If the condition changes to the abnormal state, the sensor signals the change.

Break switches are normally closed so that a current or magnetic force flows through them constantly. They are the type of sensor placed on windows and doors. The switch is placed on the door or window casing and the plate that completes the circuit in the switch is placed on the window or door. As long as the window or door remains closed, the switch plate keeps the circuit closed and the sensor remains quiet. If the window or door is opened, the plate is pulled away from the switch, thus breaking the circuit. The sensor then sends a signal indicating its circuit has been broken by the opening of the window or door.

Break sensors can be attached to any fixture where movement of one part opens the switch by separating the plate from it. Besides outside doors and windows, these sensors can protect cabinets containing valuables, rooms that should not be entered, and objects that should not be moved.

Make switches are sensors in which the normal state is open and the changed state is closed. These are most often found as **pressure pads,** which signal an alert if weight is placed on them, forcing the switch closed and completing the circuit. Pressure pads can be found in driveways where they signal a car or other vehicle moving over them. They can also signal the approach of a person stepping on them. In either of these cases, the pressure pad might not signal any alarm because of the movement. It might simply trigger the opening of a gate or start a video camera to record the arrival or departure of a person. Note that a pressure pad does not tell the direction of movement unless two are placed in line and successively tripped. Pressure pads are sometimes more useful as initiator devices than light beams or motion sensors because their normal range can be set to exclude some weight but respond to anything over a set amount. Thus, the family dog or cat walking on the pad will not open the gate or door, but the weight of a person will. Likewise, the children can play on the driveway without opening the outer gate because half the weight of a car is required to set it in motion.

Light Beam Sensors

Light beam sensors consist of two pieces: a sender and a receiver. The sender sends a beam of **laser light** toward the receiver, which receives it constantly. Receiving the light is the normal state. If something moves between the sender unit and the receiver, cutting off the light beam from the latter, the sensor sends a signal that the light beam has been interrupted, the abnormal state.

Light beams are used to sense movement, usually in places where there should be no movement. A beam can be sent across an entryway, a section of yard, a driveway, a room in the house, or any other place where the beam should not be broken during specified times. If the beam is broken, an alarm can be sounded or assistance summoned.

Light beams are limited as security devices by the fact that they can be tripped by anything that breaks the beam, and not everything that does so is necessarily an intruder. Light beams do not work well where pets roam or animals (belonging to neighbors, or living in nearby wilderness) are likely to break them. They can also be tripped by falling leaves, drifting snow, or windblown newspapers and other debris. They work best when set close to the ground (so an intruder cannot slip under them), but that is also the position where they are most likely to be set off by non-threatening events.

Figure 8-9 shows a light beam transmitter on the left and its receiver on the right. These units send a beam across a garage door opening and serve as a safety device for the automatic garage door. If anything breaks the beam between the sender and receiver unit while the garage door is descending, the door's downward motion is reversed and it rises again. This prevents the heavy door from closing on a person or on a car that has not cleared the garage entrance.

Under normal conditions, a light beam is invisible, so it is hard to detect or disable. Indoors, a light beam can be set high off the floor, where it is less likely to be tampered with or obstructed accidentally, and still function well. It can be placed so that opening doors or cabinets break the beam as well as an intruder walking in front of it. An inside beam can also be reflected in two or even more directions to increase its range. Note that the presence of a visible mirror suggests that something is being reflected and may indicate the presence of an otherwise undetectable light beam.

FIGURE 8-9   Laser Light Transmitter (Left) and Its Receiver

Heat/Motion Sensors

Most home security system motion sensors are actually **infrared devices** that detect changes in heat (infrared) radiation within their ranges. These devices sense a given amount of radiation coming from surrounding objects within their detecting range, usually about 50 feet. They then regard that amount of radiation as the normal state of their environment and do not react to it. Whenever any object that radiates heat, such as a human body or a warm car engine, enters the device's range, it radiates additional heat toward the device, thus moving the amount of radiation received out of the normal range and causing the sensor to signal that something has moved into its range. It is actually the change in heat that has been sensed rather than the motion, but the response is the same.

Infrared sensors are very good at sensing heat changes as large as a human body, but some can be set so they do not react to a smaller change, such as that made by the presence of a cat, small dog, or other animal of similar size. Once an infrared sensor has been triggered by a heat change, it usually resets itself so the current heat pattern becomes the norm, including the intruding person if he or she is still present and relatively motionless within the device's range. The sensor will not trigger again unless the heat level changes again, so the intruder can stay undetected within the range of the sensor. It does trigger again, however, when he or she moves quickly or leaves the range because again the heat level is changed.

Infrared sensors do not work well in high heat. As the ambient air temperature approaches 98 degrees, the temperature of the human body, the movement of a person into the sensor's range no longer alters the radiant heat pattern greatly. The person is radiating about the same amount of heat as the air and surrounding objects so the sensor does not signal any change. It does not "see" the intruder as a difference in heat. For this reason, infrared sensors should not be used outdoors in areas where the temperature may often reach the high 90s.

Figure 8-10 shows an infrared motion sensor mounted in a garage location with a pair of security lights. This unit turns on the lights when it detects motion inside the garage. The increased light level in turn triggers a light sensor that starts a video camcorder that records the lighted area. This is a self-contained system

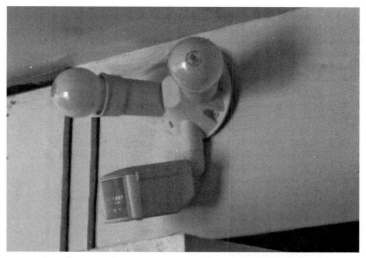

FIGURE 8-10   Heat-Sensing Motion Detector Mounted with Security Lights

not connected to a wider security system. It simply records activity whenever any movement is detected. Although the sensor is mounted near the lights, which produce considerable heat when illuminated, it is not affected by their output. Once triggered, the motion sensor has a timer that will keep the lights on for about 10 minutes. It will then turn them off unless further motion is detected.

Temperature Sensors

Temperature sensors can be used to signal either great heat or extreme cold. They measure the temperature of the air around them and have a normal range to which they do not react. If the air temperature rises above (or falls below) the normal range, the sensor signals the changed state. Some temperature sensors can record temperature changes degree by degree so that actions may be taken when certain levels are reached, but most are simply switches that signal only when the temperature moves outside their normal range. Most are set to signal a temperature drop or a temperature rise, but not both.

High-temperature sensors are used as fire alarms. They are usually mounted high in a room and signal a temperature change when the air temperature at ceiling level reaches about 140 degrees. Since heat rises in a room, a fire burning at floor level will quickly raise the temperature at the ceiling to this critical level. Heat from a stove or other warming device will not get the air up to critical temperature unless these appliances are left on too long and become a fire danger. Temperature sensors do not react to smoke, only to heat, so they make good fire detectors in kitchen areas where some occasional smoke is common and smoke detectors cannot be used.

Low-temperature sensors are mostly used for protection against freezing. Their normal range may go down to the mid-30s, but as the temperature drops to near freezing (32 degrees) they signal a temperature change. The security system may then be programmed to shut off outside water systems to prevent pipes freezing. It may also activate de-icing systems on driveways, roofs, and rain gutters. When the temperature rises again, these latter systems are shut down, although the sprinkler system may remain off until a programmed length of normal temperature time has elapsed. Low-temperature sensors can also be useful in outbuildings where they can signal danger to animals or machinery when the temperature falls too low because of failed heaters, windows or doors left open, or other mishaps.

Water Sensors

Water sensors are switches that react to the presence of water. Their normal state is dry. If they detect water, a circuit is completed and a signal sent to the central control. They are used to warn of water's presence where it should not be. The cause of such water can be leaking or broken pipes, seeping ground water, excessive lawn watering, or any number of other causes. If the water is in the home's basement, it will likely cause damage. The water sensor can prevent that with a timely signal. A water sensor should be placed at the lowest point in the drainage area it is protecting. This will allow the first water entering the area to trigger the sensor. The signal may start a pump working or merely send a warning alarm that something is amiss.

Fire/Smoke Detectors

Smoke detectors are sensors that constantly measure the purity of the surrounding air. Whenever the air becomes polluted with smoke particles, the smoke detector signals an alarm, usually an audible warning. Smoke detectors are the preferred means of fire detection because they cost less than heat detectors and thus can be more economically placed throughout the home, but they do have some limitations. They are often set off falsely in homes where people smoke or where fireplaces or cooking smoke exist to any significant extent. When this happens frequently, the smoke detector is likely to be left off to stop the false alarms and it thus becomes useless for its intended purpose. Smoke detectors do not react to heat, only to smoke, so a fire burning with relatively little smoke may not trigger them for a while.

Smoke detectors, like heat detectors, can be connected to a security system and trigger a call for the fire department when they go off, but this setup should only be created if false alarms are very rare. The fire detection system least vulnerable to false alarms may be one with smoke detectors in bedrooms, family rooms, and living rooms, and heat detectors in kitchens, furnace rooms, and wherever open fireplaces or barbecue grills are located.

Security Panels

Security panels are the central control devices for security systems. A security system control panel may also be referred to as a control box. Arming and disarming buttons (or combination codes) on the control panel allow the system to be turned on and off by the owner. In some systems, the arming and disarming controls may be located in a separate box that is also called a control box. Each of the sensors set up in the security system is wired to, or wirelessly connected to, the security panel and reports its status to the panel. If a security breach is signaled from any of the sensors, the panel can be programmed to take appropriate action in response. This may be to sound an audible alarm to notify the homeowner of the problem, or to automatically call and summon emergency assistance, as in the case of a fire alert.

The monitoring functions of a security panel are divided into zones so the location of any security breach signal can be easily located. Some panels can connect to a PC, either as a node on a network or directly through a USB port. This connection enables security status to be displayed on the PC and its display monitor to serve as a monitor for the security system's surveillance cameras.

Some security panels can record the time of any security breach anywhere in the system and make other responses as well. These responses may include turning on lights in the home to aid the occupants in getting out, starting video cameras and recorders photographing in real time, sounding an outside alarm intended to summon assistance from neighbors, or shutting off gas lines. The panel does not perform any of these actions, but sends signals activating response devices in appropriate locations.

Alarms

An alarm is a response device that sounds an audible or sometimes a visual warning of a security system breach. The alarm is connected to the security system's control panel or, in some systems, may be built in as part of the panel. When the panel receives an alert signal from a remote sensor, it directs the alarm to sound the appropriate warning. Different alarm sounds can be programmed in an alarm as responses to different warning signals from sensors. A siren sound might signal fire danger while a repeating bell might be programmed as a response to an unauthorized entry attempt. Both signals are intended to alert people in the home to the security breach, and the latter one may also scare off a potential intruder by letting him or her know clearly that his or her presence has been noticed.

Alarms are usually low-voltage powered (12 or 24 volts) and should have backup battery power in case house current is interrupted. They should always be loud enough that their sound reaches all areas of the home. They should also be impossible to sleep through. Because they are loud and sound a high-pitched noise, alarms can be very annoying if they continue to sound once their alert has been delivered. For this reason, alarms should be placed where they can be easily turned off in case of a false alert. (If the alert is genuine, it is intended that the intense sound will literally drive people out of the house and to safety to get away from it.) Alarms generally do not do more than sound an audible warning, but they are often connected to calling devices that summon emergency assistance automatically.

Call-in Devices

Call-in devices are essentially automatic telephones programmed to call one or more numbers and deliver a preprogrammed message. The simplest call-in devices dial a number and begin repeating their message. They do not detect any response at the other end of the call they have placed. More sophisticated call-in devices dial a number, wait for a pickup response at the other end, and then deliver their recorded message. If a busy signal is noted on the line, the call-in device goes into a recall mode, waiting a specified length of time and then redialing the number. It may also call a second number and deliver its message. Some devices can call up to four numbers in succession to be certain someone at one of them hears the intended emergency message. Call-in devices are also powered by low-voltage current and should have a battery backup power source in case the AC current in the house is shut off.

Summary

- Factors to consider in designing a security system for a home include: what or who the security system is going to protect; what major threats must be warded off; what level of risk exists; and what factors external to the home environment affect the risk assessment.

- Security systems can be wired, wireless, or a combination. Hardwired systems can be easily installed in new construction, but are much more difficult to retrofit in existing homes. Wireless security systems are usually the preferred choice for retrofitted installations as well as many new homes.

- Remotely accessed or monitored security systems have been offered by commercial security firms for years. For a monthly fee, the company installs a security system and monitors its sensors over a telephone line.

- Security systems are organized in zones so their protection devices can be easily monitored and the origin of any security breach quickly located. A security zone can be any area of a home or yard. The limits of a zone are really the number of devices that the security system's central control can monitor for a zone.

- A home security system should be integrated as much as possible with other technology systems in the home. It should conform to local code and ordinance requirements and provide monitoring of utility systems against failure or breakdown.

- Security devices, especially sensors, need to be placed where they cannot be easily blocked from functioning. They need to be placed where false reports will not be triggered by

Security and Access System Fundamentals • 273

events or conditions that are not security threats, where they will not be disturbed by curious (or malicious) people, and where they are protected from weather and other hazards as much as possible.

- Access devices in security systems include electronic locks, which are accessed by keypads or encoded keys. Other access controls include timed locks, passwords, and biometric identifiers, but the last of these is expensive and rarely used in home systems.

- Video cameras located around the exterior of a home and at the entrances provide a means of visually checking the area in view of the camera. The camera images can be monitored from inside the home on a television screen and recorded on tape for later review of significant events.

- Sensors are the devices that provide input to a security system. They tell whether conditions are normal, and if conditions change, they report the altered state of whatever conditions they are designed to monitor.

- Many types of sensors are available, including break and make switches, motion detectors, heat sensors, smoke detectors, light-beam sensors, temperature sensors, water detectors, and electric current monitors.

- Security panels are the central control devices for security systems. Each of the sensors set up in the security system reports its status to the security panel. If a security breach is signaled from any of the sensors, the panel can be programmed to take appropriate action in response.

Key Terms

Alarm   An audible warning of fire, illegal entry, or other security breach. Usually a siren, a bell ringing, or a similar sound.

Biometric identifiers   Unique distinguishing physical features of an individual that can be used as means of identification. They include fingerprints, eye retinas, voice prints, and facial characteristics.

Break switch   A sensor that monitors a closed switch and signals if the switch is opened and the circuit broken. Used on doors, windows, and containers that should remain closed as their normal state.

Bypass   Any means of evading a security device by making it appear to be functioning in a normal state when, in fact, conditions have changed. A window sensor, for example could be bypassed by slipping a loose plate onto its contact points to make the switch continue to signal normal (closed) even when the window is opened.

Cipher lock   An electronic lock that is opened by entering a preset numerical code. Some locks can have many access codes, some only one.

Detection devices   Sensors that discern changes in conditions that indicate the presence or passage of a person or persons.

Flow meter   A device that measures the amount of water flowing in a pipe and signals its measurement to a recording device. Used to detect leaks or breaks in pipes or sprinkling systems.

Gas detector   A device that senses the presence of natural gas, carbon monoxide, or other poisonous gas.

Heat sensor   A device that monitors air temperature and signals if it rises above or below a set range. Used in fire detection alarms (high temperatures) and freeze detectors (low temperatures).

Infrared devices   A device that functions by detecting heat (infrared-wavelength energy) or changes in the heat level of its surroundings.

Keypad   A numeric pad similar to a telephone dial pad that is used to input passwords to cipher locks and other security devices.

Laser light   Coherent light waves transmitted in parallel beams so they maintain their intensity over long distances. Used in light beam sensors and many other devices.

Light Beam sensor   A device in two pieces, one of which transmits a beam of light to the other across a distance. If the light beam is obstructed, the sensor signals its failure to arrive.

Light sensor   A device that senses the presence of light and triggers an alert, or senses the absence of the light and does the same.

Lock-down feature   A program in some security systems that allows all door locks to be deactivated during specified times so they cannot be opened from the outside at all, even with a valid access code.

Magnetic key   A key similar to a plastic credit card with its access information coded on a magnetic stripe, which can be read by an access device.

Make switch   A sensor that monitors a switch that should remain open and signals if it is closed. Most commonly used as pressure pads that are closed by weight passing over them.

Motion sensor   A device that detects any object that radiates heat moving into its range and signals the change in its surroundings.

Panning motor   An accessory device for a surveillance video camera. The panning motor slowly swings the camera right and left, allowing it to cover more area than a stationary camera could.

Power failure warning sensor   A device that detects the presence of voltage (electrical pressure) in a circuit and signals its absence to a monitor.

Pressure pads   A make switch that is closed when a set weight is placed on it. The sensor then signals its changed state.

Pressure sensor   A device that monitors gas or liquid pressure in a pipe or tank. If pressure drops too low (or, in a steam pipe, rises too high) the sensor signals the change.

Redundant protection   In a security system, having two or more sensors or detectors monitoring one area or entrance. Provides additional security to high-risk areas.

Remote access   Security system monitored by an outside commercial firm through a telephone or radio connection to the home control panel.

Security panel   The control center of a security system to which all of the system's sensors and surveillance devices report their status. The panel may also direct responses, record data, and sound alarms.

Security zone   An area in a home or its yard that is monitored by a specific group of sensors in the security system. Zones are set up with defined monitoring devices so the location and nature of any security breach can be quickly identified.

Sensors   Devices that monitor an object (door, window, floor area) or condition in or around the home and signal the security panel if its status varies significantly from normal.

Smoke detector   A device that signals the presence of smoke in the air around it. It does not react to heat or flame, only to the presence of smoke particles.

Surveillance video camera   A small video camera used in a home security system to monitor a yard or part of the home's interior. The camera's output can be displayed, recorded, or both.

Swipe slots   An input device for plastic magnetic-striped keys. The plastic key is passed (swiped) through the slot in the device that reads the key code and grants access.

Temperature sensor   A device that measures the ambient air temperature and signals it to a security system monitor that can compare it to programmed instructions and take appropriate action.

Thermocouple   A device that transforms heat into an electrical current. Used in gas appliances to monitor the pilot light. If the light goes out, the thermocouple stops generating electricity, thereby signaling a problem.

Transformer   An electrical device for stepping voltage in a circuit up or down with an inverse increase or decrease in amperage.

Water sensor   A device that senses the presence of water and signals it to a security monitor or panel.

Review Questions

1. The three traditional objectives of home security are protecting _____, _____, and _____.

2. Most security system data lines do not carry large amounts of data in comparison to network cables. True or false?

3. Why are wireless security systems usually the preferred choice for retrofitted installations?

4. Security systems are organized in _____ so their protection devices can be easily _____ _____ and the origin of any security breach quickly located.

5. Statistics show that most burglary attempts are made at what three entry points: _____, _____, _____.

6. The most convenient wired security systems are those that use existing home AC or telephone wiring rather than requiring the installation of new cables. True or false?

7. The two largest firms offering remote access service in which the company installs a security system for a nominal charge and the customer pays a monthly service fee for monitoring are _____ and _____.

8. Keypads on or near doors are used to enter _____ _____ .

9. Motion sensors' locations should take into account which factors?

10. Why is double coverage of video cameras in a yard a good security practice?

11. An encoded key contains a _____ _____ recorded on it, which opens an electronic lock.

12. Using a telephone keypad, what is the numeric code for the password "Nicholas"?
   a. 54145616
   b. 64246527
   c. 74173637
   d. 10726685

13. How can a time-coded electronic lock with keypads deactivated from midnight to 6 am be opened at 2 am?

14. Any good television set can serve as a monitor for a security system. True or false?

15. The normal state of a break switch sensor is ____ _____.

16. A make switch sensor is used for turning off lights. True or false?

17. A light beam is usually used as a _____.
   a. Water detector
   b. Motion detector
   c. Fire alarm
   d. None of the above

18. The normal condition of a water detector is ____ _____.

19. A smoke detector signals when the temperature reaches 140 degrees. True or false?

20. A motion detector actually senses changes in _____.
   a. Infrared radiation
   b. Air pressure
   c. Moisture content
   d. Weight

21. A low-temperature sensor is usually used to protect against _____.
   a. An open door
   b. A water heater failure
   c. Frozen pipes
   d. Power failures

22. The central control device for security systems is the _____.

23. If a security breach is signaled by sensors, the respective security panel can be programmed to take appropriate action in response. True or false?

24. What function do call-in devices perform?

25. An alarm device is most often used to warn the home's occupants of _____ .

Hands-On Projects

Project 8-1: Find Specifications for a Freeze Sensor

In this project, you will search the Internet to find a low-temperature sensor called a freeze sensor that signals when the temperature around it drops to the freezing point of water. You will determine the specifications of this device and what other hardware items are required for it to perform its function. For this project, you will need access to a computer connected to the Internet.

1. Log on to the computer and connect to the Internet.
2. Use your browser to go to a search engine you prefer and start your search.

3. Search for the term "freeze sensor."

4. In the results the search engine returns, find a freeze sensor that sends a signal when the air temperature falls to near freezing.

5. Carefully note the specifications of the freeze sensor you select. Is it wired or wireless? What signal does it send, and what does it send the signal to?

6. Determine what additional hardware and wiring items you would need to have to install this freeze sensor and be able to monitor its signal. Does it require a control panel to receive the signal? Could the signal go directly to an audible alarm?

7. If the purpose of the freeze sensor is to protect a sprinkling system from freeze damage, can the sensor connect directly to the sprinkler system? If not, what additional devices are needed to shut down the sprinkler system?

8. Try to diagram a complete sprinkler protection system using the information you have found. Include all the items required to get a freeze warning signal from the sensor and activate a device to shut off the sprinkling system.

9. Identify which parts of the sprinkler protection system could be used for other security functions. What are those functions?

Project 8-2: Set Up a Motion Detector

In this project, you will connect an infrared motion detector to a control module so its signal can be monitored. You will test the range of the detector in a classroom or other test area. For this project, you will need access to an X10 wireless motion sensor with a built-in transmitter, an X10 receiver module, and an X10 security control panel.

1. Connect the security control panel to its power supply, and then plug the power supply into an outlet. Follow the instructions that come with the unit to set it up. Do not use any power supply other than the one that comes with the panel. Set up one button on the control panel to receive code A-5.

2. Set up the motion sensor in an area where you want to detect movement. Follow the instructions with the sensor to set it to transmit on code A-5. If the sensor is battery powered, be sure it has good batteries. If it has a power supply, plug the power supply into an outlet in the room.

3. Plug the receiver module into a wall outlet in the home. It must be close enough to the motion sensor to be within the range of its wireless transmitter. You do not need to set any code in the receiver. It will relay the same code number it receives from the sensor.

4. Position the sensor in a corner of the room opposite the entrance door. Aim the sensor at the door.

5. Turn on the sensor, and the security control panel.

6. Walk in front of the sensor a few feet away from it. You should get a signal at the security panel. If you do not get a signal, check the setup of both the panel and the sensor. Be sure both are powered and that the sensor is set to transmit code A-5, the same code that will be received by the security panel. Test the system again, if necessary, until you get a signal from walking in front of the sensor.

7. Once you have the sensor working, test its range in the room. Try to find the edges of the area where it signals motion. Work outward from the sensor and observe how the range area fans out in a broad arc.

8. Have someone enter the room through the door. Does the sensor immediately signal motion? If it does, can a person move slowly into the room without setting off the sensor? Can a person move along the walls and not be detected in motion?

9. Find a large sheet of cardboard to use as a shield. Walk into the room with the shield held between you and the sensor. Does it signal motion? How close can you approach the sensor before it signals?

10. Slowly place the shield a foot or so in front of the sensor. Does it signal motion? With the shield in front of it, does the sensor still function as it did before? Why does the sensor not respond to motion of the shield as it does to motion of a person?

Project 8-3: Set Up a Wireless Heat Sensor

In this project, you will connect a wireless heat sensor to a control module so its signal can be monitored. You will test the detector in a classroom or other test area. Most heat sensors provide only an audible alarm when they detect heat, but in order to alert the security panel so it can summon emergency assistance, the sensor must transmit a detection signal to the panel. For this project, you will need access to an X10 wireless heat sensor with a built-in transmitter, an X10 receiver module, and an X10 security control panel. You will also need an electric hair dryer.

1. Mount the wireless heat sensor on a wall about four feet above the floor. Before doing so, be certain it has batteries. Follow the mounting and setup instructions furnished with the unit. Set the unit's transmit code to A-6.

2. Set up the receiver as you did in Project 8-2. Remember, the receiver must be within the transmitting range of the heat sensor.

3. Set up the control panel as you did in Project 8-2. Set a button on the panel to receive Code A-6.

4. If the sensor has a test button, use it to test whether the security panel can receive the sensor's wireless signal. If the test button sends the signal, proceed to the next step. If the signal is not sent and received, check the setup of the sensor and the security panel and try the test again.

5. When the sensor tests correctly with the test button, test it again with heat. Turn the hair dryer on high heat and point it at the sensor from a distance of about 1 foot away. Keep the dryer pointed at the sensor until it sends a high-heat signal to the security panel.

6. Turn off the dryer and wait until the sensor cools and resets itself.

7. Turn the dryer on again and repeat the test. Do you think the heat sensor is a reliable indicator of fire danger? Why or why not?

Project 8-4: Diagram a Security System for a Home

In this project, you will draw a diagram for a home security system using a scale floor plan of a residence as the basis for your design. Use a floor plan from a home design magazine or an architectural floor plan drawing. Assume the home is already built and that you will use a wireless security system for your design.

1. Diagram the security system on the floor plan as accurately as you can.

2. Designate areas of the home as zones. Assume you can have a maximum of four devices in each zone and a maximum of eight zones.

3. Include protection for all windows and doors. Include fire protection devices in the kitchen and at least two other rooms. Include motion detectors outside the home to cover both the front and back door areas. Include any other protective devices you feel appropriate that can be accommodated within the limits of the system.

4. When your diagram is complete, total the number of devices of each type you have included. Make a complete list of hardware required for the system.

5. Visit an X10 Web site such as http://www.smarthome.com or http://www.x10. com and price the items you need for your security system. Try to price some items together in package deals to save money.

6. Calculate the total cost of the hardware for the security system you designed.

Project 8-5: Research the Cost of a Remotely Monitored Security System

In this project, you will research how much a security system installed by a commercial company will cost, including the monthly service fee. For this project, you will need access to a computer connected to the Internet.

1. Log on to the computer and connect to the Internet.

2. Use your browser and go to a search engine you prefer to start your search.

3. Search for the term "monitored security system."

4. In the results the search engine returns, find a company Web site for a firm that provides monitored security service in your area. You may want to check several firms and decide which you like best before pricing the service.

5. For the firm you choose, note the cost of installing a basic security system and what is included in the base figure. How does the hardware list compare to the system you designed in Project 8-4?

6. Note the cost of any additional hardware items you would want added to the system if it were going into the home for which you designed a system in Project 8-4.

7. Note the monthly fee for monitoring and the minimum length of the required contract for monitoring. Multiply the number of months in the minimum contract by the monthly fee to get the total cost of the monitoring service. Add that amount to the installation cost and any extra items you have determined are needed.

8. Total all the costs to get the amount of the monitoring service. How does this figure compare with the hardware costs for the system you designed?

Project 8-6: Research the Marlock Lock and Key System

In this project, you will learn more about one type of electronic lock and key system. In Figure 8-3 of this chapter, a Marlock system key is displayed. This key is metal, but it is also electronic and has a number encoded on it rather than cuts and ridges like the traditional key beside it in the picture. For this project, you will need access to a computer connected to the Internet.

1. Log on to the computer and connect to the Internet.

2. Go to the search engine of your choice and start your search.

3. Search for the term "Marlock system" or just "Marlock."

4. In the results the search engine returns, try to find the site of the company that makes this lock system. If you cannot locate the manufacturer, find several companies that sell it and learn as much about it from their sites as you can.

5. Try to answer all of the following questions about the Marlock system:

   a. How are the keys encoded?

   b. Does every key have a unique number?

   c. What information can a Marlock lock record?

   d. What features does the lock have that a traditional key lock does not?

   e. How much does a Marlock lock cost?

   f. How much are the keys?

   g. How can key codes be changed or new keys obtained?

   h. Do you think this lock system offers better security than a traditional key lock?

Case Projects

Case Project 8-1: Research the Need for Security Systems in Your Area

You are planning to offer security systems as part of the DHTI products you install. To sell these systems, you need to know what benefits you can offer customers for installing a security system. On the Internet or at your local library, find out the following information about your area. Research your city, county, or state—whichever one you can find statistics for.

- How many home fires occurred in the past year?
- How many home burglaries?
- How many home break-ins occurred while people were at home and at risk of personal harm?
- How do these numbers compare with national figures for the same events?

Case Project 8-2: Prepare a Sales Presentation for a Security System

Two clients ask you about installing a security system as part of their home technology. They have never had a home burglary, but they travel a lot and are concerned about their children being in the home alone at night. They would like some protection and some peace of mind about their children's safety when they are away. Based on the information you learned in Case Project 8-1 and the concerns the clients have expressed, write a brief presentation about what a security system could do to reduce real risk to this family and calm perceived fears.

Case Project 8-3: Compare Emergency Response Times to Homes with and without a Security System

The typical home fire takes only a few minutes to spread rapidly and do enormous damage to a house and its contents. The more rapidly firefighters arrive at a home fire, the more quickly they can put out the fire and minimize damage. Contact the fire department in your area and find out what their average response time to a fire is after an alarm is received. Ask what the minimum and maximum time is, and whether they have any records of which home fires were in buildings equipped with call-in security systems. If this information is available, find out how the homes equipped with security systems compare with others in damage from the fires they had. How much do you think a client would spend on a security system that might reduce fire damage to his home by $10,000? By $20,000?

Case Project 8-4: Research Insurance Rates for Homes Equipped with Security Systems

Contact several insurance companies in your area that write casualty insurance (home owners insurance). Ask each of them if they offer reduced premiums for insurance on homes with security systems. If they do, find out exactly what protection the system must include to earn the discount in insurance. Ask each company what information caused them to offer such a discount (insurance companies always determine premiums on the basis of hard financial data, although they may not always be willing to share that data with you). Write a brief report on what you find and your conclusion about whether a security system can save money on insurance premiums.

Security System Installation and Setup

After studying this chapter, you should be able to:

- Describe how hardwired, wireless, and combination security system infrastructures should be installed.
- Install various types of sensors and other components in a security system.
- Install various accessory devices.
- Program a security system with appropriate settings and parameters.
- Describe the procedures for monitoring, maintaining, and servicing a security system.

Introduction

In Chapter 8, you learned about the various types of security systems and the large variety of components these systems can include. In this chapter, you will learn how to install the infrastructure needed for a security system. You will also discover how to install and set up the appropriate components to create an effective home protection system. You will learn how accessory devices can enhance the function of a security system and be able to describe how to program a security panel, keypad, and call-in device. After you understand how the system is set up and operated, you will learn how to maintain it and how to service components so their operation remains flawless.

Security System Installation

Many manufacturers offer components for security systems, both wired and wireless. Some offer complete lines of products from which a full security system can be assembled. Those whose product lines are this extensive generally also offer pre-selected packages of components ranging from so-called "starter kits," consisting of a security panel and two or three sensors, to entire house security systems with a dozen or more varied components. Remote access companies, who usually install the security systems they monitor, also offer security component packages of various sizes from which their customers can select, along with several levels of **monitoring service.**

The level of standardization in security products is not high, particularly among wireless components. Buyers should not assume that any two components, especially if they are made by different companies, are compatible with one another unless the packaging and instructions clearly indicate they are. For this reason, buying a preassembled package of security components to use in a home security system makes sense. Components sold as a package will work together and are offered by those manufacturers whose product lines in the security area are the most extensive. Both considerations are important in assuring that all elements of the security system are compatible with one another and that a range of additional products are available from the same company that can work with the package system, if needed.

Not all security systems are capable of being integrated with a home network. Some definitely are, but this fact should never be assumed any more than compatibility among manufacturers. The ability to integrate with a home network, or even a single computer, is a valuable feature and is always advertised as such. If a security panel's packaging and instructions do not clearly state that it can connect to a PC and has the necessary software included to do so, you should assume that it cannot.

In this chapter, you will learn about the installation of a wired security system and a wireless one. Each is representative of a type of system, but not all similar systems include all the features described here. Some may also include features not covered. As with other home technology products and systems, the user should become familiar with the range of products available and choose the system and manufacturer that best meets his or her needs.

Low-Voltage Wiring

As was noted in Chapter 8, the amount of data sent over most wired security system lines is minuscule when compared with the multi-megabit data capacities needed for an active home LAN. Lines from sensors to a security panel operate at a similarly low voltage like network systems, but they rarely carry a megabit of data in a month, and so there is no need for them to be installed using Cat5 cable or even standard UTP or STP cable. Wire that is less expensive and smaller in size is preferable for security system use.

Category 2 (Cat2) telephone wire, containing two pairs of wires, is adequate for most security component wiring. Only one pair of wires is needed for a sensor, so two sensors can actually be wired from a single run of Category 2 wire. If multiple sensors are to be wired to windows and doors in a room, a run of Cat2 or **Category 3 (Cat3) wire** containing four or six pairs of wires can be used. This eliminates multiple wire runs and provides individual home run wiring for each sensor.

Neither Cat2 nor Cat3 wire is armored or in any way resistant to being cut, but this is not a requirement for security system use. All **monitored** security system sensors regularly report their status to the security panel. The frequency of reporting varies with the system and is a feature that should be evaluated when selecting a system: The more frequently status is reported, the sooner any **security breach** creates an alert. Failure of a sensor to report its normal status is regarded by the security panel as a breach, the same as if the sensor had been activated by an opening window or other break-in. Consequently, cutting a sensor's wires has the same effect as opening a secured door: a security breach is noted by the security panel.

Sensors can be **bypassed** (made to appear normal when dysfunctional by a device wired into the sensor circuit), but only by a sophisticated signal generator that can match the timing and duration of the sensor's own signal and be wired into the circuit between the sensor and the security panel without triggering an alarm. This is an occurrence so rare, except in movies, as to not justify armored wiring.

Unlike high-speed data lines, telephone wire can be spliced. A single four- or six-pair line can be run to a junction box and the pairs in the line spliced to individual Cat2 lines running to sensors in different parts of a security zone. Because the signals from security components are low power to start with and attenuate with distance traveled on the wire, any splices should be carefully made and soldered or clamped under screws to assure a good connection. Interference is not normally a problem on security system data wires, but it can become an issue if bad splices and other poor installation practices cause the signals in the wires to become weakened or corrupted.

Many security (and telephone) installations using Cat2 wire have been made by running the wire on the surface of walls, usually on or just above the baseboard, and holding it in place with small staples. Cat2 wire is only a little more than 1/8 inch in diameter, smaller than Cat5, and is less obtrusive when run on the surface. Its data-carrying capacity, being much lower, is also not impaired from being squeezed by the staples as Cat5 or other high-speed cables would be.

Surface runs of security wire are still acceptable in many cases and may be necessary at the ends of runs to get wires close enough to window and door sensors so they can be connected. But concealed wiring looks better in the home and should be used wherever reasonably possible for a more professional appearance and usually a more satisfied customer. If surface runs are necessary, it is better to do them with the wire alone rather than use raceways, which are too large for Cat2 or Cat3 wire and detract more from a room's appearance than the exposed wire. Cat2 wire can be bent sharply without damage and so can be made to follow irregular contours of moldings or other obstructions closely. If the installer runs the wire so that it blends with wall surface features (along grooves in baseboards and moldings, up the sides of door casings, under window sill overhangs) as much as possible, an almost invisible finished appearance can usually be achieved.

If a house or room has wall-to-wall carpeting, Category 2 wire can be run along the inside edge (the edge away from the wall) of the tack strip that holds the edges of the carpeting, and then stapled to the floor. The tack strip prevents damage to the wire, but wire should not be run under carpet across the middle of a room since furniture or constant foot traffic is almost sure to damage it.

In large security systems, four- or six-pair wires should be used wherever needed to avoid multiple wire runs on wall surfaces. More than a single wire on a wall never looks good and will look worse over time as lint and dust become trapped between the parallel running wires. Avoid two wires next to one another wherever possible, even if it means running a single wire around opposite sides of a room to reach two window sensors.

Another solution to wiring closely spaced sensors (to two windows on one wall or to French doors in a room) is to wire them in series (not in parallel) on a single wire run. This eliminates one wire run to the security panel while still allowing both sensors to be fully active. If wired in series, both sensors are on the same connection at the security panel and there is no means of determining which sensor is tripped in case of a break-in, but this really does not matter. Through which of two windows or doors in a room a burglar makes his or her entrance is not important as long as the attempted break-in is signaled to the security panel. Figure 9-1 shows how two sensors can be wired in series so that if either sensor **(break switch)** is opened, the security panel is alerted.

Many security panels are constructed with zones that have four (or sometimes six) sensor connections in each zone. These systems can often be wired by running a single Cat2 or Cat3 wire (four-pair or six-pair) to a central point in the zone and then splitting the wire pairs into individual runs from the central point to sensors in the zone using smaller wire. This reduces wire runs to the security panel and still allows each sensor to be individually connected and monitored.

Sensors can all be wired with Cat2 wire, as can all **response devices** such as emergency call-in devices, remotely activated locks, electronic valves, automated appliances, utility controls, and intercom speakers. The only notable exceptions to this wiring standard in security systems are video surveillance cameras. These

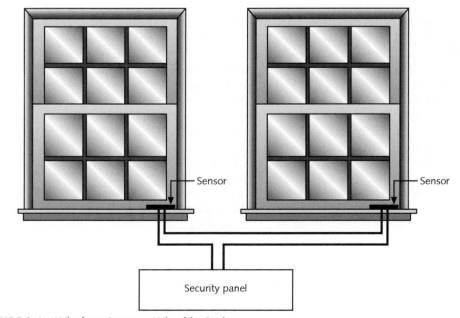

FIGURE 9-1   Window Sensors Wired in Series

require standard composite analog video cables with RCA connectors at each end. Only one cable is required for each camera because security cameras do not have any audio output as does a regular TV camera.

Video camera cables connect into a monitor or a **switcher** that can switch the monitor's view from one camera to another. Other than the different wire and connectors used, wiring for video cameras is similar to wiring for other security sensors and input devices. The video cables can be run on the surface of walls or concealed. There is no restriction on sharp bends in the cable, but it does need to be kept clear of high-voltage, AC wiring and equipment, which can introduce interference in the signal just as it can with cables to other video devices.

Wireless Security Systems

Wireless sensors do not require any wires either to transmit their signals or to power the unit. Transmission of data is handled by a transmitter and both the sensor unit and the transmitter are battery powered. High-quality sensors and other wireless devices also monitor their batteries and signal when battery power is getting low. This feature enables the user to change the batteries in a timely manner without waiting until the sensor goes dead. It also prevents changing the batteries too soon and losing a large portion of the batteries' life.

Power Usage of Wireless Systems

Wireless sensors and their transmitters consume little power and their batteries often last a year or more, but the batteries used are typically the 9- or 12-volt alkaline type and are three or four times the price of 1.5-volt C or D batteries. If the batteries are used in a device that is outdoors, or on an outside wall, battery life may be reduced somewhat by extreme temperature changes. Shielding outside devices from direct sunlight is a good idea to extend the life of the unit, as well as its batteries.

Action response devices such as audible alarms, emergency call-in devices, video cameras, utility controls, and intercom speakers can also be battery powered, but are more usually wired to an AC power outlet either directly or through a step-down transformer, which is a device that lowers AC voltage to the level required by the equipment. It is easier to power these devices with AC current because they are nearly always located near a power outlet (appliance controls, lamp modules, and utility controls) or can be located near one (audible alarms and emergency call-in devices). Some response devices also require quite a bit of power to perform their functions, and battery power would be expensive to provide for these.

Some AC-powered devices have battery back-up power in case the AC connection is lost or the power in the home is cut off. These are primarily warning devices such as alarms and emergency call-ins that absolutely must function properly in life-and-death situations. Backup power for these devices is critical.

Signal Transmission

Wireless sensors can transmit their signals to a security panel through walls and other obstructions within reasonable limits. The transmitters are low power and their signals are attenuated by distance. This means that their transmission range is limited to a few hundred feet. Heavy obstructions such as concrete walls, metal structures, and other dense materials can absorb some of the signal and further reduce its range. For this reason, all sensors in a wireless security system should be tested after they are mounted and the security panel is positioned in its permanent location. Testing assures that the system is working under the actual range and obstruction conditions of the home. If either a sensor or the security panel is relocated, the testing should be done again (if it is the security panel that is moved, all sensors should be retested) to be sure all transmissions can still be received.

Wireless video cameras are available for home security systems, but many of the transmitters on these units can only transmit to a receiver visible from the transmitter in a direct **line of sight.** For inside cameras, this is usually not a problem. It allows the camera to be located in a good surveillance position while the wireless receiver is placed in a position where it can be wired to the monitor by standard coaxial video cable.

For outside cameras, a more powerful transmitter is needed that can broadcast through obstructions (walls, etc.) to a receiver inside the home. Outside wireless video cameras can also be battery operated or connected to AC power through a transformer, if a power connection is available. Figure 9-2 shows a wireless security system that can handle both wireless and hardwired video cameras. One hardwired video camera (right) and one wireless camera (left) are shown at the front of the picture with three wireless window/door sensors, two keyless entry devices, two motion sensors and a recorder behind them, and a master control unit at the back. This system can be connected to an Ethernet LAN and has software that allows it to be monitored remotely and securely on the Internet. It also includes wireless access points and handheld wireless remote controls.

To conserve battery power, wireless video cameras usually do not operate continuously, but only on a timed schedule or when signaled from the monitoring station to turn on.

Most of the wireless security systems manufactured by SkyLink, Honeywell, and others operate on RF standards not related to any of the networking technologies. They also have no means of interfacing with any of the networking technologies,

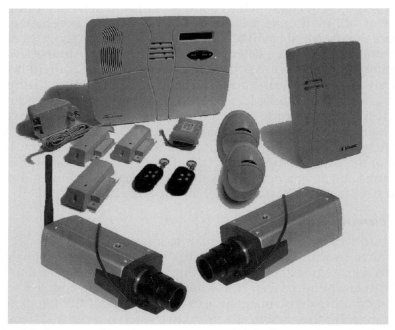

FIGURE 9-2   Wireless Security System Components Including Hardwired and Wireless Video Cameras

wired or wireless, except Ethernet. Not all security systems can interface with Ethernet, but we will examine a couple that do and see how the network connection enhances the security function of both.

WiFi Wireless Security Systems

Many WiFi-based wireless security systems are now on the market. Because WiFi has the bandwidth to transmit video data, most of the wireless security systems that include video surveillance components use WiFi technology, at least for the video segment of the systems. WiFi can also be interfaced with an Ethernet network so the security system can be an integrated as part of the home LAN.

Other components of a WiFi security system do not need large bandwidth to transmit small amounts of data, but may still use the WiFi frequency because of the convenience of sharing wireless hubs with other nodes on the LAN. Wireless WiFi sensors can be linked as individual nodes on the LAN in the same manner as any other wireless device. The only negative side to setting up a security system as nodes on a LAN is the possibility of the network going down or being disabled and taking the security system down with it. Security systems that are independent of any other technology are less likely to malfunction than integrated systems because they are simpler and less subject to interference, power failure, data corruption, and external hacking than a complex home network. Still, the ability to monitor and control a security system on the LAN more than compensates for any small increased risk of system failure.

Figure 9-3 shows a diagram of a WiFi video surveillance security system that can be controlled and monitored from a desktop computer, laptop, or a remote

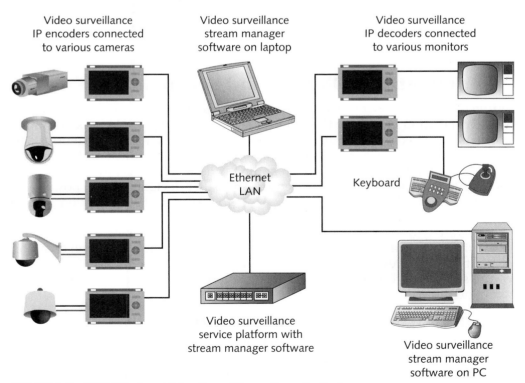

FIGURE 9-3   WiFi Wireless Video Surveillance Security System

controller. The system includes various types of cameras and several monitors that can display video surveillance images, but are not control stations. Each node in the system is connected to the LAN through digital encoders that transform the analog video images into digital WiFi data or decoders that take the digital data back to analog form for display. All the data lines shown in the diagram are not wired connections but wireless links to the various nodes and their encoder/decoders.

X10 Security Systems

Alone among the power-line technologies, X10 manufacturers have developed and marketed a wide array of security components and systems. These include sensors and security panels that operate on the traditional X10 technology of data transmission over AC power lines, but also a range of wireless devices that have greatly expanded the X10 system's capability and enabled it to interface with networks through an Ethernet port built into X10 control components.

With power-line wired and wireless components, X10 security systems are actually hybrids that utilize both technologies to provide extensive well-placed security protection and easily installed monitoring devices at convenient locations. Lighting control modules, the original automation devices for which the X10 technology was developed, can readily be incorporated as action response devices to any security alarm received. Appliance modules and other specialized devices can also be activated in response to a security signal or on a timed schedule as part of a monitoring routine.

Despite the development expertise and power of the companies promoting the other power-line and telephone line technologies through the alliances organized for that purpose, no viable home security systems have yet emerged that operate on HomePlug, HomePNA, or HomeRF standards.

Component Installation and Setup

Setting up the components of a home security system can be either the easiest part of the installation process or the most difficult, depending on the type of system selected. In a wired system, pulling the necessary wiring is the labor-intensive part of the job and, once it is completed components can be mounted and connected quickly. In a wireless system, no wire pulling is necessary (except for power connections to some devices), but the sensors and other devices must be positioned, mounted, and tested with the security panel to be certain that reception is good from every location.

The variety of components provided and available as options with each type of security system make every installation a custom design job, although common elements will be present in all of them. The three examples discussed in the following sections give a clear idea of what is involved in installing security system components.

Installing Wired System Components

A basic security hardwired system, such as the one described in the following example and offered by a large offsite-monitoring security firm, includes some devices as standard requirements and offers others as options. The standard items include:

- One multizone security control panel (to which all sensors and response devices are wired)
- Three door sensors (to detect outside doors opening)
- One motion detector (for inside or outside use)
- Two emergency buttons (installed at convenient locations to summon emergency assistance)
- One digital keypad (to **arm** and **disarm** the system)
- One interior siren (to sound an alarm in case of a security breach)
- One backup battery
- One yard sign and several window decals (noting monitored protection)
- One telephone connection link to the monitoring service
- An around-the-clock monitoring service

All of the sensors and emergency buttons are wired to the security panel to warn of an **intruder.** If a security breach is detected, the siren is directed to sound and the monitoring service is automatically contacted. The service, in turn, contacts the homeowner, sends its own personnel to check on the breach, or summons an **emergency response** from community services, according to the service contract instructions.

Note: The yard sign and window decals may seem minor pieces in the security system, but they actually contribute greatly toward accomplishing its purpose. Most burglars "case" a home before attempting a break-in, either days before the crime or a few moments before starting it. A sign or decal noting that the premises have a security system is often enough to dissuade them from attacking. This is especially true if other equally attractive targets exist nearby that do not advertise the presence of a security system. By openly stating that a security system is operating, the signs enhance the system's effectiveness by deterring burglary attempts rather than sounding an alarm when they occur. Many security companies and user-installed systems offer these signs, and they should always be used.

In addition to the basic package, the following additional wired items can be added to this security system:

- **Glassbreak detector** A wired device that detects when a window is broken. Not the same as a window sensor, which detects a window being opened.

- **Smoke detector** An alarm from this device causes the fire department to be notified immediately by the monitoring service.

- **Pet-immune motion sensor** An adjustable motion sensor that can be set to ignore infrared radiation changes caused by small animals, but still respond to any person entering its range.

- **Wireless key** An RF device, similar to the keyless entry devices on cars, which can arm or disarm the security system with the push of a button from outside the home, thus eliminating the need to use keypad codes.

Video surveillance is not available from the company that offers the just-described system and usually is not included with monitored service contracts because of the cost of connecting a video signal to the monitoring station. Video cameras could be added with monitoring done inside the home by the homeowner who could summon assistance, if needed, through the monitored system's emergency buttons when any intruders were detected.

The installation of this system, as is the case with most monitored security systems, is done entirely by the monitoring company, which maintains a staff of professional installers to do the work. The installed equipment is leased, but after the expiration of a 3-year monitoring contract, it becomes the property of the homeowner. The homeowner can then continue the service contract for monitoring at a lower rate, or drop the service and keep the equipment. The original cost for this system is about $900, including installation, but this amount is normally included as part of the monthly monitoring fee for the first 3 years of service. The monthly fee is about $48.

To install this system, the technician first determines the best locations for the control panel, the keypad, the motion detector, the door sensors, and the telephone connection point. Such a system does not include window sensors, but they could be added, if the technician determined they were necessary for the system to give adequate protection. Once component locations are identified, the wiring for each of them is run.

Security wiring can be run on the surface of walls in the same manner as telephone wire (which it actually is), but the installation detracts less from the home's appearance if as much of the wiring as possible is concealed. Concealed wires are pulled through unfinished spaces (mainly attic and basement) and within walls to a point near the location of the security device, then brought to the surface through a small hole so the wires can be connected to the device. Because many security devices are mounted on exterior doors and windows, it is often not possible to run their wiring within the walls, which are frequently solid masonry or filled with insulation. In these situations, the wire should be run through the basement, if possible, to a floor hole at the base of the wall and directly below the location of the device. The wire is then pulled up through the hole and run on the wall surface up to the security sensor. Telephone wire staples or contact adhesive hold the wire in place on the wall surface.

The technician labels the control panel end of each wire as it is installed. He or she should also label the sensor end as well with a label that is identical to the one on the control panel end. For a better appearance, the labels can be removed from the sensor ends after (and only after) the wires are connected and the system is tested and working properly.

There is often a little more flexibility in the location of motion sensors, light beams, emergency buttons, and similar devices than there is with door and window sensors, which must be placed on the openings they secure. It is best to pull concealed wiring for all components before actually fastening them in place. By doing this, the locations can often be altered to better accommodate the run of the wiring. A few inches movement one way or the other will not usually make any difference in a sensor's function, but could be a great help in getting wiring to it through a convenient path.

Concealed wiring can be run in any unfinished spaces through holes drilled in the wood framing members. Wiring can also be run on top of attic truss plates (the bottom part of the roof trusses to which the ceiling wallboard is fastened), or stapled to the bottom of floor joists in the basement. If there is any danger of damage to the wire, it should be covered or run through holes drilled in the wood framing. Security wire can also be run through conduits installed for the home's AC wiring and through heating and air-conditioning ducts. In the latter case, the NEC requires that the wire be protected from damage, and be, according to NEC Supplemental Electrical Safety Standards 3.1.1, "plenum cables that pass the NFPA262/UL 910 Fire and Smoke Test." Points where the wire enters or exits the ducts must be sealed so as to prevent smoke from leaking from them.

Surface wiring should not be run in raceways unless multiple wires are used because a single strand of exposed wire is less obtrusive than a raceway. Wiring can be run along the top of baseboards or the bottom of crown molding. It can also be run next to door and window casings and along the bottom of chair rail or similar decorative molding in the center area of a wall. Staples are the easiest method of fastening it, but adhesive makes a smoother appearance. Unlike Cat5 cable, telephone wire can be tightly bent to conform to shapes around which it is run. This is a useful characteristic when wiring on window casings and around doorjambs where the wire must often be tucked into corners and other tight areas.

Once the system wiring is in place, the components are mounted in place and connected to it. Security components generally connect with clamp screws that the wire ends are wrapped around. The screws are then tightened to make a secure connection. When wiring for networks, some slack in the cables is a good thing, but this is not true with security wiring. All wire should be trimmed as short as possible before connecting so it is flush with the windowsill or door frame on which the sensor is mounted. The less excess wire there is, the more professional the installation appears.

Door sensors should mount on the top part of the door casing on the opposite side from which it hinges so that any movement of the door inward opens the sensor switch. Window sensors can mount anywhere on the casing as long as the plate mounted on the window can be positioned so any movement opens the circuit and signals a security breach.

Emergency buttons, which are intended to be pressed by the home's occupants in case of a problem, should be conveniently located where they can be easily reached but not be accidentally pressed by someone leaning against them. Most people want the buttons placed out of sight, such as under a counter, so they can be pressed unobserved. This is fine as long as the homeowner can remember easily where the button is and get to it easily in an emergency. For elderly persons or those with disabilities, a fully visible location is often better.

Motion detectors and other sensors must be located so they function according to design and are not triggered accidentally. This often requires a compromise between the ideal placement (the location that is safe from **false alarms**) and the easiest wiring spot.

The opposite ends of the wires from the sensors are connected to the control panel according to the labels on each wire. If the security system plan calls for zones, connect the wires to make up the zones on the security panel for which you planned. If the plan does not have zones, connect the sensors in a linear fashion and leave the labels attached to each wire. Connect the telephone wire to the telephone system and hook up the alarm. Last, connect the keypad arming/disarming device to the control. When all components are wired, turn the system on, arm it, and test each sensor to be certain it is working.

Installing Wireless System Components

Some homeowners are willing to invest in the cost of hardwiring a security system, especially if the wiring can be installed as part of the home's original construction. For others, a wireless system is a better choice, both because of its lower initial cost and its comparative ease of installation. A typical basic wireless security system is offered by SkyLink Home Security and contains the following:

- One multizone security control panel
- Two door/window sensors
- One motion detector
- One keychain transmitter (for arming and disarming the system)
- One emergency dialer
- One backup set of batteries

An alarm siren is not included as a separate item in this system, but is a built-in feature of the security panel, which also activates the automatic dialer to summon assistance in case an alarm is triggered. Because calls for assistance do not go to a monitoring service, but to private telephone numbers designated by the user, up to nine numbers can be dialed in succession and the same **recorded emergency message** played to each when the line is answered. Cost for this basic system is about $300.

The security panel in this system can accommodate up to 24 different sensors. In addition to the door/window sensors and motion detectors included in the basic package, the following sensors and response devices are available and are compatible with the system:

- Smoke detectors
- Heat detectors
- Flood (water) detectors
- Light beam detectors
- Freeze detectors
- Light control modules

All of these are wireless and battery powered. They also signal their operational and battery status to the security panel. The security panel, dialer numbers, and recorded message are programmed by a coded process using the panel's keypad, but the system can also be connected to a PC and programmed from the computer screen using software provided. The monitoring function of the security panel can also be duplicated from the PC.

No video surveillance is included with such a wireless system, but wireless video cameras and transmitters, such as the one shown in Chapter 8, are available at reasonable cost for use with wireless security systems. Unlike the wireless sensors, however, the wireless transmission of video signals is usually confined to direct line of sight, and the receiver must then be wired to a monitor. If WiFi-based cameras are installed, they can communicate directly to a wireless hub on the home LAN and display their images on a computer or television node. This connection is independent of the rest of the wireless security system, which does not operate in the WiFi band.

No wiring is necessary for the described system. Installation consists of placing and testing the components. All the sensors are battery powered, so no power lines are required for them. The security panel and call-in dialer are both AC powered and must be placed near an outlet. The security panel needs to be centrally located within the home because all the sensors broadcast to it and must be within its receiving range. Because no wiring other than AC power is needed, the control panel can be placed on a bedroom dresser or a kitchen counter. After the panel is located and the sensors tested for transmission to it, it should not be moved in any significant manner unless the wireless devices are tested again afterward to be sure the new location is workable.

The call-in dialer can be connected to a telephone line through a splitter so the same telephone jack can also connect a telephone extension. The call-in dialer has a 6-foot AC cord and therefore must be fairly close to an AC outlet, but the telephone connection line can be up to 25 feet long in order to reach a jack. No wired connection is necessary between the control panel and the call-in dialer. The panel communicates with it wirelessly if a call-in is required, and the dialer is programmed independently for the numbers to call and the message to play.

Sensors are placed on windows and doors in the same positions as wired units. Other types of sensors are placed so they can function effectively, but no consideration needs to be given to their connections since all send wireless signals to the control panel.

When all sensors and the control panel are in place and equipped with batteries, the sensors should all be tested to be sure their signals reach the control panel correctly. In this type of wireless system, the sensors all broadcast on one frequency, but are equipped with individual unit numbers. This allows an alarm or low-battery signal to be identified to the sensor sending it so that appropriate action can be taken. The control panel is not set up for a zone system, but only to identify the responses of each unit. The testing should be done before the call-in dialer is connected so that accidental alarm signals will not trigger a call.

When the sensors are confirmed to be working correctly, the call-in dialer can be connected to the telephone line and programmed with the numbers it is to call and the message to be recited. The call-in dialer can be tested by programming it with the telephone number of another line to the home or to the home of a friend who can be made aware of the test beforehand. With only the one call number programmed, test the system by triggering an alarm from one of the sensors to see if the call-in dialer performs as intended. If it does, the test number can be deleted and replaced with the permanent call-in numbers the system is to use.

Installing X10 Power-Line System Components

X10 security systems are hybrid wired and wireless systems. The wired portions of an X10 system utilize the home's AC wiring, and components are connected by being plugged into an AC outlet or wired into a wall switch. No new wiring needs to be installed for an X10 system. The security panel of the system requires AC power to function and so must also have a battery backup power source in case of power failure. The components that use the AC wiring for data transmission are powered by the AC outlet into which they are plugged. No other power source is needed. The data transmission of an X10 system is not affected by whether or not AC current is flowing in the wiring: The system's sensors function with or without AC power.

Most wireless components of X10 systems are battery powered and can be installed without regard to a power source. They transmit directly to a receiver installed in an AC power outlet or to the system's security panel. When a signal is received from a sensor indicating a security breach, the system can respond by sounding an alarm or taking other appropriate action. The response commands are usually sent via X10 power-line technology over the home's AC wiring to a responder device that is connected to the AC wiring.

X10 video surveillance cameras send wireless video signals, but require a continuous low-voltage power supply. Like other video surveillance cameras, they must be wired through a low-voltage transformer to the AC power in the home. These cameras broadcast direct to a wireless receiver up to 100 feet away from the camera. The receiver does not require a power source. It functions as an antenna for the wireless broadcast of the video camera and can be connected via an RCA cable to any analog television. An X10 control panel can control up to 16 video cameras and switch between them for viewing on the monitor. By adding a wireless motion sensor coded to the video camera, and an X10 controller for a video recorder, an X10 video surveillance system can be set up to record video from any camera whenever motion is detected by the sensor for that camera.

Many X10 security system packages are available on the market. Most include only the basic components required in almost every home security system. These can be augmented with an enormous array of additional devices to customize and complete each individual system. Figure 9-4 shows a typical basic system offered by SmartHome, a large supplier of X10 systems. The system includes the following items:

- One PRO2000TM security control panel (including alarm siren and call-in function)
- One motion sensor
- One handheld remote
- One keychain remote
- One X10 lamp module (for controlling lighting)
- Two window/door sensors

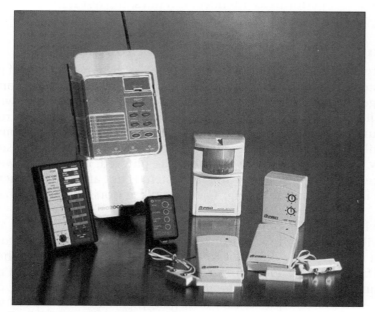

FIGURE 9-4   Basic X10 Security Components

To install the X10 security system, the security panel is connected to the AC power through an outlet (the unit's AC cord also connects it for data transmission). The remote AC units (light modules and some sensors) are similarly connected by being plugged into wall outlets in appropriate locations. The wireless components function like any wireless system and are installed where needed. All sensors and lamp modules must be set with a house code and unit number so their transmissions can be identified by the control panel. These numbers are set with selector wheels mounted on the units or by programming buttons located under the battery covers.

Wireless sensors should all be tested to be certain they are within range of the receiver in the security panel. If any are not, an auxiliary receiver can be installed at a wall outlet within range of the wireless units. The auxiliary receiver can then relay the wireless transmissions via the AC wiring to which it is connected.

In addition to the basic X10 security system components, literally dozens of other devices are available to address every possible security need. Among these are:

- AC-current detectors
- Appliance control modules
- Carbon monoxide sensors
- Electric deadbolts
- Electric door locks
- Freeze detectors
- Glassbreak sensors
- Humidity sensors
- Keypads
- Light control modules
- Light sensors
- Listening devices and intercoms
- Outside alarms
- Pressure pads
- Proximity access sensors
- Smoke detectors
- Temperature sensors
- Video surveillance cameras
- Water detectors

In addition to all these sensor and response components, X10 systems have many control options available, including touch screens, telephone links, remote wireless controls, and computer links with software so a system can be controlled and monitored on a PC.

Accessory Security Components

In addition to the sensor and emergency response security systems described in earlier sections of this chapter, several accessory devices may be installed in both wired and wireless systems. These will not be wanted by all users, but they are regarded as essential by some.

Outside Alarms

Inside alarms alert the home's occupants to danger from intruders, fire, or other problems. Outside alarms alert the neighbors to the same dangers. Not all home security systems should have an outside alarm, but they are helpful in some cases. If the security system includes an emergency response notification device, that may be sufficient in summoning professional emergency help. Most call-in devices can include multiple numbers, so some neighbors, as well as the authorities, can be notified of trouble.

In homes that have an elderly or disabled resident, an outside alarm may help summon nearby assistance to get people out of the home quickly. An outside alarm can also have some deterrent effect on a burglar, both seeing it mounted on the home and hearing it if he or she decides to attempt a break-in anyway.

Outside alarms can be wireless or wired. They require AC power or a large battery to produce a high-volume sound. Batteries are quickly drained, so most use rechargeable units. Wireless alarms are usually battery powered, while wired ones use AC current from the home. Both types are connected in the same manner as an inside alarm on the system (and can be connected to the same circuit so both alarms sound together). Wires that run through the outside wall of a home should be sealed to prevent heat loss and insect penetration.

A good feature to have on an alarm is a timed shutoff. Ten or fifteen minutes of sound is adequate to bring a response and, especially if it is a false alarm, that is all most people want to hear. If the alarm can be shut off only from the inside security panel, it is likely to fray nerves.

Controlled Access Components

Controlled access components (exterior door locks and garage door openers) can be installed independently of the home security system or as an integrated part of it. Independently installed locks and openers can be activated from outside the home by a keypad or other numeric code device mounted near the door or by a wireless access transmitter usually carried by the user. The doors can be accessed from inside only by manually unlocking and opening them.

Door access devices integrated into the security system can be accessed in the same manner as independent doors, but they can also be remotely locked down or opened from the security panel. In addition, their status (locked or unlocked) can be monitored in the same manner as the door itself can be monitored (open or closed) from the panel. This feature allows the homeowner to identify visitors via an intercom or video surveillance camera before unlocking a door. It also allows all access to the home from outside to be cut off during night hours or at other times when the occupants are away.

Independently installed access devices are most conveniently installed with the keypad or access device on the door itself next to the lock. This eliminates the need for any extensive wiring from the keypad to the lock. Most of these locks are integrated units with the keypad and lock combined in one box. These require a little custom fitting in standard doors, but no external wiring. Most are electric and require battery power (accessible from inside the door only), but a few are mechanical and work much like a combination lock with no electric power necessary. The mechanical types cannot use a remote opener, but also do not require any battery replacement and are pretty much service-free.

Electronic locks, which are operated by a solenoid in the lock mechanism, can have the keypad located on the door or on the wall beside the door. The keypad can either be wired to the door lock or can open it by transmitting a wireless signal. The latter method is now more common and also allows the lock to be remotely activated by a keychain remote. The remote eliminates the need for entering any numeric code. It opens the lock at the push of a button. The keypad is still necessary, however, for persons who visit the home and do not have a remote key.

Almost all garage door openers are activated wirelessly by remotes normally carried in the family cars. They can also include a keypad access mounted at the side of the door and this unit can be either wireless or hardwired. Wiring a keypad from the side of the door to the opener does not require the same amount of concealment of the wires as would be necessary in the home's interior. The pad can usually be wired along with the **obstruction detector** that all powered garage doors have (or should have) as a safety precaution against the door being lowered on a person or object under it. Wireless keypads transmit to the same receiver that the remotes transmit to and are no more expensive than buying another remote. A plain remote should never be used as a mounted outside access device, however, because it offers no security against unauthorized use. The keypad assures that only those with the access code can open the door.

Gate Controls

Gate openers and locks are access devices much like garage door openers and door locks, but are installed on devices at the yard perimeter. Driveway gate openers can be the rollback type that retracts a gate along the fence or wall on wheels or bearings, or the swing type that draws the gate open or closed on hinges in the traditional way. Both types are powered by AC electric motors (or motors attached to hydraulic units) and so must be wired to the home's electric service. Fence and wall gate wiring usually runs underground and so should be armored cable or similar wiring required by the local electric code.

Any remote access devices used on doors can also be used on a gate. The placement of the opening device is usually more remote from the entrance so it can be used from a car window, but the access device can be wired or wireless. A keypad for visitors and keyless remote units for family members are often the preferred methods of controlling gate access. The family's remote units can often be programmed to also function as garage door openers, thus eliminating the need for two remotes.

Because they open and close slowly (for safety reasons) and are considerably larger than a car's width, security gates on driveways do not offer much security against unauthorized entrance. Someone can easily walk through a gate while it is open to allow a car to enter or leave the home. If security is needed at such gates, a video surveillance camera should be included to monitor anyone entering while the gate is open.

Fence and wall gates, other than those on driveways, rarely warrant openers. They are not used often enough. But if high security in the home's yard is important, they can be fitted with door/window sensors to monitor them if they are opened.

Uninterruptible Power Supply

Many home security systems have backup battery power, which can be used in case of power failure, or some type of **fail-safe** system that warns if power is low or that sends an emergency signal in the event of a power failure. This backup power method is sufficient for most systems, but if the home network has an uninterruptible power supply (UPS) for use in maintaining all or part of the LAN's operational ability during a power failure, this power backup can easily be configured to include the home security system. Security panels do not require a lot of power and so will not put a heavy drain on a UPS designed to supply and protect the network.

A UPS is essentially a large battery with built-in transformers, switches, and other electronic hardware. It is connected to the home's AC power through an outlet, and components of the home LAN are connected to the UPS. When AC power is functioning normally, all the devices connected to the UPS receive their power from the home's power lines. The AC current passes through the UPS to each component without any alteration.

When AC power is cut off, the UPS detects the drop in voltage on the AC line passing through it. A good UPS will sense even a momentary voltage drop and automatically switch all the devices connected to it over to the battery power of the UPS. Batteries store power in DC form, but the UPS has hardware that transforms DC current to AC and adjusts its voltage to match the AC power supply of the home. This power supply will only last a few minutes at full power (usually 15 to 30 minutes at 1,000 to 1,500 watts), but this is enough time to shut down everything on the LAN without any data loss. Figure 9-5 shows the front and rear of a home-size UPS with 750 watts output capacity. For additional information on UPS units, see Chapter 13.

The exception to shutting down the entire LAN may be the security system. Because most security systems (without video surveillance and monitoring) draw less than 150 watts of power, they can be left functioning on a UPS power supply for 12 hours or more without depleting the UPS batteries. Crimes often multiply during blackouts and criminals may cut power to a home as a prelude to a break-in. In the event of a power failure, it is a good idea to leave the

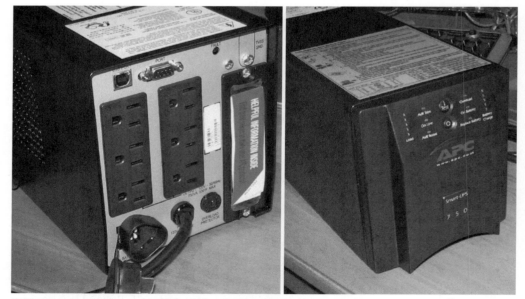

FIGURE 9-5 UPS Rear Panel (Left) and Front (Right)

security system functioning on the UPS in order to protect the home during a high-risk situation.

When power is restored, most UPS units automatically switch back to the AC power source with no input required from the homeowner. They also begin recharging their batteries in order to be ready for the next emergency.

A 12- or 24-volt security panel powered by an AC adapter can simply be plugged into one of the UPS outlets (110 volt AC) and the unit will be supplied through the UPS. Wired sensors that are powered from a security panel are also powered by the UPS through the adapter hookup. Wireless devices that are battery powered cannot be connected to a UPS, but they also are not affected by a power failure and should not need backup power. Their low-battery warnings should always be heeded so they do not lose battery power in an emergency situation.

Home UPS systems are battery powered and have limited operating times under load. They are not designed to keep large systems running when normal power fails, only to give users enough time for an orderly shutdown so no data is lost. The home security system will not tax a UPS nearly as much as the LAN, but when other devices deplete the UPS batteries, the security system shuts down as well. In the event of a power failure, the operating limits of the UPS should be noted and the homeowner should be aware that the security system ceases functioning when the battery power runs low.

Backups and Fail-Safe Systems

Commercial security systems generally hold to the rule of **double protection** wherever possible so that if one level of security is evaded or fails, a second still picks up the emergency and sounds a warning. Home security is not usually as deep as the commercial type because the risk of home fire and crime is not usually as great. Exceptions exist to that generalization, however, and some home situations may justify multiple levels of protection.

No system of security is absolutely fail-safe. The level of protection is increased by how many separate warning systems are in place and operating on the premises. If two detection systems are present and the chance of either failing is 1 in 1,000, then the chance of both failing at the same time is 1 in 1,000,000. To get that 1-in-1,000,000 failure rate, both systems must operate entirely independently of one another so the failure of one does not have any effect on the other.

Most home security systems do not have two independent warning systems, but they can have dual elements that substantially increase the protection level. Window and door sensors combined with inside motion detectors are one example of dual protection against intruders. Smoke detectors combined with heat detectors are a dual fire protection. If all sensors connect to one security panel, then the system as a whole is not stronger than the security panel. But security panels, especially with backup power, almost never fail and are almost never compromised by intruders, fire, or other hazards they protect against. Dual sensor protection may be justified for some homes, but one good panel is enough security for any system.

Other backup features commonly available in home security systems—such as multiple call-in numbers, low-battery warnings, double-function locks, and others—make even the average system virtually certain to detect danger and warn the home's occupants of it.

Security System Programming and Settings

The main items that need configuration in a security system are the security panel and the response devices it controls. All the sensors in a security system are normally set to signal any alert they detect to the security panel. When the panel receives an alert signal, it then directs the response of various devices. If the security system includes other devices, these may need to be programmed individually.

Programming may be needed for keypad access devices, security panels, timers, and notification devices. Each must be programmed so it makes the correct response to the input it receives. This section discusses these items.

Keypad Access Devices

A keypad access device must be programmed for the codes that open it. Most systems allow for multiple codes to be used and, if this option is available, it is good to program a separate code for each family member. If one person's code is compromised, only that person has to learn a new one. One code can be reserved for visitors who need temporary access to the home. That code can then be changed whenever necessary without impacting the codes of any other family members.

Some access devices allow for a **duress code** (a code to be used in the event the person is accosted or threatened at the door) that opens the door, but also summons help. This is normally a single code that everyone with access can use and should be an easily remembered combination (12345, for example, or 77777) because it is used very rarely and may be forgotten, especially when under duress.

Some access devices can also be scheduled for times of day. Electronic keys as well as keypad systems often have this function. For some times of the day or night, the access device can be set to deny outside access to everyone or to allow it only to one master code number. This feature can provide better security when the home is unoccupied and also when the occupants are sleeping. If the homeowner is away from the home for an extended period, the lockout times can be extended to cover around the clock and allow no one inside, or they can be customized to allow a house sitter in at specified times to feed pets and water plants.

Security Panel Programming

The main programming of a security panel is for the outside and **inside arming** of the security system. **Outside arming** activates all security sensors to function while the home is unoccupied. The outside armed state is usually set by entering a command code on a keypad near the home's entrance when the occupants leave. The code can also be entered on the security panel itself. In either case, after entry of the code, a delay follows during which the occupants leave the home and secure the entrances. When the delay expires, the security system activates all sensors inside and outside the home, and sounds a chirp or bell to confirm that it is functioning.

When the home's occupants return and open a door to enter the home, another delay is activated, during which the security system must be disarmed so it will not interpret the homeowner's return as an intrusion. Typically, a 30-second delay is sufficient to disarm the system. During the delay, most systems sound an intermittent beep to remind the entering person that an alarm will be triggered if the system is not disarmed. Exit codes, exit delays, and entrance delays can all be varied to suit the needs of the owner.

Inside arming activates perimeter devices in the security system and may activate some interior devices, but usually not all. Inside armed is the normal status for the security system at night while the family is asleep. All window and door sensors and any sensors and detectors in the yard are active. Some inside motion detectors, such as those that guard doors, might also be activated, but motion sensors covering broad areas of the home's interior are left inactive. These provide additional protection when the house is empty (outside armed), but create too high a risk of false alarms to use with inside arming because anyone moving in the home at night could set them off.

Inside arming is done from the security panel and involves no delays, when either arming or disarming the system, because the home's occupants can still move about freely with the security system functioning in this mode. Many systems allow inside arming to be set by a timer so it activates every night at, for example, midnight, and disarms automatically each morning at 6 am. Timing the system eliminates forgetting to arm it (or disarm it), and the schedule can be overridden manually with different settings when necessary.

If a sensor signals an attempt to enter the home, the usual programmed response from the security panel is to sound an alarm siren inside the home and direct the call-in device to telephone for emergency assistance. This response is made whether or not anyone is present in the home. The panel is normally programmed to summon the police, but many police departments will not respond to automated alarm calls because so many of them turn out to be false alarms. If the police in an area will not respond to a security system call, the security panel can be programmed to call a family member or friend who is willing to receive the call and summon assistance or take other appropriate action. A fire sensor signal triggers an audible warning (which could be the same as the intruder alarm or different) and a call-in response to the fire department.

Fire and intruders are the two most common sensor signals for which a security system is programmed, but others may also be needed, depending on the other sensor devices the system includes. Water sensors and freeze sensors may require only an audible warning if the homeowner is available to correct the problem. If not, an automated response such as shutting off a water line or starting a heater may be programmed into the security system. Other responses that can be programmed into security systems include turning lights on in the home or yard (or flashing them repeatedly as a warning), turning on speakers with recorded messages, turning on sprinkler systems, or shutting down utilities.

Some security systems have access devices with a **latchkey function.** This function programs the call-in device to notify a set telephone number (or two numbers) when a specific door of the home is opened during set hours. This feature can notify working parents that children have arrived home from school or that service personnel have arrived or left the home.

Time and Notification Settings

Security systems should always be set to the correct time of day so timed functions occur at the hours specified. Many systems record the time of sensor alarms and notifications and the accuracy of these recorded events may be important in later investigations or insurance claims.

Call numbers programmed into call-in devices should be thought through carefully. A 911 emergency call is normally the first entry, if the jurisdiction where the home is located permits automated calls. Other call numbers should include relatives or friends who can respond appropriately. Call messages should give the address of the home, even if the call is going to someone who knows it. There is no need to give the telephone number, however, since emergency services record all incoming calls and the numbers from which they originate.

Maintenance, Servicing, and Recovery

Security systems rarely break down. Their reliability is partly a function of good initial construction to meet the durability requirements of a quality system, and partly due to the fact that their components have few moving parts and a very low level of activity. They simply observe and occasionally signal conditions around them. Security system components almost never wear out. Most service and maintenance for these systems centers around their aging rather than mechanical or electronic breakdowns.

System Service and Maintenance

Security sensors are self-monitoring, sending a signal to the security panel periodically to confirm their functioning status. They also monitor their batteries and signal when power begins to drop off. Finally, if the condition that a sensor is monitoring changes, or something happens to the sensor to impede its monitoring of that condition, it signals an alarm. With this degree of self-monitoring, little external monitoring of sensors is ever needed. As long as their batteries are changed when signaled as low, sensors can function almost indefinitely without further service.

Security panels also require little in the way of service or maintenance. They are generally AC powered and have backup batteries that age over time. These should be changed every 2 to 3 years, even if never used. Like all electronic equipment, security panels can be damaged by power surges and spikes and should be protected by a surge suppressor on the AC circuit. If the panel is connected through a UPS, this will usually have built-in surge and spike protection.

The response devices in a security system are the most vulnerable to maintenance problems and should be tested annually to make sure they are functioning properly. Electronic speaker alarms rarely malfunction, but mechanical bells do and should be activated to confirm they are working. The same is true of any mechanical response device that functions only if an alarm is signaled. Electric solenoids, motor-driven actuators, and especially valves, left unused for a year or more, can corrode and become frozen in their static position. They should all be activated periodically to prove they can activate when signaled to do so by the security panel.

As is the case with all electronic systems, the enemies of electronic security systems are heat, dirt, power spikes, and water. Wherever any of these can attack a part of the system, it should be protected as much as possible and regularly monitored for possible failure. Among the specific points in a security system that should be checked are:

- Wireless window detectors (for heat damage to the transmitters from sunlight coming through the glass)
- Outside motion sensors (for dirt or obstructions that may block all or part of their fields)

- Inside motion sensors (for dust and spider webs blocking their fields)
- Outside wireless sensors (for weather damage)
- Break switches on window and door sensors (to detect anything inserted in the switch to prevent its opening)
- Make switches on pressure pads or similar devices (to detect anything inserted in the switch to prevent its closing)
- Locks (to be sure they do not stick open)

System Recovery

If a security system signals a fire through one of its sensors, that sensor should always be replaced when the system is brought back into service after the fire. Fire and smoke sufficient to set off an alarm should always be assumed to be sufficient to damage the sensor so it will not function as well in the future. These sensors are not expensive and the safe practice is to replace the possibly damaged sensor with a new one.

Fire damage to other parts of the security system, especially a wired one, should be carefully checked and, if necessary, repaired. Heavy smoke can damage some sensors even if fire does not reach them. Water damage is also a real possibility, particularly for wireless transmitters on sensors and the security panel.

Other sensors in a security system are rarely damaged by being activated. They can usually just be reset and used again. The possible exception is a water sensor, which may be affected if submerged for a long period of time. Again, this is not an expensive sensor, and to assure that it functions well, replacement may be the best policy.

If a security system is breached by an intruder, all of its sensors should be checked afterward to be certain no attempt was made to evade or disable them. The fact that the system worked to signal an intrusion does not preclude the possibility that efforts were made to get around it in some other way.

False alarms can sometimes be very troublesome in security systems. Whenever they occur, they indicate the need for a service evaluation of the system. Eliminating the causes of false alarms makes the system more reliable and also makes the homeowner and emergency services more willing to rely on it.

False alarms can be partially prevented by careful wiring and installation so that signal failures do not occur as a result of poor connections. False alarms can also be generated from a number of innocent sources that are difficult to entirely eliminate. Outside motion sensors may be triggered by debris blowing in a high wind. Light beam sensors can be broken by a wandering cat or a sleepwalking child. Window and door sensors can sometimes be tripped by strong winds flexing the door or window inward slightly, thus breaking the sensor switch.

There is no sure way to entirely prevent false alarms, but they can certainly be minimized by noting the conditions under which they occur and carefully trying to trace the cause. Adjusting or repositioning a sensor may be all that is needed to correct the problem. If not, changing to a different type of sensor may be an answer, or simply eliminating one troublesome sensor and covering the area with other devices in different locations might resolve the issue. If false alarms happen only at night or only with sensors that cannot be easily observed by those inside the home, the possibility that someone is deliberately testing the security system's effectiveness should not be dismissed. False alarms should not be used as a reason to turn off a security system.

Summary

- Security systems are manufactured by many companies and are not standardized to any great degree. Systems bought as a package of components are more likely to function together better than those purchased individually from different manufacturers.

- Wired security systems have low data-transmission speeds and can be wired with Category 2 (two-pair) wire or category 3 (four- or six-pair) wire. This wire is smaller than high-speed network cable and can be concealed or run on the surface of walls without the need to use raceways.

- Wireless security system sensors are generally battery operated and require no data or power wiring. The security panels of wireless systems are AC powered, but can always be located near a power outlet and connected to it. They may include both a wireless receiver and a transmitter.

- Except for WiFi-based systems, wireless security systems do not operate on the same frequencies used by wireless networks.

- None of the power-line network or telephone-line network technologies have developed security systems or products, except for the X10 system, which has a wide array of security systems and devices.

- Most security systems, both wired and wireless, are marketed as packaged sets of compatible products to which accessory devices can be added according to need.

- X10 technology offers the most extensive array of security-related products in both power-line wired devices and wireless systems. All X10 devices in both technologies are compatible with one another.

- Accessory security components that can be added to basic systems include video surveillance, alarms, access devices, gate controls, uninterruptible power supplies, and backup devices.

- Installed security systems must be programmed for access control, arming (outside and inside), disarming, timing, and notification schedules.

- Security systems are low-maintenance systems with few moving parts and low activity rates. Their components should be tested periodically to be certain they can work if activated, and their power supplies should be regularly maintained.

- If a security system is breached either for a fire or a break-in, all its components should be carefully checked for damage before being placed back in service.

Key Terms

Arm Activate a security system so its sensors are functioning.

Break switch A type of sensor that consists of a switch whose normal state is closed. If the switch opens (breaks), a security breach is signaled.

Bypassed In security systems, a situation where a sensor is made to appear in normal status, but in reality is not.

Category 2 (Cat2) telephone wire A low-speed data wire containing two pairs of solid core insulated wires. Named for its widespread use in wiring analog telephone lines.

Category 3 (Cat3) wire A low-speed data wire with the same composition as Category 2 wire but having either four or six pairs of wires.

Disarm Deactivate a security system so its sensors are not functioning.

Double protection In security systems, at least two devices monitoring a means of entry into a home so that if one is disabled or fails, the other still detects an intruder.

Duress code An access code (to be used in the event a person is accosted or threatened at the door) that opens the door, but also summons help.

Emergency response   Any assistance delivered as a result of a security system call-in. Fire department, police department, and medical teams are all emergency responses.

Fail-safe   A theoretical term for a security system that cannot be disabled or bypassed. Not possible to achieve in actual systems.

False alarms   Any security breach not produced by a genuine security threat.

Inside arming   A code or command that activates perimeter devices in the security system and may activate some interior devices, but usually not all. Inside armed is the normal status for the security system at night while the family is asleep.

Intruder   Any unauthorized person trying to enter a secured home or yard.

Latchkey function   A command or code that programs the call-in device to notify a set telephone number when a specific door of the home is opened during set hours.

Line of sight   A term used to describe the location of wireless transmitters and receivers. It means that the receiver must be visible when viewed from the transmitter for the signal to be received.

Monitored   In security systems, watched either by a person at a control console or by the electronic security panel itself, so that any change in status can be responded to.

Monitoring service   A commercial service in which a company's staff continually watches a home security system via a telephone line linked to the home.

Obstruction detector   A safety device that detects anything unusual under a descending garage door and stops the door's downward movement.

Outside arming   A code or command that activates all security sensors to function while the home is unoccupied.

Recorded emergency message   A message recorded on a security system call-in device; it plays after a device calls a preset telephone number and gets a connection. Some systems dial multiple numbers in succession and some can play multiple messages.

Response devices   Security devices that can perform an action when commanded to do so by the security panel. They may sound alarms, call for assistance, activate systems, or perform other tasks.

Security breach   An event that occurs any time a sensor signals a change of status to a security panel.

Switcher   A multiple video input panel with a single output to a monitor and a switch that allows the monitor to display any video input selected.

Review Questions

1. Companies that provide remote access service with a security system usually provide 24-hour-a-day _____ of the home security system.

2. Security systems can be wired with what type of wire?
   a. Category 2
   b. Category 5
   c. Category 6
   d. Category 7

3. Category 3 wire is like Category 2 wire except it contains _____ or _____ pairs of wires instead of two.

4. Bypassing a sensor means making it appear to be working normally when, in fact, it is not. True or false?

5. Security system wiring can be bent sharply, stapled to the surface of walls, and spliced, but network data lines cannot. Why are the security system wires not as fragile as network data lines?

6. Two window or door sensors can be wired together to one security panel connection, provided they are wired in _____.

7. What is a zone in a security system?

8. A video camera in a security system is wired to a monitor or switch panel using what type of cable and connector?
   a. Category 2 with RJ-45 connectors
   b. Composite video cable with RCA connectors
   c. Category 5 with RJ-45 connectors
   d. MIDI cable with USB connectors

9. Wireless sensors can transmit through walls and obstructions to a security panel, but wireless video transmitters usually must transmit on a _____ to their receivers.

10. Wireless security systems work on which of the following technologies?

    a. HomeRF

    b. WiFi

    c. Bluetooth

    d. None of the above

11. What does arming a security system for outside or away status mean?

12. Wired security systems work on which of the following technologies?

    a. HomePlug

    b. HomePNA

    c. X10

    d. All of the above

13. Why is backup battery power not needed for sensors?

14. A duress code is _____.

    a. A backup access code to be used if you forget the main one

    b. An access code that reads the same in either direction

    c. An access code that opens the door and summons emergency help

    d. An access code for visitors so they do not learn the main one

15. Timed access devices may exclude some or all persons with valid access codes during certain hours. True or false?

16. An exit delay on a security system allows the user to _____ before the security system arms.

17. Inside arming of the security system is used when people are not inside the home. True or false?

18. When a security panel receives a breach signal from a sensor, what does it normally do?

19. What is a latchkey function on an access device?

20. To prevent damage from electrical surges and spikes, a security panel should be powered through an _____.

21. If a fire sensor detects a fire, what should be done with it when the system is reset?

    a. The homeowner resets the sensor.

    b. The homeowner relocates the sensor.

    c. The homeowner replaces the sensor with a new one.

    d. The homeowner replaces the sensor's battery.

22. X10 security systems are usually hybrid systems. What does this mean?

23. Raceways are not usually used in wiring security systems because _____.

    a. They are too large for security system wires

    b. They interfere with the security system's transmissions

    c. They are too small for the security system's wires

    d. They cannot be installed on windows

24. Security system wiring is not armored because the risk of it being cut is very low. True or false?

25. An independent access device may use a keypad, but it is not connected to the _____ _____.

Hands-On Projects

Project 9-1: Set Up a Motion Sensor

In this project, you will set up a wireless motion sensor and connect it to a lamp so when the motion detector signals motion, the lamp turns on. For this project, you will need access to an X10 MS14A or MS13A Motion Sensor, an X10 RF transceiver (model RR501 or TM751), an X10 lamp module (any model), and a lamp.

1. Set the X10 lamp module to house code A and the unit code to 1.

2. At one end of a classroom or other large room, plug the X10 lamp module into an AC outlet.

3. Plug the lamp into the module.

4. Plug the RF transceiver into an AC outlet at the opposite end of the room.

5. Remove the holding screw from the battery cover of the X10 motion sensor and then remove the battery cover. If the sensor does not have batteries installed, put two AAA batteries in it, but do not replace the battery cover.

6. The sensor's default setting is A1, so it does not need to be reset. Test the sensor by pressing the House button once. The unit transmits "device on" and the red light on it flashes. Press the Unit button once. The unit transmits "device off" and the red light flashes.

7. Place the motion sensor in a position at least 6 feet from the floor and facing out into the room.

8. Stand aside from the sensor for a full minute to let it arm itself and set a heat pattern.

9. Walk past the sensor and observe the lamp at the opposite end of the room. It will turn on and remain on for 1 minute (the default setting on the sensor).

10. Walk in front of the sensor again and watch the light come on. Stand as still as you can for one minute and see if the light goes off with you still in front of the sensor, but not moving.

11. If the light goes off, move again and see how quickly it comes back on as the sensor detects movement.

Project 9-2: Change the House and Unit Codes and Light Delay on a Motion Sensor

In this project, you will use the same equipment as in Project 9-1 but will change the house code and unit code to which the sensor broadcasts, and the length of time it transmits.

1. Unplug the lamp from the X10 module and the module from the AC outlet.

2. Set the house code on the lamp module to D and the Unit Code to 3.

3. Plug the module back into the outlet and the lamp into the module.

4. On the motion sensor, press and hold the House button until the red light flashes and then blinks the current setting (A = 1 blink).

5. Immediately release the button and then press it four times to set the house code to D. On the last press, hold the button down for 3 seconds. The red light blinks the number of the house code you set. Release the button.

6. Press and hold the Unit button until the red light flashes and then blinks the current setting (1 = 1 blink).

7. Immediately release the button and then press it three times to set the unit code to 3. On the last press, hold the button down for 3 seconds. The red light blinks the number of the unit code you set. Release the button.

8. Set the delay on the sensor for 2 minutes. To do this, press the House button once. The red light flashes. Immediately press the Unit button and hold it until the green light turns on. After 3 seconds, the red light blinks the current delay setting (one blink for 1 minute).

9. Immediately release the Unit button and press it twice to set the delay to 2 minutes. On the last press, hold the button down for 3 seconds. The red light blinks twice, showing the delay is set at 2 minutes.

10. Follow Steps 7 through 11 in Project 9-1 again to test the sensor and determine if it turns the lamp on using the new house and unit codes, and whether the lamp stays on for 2 minutes.

Project 9-3: Set Up a Wireless Video Receiver with a TV

For this project, you will need access to a television set with a video input jack (on the front or back of the television), and an X10 wireless camera system (Model VR36A or equal), which contains a wireless video camera with a power supply and a wireless receiver.

1. Set the television up in one end of a classroom or other large room. Plug it into an AC outlet.

2. On the bottom of the wireless receiver, find the channel switch and set it to C.

3. Connect the receiver to the television using a video cable with RCA connectors at both ends. Plug one end of the cable into the receiver and the other into the video jack of the television.

4. Turn the receiver on with the switch on the side of the unit.

5. Raise the antenna panel and turn it so the receiver side faces toward the opposite end of the room where the camera is placed.

6. Turn the television on and select video input.

7. Adjust the antenna of the receiver, if necessary, to get the least interference on the screen. You will know that the interference is decreasing when the picture quality is more pleasing to the eye.

Project 9-4: Set Up a Wireless Video Camera

For this project, you will need access to an X10 wireless camera system (Model VR36A or equal), which contains a wireless video camera with a power supply and a wireless receiver.

1. Set up the camera tripod on a table or other support at one end of a classroom or other large area. Mount the camera on the tripod using the screw mount on the bottom of the camera base.

2. Plug the power supply jack into the adapter cable attached to the camera. Plug the power supply into an AC outlet.

3. Carefully remove the rubber plug on the base of the camera. Set the channel switch under the plug to C using a pen point or other pointed object. The switch position at the bottom of the base (where the cord exits) is A. The position at the top is D.

4. Adjust the camera antenna so its broadcast side is toward the opposite end of the room where the receiver is.

5. Point the camera where you want its image to be. The camera is now transmitting.

6. Check the television and receiver you set up in Project 9-3. Is the television showing a picture from the camera? If it is, adjust the antennas to get the best quality picture you can.

7. If there is no picture, check the settings for the camera and receiver to be sure they match and also check all connections to be sure they are secure. Check the television to be sure it is set on the input channel. Troubleshoot the system using the instructions provided with the equipment until you get a clear picture on the television screen.

Project 9-5: Configure a Motion Detector to Trigger an Alarm

In this project, you will set up a wireless motion sensor and connect it to a control panel so that when the motion detector signals motion, the control panel signals an alarm. For this project, you will need access to an X10 MS14A or MS13A Motion Sensor and an X10 Wireless Security Console (Model PS561 or similar).

1. Check to be sure the control panel has a backup battery (9 volt) installed. Connect the panel to an AC power outlet that is always on. Extend the unit's antenna fully and place it on a table or desk.

2. Set the panel's house code dial to C.

3. Set the slide switch on the upper right of the panel to Install.

4. Press the House button on the motion detector once. The unit transmits "device on" and its red light flashes. The control panel's first available zone indicator light goes on, and the panel sounds a tone. This indicates that the panel has received the signal from the motion detector and assigned the detector to a zone.

5. Set the panel's slide switch to run.

6. Place the motion sensor in a position at least 6 feet from the floor, facing out into the room. Face it away from the control panel.

7. Stand aside from the sensor for a full minute to let it arm itself and set a heat pattern.

8. Walk past the sensor and listen for the control panel. It will chime a tone indicating that the motion sensor has detected motion. You can repeat the test after a minute. The sensor has a built-in delay so it will not transmit an additional motion signal for about 20 seconds. This helps extend the unit's battery life.

9. Stand away from the motion detector for a minute. Press the Arm button on the control panel to arm the system.

10. Walk in front of the motion detector again. The alarm will trip and sound.

11. Press the Disarm (bypass) button on the control panel to shut off the alarm.

Project 9-6: Configure a Call-In Dialer to Make a Programmed Call

In this project, you will configure a call-in dialer to call a phone number in response to an alarm warning triggered by the wireless motion sensor and control panel you set up in Project 9-5. For this project, you will need access to an X10 MS14A or MS 13A Motion Sensor, an X10 Wireless Security Console (Model PS561 or similar), and a telephone line with RJ-11 plugs on both ends.

1. Check to be sure the control panel has a backup battery (9 volt) installed. Connect the panel to an AC power outlet that is always on. Extend the unit's antenna fully and place it on a table or desk.

2. Connect an accessory phone cord with an RJ-11 phone line plug into the back of the control panel. Connect the other end of the line to a live telephone jack. If the only jack available already has a phone connected to it, use a splitter to divide the jack connection so both the phone and the control panel can be connected to the same line.

3. Set the slide switch on the upper right of the panel to Install.

4. Press the Prog (program) button on the panel. Enter the telephone number you want the dialer to call on the keypad at the lower left of the panel. When you have entered the seven-digit number, press the Mem (memory) button, then press the 1 button on the keypad to store the number in the first call position.

5. If you want to enter a second number, press the Program button again and follow the procedure in Step 4. When the number is entered, press the memory button followed by the 2 button to store the second number. (Caution: Only program a phone number if the owner is aware you are doing so and expects to receive the called message. Do not program a police or other emergency number.)

6. Connect an earphone to the control panel's earphone jack on the side of the unit. Keep the slide switch set to Install.

7. Press the panel's Record button and watch for the Record light at the bottom of the panel to come on. When the light comes on, record the message you want the dial-in caller to give when it dials the phone number. For this test, the message might be, "This is a test of the Smith's security system. Please call back 555-1234 to confirm that you received this call."

8. Set the control panel slide switch to Run. The message you recorded plays back. If you need to rerecord it, repeat Steps 6 and 7 to do so.

9. Press the Arm button on the control panel. Wait a full minute for the system to arm itself.

10. Walk in front of the motion detector. The alarm sounds and the call-in dialer calls the number you entered and delivers its message when the phone is answered.

11. Disarm the system and wait for the person you called to call you back with confirmation that he or she received the security message.

Case Projects

Case Project 9-1: Design a Security Response for a Hearing-Impaired Person

You have been asked to install a security system in a home occupied by a family with a child who is severely hearing impaired. The child has her own bedroom on the second floor of the home. The security system is configured to warn of fire and intruders with a warning alarm and emergency response call-in, but this child probably would not hear the warning alarm. Research what other visible warning device could be installed in her room or in other parts of the home that would give her warning of an emergency. Describe in a short report what you would do to set up such a warning response.

Case Project 9-2: Diagram a Video Surveillance System

In the Hands-On Projects, you set up a motion sensor and a video camera. These devices can work together. If you were installing a security system with a video surveillance camera, describe or diagram how you could connect the motion sensor, X10 wireless receiver, X10 module, video camera, video receiver, and television so the video camera would turn on only when someone walked in front of the motion sensor. Would any additional equipment be needed or could the video system be set up with only the items noted?

Case Project 9-3: Use a Security System to Identify an Intruder

A client with an installed security system contacts you with a problem. He believes that someone is entering his home while the security system is activated, turning off the system, and searching for confidential information in his home office. The intruder then resets the security system and leaves the home. This could only be done by someone who knows the arming and disarming codes to the security system. Describe what could be done to the programming of the security system to prevent such an intrusion from happening again. Also describe what could be done to identify the intruder without permitting him to actually enter the home again.

Case Project 9-4: Design a Swimming Pool Security System

You have installed a security system for a client who has a swimming pool. The homeowner is often away, leaving the home and pool unattended. Many small children live in the neighborhood and often ask to swim in the pool. When he is home, the homeowner allows this, but he fears that one of the children could come while he is away and drown in the pool. Standard motion sensor alarms do not work well around the pool because heat reflected from the water causes continual false alarms. What type of device can you recommend installing in or around the pool that could warn of someone about to enter the pool?

Telecommunications Fundamentals and Installation

OBJECTIVES   *After studying this chapter, you should be able to:*

- Describe the characteristics of analog and digital telephone communication systems and the differences between them.

- Define the characteristics of the various local telephone systems.

- Identify the components of telephone systems and describe their functions.

- Describe the installation of wiring and components of various telephone systems.

- Define external services offered through local and central telephone systems.

OUTLINE   Introduction

Telecommunications System Types and Characteristics

Local Telephone Systems

Telephone Components and Features

Telephone Installation and Configuration

External Services

Introduction

Telecommunications systems are both an independent voice communication system in the home and a data transmission system that may form an integral part of the overall home technology network. In this chapter, you will learn about analog and digital telephone communications and the capacities of each. You will understand the various types of local telephone systems and how they are integrated with home networks. You will learn about the basic components from which telecommunications systems are assembled and how these devices function. You will also learn how to install telephone systems and set up their operation for maximum efficiency. Finally, you will learn how to configure the services that are offered by local and central telephone systems.

Telecommunications System Types and Characteristics

Telecommunications began with an analog device that quickly developed into a worldwide analog system. The telephone was invented in 1876 for the transmission of speech over wires. Speech is analog sound, and calls on telephone lines were exclusively analog signals from the 1870s until the 1960s.

Telephones were the first means of transmitting voice messages over wires. They became the first mass communication system, quickly surpassing the first digital communication system, the telegraph, which never achieved wide consumer use because of the complexity of its Morse code. Telephone use grew until the demand for ever greater voice transmission began to exceed the capacity of analog lines. By that time, digital transmission of voice and other data had been developed to a point where it could provide the additional capacity needed in the telephone transmission system. Digital telecommunications systems now have taken over much of the transmission load for both voice and data, but analog systems continue to function in many areas.

Analog Telephone Communication Systems

What we know as Plain Old Telephone Service (POTS) is an analog system of phone lines and equipment. Even today, almost 50 years after the introduction of digital voice transmission, most home telephones are still analog devices that are connected to analog phone lines. Once the phone line reaches from the home to the nearest telephone exchange, it likely connects to a digital transmission line, but the local "last mile" of phone line from the telephone exchange to the typical residence is still analog.

As the need for more and faster communication accelerated in recent decades, the analog phone system was taxed to provide the necessary service. Analog signals move relatively slowly when carried in physical media such as copper wire and cable. Telephone voice transmission signals travel at speeds from about 300 Hertz (cycles per second) to 3,300 Hertz. Although this might seem relatively fast when compared with, for example, the speed of an electric motor, it is downright snail-paced when compared with digital data speeds over a network that are measured in millions of Hertz (MHz) or wireless data transmissions that are measured in billions (GHz).

Analog voice communication signals are also subject to weakening due to resistance in the wires on which they travel, and to interference (noise) caused by electrical equipment, power lines, and even some electric lights. In a telephone conversation, this noise is heard as **static** on the line. The weakened signal of an analog transmission must be strengthened periodically by an amplifier. Unfortunately, an amplifier cannot separate the noise from the voice part of the signal and it consequently amplifies both. Voice conversations that have been amplified several times usually can still be understood because the human hearing system is very adept at extracting the correct meaning of spoken words from background sounds. Data transmissions sent over analog lines do not always fare as well because the data is interpreted by machines that do not "hear" as well as people. The result is that noise in analog data transmissions frequently causes errors.

Data transmission using analog modems is limited to about 56,000 bits per second when receiving and about 33,000 bits per second when sending data. The analog public switched telephone network operates at 2,400 **baud.** A baud is one complete cycle or wave in a transmission signal. A baud starts at zero voltage, goes up to maximum positive voltage, comes back to zero, goes to maximum negative voltage, and finally returns to zero voltage. A 2,400-baud line carries current that completes 2,400 of these wave cycles each second. The baud rate refers only to the frequency of the current. It does not indicate the amount of data that is transmitted on the line. A 56,000-bit modem uses technology that allows it to put 24 bits of data on each wave of a 2,400-baud line (56,000 divided by 2,400 = 24). The line current is still running at 2,400 baud even though 56,000 bits of data per second are being transmitted on it. That is about the limit for an analog data line, however. To send more than 56 Kbps of data requires a shift to digital transmission technology.

Voice and data can both be sent on an analog line at rates up to the limits that analog lines can transmit. Within a home, the requirements for voice communication by telephone will almost never exceed the limits of analog technology. **Frequency division multiplexing** even allows analog lines to carry multiple conversations or data streams simultaneously. This is accomplished by dividing the available analog frequencies on a line and sharing them among different voice and data transmissions. Thus, one voice data stream might be running on a line at 2,400 baud, while a second runs at a higher baud rate, and a third at a still higher rate. **Multiplexing** allows multiple telephone lines to be installed in homes and businesses without the need to install additional wiring from the telephone company office to the home.

If voice links were all that were required of phone lines, analog alone would probably be sufficient, at least within the home. But most homes now also require data transmission over the same telephone lines that handle conversations on the telephones. If the phone lines are even occasionally filled by analog voice transmissions, which cannot be delayed or interrupted without interrupting or garbling a real-time conversation, they cannot also meet data transmission requirements.

To have enough capacity for the system to handle both voice and data needs, additional lines must be installed, or the capacity of the existing lines must be increased by transmitting digital data on them. Not only must the data be sent digitally, but the voice transmissions must be sent digitally as well. If either were left in analog form and sent with that technology, they would consume so much of the line's capacity there would be insufficient capacity left for the digital transmissions.

Changing both voice and data transmission to digital format provides a huge increase in capacity for a telephone line. The majority of homes are still using analog lines for telephone voice transmission and for data lines that connect analog 56-Kbps modems to ISPs, but the shift to digital transmission lines that can carry anywhere from three times to dozens of times the data of an analog line is accelerating. The demand for increased data transmission rather than additional voice lines is driving this transition, although voice traffic is also increasing.

Digital Telephone Communication Systems

Digital data transmission over telecommunications wires has several advantages over analog, the two greatest being that it is much faster and much less error prone. As it does on a network, digital data on a telephone line travels in packets. It follows the rules of established protocols so it can be controlled and sent to the correct address and interpreted when it arrives there. Voice data and other analog sounds are converted to digital form by sampling techniques similar to those used for digital sound recording. The coded audio signals are sent as digital data and decoded at the receiving end so they can be "played" through a speaker as sound. Data sent by digital transmission (whether it is digital audio data, text data, or video data) travels far faster and with far fewer errors than it would as analog data.

As with analog signals, digital transmissions also weaken and pick up noise as they travel, but the amplification process used to strengthen digital signals is different from that used for analog. An analog amplifier must strengthen an existing signal including the data tones, variations, pauses, and noise that occur in it. The amplifier cannot change the data signal or any corruption that has come into it; it can only amplify it and send it on. The distance that analog signals can travel is limited by the amount of noise the signal picks up. Eventually, the noise level becomes so high that the analog voice or data signal can no longer be understood.

Digital signals consist only of zeros and ones (binary numbers), which are very simple to read and duplicate. A digital amplifier in a data transmission line is called a **regenerator.** Instead of amplifying the existing signal, the regenerator simply reads the signal and generates a duplicate signal of all the data in it, but none of the noise. The clean duplicate signal is then sent on the line in place of the weakened one. Digital signals thus move more cleanly with less chance of error than analog signals because they can be regenerated periodically without any noise being present in the regenerated signal.

Capacity in telecommunications is described as **bandwidth.** Analog transmission or POTS, whether voice or data, is **narrow band,** each line being limited to one voice channel or one data channel at a maximum speed of 56 Kbps. By contrast, a BRI ISDN line, though still narrow band in the digital data transmission scale, provides for two voice or data channels, each capable of carrying 64 Kbps, and a third channel for control signals that can carry up to 16 Kbps. A **T-1 line,** which can be carried on two pairs of wires such as Cat5 or similar cable, can carry up to 1.54 Mbps of digital data. This means that it can simultaneously transmit 24 data signals or voice conversations, each containing 64 Kbps of data. T-3 lines and other broadband technologies, transmitted over fiber-optic cables or radio frequencies, can reach capacities ranging from 44.7 Mbps to 13.22 Gbps.

For most home telephone users, a digital subscriber line (DSL), which provides digital data capacity from 128 Kbps to 7 Mbps or even more, is adequate for both voice and network data requirements. The cost for a DSL line is usually less than for a T-1 connection, and DSL service is now available in most urban areas. As demand for DSL lines continues to grow, telephone companies are installing additional lines. Where DSL service is not available, users generally retain their analog telephone service and obtain digital data service through a cable company or a satellite hookup. Both cable and satellite data transmission services are still more widely used than DSL because neither of these technologies is subject to the distance restrictions from a central telephone exchange that limit digital subscriber lines.

TABLE 10-1
Types and Characteristics of DSL Telecommunications Service

Type of Service	Data Rate Upstream	Data Rate Downstream	Central Office Distance Limit	Voice	Comments
Symmetric DSL	1.1 Mbps	1.1 Mbps	12,000 ft 2.25 miles	No	Uses one wire pair
DSL Lite	384 Kbps	1 Mbps	18,000 ft 3.4 miles	Yes	Uses one wire pair
Asymmetric DSL – downstream	640 Kbps	7.1 Mbps	12,000 2.25 miles	Yes	Uses one wire pair
Asymmetric DSL – upstream	1.76 Mbps	1.54 Mbps	18,000 ft 3.4 miles	Yes	Uses one wire pair
G.SHDSL	192 Kbps to 4.62 Mbps	192 Kbps to 4.62 Mbps	6,500 ft (1.2 miles) 40,000 ft (7.5 miles)	Yes	Uses one or two wire pairs
HDSL	1.54 Mbps	1.54 Mbps	12,000 ft 2.2 miles	No	Uses one or two wire pairs

Several varieties of DSL are available, not all of which are available in some areas. Rural areas are unlikely to have DSL service available because of their distance from a central exchange. Distance of the user from the telephone company central office is a determining factor in the availability of all DSL lines. Table 10-1 shows different DSL line types and the varying distance limits, typical speeds, and characteristics that each has. Users should check carefully on the type and speed of service they will receive when purchasing a DSL line.

Local Telephone Systems

The first telephones were wired to each other so that a conversation could be transmitted between them. A few telephones could be wired in this manner, but the cost and labor of wiring every telephone user to every other user quickly became excessive. By the year after the telephone was invented, **switchboards** had begun to appear. Individual telephones were not wired to one another any longer, only to the central switchboard. An operator there answered every call and connected the requesting party to another telephone subscriber through the switchboard. By 1891, automatic switching gear was patented that could automatically connect calls through relays according to the number dialed on the rotary phone. No operator was required.

This section describes the various local telephone systems that provide local area switching services. At first, this switching service was provided by the telephone company, but competing systems now provide most local switching.

Centrex Systems

The central switching exchange or **Centrex** was the forerunner of **private branch exchange (PBX)** telephone systems that are now located on site at many businesses. When switching was done at the telephone company's central exchange, every telephone line coming into a home had to connect to that exchange. A line could have several **extensions,** but a call could be placed on only one of them at a time because all the extensions had only one line connecting to the exchange where it could be switched to the number called. PBX systems have become very common in business establishments and have taken over much of the local switching once done by Centrex systems. In turn, PBX systems now face strong competition from another control method known as a **key system.**

Centrex service, which uses telephone company–supplied and –maintained switching equipment, is still used on 10 to 15% of U.S. telephone lines. Centrex switching has some advantages, especially for large organizations with facilities spread out over a wide area. None of the Centrex equipment needs to be located on the customer's premises, although some of it can be if there is a cost or efficiency advantage in doing so. All switching can be done from the telephone company's central exchange. This allows incoming calls to be routed directly to individual extensions without passing through an operator or receptionist. Likewise, outgoing calls can be placed directly without an operator or need to dial for an outside line. All outgoing calls are also identified and billed to the extension that made them. There are no "common line" calls.

Centrex systems are best used by organizations that have many telephones in many locations. The telephone company can provide the wiring infrastructure to connect separate locations without any need to install private lines. It can set up the organization's service so internal calls are routed to other extensions using an internal dialing sequence and outside calls are routed correctly using the public telephone number system. Local governments, hospitals, large educational institutions, and businesses with multiple locations or divisions are among the types of organizations that find Centrex systems the most efficient and economical for their needs. Smaller users favor PBX systems for the same reasons.

PBX Systems

A PBX is a private on-site switching system that routes calls within a single entity in the same manner that a central exchange routes calls on the public telephone system. The site for a PBX can be an entire organization such as a business or educational institution occupying one building or several. It can also be a small business occupying only a suite of offices or an individual home.

A PBX system controls telephone lines and switching from the **demarcation point** or demarc, the point where the incoming lines from the telephone company connect to the internal telephone wiring of the home. All calls to and from the telephone extensions and other devices throughout the house are controlled by the PBX. Figure 10-1 shows the inside (user side only) of a demarc with three incoming telephone lines wired in it. The lower two lines are voice lines and are wired with old-style single-strand wire (see the section "Connecting Telephone Equipment" later in this chapter for details about old- and new-style wiring). The top line is a data line and is wired with new-style wire.

Telephones inside the home are connected to the PBX rather than directly to outgoing telephone lines. The PBX is connected to the demarc point and there its lines connect to the outside lines of the public telephone system. Calls from one telephone to another within the home are connected through the PBX without ever passing outside the home. Calls going out to the public telephone system are connected to an outside line through the PBX. Incoming calls on any line can be routed to any extension telephone within the home, again through the PBX.

FIGURE 10-1   An Open Demarcation Box with Three Lines on a Residence

Mid-size businesses and other organizations find PBX switching meets their needs more effectively and economically than Centrex. Outgoing calls from any extension go through the PBX and out to the demark point. From there they are carried on **trunk lines** to the central exchange and are routed on to their destination. Incoming calls travel the other direction on the same trunk lines to the PBX, which then routes them to the correct extension. Most organizations with a PBX use a T-1 line for their trunk. A T-1 line can carry up to 24 voice or data lines simultaneously. Those 24 lines can be shared by 100 or more voice users (only a few of whom would be making calls simultaneously) and still provide data connections for the business as well.

PBXs are **ground start systems.** This means that when a telephone receiver connected to the PBX is lifted to make a call, the PBX responds by sending a **dial tone** to the phone and also by requesting that an outside trunk line from the PBX to the central station be reserved or "seized" for the call. If the extension dials an access code (usually 9) for an outside line, it is connected to the seized trunk line, the caller hears another dial tone provided by the telephone company over the trunk line, and the call can be dialed. If the call is to an internal extension on the PBX system, it is routed directly to the called number and the seized outside line is released. Incoming calls are also sent over a "grounded" line seized by the central exchange when it receives the incoming call destined for the PBX.

Few homes have enough telephones or enough user demand to warrant installation of a PBX, even a small one. A large home having one or more offices with multiple lines and several additional home lines might want a PBX that could monitor outgoing calls from the business extensions in order to track expenses for tax purposes. Trunk-line service of less than a T-1 line, however, would hardly ever justify a home PBX, especially when another type of switching system is available for smaller users. Figure 10-2 shows the back (connection panel) side of a "home-size" PBX system with a **voice mail** attachment on top of it. This system, made by DataLabsUSA, allows for three incoming trunk lines and up to eight residential extensions. It provides all standard PBX features, including fax detection, which routes incoming fax calls to a designated extension where the fax receiver is located.

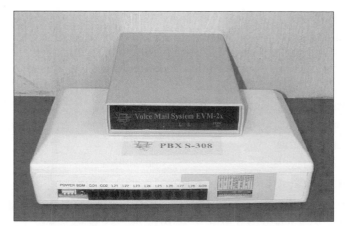

FIGURE 10-2 A Small PBX and Voice Mail System Suitable for Home Use

Key Systems

Key systems are designed for smaller organizations with fewer users than a mid-size business. Key systems began when the old AT&T Company began installing individual telephones connected to multiple lines. These phones had buttons or keys that could connect them to any one of the several telephone lines coming into the home. They also had a hold button that kept a line that was in use connected while another line was used by the same telephone extension. The line buttons each had lights to show when their line was in use so that another extension user would not push a button for a busy line, unless of course, he or she wanted to listen in on that conversation.

On these early key systems, all the extensions could connect to all the lines simply by pushing a connect button to access the line. No access code for an outside line was required because there was no PBX to request that an outside line be seized for a call. Key systems are **loop start systems** rather than ground start, like with a PBX. This is the major difference between the two systems.

In a loop start system, when a telephone receiver is lifted to make a call, a dial tone is sent from the central office to show that a path is open for the call, but the line is not seized until the call is actually placed. Likewise, an incoming call is sent to the destination number by any available path, with no line being seized from the central station to the receiver in advance. This means that it is possible, though unlikely, that a user on a key system could pick up a telephone to make a call and find someone already on the line calling her or him, even though the telephone had not rung. The incoming call simply arrived over the available path slightly before the outgoing call could be placed. The loop path was filled first by the incoming call which would have to end before the outgoing call could be completed.

Key systems with multiple loop start lines are well suited for small business users and homes with more than one line. All phones have access to all lines and can use any available line for outgoing or incoming calls. Key systems have an **intercom** button that allows them to make internal calls to other extensions without accessing an outside line. The internal call simply goes to the called extension over wiring inside the home.

Hybrid Systems

Newer key systems have most of the features of a PBX system and in fact may be hybrid systems that offer multiple-line buttons with a holding feature as well as the grounded trunk lines with a dialed outside access code used in PBX systems. These hybrid systems also offer many convenient features, including:

- Hold buttons for keeping a line open
- Speed dial and redial buttons for rapid dialing of frequently called numbers
- Voice-mail alert lights
- Caller identification screens
- Cordless telephone sets

These features and the multiple-line capacity of key systems and hybrids make them ideal for home use where multiple-line service is desired.

Both PBX systems and key systems carry voice calls over **circuit-switched lines.** This means that all the data in a call follows the same path on the telephone network, and the line over which it travels is held open for that data for the full duration of the call. Telephone systems have now converged with computer networks over which data is sent in packets that travel by different routes to the same destination. Manufacturers of PBX systems and key systems have now equipped their products with data transmitting capabilities using Internet Protocol. The result is **Voice over Internet Protocol (VoIP)** telephone service.

Voice over IP (VoIP)

Data moving on a LAN or other network travels in packets that adhere mainly to the Internet Protocol (IP). Voice data packets can also travel on the same networks, provided they follow the same protocol.

VoIP systems must perform a number of tasks rapidly and flawlessly in order for voice telephone conversations to be sent over a data network. When a voice call is made over IP, the sending system must encode the analog voice signal into digital form, compress the digital data to reduce its size, assemble the compressed data into packets with appropriate headers and trailers, and transmit the data onto the LAN. At the receiving end, the data packets must be received on the LAN, reassembled into the order they were sent in, decoded (decompressed and converted from digital to analog voice), and sent to a speaker (in a telephone or in a computer). All of this must be done as nearly as possible in real time because the telephone conversation is a two-way process and any appreciable delay in sending or receiving the transmissions results in periods of silence between every statement of the two speakers.

Data packets sent over a network do not need to arrive in sequence and can easily be sent again if a collision causes one to be lost. Voice data packets, however, are made up of parts of a real-time conversation and need to be reassembled at the receiving end in proper sequence and in real time. A lost packet or delayed transmission often causes the reconstructed voice to sound choppy or "clipped" and may make it difficult to understand. The biggest challenge for VoIP telephone transmission is to prevent the telephone conversations from being degraded significantly below the quality of analog voice transmissions.

To avoid significant degradation of VoIP transmissions, the network must have sufficient capacity so that only a small part of its data packets suffer collisions and must be sent again. The voice data packets must also be sufficiently compressed (uncompressed sound files are large) so that they can be sent and converted back into sound in real time, despite the delay caused by any lost packets. Providers of VoIP equipment are continually working on these two requirements in order to improve the quality of VoIP transmissions. New electronic chips that more efficiently compress audio and put it into packets for network transmission are being developed. Some of these new systems for VoIP also provide a means for voice data packets to have priority over other data packets so the continuity of transmissions can be better maintained when traffic on the network is heavy.

Most home network users with IP-based phone systems still make the majority of their calls over the **public switched telephone network (PSTN).** Intercom calls between nodes on the LAN are made over the network, and long-distance calls can be made to a receiver node on another LAN that has appropriate software for decoding the data into voice.

Several companies, the largest of which is Vonage, now provide VoIP digital telephone service for long-distance calls. These companies have proprietary VoIP software installed on Internet-connected computers around the world. When a homeowner buys their service (for a monthly fee), the VoIP provider installs its software on the customer's computer. When the customer makes a telephone call, the analog data is digitized by the computer, which then routes it over the Internet to a VoIP-equipped destination computer located in the area of the person being called. The receiving VoIP computer decodes the call and connects it via a local telephone line to the person called.

Although the installed number of independent IP-telephone-equipped home LAN nodes is not yet large, the commercial services offered by VoIP companies are growing rapidly. Interest in VoIP technology is high because of its low cost for long-distance phone service and its ability to make greater use of network infrastructure without the need for new telephone wiring. The reliability of VoIP telephone systems is not yet as high as the PSTN, which does not crash nearly as frequently as the typical computer system, but it is improving rapidly. The use of VoIP in business and in home LANs is increasing.

Remote Access Methods, Standards, and Protocols

VoIP systems can deliver telephone service to standard analog telephones, to PBX- or key system-connected multi-featured phones, or to **softphones** built into a computer with software. Softphones are software programs that display phone features (hold button, **caller ID,** message waiting) on the computer screen and route the calls through a handset or an earphone and microphone wired to the computer. In addition to enabling VoIP phone capability with a minimum of new equipment (softphone programs work with standard PC peripheral audio components), softphones can also provide remote phone access. If the computer is a wireless laptop, then the softphone system goes with it anywhere within the range of wireless reception. For the home user, this extends the telephone system to the full limits of the wireless hubs on the LAN without the need to install **cordless phones** or carry any additional equipment.

Another means of extending telephone service to all parts of a home or yard is through the use of limited-range cordless phones. These phones are transceivers (transmitters and receivers) that operate on radio frequencies in the 900 MHz and 2.4 GHz bands. Calls on these phones are relayed back and forth between the wireless handset and the base unit in a two-way radio broadcast. The handsets are battery powered (usually by rechargeable batteries), but most base units require

AC power through a low-voltage transformer in addition to the line voltage in the telephone system.

Neither cordless phones nor wireless VoIP softphones are secure from eavesdropping by significantly interested parties. Calls on both types of phones can be received by anyone within range who has suitable equipment. The home user of wireless telephone equipment should be as careful as the business user not to give personal information, identification numbers, or any other nonpublic data over a wireless phone call.

Cell Phones

Cell phones can integrate with a home LAN either through their wireless connection to the home telephone system or via a connection to the Internet. A telephone link allows a cell phone to connect to the home LAN by calling a home land telephone line and connecting through that call to other segments of the network. With this link, the cell phone user can download information from the network's other nodes and forward command or control instructions to the security system, utility control systems, or other parts of the network.

The same objectives can be achieved by connecting to the home LAN through the Internet. Most cell phones now have a direct wireless link to the Internet through the service provider. Using this path, the home owner can link to the home LAN and perform all of the functions that would be available by linking to the LAN through a remote computer. Cell phone Internet connections are not as fast as a computer link, nor do they have the storage capacity or ease of data entry available on most laptops, but they offer great convenience for contacting the home LAN from any remote point where cell phone service is functional.

Telephone Components and Features

The telephone is a fairly simple electronic device, especially when compared to modern computers, audio systems, and other devices used in communications. Telephone service, however, is one of the most universally available communications systems on Earth and its widespread use has resulted in many refinements in the basic unit over the years. Many pieces of peripheral telephone equipment providing enhanced telephone service or adding additional features to its basic communication function have also been developed. The basic telephone's design and some of the important improvements made in telephone service and equipment since its invention are described in this section.

Telephone Fundamentals

A telephone delivers (plays) audible sound derived from electronic signals received over wires and also converts audible sound into electronic signals that are then sent over the same wires to another telephone. Figure 10-3 shows the main parts of a modern telephone, which are still fundamentally similar to those that would be found in an early telephone. In fact, most antique telephones from the 1900 era, if plugged into a present-day analog telephone circuit, would still work.

The handset of a telephone contains a diaphragm attached to an electromagnet at its upper end. Together they form a speaker that reproduces sound from the signals coming to it on the electric circuit in the handset. At the bottom of the handset another diaphragm set in a carbon-filled chamber acts as a microphone to

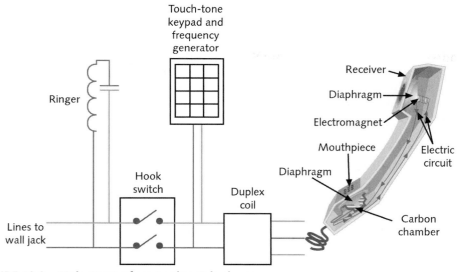

FIGURE 10-3   Main Parts of an Analog Telephone

convert sound waves into analog electric signals, which travel to the base circuitry of the telephone.

The base of the telephone contains a duplex coil, through which the wires containing both the incoming and outgoing speech signals pass. This device blocks outgoing speech signals from feeding back into the receiving speaker so the telephone user will not hear his or her own voice echoing in the phone's speaker when talking. Other devices inside the telephone base are the hook switch, which disconnects the telephone from the outgoing line when the handset is hung up on it; the ringer, which remains connected to the outgoing line and rings when an incoming current (call) on the line activates it; and the touch-tone keypad and frequency generator, which generates the tone signals to dial numbers as its keys are pressed.

Many telephones have additional features added to them for enhanced service, but all telephones have the basic components previously described and all work in the same manner. Telephones are analog devices only. If their voice signals are sent in digital form over networks, they must be encoded (converted) into digital form after leaving the analog telephone at the sending end of the call, and decoded (reconverted) into analog signals again at the receiving end of the call before the signals are delivered to the receiving phone.

Extension Dialing

When telephones were all switched through Centrex systems and each telephone was connected to an outside line, extensions could be added so that one line was connected to several phones simultaneously, but conversations between extensions on the same line were not possible. There was no way to switch one extension to connect directly with another without also being connected to the outside line. Since both were already connected to the same line, if two people picked up extensions on a line, all either would hear was the dial tone sent from the Centrex, denoting it was ready for them to make a call.

PBX systems, which are connected to all the extensions in a home, allow any extension to be connected to any other without the need of also being connected to an outside line. The PBX simply switches the two extensions together as an internal phone call. Each extension has its own internal number that can be used only by another extension on the system. Outside calls come from the PSTN using the assigned number of the telephone. Internal calls use the extension number (usually one or two digits in a home system) to connect one extension to another.

Key systems are like the early Centrex systems in that they cannot connect one extension to another directly because all the internal lines in a key system are also connected to outside lines. If a caller on an extension pushes a line button, he or she gets the dial tone from the Centrex. The caller can then call another extension by calling through another outside line to the extension. This amounts to placing an outside call to the Centrex and then having it routed back to the home on a second line as an incoming call. Such calls require relatively long travel for the signal and tie up two phone lines to make one extension call; it is not a very efficient way to talk to someone at the other end of the house. A more effective solution was clearly needed.

Intercoms

To get around the problem of Centrex-routed extension-to-extension calls, all key systems have intercom capability. An intercom link is simply an internal method of connecting two telephones in a home without connecting either to an outside line. If there are four incoming lines to a home with a key system, each phone has four line buttons so it can connect directly to any of the incoming lines. It also has a fifth button for intercom calls. This button does not connect to an outside line, but to a switch panel that is also connected to each of the other intercom buttons on the other telephones in the home. As with the PBX system, each extension has a number and any other extension can dial it directly, but only on the intercom line. If an outside line button is pressed for an intercom call, it cannot connect because those buttons only connect the phone to an outside line, not to the inside switch that links all extensions.

Intercom service is very useful in letting users in different areas of the home talk to one another without either person moving to the other's location. Some systems allow conference intercom calls (three or more participants) and many include a paging feature.

Paging allows one extension to call another and talk over a speaker on the called extension before the extension handset is picked up by anyone. A person looking for someone can thus page them at various extensions around the house and yard (some paging features allow all extensions to be dialed for a page simultaneously) without having to shout their name or conduct a physical search for them.

Phone-Line Extensions and Splitters

Extension phones can also be added on a single line in a home by using **splitters.** Splitters are devices that allow two (or even three) telephones to be connected to a single wall jack. In this arrangement, both extensions ring when a call is received and either or both can answer it and hear the incoming call since they are connected to the same line. Only one splitter extension at a time can make an outgoing call, however, because there is only one outgoing line on which it can travel. The other extension can listen in on the outgoing call, but cannot make another until the first is completed and disconnected.

In addition to phone extensions, a splitter also allows a line to be connected to a phone and a peripheral device such as a fax machine. In this arrangement, all incoming calls are answered first by the fax machine, which must be equipped to determine which calls are fax messages. If the incoming call is not a fax message but a voice call, the fax machine allows it to ring through to the phone. Outgoing fax messages are also sent over the line from the fax machine. The extension phone is only used for outgoing calls when the fax machine is not in use.

Another peripheral that can be attached as a splitter extension is a computer, giving it a primary or backup connection to an ISP, or allowing it to receive and record voice mail and fax messages on its hard drive. In the former use, the computer only accesses the phone line when an ISP connection is needed. At all other times, the line is available for use as a regular phone extension for placing and receiving calls. Such a computer extension hookup might also be used for an emergency assistance call by the computer as part of the home's security system. An emergency link would rarely be used, but needs to have constant availability of a phone line in case a distress call is required.

If the computer is used for messages (voice mail or fax), then incoming calls are answered by the computer so it can process messages. When the computer is on the line, the other extension should not be picked up or it will interfere with data transmission to the computer. A line-in-use light should be placed on the phone line in this arrangement so users can avoid interrupting data transmissions.

Caller-Line Identification

The rise of telemarketing and other unsolicited (and unwanted) telephone calls has made caller line identification (caller ID) a high-demand feature of home telephone service. Caller ID is a feature made possible by a signal sent by the telephone company with an incoming call that identifies the calling number and its registered owner. Telephones with electronics capable of reading this signal display the caller ID information on a small LCD screen. The called party can then decide whether to accept the call or not.

Call Conferencing

PBX and key systems both permit conference calls between three or more persons. Both systems accomplish this feature in similar ways. In either system, by connecting the three outgoing lines to one another, the three parties are all joined on the same circuit and can talk to one another. Conference calls involving more than three persons simply repeat the steps needed to add another line and caller, using the PBX or key system to connect the new individual into the call.

Videoconferencing

Videoconferencing, sending full-motion picture images with a telephone call, requires a lot of bandwidth, high-level compression of the video data, and multiplexing (combining different types of data from multiple sources on a single transmission path). Videoconference software has to transmit data at rates of 128 Kbps up to 768 Kbps in order to provide acceptable quality video reproduction at the receiving end. Analog telephones cannot handle this quantity of data, and so all videoconference software is based on digital phone technology.

The compressed video in conferencing systems still is not equal to the quality found in broadcast systems. It travels at a slower pace and runs at slower speeds (fewer frames per second) than commercial video. The audio portion of a videoconference is also compressed by using slower sampling techniques, skipping silent times, and predicting future sounds based on previous ones. These improvements in compression techniques and the use of slower video make videoconferencing practical using software that requires transmission speeds of only 128 Kbps to 384 Kbps.

Videoconferencing is now available to home network users who have adequate bandwidth to support the software. All those participating in a videoconference must have common software and be able to connect to one another on the Internet or via telephone lines. The high-speed data rates of videoconferencing require multiplexing the voice and video portions on the same transmission line in order to maintain the synchronization and timing of audio and video portions. Multiplexing can be done with both analog and digital data, but the compression and speeds necessary for videoconferencing require both the audio and video to be digital.

Call Restriction

Most PBX and key systems provide **call restriction** options for outgoing calls, and some can also block specified incoming calls. Outgoing restrictions can be applied to any or all of the phones internally connected to the system. Incoming call restrictions are nearly always universal (blocked from entering the phone system, hence not available to any extension). A few systems allow specified incoming calls to automatically route to a specified number in the system. Among the outgoing call restrictions available are:

- Total restriction (no outgoing calls permitted)
- Long-distance calls restricted (all calls out of the local area code are restricted or only specified codes permitted)
- Call-time restrictions (outgoing calls permitted only during specified hours)
- International calls restricted
- Specific telephone numbers restricted

Those systems that can restrict incoming calls usually do so with a recorded message such as "This telephone subscriber does not accept calls from . . ." or similar wording. For the restriction message to be played, the call must be connected to the PBX or key system. On long-distance calls, this results in a minimum charge to the calling party, even though the call did not actually reach the called party. Among the call-in restrictions that some systems can invoke are:

- Total restrictions (no incoming calls accepted for specified extensions)
- Calls without caller ID (or on which the ID has been deactivated or blocked)
- Long-distance calls (or calls from specific area codes)
- Calls from specified telephone numbers

Voice Mail

As with call restriction features, voice-mail functions were originally offered by the central phone company. All digitally processed and stored voice-mail systems trace their beginning to the VMX Corporation's original system first installed in 1980. All current manufacturers of Centrex, PBX, and key systems with voice mail still license their software from the successors of the first patent holder.

Voice mail is usually compressed in order to conserve storage space on the hard drive. Messages are indexed in sequential fashion for playback, but the data is stored randomly as files on the hard drive. If the drive fills, no more messages can be stored until space is cleared by erasing some of the older ones.

PBX and key systems performed voice-mail functions as a stand-alone telephone system feature until recently. Now voice mail is changing from an independent telephone feature to **unified messaging** in which storage is located on a PC's hard drive and the messages include not only voice mail but also fax and e-mail. Cisco System's Active Voice messaging software is a unified system that operates on Microsoft Exchange. Other manufacturers are developing or have announced unified messaging software. The convenience and efficiency of unified messaging allows the home user to review all messages at one time. The systems can perform several levels of sorting to bring up the most time-sensitive messages or the most important first.

Most new PBX and key systems now offer unified messaging as an option, if not a standard feature, and use a PC as a storage and display console. Unified systems store calls as voice files, faxes as graphic files, and e-mail as text files. The voice mails can be played on the PC's sound system and the faxes and e-mail displayed on the computer screen and then printed, if necessary.

Fax Machine Communications

Fax machines are digital devices that operate on analog telephone lines. The analog telephone line's capacity limits the speed of a fax machine to about 54,000 bits of data per second, and at this speed, some errors in data transmission are likely. However, the top speed of current machines is 33.6 Kbps.

A fax machine views a page of printed material (text or pictures or both) as a bit-mapped graphic. It divides the page into tiny dots (about 300 dots per inch arranged in rows across the page. It scans the page to determine whether each dot on its surface is white or black (fax transmissions have no grays or other colors), then compresses the digital data it derives and sends it over the telephone line to another fax machine. The receiving machine reverses the process: It decodes the digital data back into white or black dots on the page and then prints the black ones in their correct positions on the bit-mapped graphic, thus reproducing the original printed page.

All fax machines scan documents in the same manner, but some have memory chips that enable them to store the scanned data before sending it. These machines can scan documents more quickly than those that do not have memory chips because they are not limited by the speed at which the data can be compressed and sent over the telephone line. The scanned information is simply stored in memory, and later sent on the telephone line as rapidly as the line can carry it. Fax machines can have memory capacity for a single-page document or for many pages. The larger the machine's memory capacity, the faster it can scan a document and store the data, but the speed of the fax transmission is still limited to the capacity of the analog telephone line over which it is being sent.

Fax machines print the documents they receive in different ways. The least expensive printing process is called thermal printing. It uses a heat-sensitive paper on which the fax machine prints by using a row of heated elements across the width of the paper. The heating elements do not actually touch the thermal paper, but they pass very close to it and the heat transfer is sufficient to print on the paper. Plain paper fax machines use the same printing technology found in laser printers or ink jet printers. Instead of receiving printing data from a computer, the fax machine receives its data over the telephone line and with it creates the image for the laser printing module or ink jet technology to print.

When sending or receiving a document, a fax machine uses the full capacity of a telephone line. No simultaneous voice transmission is possible. But fax machines do not operate constantly, especially in home environments, and this means they can usually share a telephone line that is used for voice calls when not occupied by a fax transmission.

Many fax machines are designed for this type of line sharing. They have built-in circuitry that determines whether an incoming call is a fax transmission or a voice call. If it is a fax message, the machine activates to receive and print it. If not, it allows the call to ring the attached telephone so the voice call can be answered. In many fax machines, a voice telephone is also built in so incoming calls that are not fax transmissions can be answered on it and outgoing calls can be made. Fax machines that do not have a built-in telephone can still share a telephone line by being connected to a telephone jack through a splitter that also has a telephone attached.

Telephone Installation and Configuration

Since telephone service began, the equipment used has continually increased in complexity and capacity to meet the demands of users. Looking back, it is easy to see with perfect hindsight that early wiring and infrastructure did not provide for nearly enough future expansion. This is especially true in residential telephone service where single-line analog telephone service wired with two-wire telephone line is the norm. Wiring and installation of telephone systems today should benefit from past experience and provide not only for more capacity than is currently needed, but more than current technology levels suggest will ever be needed. Experience has shown that the use of data transmission systems (telephone, network, radio, and television) has continually increased at a pace that far exceeded the most optimistic predictions. There is no reason to think that home telephone service 10 years from now will not need a hundred times the capacity it needs now.

Telephone equipment is wired and installed in much the same manner as network and other electronic equipment. The same care should be taken in wiring to use adequate cabling and install it to standards, make data-secure connections, and test systems carefully to be certain they function correctly.

Literally thousands of devices are available for use in home telephone networks. This section discusses only a few of the basic installation procedures you need to know for all home systems. The installation instructions for any telephone system device should always be studied and followed precisely to ensure that the individual requirements of the equipment are met.

Connecting Telephone Equipment

Analog telephones have long been wired with cable consisting of two or four copper wires. The wires are single strand and are not twisted into pairs. A single telephone line requires only two wires, and so a four-wire cable could connect two lines. Because the wires were not twisted, cross-talk interference was common. Persons speaking on one telephone line could often hear a low-volume background conversation, which was usually someone talking on the other line.

FIGURE 10-4   An Old-Style Telephone Block Wired for Two Lines

Figure 10-4 shows an old-style wiring block wired for two lines with single-strand wire. The round four-wire cable, secured with staples, can be seen coming in from the right side of the photo. The individual wires lead out to an extension phone (not visible in the picture) above the block.

This old-style wire has a color code that is slowly disappearing in favor of the color coding used for twisted pair cables that connect high-speed data networks. Table 10-2 shows this older wiring color scheme for a four-wire or a six-wire cable. This color code is still important to know since you will likely encounter many existing telephone systems wired with it.

The designations "tip wire" and "ring wire" come from the old ¼-inch phone plugs (see Table 10-2) that were used to connect phone lines in operator-run PBX boards. Each jack connected a phone line and had two wires running from it. The tip wire was the positive wire and was connected to the tip of the plug. The ring wire was the negative wire and was connected to the ring or side of the plug. This traditional designation has remained in use even though its original meaning no longer applies.

The old color code has now been replaced by one that includes up to five pairs of wires, but is nevertheless easier to remember. It is shown in Table 10-3.

Additional pair colors exist for this color code, just like for the older one, but home telephone systems are not likely to use cables with more than five pairs of wires.

TABLE 10-2
Old-Style Wire Colors for Telephone Wires

Wire Pair	Wire 1 Color	Tip/Ring	Polarity	Wire 2 Color	Tip/Ring	Polarity
1	Green	Tip	Positive	Red	Ring	Negative
2	Black	Tip	Positive	Yellow	Ring	Negative
3	White	Tip	Positive	Blue	Ring	Negative

TABLE 10-3
New-Style Wire Colors for Telephone Wires

Wire Pair	Wire 1 Color	Tip/Ring	Polarity	Wire 2 Color	Tip/Ring	Polarity
1	Blue/White	Tip	Positive	Blue	Ring	Negative
2	Orange/White	Tip	Positive	Orange	Ring	Negative
3	Green/White	Tip	Positive	Green	Ring	Negative
4	Brown/White	Tip	Positive	Brown	Ring	Negative
5	Slate/White	Tip	Positive	Slate	Ring	Negative

Today, telephone wiring is installed using twisted-pair cable. Category 3 cable is adequate for analog telephone lines and contains at least two pairs of wires, which can be used to wire multiple telephone lines. In keeping with the practice of installing excess capacity, new telephone wiring, especially if it is to carry any digital transmissions, is usually done with Cat5 cable. Basic telephone wiring is not difficult, but should be performed carefully to ensure good connections and make certain lines will meet their data transmission requirements.

Pulling Cat5 or similar cable for telephone connections should be done using the same techniques and follow the same cautions used when cable is pulled for a network. Even though the data transmission requirements for the telephone lines may not be as high as for the data lines, installing the cable with the same (maximum) adherence to standards ensures that it performs with peak efficiency and can be used in upgraded systems with higher data requirements later.

RJ-11 Connections

The Bell System telephone company developed the Universal Service Ordering Codes (USOC) specifying the wiring configurations for the series of Registered Jack (RJ) types used to connect residential telephone equipment to the public network. Figure 10-5 shows these wiring configurations for one line, two lines, three lines, and four lines. An **RJ-11 jack** is the standard single-line telephone connector. An **RJ-14 jack** is the standard two-line connector. RJ-25 and RJ-61 Jacks are for three- and four-line connections, respectively.

These color codes for telephone wiring differ from those used for RJ-45 jacks and plugs that connect network wiring, but the procedures for terminating cables to jacks and plugs are the same. Wires must be set in the plug according to the color pattern shown in Figure 10-5, but all the other steps in the terminating process are the same. Wires not used in a multiple-pair cable when terminating it to a jack or plug should be cut off at the plug or jack entrance so they do not interfere with connected wires or produce short circuits in the cabling. Figure 10-6 shows some of the major parts used for wiring telephone lines. Moving clockwise from the upper left, the photo shows an RJ-11 plug, four-pair wire, a splitter for doubling extensions from an RJ-11 plug, and an RJ-14 four-wire jack with its cover removed.

If wiring is being installed for a network and for telephone lines, a single Cat5 or similar cable can be used to accomplish both connections. Only two of the Cat5 cable's four pairs of wires are used for network connections. The remaining two

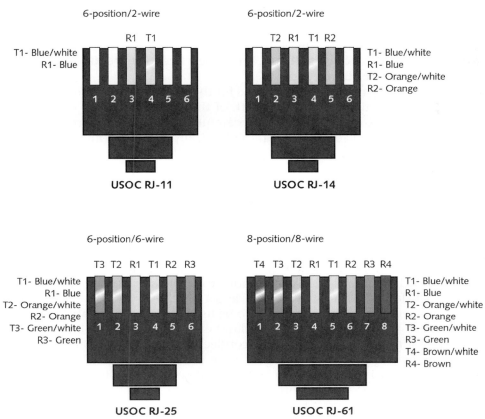

6-position/2-wire

R1 T1

T1- Blue/white
R1- Blue

1 2 3 4 5 6

USOC RJ-11

6-position/2-wire

T2 R1 T1 R2

T1- Blue/white
R1- Blue
T2- Orange/white
R2- Orange

1 2 3 4 5 6

USOC RJ-14

6-position/6-wire

T3 T2 R1 T1 R2 R3

T1- Blue/white
R1- Blue
T2- Orange/white
R2- Orange
T3- Green/white
R3- Green

1 2 3 4 5 6

USOC RJ-25

8-position/8-wire

T4 T3 T2 R1 T1 R2 R3 R4

T1- Blue/white
R1- Blue
T2- Orange/white
R2- Orange
T3- Green/white
R3- Green
T4- Brown/white
R4- Brown

1 2 3 4 5 6 7 8

USOC RJ-61

FIGURE 10-5   Telephone Plug and Jack Wiring Standards

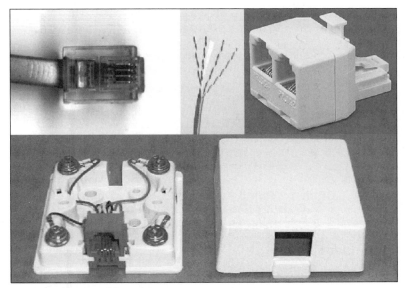

FIGURE 10-6   Telephone Wiring Components

pairs can each be used to wire a telephone line in a combination jack. If one cable is used for both types of connections, the RJ-45 jack must be wired using only the two pairs of wires actually needed for the data connection. The other two pairs are not included in the jack connection, but are left free to be wired to an RJ-11 or RJ-14 jack as telephone connections.

The colors of the pairs that must be used for the telephone lines will not follow the USOC standard, but will be one or both of the twisted pairs not used for the two-pair data connection. You must keep careful track of which pair is used for each telephone line, and connect the lines identically at both terminations.

Connection Blocks

Multiple telephone lines can be connected to a centrally located **punchdown block** or patch panel that is connected by a trunk cable to the telephone company's lines at the demarc point, which is usually located on the outside of the home.

Punchdown blocks are the usual means of centrally terminating telephone lines and connecting them to the trunk lines, but these blocks are used primarily for commercial installations, which have many lines. Figure 10-7 shows a large 110-type punchdown block with connections for 100 pairs of wires, as well as a 110C-4-4 block designed for four wire pairs and a 110C-5-5 block designed for five pairs. The smaller two are the ones likely to be used in home installations.

For home installations that usually will not exceed three or four lines, a patch panel termination may be preferable to a connection block. Patch panels provide for easy reconfiguration of the wiring because they are connected through RJ-11, RJ-14, or similar jacks. They also assure good connections for the multiple-stranded wires found in Cat5 or other multiple-pair cables. Punchdown blocks connect by insulation deformation when each wire is punched down into the block using a specialized punchdown tool. They work best with single-strand wire, which is strong enough to accept this type of connection without damage to the conductors.

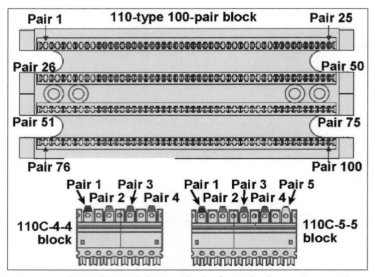

FIGURE 10-7 Large and Small Punchdown Blocks for Telephone Connections

External Services

Some telephone services available on PBX or key systems still require some participation by the telephone service provider. For homes that do not have a local phone system, the central telephone company may provide the entire service. Telephone companies charge a monthly fee for each of these services, so obtaining all of them can eventually cost more than purchasing a local system that has the features built in. This is especially true if the home has multiple incoming telephone lines because, unlike a local telephone system that can apply its features to all the home's telephone lines, the outside telephone provider charges a fee for each line that subscribes to these services.

Caller ID

Caller ID is always provided in part by the telephone service company because it is the only entity that can supply the caller information on incoming calls. Local equipment must have the necessary electronics to read and display the caller ID signals. If a local system has caller restrictions on incoming calls, the incoming lines must have caller ID signals on them because these signals are the means which the local system uses to determine if a call should be allowed through or blocked.

Call Blocking

Central telephone company call blocking is limited to an inserted message that states which calls are not accepted by the subscriber. The law requires that callers who are included in a blocked group hang up when they hear the blocking message, but if they do not, the call goes through and the subscriber has to pursue the violation through a complaint to the service provider. A local system also receives blocked calls, even after playing a recorded message that they are not accepted. But a user will not ever actually receive one of these blocked calls because the local system simply does not put them through to any phone extension where they could be answered. The calling party will hear a ring signal on his phone, but no ring sounds in the home. The local system either disconnects the call or holds it until the caller hangs up.

Three-Way Calling

Three-way calling provided by the telephone company has the advantage of only using one line from the home that originates the call. If a three-way conference call is set up by a local system (PBX or key), it uses two lines to bring both of the called parties to its home connection. If more than three parties are included in a call, a local system must have a new line available to add each one to the call. The telephone company has no such limitations. Through its central switching facilities, it already has a line connected to every phone and can connect any number of these lines into a single call.

Call Waiting

Call waiting on a single telephone line can only be provided by the telephone company. The company can keep both lines open to their station until the user disconnects them, but only one conversation at a time can reach the destination phone.

Voice-mail service is also available from the phone company on both lines of a call-waiting setup. Because the telephone company's voice-mail system is located in the central office, it does not need to route a waiting call to the home; it simply redirects it to the subscriber's voice mail located in its offices.

Emergency Response System

A home with a security system that has an emergency response function should always have an available outside line on which an emergency call can be sent. Whether the call goes to a commercial security service or to a family member or friend, the security system must always be able to get a clear line to make the call or calls immediately. In rare cases, this may mean that the local telephone system needs to disconnect another call in order to free a line for an emergency.

Since some incoming calls cannot be immediately disconnected by the receiver (the line is seized by the originator of the call and will not be released until he or she hangs up), the safest line to put emergency call-in equipment on is one that is used exclusively for that purpose and no other. If keeping an open telephone line for only one possible call in years seems a needless expense (as it will to most), the next safest line for a call-in response is one on which only outgoing calls are made. If the home has a separate line for the home LAN ISP connection, that's the line to also put the emergency call-in device on as well. It is possible that an emergency call might interrupt a data transmission one time, but that small risk is more than offset by having a line that can be instantly cleared when the emergency response call is activated.

If an independent line or an outgoing-only line is not available, emergency response devices can be connected to any other line coming into the home. The only security risk in using a regular voice line is that an incoming call, which the emergency response device cannot override, might be on the line and delay an emergency call for a time. A few cases have even been reported in which burglars called into a home in order to occupy the phone lines so an emergency call-in could not get through, but criminals this clever are rare.

Summary

- POTS is an analog voice telephone system which has been built as a worldwide network over a period of about 130 years. Its lines can also transmit data at a maximum speed of 56 Kbps.

- Digital telephone transmission of voice and data began in the 1960s and vastly expanded the data-handling capacity of POTS. Digital transmission of data, including voice, text, and video, can now be achieved at speeds ranging from 128 Kbps to 7 Mbps or more.

- The public switched telephone network (PSTN) provided central exchange or Centrex services to connect telephone calls nationwide. Centrex handled all switching until local systems were developed to provide organizations with internal switching.

- Local PBX and key systems now handle 85% of telephone switching. PBX connects all local phones through a switchboard to trunk lines that go to the central exchange. Key systems connect multiple lines directly to Centrex and also provide intercom service.

- Digital telephone transmission is now expanding over the Internet as VoIP that provides fast and economical service to customers who have suitable transmitting and receiving software. Some users have integrated softphones on their PCs and receive all telephone service by computer.

- Telephones are analog voice devices with a microphone, speaker, and digital keypad for tone dialing. In addition to basic calling service, other functions offered on telephones include intercom, call waiting, caller ID, split lines and extensions, conference calling, call restrictions, voice mail, and video-conferencing.

- Fax machines are digital devices that operate on analog telephone lines. They send printed pages containing text and graphics at speeds up to 54 Kbps.

- Telephones are wired with two wires per line and do not require a high-speed data cable.

Most telephone wiring now being installed uses high-speed cable because digital transmissions do require it and telephone systems are becoming more digital and high speed.

- RJ-11 and RJ-14 jacks and plugs are the standard terminators for telephone cables. They are wired according to a color code, which is not the same as that used for network cables and RJ-45 jacks.

- Some telephone services are still provided wholly or partially by the central telephone system. These include caller ID, call blocking, three-way calling, call waiting, and emergency response call-in.

Key Terms

Bandwidth In telecommunications, the amount of data transmission capacity in a line.

Baud One complete cycle or wave in an analog transmission signal. A baud starts at zero voltage, rises to the maximum positive voltage, comes back to zero, goes to the maximum negative voltage, and finally returns to zero voltage.

Call restriction A telephone system feature that allows certain calls to be blocked from incoming or outgoing lines.

Caller ID A system in which a signal sent by the telephone company with an incoming call identifies the calling number and its registered owner. Telephones with electronics capable of reading this signal display the caller ID information on a small LCD screen.

Centrex Central exchange, the main switching point in the telephone company where calls are connected by switching them to the line called.

Circuit-switched lines The normal method of completing an analog call in which the calling line is physically connected to the called line by a defined path, and the path is kept open for the duration of the call.

Cordless phones Phones with transceivers (transmitters and receivers) that operate on radio frequencies. They have a limited range from the base unit, which contains another transceiver.

Demarcation point The physical place where the incoming telephone lines from the telephone company

connect to the internal wiring of the home; also known as demarc.

Dial tone A continuous tone sent on a telephone line by the PBX or the central exchange to indicate it is available for a call.

Extensions In telephone systems, an additional phone wired in parallel into a single line.

Fax machines A digital device that operates on an analog telephone line. It sends and receives printed documents by transmitting them as bit-mapped images.

Frequency division multiplexing An analog technology for carrying multiple data streams, voice or data, on the same wires.

Ground start system Local telephone system that seizes a telephone line as soon as a receiver is lifted so no other call can transmit on the line.

Intercom A line in key systems that allows any extension to connect to any other for an internal call without accessing an outside line.

Key system A local telephone system in which multiple phones are connected to multiple lines by switching buttons so any telephone can use any line.

Loop start system Local telephone system in which a Centrex-supplied dial tone indicates when a line is available for a call, but no line is seized until a call is actually placed.

Multiplexing Combining different types of data from multiple sources on a single transmission path.

Narrow band In telecommunications systems, bandwidth of 128 Kbps or less.

Private branch exchange (PBX) A local switching point where phone lines within an organization can be connected to one another or to outside lines.

Public switched telephone network (PSTN) The commercial network of telephone lines and transmission facilities over which most telephone calls are made.

Punchdown block The usual means of centrally terminating telephone lines and connecting them to the trunk lines.

Regenerator A digital data amplifier that reads weakened data signals and re-creates them at full strength and without noise on the transmission line.

RJ-11 jack The standard single-line telephone connector wired with two wires.

RJ-14 jack The standard two-line telephone connector wired with four wires.

Softphones Software programs that display phone features (hold button, caller ID, message waiting) on the computer screen and route the calls through a handset or an earphone and microphone wired to the computer.

Splitters Devices that allow two or more telephones or peripherals to be connected to a single wall jack.

Static Noise or interference on a telephone line that is typically heard as a background crackling sound.

Switchboard A device to which a number of telephone lines are connected and which can switch any line to connect to any other.

T-1 line A high-capacity telephone trunk line capable of simultaneously handling up to twenty-four voice lines or 1.5 Mbps of data.

Trunk line A telephone line from a PBX to the telephone company that can be used by any of the telephones connected to the PBX.

Unified messaging A computer-based system for storing messages from multiple sources. Storage is located on a PC's hard drive, and the messages include voice mail, fax, and e-mail.

Videoconferencing A technology for sending full-motion picture images with a voice telephone call so callers can both hear and see one another.

Voice mail A digital system that allows telephone callers to record messages that the system stores, retrieves, and plays on demand.

Voice over Internet Protocol (VoIP) A system whereby analog phone calls are digitized and sent in packets over the Internet. At the destination, the digital data is decoded to analog form and delivered to the receiver.

Review Questions

1. Plain Old Telephone Service (POTS) is an analog network of telephones and connecting lines. True or false?

2. Data transmission over analog modems is limited to about _____ for receiving information and about _____ for sending information.

3. A complete cycle in the frequency of an analog transmission line is known as a _____.
   a. Bit
   b. Baud
   c. Byte
   d. Bend

4. Give one reason why digital transmissions over long distance are less prone to errors than analog transmissions.

5. In telecommunications, "bandwidth" means ___ _____.

6. What does a regenerator do?

7. A T-1 line can carry data at speeds up to _____ _____.
   a. 128 Kbps
   b. 256 Kbps
   c. 1 Mbps
   d. 1.54 Mbps

8. What is the unique feature of an ADSL compared to other types of DSL lines?

9. All DSL lines are limited by a maximum _____ _____ from the central exchange.

10. A PBX can connect _____ directly without the need of using any _____.

11. A PBX controls local telephone lines from the _____ to the end user extensions.

12. A trunk line runs from the central exchange to a microwave station. True or false?

13. What is a ground start system?

14. A key system is a local telephone system that does not require callers to dial a code for an outside line. True or false?

15. An intercom feature is available on which local system(s)?

    a. PBX

    b. Key

    c. PBX and key

    d. Neither

16. What is a loop start system?

17. The weakened signal of an analog transmission must be strengthened periodically by an _____.

18. Why is digital voice data more difficult to send over the Internet than ordinary numerical data?

19. A softphone is a _____.

    a. Padded telephone

    b. Wireless telephone

    c. Computer-based telephone

    d. Telephone with no speaker

20. The PSTN includes all the public telephone network up to a home's individual _____ _____.

21. A splitter is used to divide T-1 lines into multiple trunks. True or false?

22. For a PBX or key system to set up a three-way conference call requires how many outside lines?

    a. One

    b. Two

    c. Three

    d. Four

23. The standard plugs of terminating one- and two-line telephone connectors are _____ and _____.

24. In the old-style telephone wire color code, what colors are paired?

    a. Blue/green   red/yellow

    b. Green/red   black/yellow

    c. White/blue   black/green

    d. Yellow/green   black/white

25. Call-waiting service can only be provided by the _____ system.

Hands-On Projects

Project 10-1: Terminate a Four-Wire Telephone Cable with an RJ-14 Jack

In this project, you will terminate one end of a 10-foot cable to an RJ-14 jack. The cable length will also be used in Project 10-2 and so should not be shorter than 10 feet. Use two-pair Category 3 telephone wire for this project. The cable can be round or flat. You will also need an RJ-14 jack, a cable stripper, and a punchdown tool.

1. Use a CatX cable stripper to strip the cable jacket 1.5 inches.

2. Place the stripped end of the cable in the jack and arrange the wires in the correct color code for an RJ-14 jack, as shown in Figure 10-5. Be sure the stripped wires are pulled as far into the jack as the sleeving allows, so the excess stripped wire can be cut off when the wires are set.

3. Use an RJ-14 punchdown tool to set the cable, connect it, and cut the wires.

4. After your instructor inspects your work, cut off the jack from the cable end, leaving a 9-foot piece of cable that can be used for Project 10-2.

Project 10-2: Make a 9-Foot RJ-14 Two-Line Telephone Cable

In this project, you will terminate both ends of a 9-foot telephone cable with RJ-14 plugs. Use the same length of cable you used in Project 10-1 for this project. The cable should be two-pair Category 3 cable, round or flat. You will also need two RJ-14 plugs and a combination cable stripper and crimping tool.

1. Use a CatX cable stripper to strip a half inch of the cable jacket.
2. Untwist and flatten the wire pairs in the correct color pattern for a two-line (four-wire) RJ-14 cable.
3. Cut the stripped wires to a maximum half-inch length, if necessary.
4. Place the plug on the cable end with the flattened wires in the correct position. Be sure the wires are fully inserted in the plug.
5. Use an RJ-14 crimping tool to crimp the plug in place.
6. Repeat the preceding steps to terminate the other end of the cable with an RJ-14 plug. Use the same wire pattern so the telephone cable is a straight-through cable.

Project 10-3: Connect a Fax Machine to a Phone Line through a Splitter

In this project, you will connect a fax machine and an extension telephone to a single phone line using a splitter. You will then test the line to see if the fax machine can separate fax calls from voice calls. You will need a fax machine, a telephone extension, a splitter, and a working telephone line for this project.

1. Find the wall jack where the telephone line you are using connects to the extension telephone. Disconnect the phone extension plug from the jack.
2. Insert the splitter into the wall jack.
3. Connect the phone extension's telephone line into one side of the splitter.
4. Connect the fax machine's telephone line into the other side of the splitter. Plug the fax machine into an AC power outlet and turn it on. Check the fax machine to be sure it has paper in it and that it is ready to operate.
5. From another telephone connected to a separate line, call the fax machine telephone number (voice call). Does the telephone extension ring or does the fax machine pick up the voice call?
6. If the fax machine picks up the voice call, check its instruction manual to see if there is any setting on the machine you can configure so it does not pick up voice calls, but passes them through to the telephone extension.
7. If you can set the fax machine to separate voice calls, test it again to be sure it actually works that way.
8. If the fax machine cannot separate calls, is there any way you can set up the phone extension and the fax on one line so some voice calls can still be made to the extension? How would you do it? What calls could be made?

Project 10-4: Test the Range of a Cordless Phone

In this project, you will test the range of a cordless phone to see how far it extends and what happens to a call when the range is exceeded. You will need access to a cordless telephone set that is connected to a working telephone line.

1. Using the cordless phone, call someone who is willing to assist you with this test. The person does not need to be nearby, just willing to stay on the phone until you complete the test.

2. When your call is connected and answered, talk on the phone with the other party while you move away from the base of the cordless phone. Keep moving away until you can no longer hear the other person talking and they can no longer hear you.

3. As soon as you are sure you are beyond the cordless phone's range, turn around and walk back toward the base unit. As you move closer to it, can you talk to the other person again or has the call been disconnected?

4. If you were able to talk to the other person again without placing another call, why do you think the call did not disconnect when you walked out of range? Where do you think the control for disconnecting a call is located in this phone?

5. If the call did disconnect, where do you think the control for disconnecting a call is located in this phone set?

6. Ask the person who is assisting you to call you at the number of the cordless phone. Repeat your range experiment while talking on the call that was placed to you.

7. When you go out of range on this call, does the line disconnect? Does the system act the same for an incoming call as for an outgoing call? Why do you think it does, or does not?

Project 10-5: Hardwire Two RJ-14 Jacks to One Another

In this project, you will wire two telephone jacks together so a telephone line can be connected through one to the other. For this project, you will need two surface-mounted RJ-14 hardwired jacks, enough telephone wire (two-pair or more) to connect the jacks, a telephone cable with RJ-14 plugs on both ends, and a telephone that can be used to test the finished line. You will also need a wall area or upright board on which to mount the jacks, a flat-bladed screwdriver, and a wire stripper or knife for stripping wire.

1. Mount the two RJ-14 jacks on the wall a few feet apart and 1 foot above the floor using the screws that come with the jacks. Both jacks should mount with the plug insert opening facing down toward the floor.

2. Strip 1.5 inches from the telephone wire's outer jacket on one end. Do not strip the wires, just the outer jacket.

3. Strip a half inch of insulation from two pairs of exposed wire in the length of telephone wire. Strip the first two pairs noted in the color codes of Table 10-2 (old style) or Table 10-3 (new style), depending on what type of wire you are using. If the cable contains other pairs besides one and two, do not strip them. Cut them off at the point where the jacket is stripped from the cable.

4. Insert the stripped end of the cable into the left jack through one of the cable channels in its side and bring the wire pairs up through the center of the jack. Pull the cable into the jack until the outer sheathing is at or inside the cable channel.

5. Keep the cable in place with one hand, and place a staple in the wall over the cable to hold it from slipping out of the jack.

6. Connect the four wires to each of the jack's connecting screws. Match the wires color for color with those in the jack, or match the correct pair 1 and pair 2 leads by color code to the pair 1 and pair 2 leads, if the color codes are not the same on the jack and the cable.

7. Measure the length of cable needed to reach the other jack, and the additional amount needed to connect to it (1.5 inches). Cut the cable at that point.

8. Follow Steps 2 through 6 to strip the cable, insert it in the other jack, and then wire it to the same color-coded connecting screws. Pull the cable snug between the two jacks so it forms a straight line between them. Staple the wire in place.

9. Use a double RJ-14 plug patch cord to connect the jacks you have wired to a live phone line. Disconnect the telephone set from the wall jack by pressing the RJ-11 plug's holding tab and pulling the telephone wire and plug out of the jack. Insert one end of the patch cord into the live telephone jack and the other end of the patch cord into the right RJ-14 jack you have just wired.

10. Insert the telephone cord's RJ-11 plug into the left RJ-14 jack you have just wired. Pick up the telephone's receiver. You should hear a dial tone and be able to make a call. If you cannot, you have wired the jacks incorrectly. Go back and check your work for errors.

11. Keep testing and correcting the jacks' wiring until you can make a call on the telephone.

Project 10-6: Connect Telephone Extensions and Lines to a PBX

In this project, you will connect a telephone trunk line and two extension phones to a PBX so calls can be made and received on the system. For this project, you will need access to a PBX (DataLabsUSA Model S-308 or similar), two telephone extensions, and three RJ-11 patch cables, each at least 8 feet in length. You will also need access to one working telephone line that can be accessed through an RJ-11 jack.

1. Connect the power supply cable to the PBX unit power port. Connect the PBX system's power supply to an AC power outlet.

2. Using an RJ-11 patch cable, connect a telephone extension to the L21 jack on the PBX.

3. Using an RJ-11 patch cable, connect the second telephone extension to the L24 jack on the PBX.

4. Follow the instructions with the PBX to set the dip switches on the unit so incoming calls from trunk line CO1 ring to extension L21 (operator extension).

5. Using an RJ-11 patch cable, connect the CO1 jack on the PBX to a working telephone-line jack.

6. Pick up an extension telephone receiver and listen for a dial tone. If you hear the dial tone, proceed with testing the unit; if you do not hear a dial tone, check the dip switch settings and connections of the telephone extensions and line hookup.

7. Using Extension 24, dial "0" (for Operator). Extension 21 should ring. If it does not, check the settings on the unit and change them so Extension 21 is the operator extension. Test the system again.

8. Using Extension 24, dial "9" for an outside line, and then complete a call to an outside telephone number. Ask the party you call to call back using the number of the telephone line you have connected to the PBX.

9. When the return call comes in, it should ring on Extension 21 (the operator extension). When it does, answer the call on Extension 21.

10. Press the * key to put the incoming call on hold. Dial extension 24 and press the * key again to transfer the call. Extension 24 should ring with the call. Answer the call to complete testing the system.

Case Projects

Case Project 10-1: Design a Home Telephone System

You have been asked to recommend a telephone system design for a home user. The home contains a business office that includes two computers needing an Internet connection, a fax machine, and two telephone extensions. The rest of the home has eight telephone extensions scattered through it, another phone extension in the backyard, two computers that both need Internet connections, and a security system with an emergency call-in device that must have a phone-line connection. The home is too distant from the telephone company offices to get any type of DSL connection, but the owner wants at least a 1.5 Mbps Internet connection for his business. Write a short report noting what local phone system, what ISP connection, and the number of telephone lines you recommend for this installation. Give your reasoning.

Case Project 10-2: Recommend Changes in a Home Telephone System

A friend asks you about improving the telephone service in her home. She presently has three telephone lines coming into her home. One is connected to the modem that provides her ISP service, one is connected to a phone in the kitchen, and one has two extensions in the home's two bedrooms. She does not use the computer much for Internet connection, but now her son is asking that another (fourth) line be installed so his computer can have an Internet connection. Even though the expense of another line concerns her, she is about to say yes because it seems whenever she wants to make a call on one of her present phones, her son is on the line. If she uses one of her existing lines for her son's ISP connection, then she will have even more trouble finding a free phone to make a personal call. What would you recommend? Write a short report noting how this person could get more service out of the phone lines she already has and whether you think she really needs that fourth line.

Case Project 10-3: Solve a Telephone Traffic Problem

A home with three telephone lines has a continuing problem with their use. Each line has two extensions connected to it in the home. In addition, one line is connected to the home computer's modem, one is connected to a fax machine, and the other to an answering machine and voice mail recording device. The problem with each line occurs when the machine on the line is in use and someone in the home picks up one of the line's extensions to make a call. When this happens, the call cannot be made because the line is already in use, but picking up an extension while a line is in use also interrupts the call in progress. If the fax is running or the computer is online, their connections are broken by the pickup. Even the answering machine disconnects from a call when a new extension is picked up. Despite having three phone lines, the family in this home is getting very poor service from its telephones. They do not want to put in more phone lines, but would like some means of improving service on the ones they have. Write a short recommendation of what they could add to their telephone system that would help the interrupted call problem and get more use out of their phone lines.

Case Project 10-4: Troubleshoot a Telephone Extension

A friend asks you for help with a telephone extension he just installed. He has two telephone lines in his home and wants an extension phone on the second line in his garage where he has a workshop. He installed a two-wire line to the garage and wired it to the black and yellow wires of the second line coming from his demarc box. He wired the other end to a new RJ-14 jack in the garage, again connecting the wires to the black and yellow connectors in the jack. He bought a used single-line phone and connected it to the new jack, but the phone will not work in the new jack. He tested the telephone on the first line inside the home and it worked all right, so he thinks the new wiring must be wrong. You ask him to look at the plug on the end of the phone cord. He says it is just an ordinary RJ-11 plug with two wire contacts for a single line. Why will the phone not work and how can he easily correct the problem?

Home Lighting Control

After studying this chapter, you should be able to:

- Describe the basics of lighting control design and planning.
- Define the major types of automated home lighting control systems available.
- Describe the major components of an automated lighting system and how they work.
- Describe how to install and set up an automated lighting system in a home.
- Describe how to program an automated lighting system in a home.
- Describe troubleshooting techniques for eliminating problems in lighting systems.

Introduction

Automated Home Lighting Design

Types of Automated Lighting Systems

Lighting Control Components

System Installation and Setup

Programming the Automated Lighting System

Troubleshooting the Automated Lighting System

Introduction

In this chapter, you will learn how data transmission can be used to control the lighting in a home through a system of **command modules** and **control modules** linked in a wired or wireless network. An automated lighting control system can enhance the visual impact of a home's architectural design and décor, increase the aesthetic influence of lighting, and make the **home lighting system** more economically efficient. You will learn to define the basic types of home lighting control systems and their characteristics and will find out about the basic components that make up these systems. You will also discover how to install and set up an automated lighting system and program it to function according to design parameters. Finally, you will learn how to troubleshoot the common problems that can occur in home lighting control systems.

Automated Home Lighting Design

Every home has lights that provide basic illumination. Some homes have elaborate lighting systems composed of **light scenes** (illuminated rooms or areas within a room) that not only provide light, but also beautify the room décor or invoke a mood in those who see the scene or are within it. Whether a home's lighting system is minimal or extensive, it can be significantly enhanced by being automated. **Automation** enables lighting to respond quickly, and with relatively little effort on the homeowner's part, to changes in natural lighting, specialized illumination needs, social situations, and even the moods of the home's occupants. At the same time the utility of the lighting system is increased, the electrical cost of operating it should not rise and may even decline due to greater efficiency in the timing and use of energy.

In this section, you examine the main considerations that go into an automation design for a home lighting system. This discussion assumes that the lighting system of the home is already installed and working and that the homeowner wants to automate it.

Note: Before the automation design process begins, some homeowners may want to consider installing additional lighting or changing some of the lighting fixtures already in the home. If lighting is to be upgraded or altered, that should be done before automation is installed. Lighting design and installation is a large field in itself and a subject on which books have been written. This chapter will not discuss lighting design and installation except where it bears directly on our primary focus, the automation of lighting systems.

Caution: High-voltage wiring and fixture installation should be performed only by a licensed electrician or under the direct supervision of one. Wiring and fixture replacement descriptions in this chapter are informational only and should only be performed by persons trained and licensed to do so. Always shut off electric power in the area where wiring or other electrical work is being done.

Load Requirements and Grounding

An automated home lighting system, whether it is wired or wireless, will not add any **load** to the home's AC electric wiring, but the adequacy of the in-house wiring should still be confirmed before any automation is installed. Both wired and wireless automation controls are subject to interference or "noise," which can be caused or increased by overloaded wiring.

Lights are usually not the cause of **overloading** in a home electric system because they are not the largest consumers of electricity in the system when they are used individually. If acting as a group, such as when an automation system turns all the controlled lights in a home on or off simultaneously, lights can have a significant impact on the power system, but individual lights being turned on or dimmed will not make much of a difference in the overall power consumption or load. Unlike electric motors, which draw considerably more current when starting, lights consume only slightly more electricity when starting than they do when operating.

A drop in line voltage will not cause a light to consume more power either. If line voltage drops, the light simply shines less brightly; that is how dimmers work.

Chapter 5 described in detail how to determine if a home's wiring is inadequate before attempting to install a network. The same evaluation should be made before installing home lighting controls. If new circuits, wiring, or fixtures are needed, they should be installed before the automation or simultaneously. New circuits, if necessary, can be used not only to reduce the load on older wiring, but also to connect additional new lighting that might not have been installed at all if the new circuits were not available. Automation devices like controlled outlets and switches can be installed in new circuits at the beginning, instead of replacing manual units as is necessary in existing circuits.

All these considerations point to the need for a plan for the lighting system itself and for the automation system that will control it. The plan should start with a map of the existing electrical system, which can be researched and drawn according to the instructions in Chapter 5. Make the diagram of existing wiring as complete and accurate as possible on a floor plan of the home. Add any new circuits that are required and design how the wiring for these will be installed. Finally, design the control system noting exactly which lights you plan to control, which lights will be controlled as **zones** (groups of lights controlled as single units), and where the **controllers** will be located. Make the plan as detailed and complete as you can; every item planned before starting the hands-on work will enable you to avoid problems during the installation phase.

Label the plan with a unique label for each item. As you later install each device, it should be marked with the label you gave it in your plan. The plan can then serve as a schematic drawing of the completed lighting automation system, which you can use in the future to troubleshoot and maintain it. Figure 11-1 shows a plan of the first floor of a home with all the electrical fixtures noted. Because the plan has been reduced in size for publication in this book, the fixture symbols have been enlarged and only those in the family room are shown labeled. Later, you will see Figure 11-2, which shows the family room of this plan enlarged and notated with the zones and control devices that will be used in it.

All the high-voltage fixtures in the home should be grounded, as should those installed as part of any new circuits added to upgrade the home's electrical system. Non-grounded fixtures are almost never found except in old homes constructed before grounding was a code requirement, but it is still wise to check that everything is in fact grounded, as a safety measure, and also because any ungrounded circuit is likely to cause interference in the automation system you install. This is true whether it is a wireless system or uses power-line technology. **Fluorescent lighting** is especially prone to generate interference in the automated lighting system, as well as in other home technology installations.

Lighting Zones and Scenes

As you develop the lighting automation plan, you need to decide whether each of the lights will be controlled individually or be part of a zone. A lighting zone is simply a group of lights that are related by function (not necessarily by location, although they may be near one another as well). A zone, for example, could consist of all the lighting in the family entertainment room. These lights would nearly always go on or off together or have their light levels adjusted as a group to fit whatever activity (watching a movie, reading, playing games, etc.) is underway in the room. Each of the lights retains an individual manual control, so the occupants

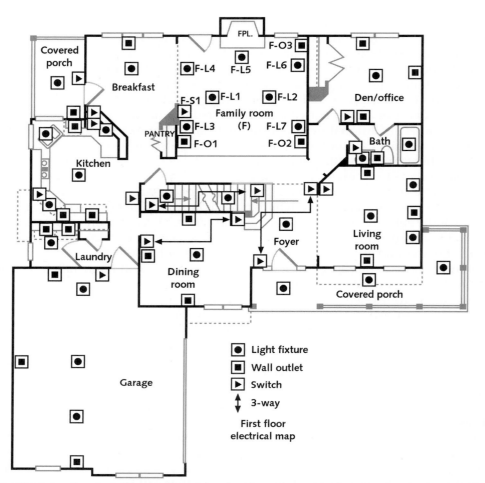

FIGURE 11-1   Electrical Fixtures Noted on the Floor Plan with Those in One Room Labeled

can override the automatic controls should there be a need, but most of the time, they function together as a zone.

A lighting scene might consist of the lights in a single zone, but can also be any set of lights designed to accomplish a specific lighting objective. A dining scene, for example, might dim the overhead dining room chandelier while turning on small spotlights to illuminate the paintings on the room's walls. Another lighting scene in a family room might dim the overhead lights to eliminate reflections on the glass while turning on backlighting behind the aquarium to make observation of its occupants easy. Yet a third scene might turn off floodlights in the backyard garden while illuminating the flowers, fountain, and footpaths with indirect lighting for a romantic effect.

Possible zone and scene configurations for home lighting include:

- Bedrooms
- Dining areas
- Indoor safety lighting (hallways, stairs, entries)
- Emergency lighting (battery-powered lights that activate during a power failure)
- Living room

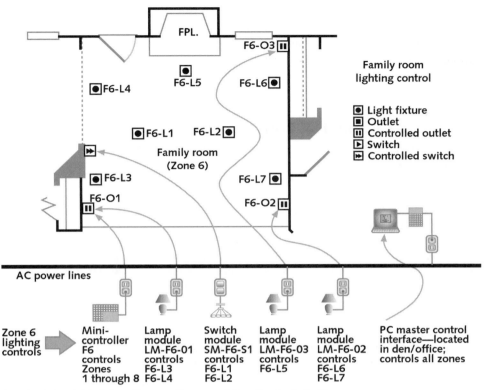

FIGURE 11-2 A Lighting System Zone and the Controls in It

- Interior decorative lighting areas
- Outside security lighting
- Outside safety lighting (sidewalks, steps, and outside entrances)
- Outside decorative lighting
- Outside special areas (swimming pool, hot tub, game area)

If any of these or other areas fit well with the automation you want for the home, plan the appropriate zone and the controls needed for it. Lights that always function together can be controlled by one device if they are on the same circuit or can be wired to a single circuit. This is especially true of lamps and other moveable lights, which can often be set up so multiple lights connect to a single outlet. This allows one module to control all of them, up to the limit of the device's power capacity.

A light can belong to more than one zone, but this requires that it have an individual control, which can be directed for the lighting requirements of each zone. Lights that are part of a room lighting scene, for example, might also be part of a safety lighting zone for the home's interior core. Outside decorative lighting might also be part of safety lighting that turns on when someone leaves the home late at night, even though the rest of the decorative lighting remains off. Security lighting might override all other zones and turn every exterior light up to full power to give maximum illumination.

Note each zone on the lighting plan and add a zone code to the label of each light that is included in a zone. A light that illuminates a painting hanging in the living room might be labeled LVRM5-Z1, showing that it is the fifth light in the living room and belongs to Zone 1. A light on a stairway descending to the basement family room might be designated STRS2-Z3-Z7, denoting it is a stairway light and belongs to both the living room zone (Zone 3) and the safety zone (Zone 7). Figure 11-2 shows the electrical fixtures and controls in a family room lighting zone. Each control module is labeled.

Home Run–Connected Lights and Daisy Chains

When a light is connected directly to its power supply, it is said to have a **home run connection.** A lamp plugged into an outlet receptacle is an example of a home run–wired light, as is an overhead light connected to a switch on the wall. The light is wired directly to a power connection (a home run wire) inside the walls. Most, but not all, hardwired lights (those that are wired as part of the home's electrical system) are home run lights. They are connected to the electrical system through a switch that allows them to be turned on and off.

Sometimes a switch controls two or more lights. In this case, not all the lights may be wired to the electrical system directly through the switch. One light may be wired to the switch, while the second light is connected to the first light, getting its power from the first light's connection to the electrical system (if there is a third light, it would be connected to the second light). This arrangement of connection is known as **daisy-chain wiring** (from the now little-known art of joining daisies by twisting the stem of one around the blossom of the next). Other examples of daisy-chain electrical connections are as follows: a series of lamps in which the first is plugged into a wall outlet, the second is plugged into a connection in the first, and perhaps a third into a connection in the second; or a television that is plugged into a wall outlet and a small backlight lamp that is connected to the television. Figure 11-3 shows two lamps with home run connections on the left and three lamps connected in a daisy chain on the right.

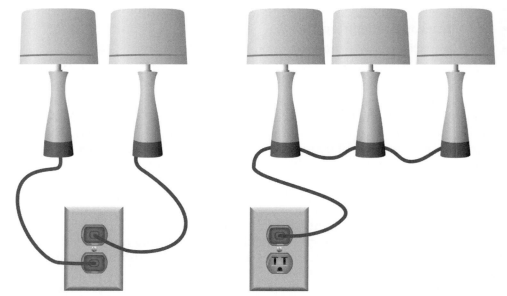

FIGURE 11-3   Lamps with Home Run Connections (Left) and Daisy-Chain Connections (Right)

Lights with daisy-chain connections always work together from a single control device because all the lights in the chain receive their power from one connection. They are wired in parallel, however, not in series, so that if any one light fails, the others in the chain will continue to work. Daisy-chained lights are useful because the arrangement permits several lights to be placed in one location without using all the wall outlets. Multiple track lights and indirect fluorescent lights are often connected in daisy chains. Decorative lighting, both indoor and outdoor, is usually in a daisy chain in multiple light segments. Parallel-wired holiday lights are daisy chains, and several strings of them can be connected one to another in an even larger chain. Only one control device can be used on an entire daisy chain, however, because it is all powered from only one controllable point.

Home run–connected lights that are part of the home's electrical system can be controlled individually from the switch that is always wired into their circuit. They can also be controlled from another switch wired into the circuit closer to the light than the first switch, if the first switch is left in the on position. Home run lights plugged into a wall outlet can also be controlled at the outlet or by another switch wired into the light's cord. Because each home run light has its own power connection, it must be controlled by a separate controller. Most plug-in controllers have two or more outlet connections, but all of these are usually controlled by the same commands and so all the lights plugged into the module work together as if they were one. Figure 11-4 shows two lights added in one corner of a room

FIGURE 11-4   Added Lights Controlled by an Automated System

that has no built-in overhead lights (a matching pair was placed in the adjacent corner). These lights connect to a wall outlet and have no manual switch on their circuit. They are controlled, however, by a control module plugged into the outlet and commanded by a controller.

Types of Automated Lighting Systems

Automated lighting systems are specialized networks that use different protocols, hardware, and frequencies than computer networks or telecommunications networks. Automated lighting networks have very narrowly defined functions. These functions consist of turning lights off and on and adjusting their brightness through a process called dimming. Automated systems can control many lights and be programmed for complex lighting patterns, but the control commands to each light are simple and do not require large amounts of data transmission. Automated lighting systems consist of either power-line control networks or wireless control networks. Each type of control system is discussed in this section.

Power-Line Control Systems

Power-line control technology uses the AC power lines to transmit control signals at frequencies higher than the 60-cycle AC electricity flowing in the same lines. This technology can therefore send lighting control signals over the same wires that provide electric power to the lights and other electrical appliances throughout the home. The most common and extensive method of lighting control and automation uses X10 **power-line carrier (PLC)** technology, which was described earlier in Chapter 3. Several other power-line technologies are now available, some of them more robust and sophisticated than X10 products, but none to date as inexpensive. The most prominent of the newcomers is **Pulseworx,** manufactured by Powerline Control Systems. It uses a signal standard known as **Universal Powerline Bus (UPB)** to provide control data over AC wiring. Some other manufacturers have adopted the UPB standard and produce products for lighting control using it, but Pulseworx is the leading control system employing UPB.

X10 Lighting Control Systems

X10 devices were once manufactured by a single company, but the patent on the original technology expired several years ago and many firms now manufacture and market X10 equipment. With more than 35 years of sales, X10 systems are now installed in an estimated 12 million homes in the United States and other countries, and the basic technology has been greatly expanded and improved. The original company, X10 Limited, and its subsidiaries are still major manufacturers of X10 equipment, but they have now been joined by many other firms who have produced a large array of control devices and controllers to command them.

Many types of command modules can be used with an X10 system. All of them connect to an AC outlet and send out signals at a frequency of 120KHz on the home's electric wiring. This frequency is way above the AC current frequency of 60Hz and well below radio frequencies (RF) so that neither interfere with the X10 signals on the wiring. Other interference does occur (which is discussed later), but the X10 signal frequency is as well placed in the electromagnetic spectrum as it could be to avoid both high- and low-frequency interference.

TABLE 11-1
X10 Basic Commands and Actions

Number	Command	Action
1	On	Turns on any control module with the code specified
2	Off	Turns off any control module with the code specified
3	All on	Turns on all control modules in the system
4	All off	Turns off all control modules in the system
5	Dim	Dims any control module with the code specified
6	Brighten	Brightens any control module with the code specified

In an X10 system, all the command signals are sent over the home's wiring and are received by all the control modules. Each module responds only to the signals that have its unique address in the header of the data packet. X10 control modules in a system are each assigned a letter (house) code and a number (device) code. There are 16 possible letter codes (A through P) and 16 number (unit) codes (1 through 16) for each letter, giving a total of 256 possible different control modules in an X10 system. (Even more control modules are possible if several are assigned the same codes and function together in a zone.)

Command modules have no code designation because they do not receive commands. They only send them. An X10 system can have several command modules located in different areas of the home, and the control modules respond to commands from any of them so long as the commands are correctly addressed to the module. Table 11-1 shows the six possible commands that basic X10 lighting control systems use. Not all control modules respond to all commands.

X10 control modules come in many varieties. Table 11-2 shows those intended for lighting control, their power ratings, the commands to which they respond (as noted in Table 11-1), and the function of each type of module. Note that lamp

TABLE 11-2
X10 Control Modules and Their Functions

Control Module	Power Rating	Command Response	Function
110v lamp	300 watts	1, 2, 4, 5, 6,	Plugs into outlet, controls lights plugged into it
110v lamp	600 watts	1, 2, 4, 5, 6,	Plugs into outlet, controls lights plugged into it
110v appliance	800 watts	1, 2, 3, 4, 5, 6,	Plugs into outlet, controls appliances plugged into it
110v outlet	500 watts	1, 2, 4, 5, 6,	Replaces outlet, controls lights plugged into it
110v switch	500 watts	1, 2, 3, 4, 5, 6,	Replaces switch, controls lights connected to it
110v three-way switch	500 watts	1, 2, 3, 4, 5, 6,	Replaces switch, controls lights connected to it
Inductive switch		1, 2, 3, 4, 5, 6,	Controls low-voltage lighting

and outlet modules do not respond to the "all on" command because they may be used to control non-light devices that should not be activated with this command.

The following types of command modules (controllers) are available for X10 systems. All send commands to X10 control modules. The differences between them consist mainly in how many devices the command module can control and how the command module itself is programmed to perform its work.

- **Mini-controller**   Usually controls up to eight control modules. Plugs into a wall outlet via a cord and sends commands on the AC wiring. Works like a handheld remote with buttons for each control module.

- **Wireless mini-controller**   Similar to a mini-controller, but wireless. Sends radio commands to an outlet-mounted receiver, which transmits them on the AC wiring.

- **Tabletop controller**   Larger than a mini. Manual remote control. Controls up to 16 control modules. Some can control multiple house codes, providing control for up to 256 devices.

- **Programmable controller**   Controls up to 16 modules and can use multiple house codes. Allows manual remote control and timed sequences of lighting actions.

- **Telephone interface**   Connects to telephone line and AC wiring. Controls up to 16 modules and can use multiple house codes. Allows manual remote control from any phone.

- **Touch-panel interface**   Controls up to 16 modules and can use multiple house codes. Allows manual remote control and timed sequences of lighting actions.

- **Computer interface**   Controls up to 16 modules and can use multiple house codes. Allows manual remote control, timed sequences of lighting actions, or sensor response actions.

Standard X10 command modules and control modules use one-way communication only. Signals go from the command module to the control module, but nothing comes back from the control module to the command module. This means no confirmation is sent to the command module stating its instructions were received and carried out. Most of the time they are, but this lack of confirmation signal was the greatest weakness of X10 technology.

One of the best advances in recent X10 products is the development of bidirectional (two-way) X10 modules. These modules confirm receipt and execution of commands and are available (at higher cost) for most X10 devices. They are worth the money, however, because they provide much more reliability in programmed lighting sequences. If a command is not confirmed within a short time, it is re-sent and the program is kept accurate through its timed sequence. Many of the touch-panel interfaces and nearly all of the computer X10 interfaces use bidirectional modules.

UPB Lighting Control Systems

The other major power-line technology was developed by Powerline Control Systems. It uses a signal standard known as Universal Powerline Bus (UPB) to provide control data over AC wiring. The Powerline Control Systems products that use UPB technology are marketed under the brand named Pulseworx. Some other

TABLE 11-3
Comparison of UPB and X10 Control Products

Feature	UPB Product	X10 Product
Pulse-position modulation	Yes	No
Spread spectrum	Yes	No
Two-way signals	Yes	No
Bps speed	480	60
Commands per second	5	1
Settings stored in device	Yes	No
Programmable	Yes	No

manufacturers have adopted the UPB standard and produce products for lighting control using it, but Pulseworx is the leading control system employing UPB.

Two principal differences exist between UPB power-line standard control devices and those using the X10 standard: UPB devices send signals at a much lower frequency than X10 devices (4Khz to 40Khz, as compared to X10's 120Khz frequency), and all UPB devices have two-way communication capability so command signals can be confirmed and the status reported on any device.

The UPB standard also uses spread spectrum signal hopping (see Chapter 4) and sends its signals at a strength of 40 volts (X10 signal strength is typically about 2 volts). These features make the UPB control signals very reliable and robust, even in power-line systems that have considerable noise and interference. UPB is also faster than X10. It can send command signals at the rate of 5 per second, as opposed to 1 for X10 systems. Table 11-3 gives a comparison between the features of typical X10 and UPB products. The table shows only standard X10 product features. Some newer X10 products, especially hybrid power-line/wireless units, have been enhanced and may include features missing from the older versions.

The Pulseworx lighting control product line includes all the standard devices available in X10 technology plus a few extras. In order to meet the demands of larger lighting scenes and multibulb fixtures, Pulseworx offers 1,500 and 2,000 watt wall-mounted dimmer modules and an 800 watt plug-in dimmer. These are larger than the X10 maximum of 1000 watts for wall-mounted dimmers and 300 watts for plug-in dimmers.

Other Power-Line Systems

A hybrid power-line lighting control system, employing both wireless and power-line technology is now on the market from Insteon. It uses proprietary technology and is not compatible with X10, UPB, or other wireless systems. **LonWorks** is another PLC lighting control system developed by Echelon and has a large market base in Europe and Asia. LonWorks products have recently been introduced into the U.S. to compete with UPB and X10 products but have not yet achieved a large market share. Lastly, the HomePlug consortium has announced the availability of a new PLC standard for home lighting control and related functions. It is known as **HomePlug Command and Control (HPCC),** but few products incorporating this technology have yet reached the market.

HomePNA, the telephone line technology described in Chapter 4, has not been used for lighting control or home automation systems to date. None of the HomePNA companies have viewed automation as a significant enough market to warrant a development effort despite a multimillion home market base already established by X10, UPB, and other power-line products, as well as the wireless systems described next. Fortunately, all of the existing technologies are well developed and fully capable of meeting the needs of almost any lighting automation plan.

Wireless Lighting Controls

Nearly all the lighting control devices available in X10 technology are also available using wireless technology. The largest wireless product supplier is Lutron, which offers a full range of home lighting control products that use its proprietary **RadioRA** technology. Lutron control modules do not offer the "all on" and "all off" commands available in X10 systems, but other than that, just about every feature of X10 control can be duplicated using RadioRA. Because its RF signals are much faster than X10 signals, RadioRA controllers can approach the speed of an X10 "all on" or "all off" command by sending individual signals in sequence to each module.

All RadioRA modules have two-way communication and offer automatic confirmation of commands from control modules. This wireless system can also be interfaced through an RS232 port to a computer-based control system. RadioRA components are somewhat more expensive than X10 modules, but they offer high quality and reliability and, while still not as widely used as X10 products, are making a definite place for themselves in the market.

In addition to the Lutron RadioRA wireless system, Vantage and Watt Stopper also offer wireless lighting control systems using proprietary radio technologies that are different from, and not compatible with, either RadioRA or X10 wireless systems.

Leviton is the largest of several companies that use X10 wireless technology as the basis for their lighting control systems. Recently, Leviton also started marketing lighting control products using the **Z-Wave** wireless technology originally developed by the Danish company Zensys. Z-Wave has become popular in Western Europe as a lighting control solution and is now moving into the U.S. market. In addition to Leviton, Z-Wave lighting control devices are made by Sylvania and Intermatic for the U.S. market.

Wire Runs and Wireless Zones

New wiring should be minimal for an automated lighting system, unless additional circuits are being installed. If that is the case, the new wiring should be integrated as part of a full home upgrade designed to bring the home's electrical system up to current electrical standards for all the technology that will be operating in it. For new circuit wiring, refer to the instructions given in Chapter 5 on high-voltage wiring and be certain all such work is done only under the supervision of a licensed electrician.

Additional track lighting (lights mounted on metal tracks that contain their wiring and can be attached to the surface of walls or ceilings with screws). Although it connects to the home's AC wiring, most track lighting is low voltage (usually 24 volts) and power is supplied to the lights through a step down transformer.

Some additional AC fixtures can often be added to a room without the need for more infrastructure wiring as long as the home's circuits are not overloaded

FIGURE 11-5 **Multiple Lights Wired from a Single Switched Connection Point**

by the increased power requirements. Such fixtures can usually connect to a wall outlet through a power cord run from the mounted fixture in a track or a raceway. This gives a permanent finished appearance to the lighting installation without requiring any new wiring within the walls.

Inadequate lighting in a room can often be remedied with new multiple bulb fixtures wired to the original junction box and controlled by the same switch(es). Figure 11-5 shows an example of a single overhead light connection that has been upgraded with multiple lights, which provide improved appearance and better illumination.

Figure 11-6 shows a single overhead connection that has been expanded into two chandeliers spaced 10 feet apart without any new structural wiring.

FIGURE 11-6 **Dual Chandeliers Wired from a Single Connection Point**

Lighting Control Components

Each type of lighting control component is designed for a specific automation function. After completing the lighting automation design, you must determine the minimum number of controllers and control modules you need for the lighting automation system. This section describes the function of each piece of equipment and any specific limitations you need to be aware of when using it. Whether the lighting automation system uses X10 or UPB technology power-line components or wireless modules (RadioRA, Z-Wave, etc.), the various control devices function in much the same manner.

Many lighting control systems are sold most economically as packages of components or as kits containing all the components needed for a basic home lighting control system. Additional components can be added as required for a particular installation, but the best cost savings will be realized by obtaining as many components as possible in a single package.

Some lighting control packages also provide for security system monitoring, utility control, or other types of home automation integrated in the same master control unit or computer software as the lighting control. These integrated systems also provide large cost savings over the purchase of individual systems or components. They also provide a more unified and efficient home automation system in which all the control elements can be programmed to work together. Figure 11-7 shows the main elements in the Home Automation, Inc. (HAI) Omni-Pro II control system, which is a popular universal control system. Components shown in the picture include two sizes of control panels and dimmer switches on the left, a programmable thermostat in the center, a laptop programmer/controller, Web interface and handheld controller on the right, and a master control unit at the back.

For lights in a single zone, try to connect as many as you can in daisy chains to minimize the number of control modules you need. If a light is in two zones, it

FIGURE 11-7   Omni Pro II Integrated Security and Control System

must have its own control module (or share one only with other lights in the same zones) that can respond independently to control commands for both zones. For the most professional appearance in a lighting automation system, use modules that are concealed in standard electric fixtures or placed out of sight.

A combination of power-line and wireless components can be employed to create a home lighting system, but only one technology of each type can be used. X10 or UPB devices are not compatible with one another, so you must choose which type you want to use. Likewise, the wireless lighting control technologies are all independent and noncompatible, so only one should be used in a system. Equipment that is nearly identical in function is available in each technology, so the decision of which to use comes down to finding the most cost effective and easiest to install for your particular application.

Each technology has a few unique items that address special needs. If your system has one of these needs, that may be the determining factor in deciding which lighting solution to adopt. An X10 socket module, for example, allows any standard light socket to be converted into a controlled light. The socket module screws into the regular socket and the light bulb screws into the module. The light can then be controlled by commands sent on the AC wiring. This solution is an alternative to using a controlled wall switch. It is useful for controlling lights that do not have remote switches but are turned on and off only by pull-string switches in the light fixture itself. None of the other technologies offer controlled sockets.

X10 manufacturers have developed numerous wireless devices to augment the X10 power-line products they sell. Not all the wireless components of the different companies will work together even though all can interface with X10 power-line modules. For this reason, it is best to choose a single manufacturer for all your X10 lighting system components, both power-line and wireless.

The same is true if you are using a UPB hybrid lighting control system with power-line and wireless components. Get everything for your system from one manufacturer to be sure the whole system will work together. If you must use a device from another company, be certain before you install it that it will function with the rest of your system.

Command Modules and Controllers

The smallest command modules are similar in appearance to a television remote control. They can control up to eight control modules, but are not programmable. Commands for each control module must be entered manually by pressing the appropriate buttons on the command module. An X10 command module must connect to an electrical outlet in order to transmit its signals on the AC wiring.

Wireless X10 command modules are battery operated and send their signals to a separate receiver nearby, which is plugged into an AC outlet. The receiver regenerates the command signal and transmits it on the AC wiring. In homes with more than 2,500 square feet of floor space, an additional X10 signal regenerator may be needed both to strengthen the X10 command signals and to eliminate noise from the signals so they can be more easily understood by the receiving control modules.

Command modules and controllers capable of controlling more than eight devices or of being programmed for timed lighting sequences are usually tabletop units or wall mounted. X10 command modules range in size up to those that can control the full 256 devices this technology is capable of handling. The larger command modules also operate manually so that individual commands are entered by pressing the appropriate buttons on the module to send the desired command to the control module.

The most elaborate X10 command modules have touch screens and a graphic interface that allows them to be programmed by touching appropriate entry commands on the touch screen to store the commands in memory.

Some X10 command modules also can interface with a computer, and command sequences can be programmed and stored on the computer. As with RadioRA technology, the X10 computer software allows controls to be implemented based on time-of-day settings, on sunrise and sunset times, or on sensor input from lighting sensors or security system components.

RadioRA and other wireless handheld command modules are also battery operated and transmit their command signals directly to the various control modules or to a repeater unit located nearby, which then repeats the commands in an amplified signal throughout the home. Homes with more than 3,500 square feet of floor space may require a second repeater in order to cover the full home.

Many wireless command units are also programmable, with timed sequences of lighting instructions that are sent at preset times to adjust lighting. Several can also store **macros** (sequences of commands in a specified order) in memory, which can then be sent at a programmed time or sent manually by pressing a few recall buttons on the command module.

RadioRA offers large command modules that can be tabletop units or wall mounted. These can be preprogrammed with macros for specific lighting control sequences, but these do not have timed functions unless combined with an astronomic clock interface unit, which allows the command modules to control devices based on regular time-of-day settings or according to sunrise and sunset in the local area.

RadioRA command modules also can be interfaced to a computer using an RS-232 interface. Automation control programs on the computer can then control all RadioRA functions based on time-of-day settings, sunrise and sunset times, or on sensor input from light and motion sensors. The other wireless technologies can all interface with a computer for programmed control functions.

Remote-Access Controllers

RadioRA technology has wireless remote controllers that can operate the automated lighting system from a television universal infrared remote device, from a car visor remote unit, from a telephone, or from a security system keyless entry device. Several of these controllers are more limited in function than a master control unit, but they are all adequate to command basic lighting sequences when away from the home or approaching it in the dark. The keyless entry control and the car visor unit can also be programmed to control security gates and garage doors.

X10 remote-access control devices are slightly more limited, but still include a telephone interface as well as battery-operated remote controls, keyless entry units, and car visor controls. The latter two, like their RadioRA counterparts, can also control gates and garage doors.

Outlets

Outlet control modules for X10 installations can be obtained either as plug-in units, which are about the size of a small 9-volt power transformer used with many electronic devices, or as replacement outlet receptacles that have the X10 control module built into them and must be wired into the AC wiring in the same manner as an ordinary receptacle. Control module outlet receptacles are identical in appearance to regular outlets and are available with one of the duplex outlets controlled or both. The lights to be controlled are simply plugged into the control module and receive their power through it. Plug-in outlet control modules accept all six X10 commands (on, off, dim, brighten, all on, all off); the wired-in outlet control modules accept five of the six X10 commands (on, off, dim, brighten, all off) in keeping with UL safety requirements for this type of device.

The receptacle devices are preferable both for appearance and durability, but are naturally more expensive. The newer X10 two-way communication modules, which confirm receipt and implementation of signals, are presently available only as wired-in receptacles, but will probably soon appear as plug-ins, too.

RadioRA outlet control modules are tabletop units that plug into an outlet by means of a 6-foot cord. The module's plug has a "piggyback" outlet on its back side and the lamp that the module controls is plugged into this outlet in the module's plug. The module with its manual dimmer switch sits on the table next to the lamp. RadioRA replacement wall outlets are not available at present.

Light Switches and Dimmers

Both X10 and RadioRA lamp control modules are capable of dimming lights as well as turning them on and off. Both technologies also have replacement wall switches available that, in addition to serving as manual dimmer switches, are controlled for automated dimming and brightening. As with X10 controlled outlets, these switches must be wired directly into the AC circuits because they replace the standard manual switches. Both systems have **single-pole switches** and **three-way switches**. X10 switches are also available for three-switch locations while RadioRA can go as high as nine switches controlling a single lighting zone. Switch-type control modules accept all six X10 commands (on, off, dim, brighten, all on, all off). RadioRA controlled switches accept on, off, dim, and brighten commands.

Fixtures (Luminaires)

Some X10 light fixtures and lamps are available in which the X10 control module is built into the device rather than attached outside at the wall. These lights must have continuous power supplied from an outlet or a switched circuit in which the power is always left on. The internal X10 control module then turns the light on or off and dims it.

Built-in control modules are available with only a very limited selection of lamps and lights, but they do offer additional flexibility in locating controls. They are especially useful for lights that have inconvenient switches, such as basement or garage lights that are turned on by pull strings that are hard to find in the dark.

Built-in control modules are not available in RadioRA technology except as controls on low-voltage lighting sets. The control module can be attached to these and serve as an internal switch and dimmer. X10 systems also have low-voltage control modules.

Automated Window Treatments

RadioRA systems do not provide any window shade control mechanisms, but these devices are available with X10 technology. While not really lighting controls, these modules can be very useful in controlling sunlight through windows, darkening entertainment centers in daylight hours, and other functions. X10 control modules connected to motorized pulls can open shades, blinds, or curtains, close them, or partially open or close them.

Manual remote-operated window coverings are mostly used where the windows are high and are inaccessible for hand pulls. An automated shade or curtain system can be programmed like a lighting system to respond to time-of-day commands, sunrise and sunset macros, or other command sets. Automated window treatments are more fully discussed in Chapter 14.

Sensors

Light and motion sensors can be used for lighting control input on both RadioRA and X10 systems that are programmed and controlled through a command module or a computer. Among the macros that can be programmed for both systems using sensor input are the following:

- Selected lights turn on when motion is detected by a sensor.
- Selected lights turn off when no motion is detected by a sensor for a specified time interval (unoccupied rooms).
- Selected lights turn on or brighten when the ambient light level falls.
- Selected lights turn off or dim when the ambient light level rises.
- Selected lights turn on when a light beam is broken (more selective than motion detector).
- Selected lights turn off when a light beam is broken (closed door or departing vehicle).

System Installation and Setup

Once you have the automated lighting system planned and the necessary components on hand, you are ready to install the control modules and the command modules, set up the control program, and test the system. If you do these steps for one lighting zone at a time, even a large automated lighting system can be installed and working in a short time. This section discusses the installation process and how to make it as efficient as possible.

Cautions while Wiring and Connecting Lighting Components

Caution: As you go through this part of the chapter, note that high-voltage wiring and fixture installation should be performed only by a licensed electrician or under the direct supervision of one. Wiring and fixture replacement descriptions in this section are informational only and should only be performed by persons trained and licensed to do so. If you are using X10 or UPB plug-in modules, these can be installed by someone who is not an electrician, but the same safety precautions should always be observed when working around high-voltage electrical fixtures.

Before doing any installation, shut off the AC power in the area where you are working. This usually means shutting off some house circuit breakers while leaving others functioning in non-work areas. Working in one power-off area at a time minimizes the inconvenience for others of having the AC power cut off and also allows you to run an extension cord from an outlet where the power is still on to the work area so you have light to work with while doing the installation. One final caution: If only some circuits in the house are turned off, do not assume every outlet or fixture you are working on is off. Test each one before doing any work to be sure the power is off. A small two-prong tester or a lamp can tell you whether a fixture is safe to work on.

Installing Interior Lighting Zones

Start with one lighting zone defined on your plan and install all the control modules required for controlling its lights. Wired-in outlets and switches should be installed just as if they were standard replacements for the devices already in the circuit. The various ways of wiring outlets and switches must be considered when installing control devices. Some points to remember include:

- **Always-live duplex** outlets should normally be replaced with matching X10, UPB, or RadioRA control modules in which both receptacles are controlled. If one receptacle needs to remain always on, use a control module in which only one is controlled. The other remains always on by default.

- **Switched duplex** outlets should be replaced with control modules, and the switch which controls them left in the on position. UL listing requirements for RadioRA, UPB, and X10 wall-mounted control module switches and dimmers state that they must be connected only to permanently installed light fixtures and not to lamps or plug-in devices.

- **Split duplex outlets,** in which one receptacle is switched and the other always on, should be treated like switched outlets. If only the switched receptacle is to be controlled, the control module should be placed on the outlet, not on the switch, and the switch left on. If both receptacles need to be controlled, then the outlet should be replaced with a control module and the replacement control module should be wired so that both receptacles are always live. This disables the switch that formerly controlled the receptacle and leaves the control module able to control both.

- **Single-pole switches** should be replaced with single-pole control modules switches wired the same as the original switch.

- **Three-way switches** should be replaced with three-way switch combinations. For X10, UPB, and RadioRA, three-way switch combinations consist of one control module switch and one companion switch, which does not have a control module in it but still allows manual operation of the switch. Both switches must be replaced in a three-way circuit, not just the one that has the control module in it.

- **Multiple switch** sets above three-way should be replaced with one control module and the other switches as companion switches. X10 and UPB can only handle up to four switch combinations, but RadioRA can go to nine.

As you install control modules, remember to check loads on each one. RadioRA and UPB devices are built for heavier duty than most X10 devices. They can accommodate 500 watts of power on switch modules and lamp modules. Standard X10 devices have 300-watt power load maximums, although units are available with 500-watt capacity and more. Be sure you install control modules with enough current capacity to handle the load you are putting on them.

Also note that control modules should not be placed where a high-current appliance is likely to be plugged into them. Heating appliances like hair dryers, waffle irons, space heaters, and toasters typically use from 800 to 1,500 watts of power when operating. This exceeds the load capacity of RadioRA, UPB, and X10 control modules and will burn them out in a short time. Plug-in modules can be removed when a high-current appliance is used and then put back into the outlet with no loss of control function. Wired-in control modules cannot be taken out, but will definitely burn out if overloaded, so be cautious of where you install them.

Low-voltage light sets require the special low-voltage control modules available in X10, UPB, and RadioRA systems. These control modules will normally be wired into the light circuit directly rather than being installed as an outlet or switch replacement.

When the control modules are in place and the lights connected to them, set up the master command module or controller in its assigned location. Program the zone and test it to be sure everything works as you want it to, then proceed to the next zone and continue the installation process.

If the automated lighting system has multiple command modules or controllers, each should be tested with at least one zone of lighting control to be sure it functions correctly in its assigned location. For portable remote controllers, this means testing from several locations to be sure the radio transceiver unit can pick up commands from anywhere in the house. For car visor units and keyless units, it means testing from outside the home to be sure the wireless transceiver is still within range.

Installing Exterior Lighting Zones

Once you have set up all the interior lighting zones and programmed the light scenes you want for each of them, install the exterior lighting zones. These usually consist of a decorative lighting zone and a security zone. Set up the security lighting zone first as it probably is the larger zone and requires more power. You want to be sure you have enough capacity for controlling all the security lighting before going on to decorative lighting. In the latter, some compromise can be made with bulb sizes and numbers if the load on the electrical system appears to be getting high. In the security system, you want to have all the illumination necessary.

If any new circuits or lights are being added outside, check your state and local code requirements to find out whether they need to be wired with armored cable (AC) or metal clad cable (MC). Both these types of cable, along with nonmetallic Romex-type cable were described in Chapter 5. Be sure you follow the NEC requirements as amended by any local or state ordinances when installing outside cable. Also be careful about grounding outside circuits. Do not assume that because the cable is buried in the ground, it does not need to be grounded. That is not true. The same rules apply for grounding outside circuits as inside.

Installing Security Lighting Zones

Security lighting is usually programmed with time-of-day settings or to come on and go off at sundown and sunrise. If security lighting control is augmented by sensor input, then these parameters need to be programmed into the system in addition to, rather than in place of, the time settings. Security lighting should not come on during daylight hours, even if a motion sensor detects movement of someone approaching the house. Other security features, such as video surveillance cameras, may activate in daylight from the same sensor input, but the lights should probably stay off during non-nighttime hours.

TABLE 11-4
Automated Lighting and Shade Programming Sequence

Time/Event	Lighting Action
Sunrise	Outside security lights turn off for daylight hours; all window shades open.
Sunrise + 30 minutes	East side shades close; south side shades activate close cycle shutting 20% hourly until closed.
Sunrise + 240 minutes	West side shades activate close cycle shutting 20% hourly until closed.
Local noon	East side shades activate open cycle opening 20% hourly until fully open.
Sundown	Outside safety/security light zone turns on.
Sundown + 15 minutes	All shades close. All inside light zones turn on if room is occupied. Family room set to activity scene lighting level. Living room set to entertainment scene lighting level. Decorative interior and exterior lighting turns on. Pool lights turn on.
8:00 pm	Family room zone set to entertainment scene lighting.
11:00 pm	Pool lights turn off.
11:30 pm	Outside safety light zone turns off (turns on again if motion detected in zone). Outside and inside decorative light zones turn off.
11:45 pm	All inside light zones turn off unless individual rooms are still occupied.

Programming the Automated Lighting System

Automated lighting programs are based on time-of-day schedules, event schedules (sunrise, sundown, etc.), sensor input (motion detected, light level, etc.), or combinations of all three. Every program is a unique sequence developed for the individual home and its occupants. A typical program using time schedules, events, and sensor input is noted in Table 11-4. The programs you develop may include some of these features and many others.

Note that lighting control systems timed to events such as sunrise and sundown vary daily by as much as 3 hours from winter solstice to summer solstice for areas located in the middle latitudes of the continental U.S. Northern latitudes vary even more, and southern ones somewhat less. These events, as well as local noon, are not affected by time changes (Daylight Savings Time) or by a location's position within a time zone. The data tables from which they are calculated are based on Greenwich Mean Time (GMT) and the geographical location specified.

Time schedule events follow the clock, including changes into and out of Daylight Savings Time. In cases where event programming and time schedule programming are used together, some lighting changes may need to be adjusted during summer or winter time. In Table 11-4, for example, if the home using this program was located in Seattle or Boston, the pool lights would only be on for an hour or so in midsummer because sundown occurs late in these northern cities. For situations like these, a program might include a few time-of-year adjustments to make allowance for longer or shorter days.

In addition to a timed program, such as the example in Table 11-4, command controls in rooms throughout the home can also be used to activate lighting scenes and levels for each room. Sensors can control individual lights or zones.

Sensor input to a lighting program and event scheduling of commands both require a computer-based master control system. Time schedule commands can be entered manually on many of the larger command modules in X10, UPB, and RadioRA systems. Complex time schedules are much more easily entered on a graphic interface program such as the X10 touch screens or computer programs.

Another lighting strategy that can be programmed into an automated system is a load reduction strategy. This strategy directs the lighting control system to minimize light use during peak electrical demand hours (usually during warm afternoons and early evenings) in order to reduce the home's peak use of power. This strategy may be one to consider seriously in areas where utility companies have begun to charge for electric service based on peak demand rather than average use.

The easiest and most sophisticated programming of light control and other automation systems is performed using PC-based software programs specifically designed to work with the major control technologies. New and more powerful programming software has recently become available that enables the home user to get better service from a home automation system than was previously possible. Figure 11-8 shows two screens from the PowerHome Home Automation Software package. This software is designed to work with X10 and Insteon lighting controls, as well as infrared controls for audio and video systems. The program can be managed on the computer, over the Internet, or by cell phone.

Troubleshooting the Automated Lighting System

Problems with automated lighting systems, whether X10, UPB, or RadioRA, come mainly from interference or attenuated signals. For X10 and UPB, the separation of the two legs of the home's electrical system can also be a problem.

AC electrical service to homes arrives as a 220-volt current that is divided once it enters the home into two "legs" of 110 volts each. One of these legs powers each of the home's 110-volt circuits, and both power the 220-volt circuits to the stove, clothes dryer, and other 220-volt devices. Both legs of the AC power work together in most homes, but they may not be actually joined in a manner that allows the higher-frequency X10 or UPB signals to travel easily on both legs. If the signals cannot move freely on both legs, they are likely to not reach some control modules on one leg or the other, which results in some lighting control commands not being implemented.

The solution to this problem is a bridge installed on the in-house side of the service line, which connects the hot wires of both legs of the house AC current to one another. The bridge can be installed at the electric service box by an electrician, but a new model is now available that can be installed by a homeowner or technician on any 220-volt outlet (such as for a stove or clothes dryer). The bridge, which is available as an X10 accessory from several manufacturers, provides a path for X10 or UPB signals to reach control modules on both legs and should eliminate the problem of one leg transmitting no signals or signals too weak to be read by the control modules.

Weak X10 signals can also occur in a home with a large AC electrical system. X10 signals are only about 2 volts in strength and attenuate with distance. Eventually

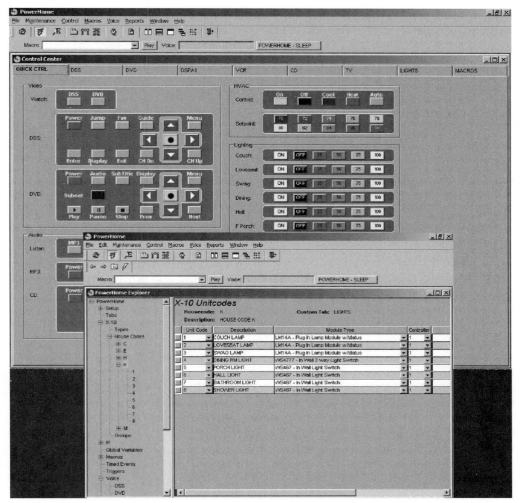

FIGURE 11-8   PowerHome Software Control Center Screen (Top) and X10 Explorer Screen (Bottom)

they become too weak to be read. A regenerator, placed in an outlet some distance from the command modules but close enough to read their signals clearly, can solve this problem. It reads all command signals on the AC wiring and regenerates them at full strength without any interference or noise they may have picked up in transit. In very large homes, two regenerators, placed some distance apart from one another, may be needed to keep signals strong throughout the residence, but in most houses a single unit is enough.

Even with a bridge and a signal regenerator in place on the AC wiring, some X10 control modules may have trouble receiving commands from controllers. Among the causes of specific problems with individual control modules are the following:

• Reverse-wired outlets with the hot wire connected to the neutral side of the outlet and the neutral wire to the hot side. Correcting the wiring often solves the problem.

- Switched circuits where the switch is on the neutral wire instead of the hot wire. Again correcting the wiring often solves the problem.

- An AC suppresser or filter that blocks signals. Removing the blocking device is the only solution. Lighting circuits usually do not need suppressors or current filters, but if other equipment that does need one or both is on the same circuit, some rearrangement of connections may be necessary.

- Noise from other electrical equipment that interferes with the X10 signals. You can locate noise-producing electrical equipment using a radio tuned to a blank (no station) area of the broadcast band. A filter on the offending equipment often cures this problem, but if not, moving it to another circuit may be necessary.

- General noise that comes from an outside source that cannot be eliminated. This type of interference is tough to combat. A filter on the incoming AC service line may help. Sometimes, an extra regenerator to keep the command signals well above the noise level works. If neither solution solves the problem, the only other option is to locate the source of the noise and somehow shield the home's wiring from it.

UPB signals are also subject to interference and noise, but much less so than X10 signals. UPB signals are 40 volts strong and are set at a lower frequency than X10. Both of these features make them less subject to interference than the weaker X10 signals. If interference is a problem in a UPB control system, the same possible causes and cures as described earlier for X10 systems may apply.

Wireless RadioRA and X10 systems can also be troubled by noise. RF frequency noise is difficult to filter out. Filtering each wireless receiver is simply too expensive and complex. The only workable answer is to find the source of the noise and filter that device.

Minor noise levels in wireless systems can usually be overcome by simply boosting the wireless signals so they can be clearly read over the noise level. Both RadioRA and X10 signal regenerators are routinely used in large homes where signals attenuate, but these can also be used in smaller installations where their signal boost may be enough to overcome background noise.

Summary

- Automated lighting systems can be designed to control any number of lighting fixtures according to manual instructions or stored programs. Among the objectives of automated lighting systems are greater convenience in lighting control, more aesthetically pleasing lighting, better energy efficiency, and more effective safety and security for the home.

- Automated lighting systems should be planned in zones according to the function of the lighting as well as its location. Zones can be controlled by a single device or by a number of devices working a unit.

- Lights with home run wiring require individual control modules in order to be automated. Lights that can be wired in daisy-chain fashion can all be controlled by a single control device at the head of the chain. Control of multiple lights in a zone is more economical using daisy chains.

- Automated lighting systems use X10 or UPB power-line technology or wireless X10 and RadioRA technologies to communicate on the automation network between command modules and control modules.

- Control modules come in several varieties and respond to signals from command modules to turn lights on or off, or to dim or brighten them. These modules plug into existing outlets or can replace existing outlets and switches in the AC wiring system. Modules hardwired into the AC system should be installed by an electrician. Those connected to existing outlets can be done by a DHTI technician.

- X10, UPB, and RadioRA lighting controls are installed by setting up control modules in separate zones, which can be individually controlled from the command modules in the system.

- Command modules can be small handheld devices that control only a few lights, or large computer-based modules that can be programmed to control hundreds of lights and sensors.

- In addition to in-home command modules, both power-line and wireless technologies have wireless remote units and telephone interface units for communicating with the lighting system at a distance.

- Control modules must be installed that interface correctly with the light fixtures and wiring setup they are designed to control. Several types of outlet control modules and switch control modules are available, as well as low-voltage modules and relay activators.

- Automated lighting systems can be programmed according to a time-of-day schedule, an event schedule, or a combination of both. It can also be used to reduce peak load on the home's electrical service.

- Attenuation of signals, noise, and RF interference are the main problems encountered in automated lighting systems. All of these can be overcome in most cases.

Key Terms

Always-live duplex outlets An AC electrical outlet that is not controlled by a switch and always has current available.

Automation The process of controlling a lighting or other system remotely, either with manual commands from a controller or by a programmed set of instructions from a computer.

Command module A power-line or wireless device that sends commands (manually generated or programmed) to control modules that control lights and perform other automated functions.

Control module In lighting control systems, any of the devices that receive lighting control instructions from the controller and implement them by adjusting the lights.

Controllers Lighting control devices that send lighting commands to control modules in an automated lighting system. *Same as* command module.

Daisy-chain wiring Wiring light fixtures in parallel one to another with only the first light being connected directly to a power supply, and each of the others receiving power in succession down the chain.

Fluorescent lighting Lighting tubes that contain phosphorescent material and glow when a high-voltage current is passed through them. The tubes are made in various lengths up to 8 feet long.

Home lighting system The entire electrical light configuration in a home, including all interior and exterior lights and the controls (manual or automated) that direct them.

HomePlug Command and Control (HPCC) A power-line control system recently introduced by the HomePlug Alliance. Only a few products using this technology are currently available.

Home run connection A wired connection that links a light directly to its power supply without the current to it passing through any other fixture except a switch.

Light scenes Illuminated rooms or areas within a room that not only provide light but beautify a setting or invoke a mood.

Load In electrical systems, the amount of current (amps) flowing in the wiring at a given time.

LonWorks A power-line control system common in Europe that has recently been introduced in the U.S.

Macros Sequences of commands in a specified order that a programmable device stores in memory and executes at preset intervals.

Overloading   Placing a current demand on an electrical circuit above its capacity to carry safely.

Power-line technology   Network data transmission method in which data signals are sent on AC wiring using a different frequency and voltage than the regular current flowing in the circuits.

Power-line carrier (PLC)   The general term for control systems that operate by sending command signals over AC wiring.

Programmable   Capable of storing instruction sets for later execution in sequence.

Pulseworx   The UPB-based control system developed by Powerline Control Systems.

RadioRA   A proprietary wireless lighting control technology developed by Lutron and used for home lighting control.

Single-pole switch   An AC basic switch that opens or closes a circuit to control power to a fixture.

Split duplex outlet   An AC electrical outlet in which one receptacle is always live and the other is controlled by a switch.

Switched duplex outlet   An AC electrical outlet controlled by a switch that turns current to it on or off.

Universal Powerline Bus (UPB)   A power-line technology for controlling lighting that uses 4 to 40 Khz signals. UPB control devices are made primarily by Pulseworx.

Three-way switch   A type of AC switch that is used in pairs to turn a circuit on or off from two locations.

Zone   A group of lights controlled together to provide a specific scene or accomplish a specific task such as security lighting.

Z-Wave   A wireless lighting control system developed in Denmark and now available in the U.S.

Review Questions

1. A light scene in a home is the same as a lighting zone. True or false?

2. Overloading an electric circuit means _____.
   a. Using wire that is too large for the amount of current being carried
   b. Putting more than one circuit breaker in the circuit
   c. Connecting more electric devices in the circuit than it can safely carry current for
   d. Grounding the circuit to a lead pipe rather than to a steel rod

3. Power-line networking technology uses _____ _____ to carry its data signals.

4. A lighting zone is a group of lights that are related by _____, and may also be in or near the same _____.

5. One automated lighting zone that would not be part of the home's AC wiring system are battery-operated _____ lights.
   a. Yard
   b. Decorative
   c. Emergency
   d. Security

6. Explain how a single light could belong to more than one zone and what kind of control it would need to have.

7. Home run–connected lights are wired directly to a(n) _____.
   a. Outlet
   b. Controller
   c. AC circuit
   d. Transformer

8. A string of parallel-wired holiday lights is an example of _____ wired lights.

9. The most common automated lighting control technology is _____, which uses AC wiring to transmit control signals.

10. The two wireless technologies for automated lighting are _____ and _____.

11. X10 command modules _____
   a. Send commands to control lights
   b. Receive commands to control lights
   c. Strengthen commands that control lights
   d. Block commands that control lights

12. X10 systems can use a maximum of _____ house (letter) codes, each of which can have a maximum of _____ unit (number) codes for individual modules.

13. X10 control modules _____

   a. Send commands to control lights

   b. Receive commands to control lights

   c. Strengthen commands that control lights

   d. Block commands that control lights

14. An X10, UPB, or RadioRA three-way control module switch uses what type of additional switch for wiring the three-way circuit?

   a. A three-way control module switch

   b. A single-pole switch

   c. A companion switch

   d. A low-voltage switch

15. What does a telephone interface for an automated lighting system do?

16. RadioRA signals differ from standard X10 signals because _____.

   a. They are wireless rather than power-line signals

   b. They are two-way rather than one-way signals

   c. They are faster than X10 signals

   d. They are all of the above

17. A programmable control module differs from a manually operated controller because it can ____ _____ command sequences for later use.

18. An outlet controlled by a switch in an AC circuit should only be replaced with a control module at _____.

   a. The switch

   b. The outlet

   c. Either one

   d. Neither one

18. In addition to controlling lights, an automated lighting system may also control _____.

   a. Motion detectors

   b. Door locks

   c. Window coverings

   d. Pool covers

19. In a lighting-controlled room, what two functions could a motion detector serve in controlling the lights?

20. A split duplex outlet has one receptacle _____ _____ and the other controlled by a _____.

21. Outside circuits often must be wired with _____ _____ cable or _____ cable for greater protection from damage.

22. Attenuated (weakened) signals in an automated lighting system can be strengthened by a _____ _____.

23. In a RadioRA lighting system, RF interference might be reduced by what means?

24. On July 4, does the sun set later in Houston or Minneapolis? How about on New Year's Day? (Hint: Houston is in southern Texas; Minneapolis is in northern Minnesota.)

Hands-On Projects

Project 11-1: Set Up Two Control Modules that Turn Lights On and Off

In this project, you set up and test a basic light control system consisting of a command module and two control modules, each controlling a light. For this project, you need access to an X10 wireless remote controller module (Model 4000 remote and base or similar), two plug-in control modules (Model 2000 or similar), two lamps or other plug-in lights, and a room with at least three wall outlets.

1. Set the house code and unit code on one control module to A3 using the wheel settings or buttons on the X10 control module you are using.

2. Set the house code and unit code on the other control module to A4.

3. Set the house code on the controller remote and the receiver base to A. Be sure the remote unit has working batteries.

4. In a room with at least three wall outlets, plug the two control modules into two of the outlets.

5. Plug a lamp or light into each control module. Be sure the lamps are turned on. If you are using control modules with a controlled outlet and an always-on outlet, be sure you plug the lamps into the controlled outlet and not the always-on outlet.

6. Plug the controller receiver module into the third outlet. Turn it on.

7. On the controller, set the lower select switch to the left (modules 1–8 setting).

8. Press the 3 on (left) button to turn the first lamp on. Press the 4 on (left) button to turn the other lamp on.

9. Press the 3 off and 4 off (right) buttons to turn the lamps off.

10. Turn both lamps on again.

11. If your controller has dimmer buttons, press the 3 button followed by the dim (right) button to dim the first lamp.

12. Dim the second lamp with the 4 button and the dim button.

13. Turn both lamps off.

Project 11-2: Program a Timer Module with a Preset Program

In this project, you use the basic light control system you set up in Project 11-1 to run a timed command program from an X10 timer controller. For this project, you need access to an X10 mini-timer controller module (Model 1100X or similar), two plug-in control modules (Model 2000 or similar), two lamps or other plug-in lights, and a room with at least three wall outlets.

1. Repeat Steps 1 through 5 in Project 11-1 to set up the control modules and lamps for this project. Set the house and unit codes for the control modules to A3 and A4 as specified in the instructions.

2. Plug the timer controller into the third outlet in the room.

3. Move the upper-left selector switch on the controller to the set clock position, and use the timer arrow buttons to set the time on the controller clock.

4. On the right side of the controller, set the house code wheel to A. Set the selector switch left for unit numbers 1–4.

5. Move the upper-left selector switch to set program.

6. Set the time with the time arrow buttons about 15 minutes ahead of the current time.

7. Press the 3 unit on button. Press the 4 unit on button.

8. Move the time ahead 1 minute.

9. Press the 3 unit on button and then the dim button.

10. Move the time ahead 1 minute.

11. Press the 4 unit off button and then the brighten button.

12. Move the time ahead 1 minute.

13. Press the 3 unit off button. Press the 4 unit off button.

14. Set the upper-left selector switch to run and watch for the program to execute when the time reaches the preset times you entered for turning on the lamps

Project 11-3: Program Lights to Turn on When Motion Is Detected

In this project, you set up and test a basic light control system consisting of a motion sensor and two control modules, each controlling a light. For this project, you will need access to an X10 wireless remote controller module (Model 4000 remote and base or similar), an X10 wireless motion sensor (Model MS13A or similar), two plug-in control modules (Model 2000 or similar), two lamps or other plug-in lights, and a room with at least three wall outlets.

1. Repeat Steps 1 through 5 in Project 11-1 to set up the control modules and lamps for this project. Set the house and unit codes for the control modules to A3 and A4 as specified in the instructions.

2. Plug the controller receiver module into the third outlet. Turn it on.

3. On the controller, set the lower select switch to the left (modules 1–8 setting)

4. Check the motion sensor to be sure it has working batteries.

5. The default settings of the motion sensor are house code A and unit 1. Follow the instructions that come with the sensor to reset its code to A3. Put the cover back on the sensor.

6. Test the sensor by pressing the house button once. The sensor sends an on command and the A3 lamp should turn on. Press the unit button once. The sensor sends an off command and the A3 lamp should turn off.

7. Place the motion sensor at least 6 feet off the ground in one corner of the room in which you are working.

8. Wait a minute or so without moving, then walk in front of the sensor. Lamp A3 should turn on, but lamp A4 should remain off. After 1 minute, lamp A3 should turn off in response to an off command from the sensor because it detected no further motion.

9. Use the command controller to turn both lamps on and wait for a minute without moving. Does lamp A3 turn off in response to an off command from the sensor? Why do you think it did or did not?

Project 11-4: Program Lights to Turn on When Darkness and Motion Are Detected

In this project, you set up and test a basic light control system consisting of a motion sensor and two control modules, each controlling a light. For this project, you need access to an X10 wireless remote controller module (Model 4000 remote and base or similar), an X10 wireless motion sensor (Model MS13A or similar), two plug-in control modules (Model 2000 or similar), two lamps or other plug-in lights, and a room with at least three wall outlets.

1. Repeat Steps 1 through 5 in Project 11–1 to set up the control modules and lamps for this project. Set the house and unit codes for the control modules to A3 and A4 as specified in the instruction.

2. Plug the controller receiver module into the third outlet. Turn it on.

3. On the controller, set the lower select switch to the left (modules 1–8 setting).

4. Open the cover of the motion sensor and set it to transmit motion detected signals only if it is in darkness. To create this setting, do the following:

   a. Press the Unit/off button once.

   b. When the red light flashes, press and hold down the House/on button. The green light turns on.

   c. After 3 seconds, the red light blinks twice, indicating the motion sensor is set for darkness transmit only.

   d. If the red light blinks only once, the sensor is still set for daylight and darkness operation. Repeat Step 4 until the red light blinks twice.

5. Place the motion sensor at least 6 feet off the ground in one corner of the room in which you are working.

6. Wait a minute or so without moving, then walk in front of the sensor. Lamp A3 should not turn on because the room is lit. If it does, reset the sensor for darkness-only operation by repeating Step 4. If lamp A3 does not turn on, continue to Step 7.

7. Darken the room by turning off the lights and covering any windows.

8. Wait a minute or so without moving, then walk in front of the sensor. Lamp A3 should turn on, but lamp A4 should remain off. After 1 minute, lamp A3 should turn off in response to an off command from the sensor because it detected no further motion.

9. Use the command controller to turn both lamps on and wait for a minute without moving. Does lamp A3 turn off in response to an off command from the sensor. Why do you think it did or did not?

Project 11-5: Install a Split X10-Controlled Receptacle

In this project, you install an X10 split receptacle into an outlet box that has been roughed in for new wiring or already contains a standard outlet receptacle. For this project, you need access to a simulated electric circuit set up in a classroom or an existing AC circuit with an outlet receptacle that you can replace. You also need a Leviton DHC split receptacle or similar, a cover plate for the receptacle, two lamps or other plug-in lights, and an X10 mini-controller.

Caution: High-voltage wiring should be performed only by a licensed electrician or under the direct supervision of one. Do not attempt to complete this project without the assistance of a trained and licensed professional.

1. Turn off the power to the circuit you are working on.

2. Test the circuit with a two-prong circuit tester to confirm the power is shut off.

3. If the circuit has a receptacle you are replacing, remove the cover plate from the receptacle and then take the receptacle out of its wall box by removing the holding screws at the top and bottom of the receptacle.

4. Disconnect the wires from the receptacle. If they are held with screws, loosen them and remove each wire. If they are the push-in type, use a flat-bladed screwdriver in the release slot of each wire to depress the retainer clip and release the wire.

5. Connect the green lead from the X10 receptacle to the ground wire or a grounding screw in the outlet box. Secure the connection with a twist-on connector if necessary.

6. Connect the white lead from the X10 receptacle to the white wire or wires in the outlet box. Secure the connection with a twist-on connector. A second white wire leading to another outlet can be connected to the receptacle without affecting the operation of the X10 receptacle or the other one connected to it.

7. Connect the black wire or wires to the black lead of the X10 receptacle. Secure the connection with a twist-on electrical connector.

8. Install the receptacle in the outlet box using the screws provided with it. Install the cover plate.

9. Set the house code on the left set wheel of the receptacle to C. Set the unit code on the right set wheel to 3.

10. Turn the power in the circuit back on.

11. Connect a lamp to the lower (always-on) outlet of the X10 receptacle. Turn the lamp on and leave it on while you unplug the lamp. Plug the lamp into the top (controlled) outlet of the receptacle. The lamp may come on again or remain off since it is now being controlled by the X10 receptacle. Plug a second lamp into the always-on outlet of the receptacle and turn it on. Both lamps should now be illuminated.

12. Plug the mini-controller into another receptacle on the same AC power system as the circuit you have been working on.

13. Send a command from the mini-controller to turn C3 off. The controlled lamp should go off and the other should remain on.

14. Try a dimming command. Does it work? Why or why not? How about an All On command? All Off? Is the second uncontrolled lamp affected by any of these commands?

Project 11-6: Install Three-Way X10 Switches

In this project, you install an X10 three-way master switch and slave switch into an existing roughed-in, new wiring, three-way circuit or an existing three-way circuit. For this project, you need access to a simulated three-way electric circuit set up in a classroom or an existing three-way AC circuit with switches you can replace. You also need a Leviton 2202W DHC X10 500-watt dimmer switch, a Leviton 22081 slave switch, two cover plates for the switches, and an X10 mini-controller.

Caution: High-voltage wiring should be performed only by a licensed electrician or under the direct supervision of one. Do not attempt to complete this project without the assistance of a trained and licensed professional.

1. Turn off the power to the circuit you are working on.

2. Test the circuit with a two-prong circuit tester to confirm that the power is shut off.

3. If the circuit has three-way switches you are replacing, remove the cover plates from the switches and then take the switches out of their wall boxes by removing the holding screws at the top and bottom of the switches.

4. Disconnect the wires from the switches. If they are held with screws, loosen them and remove each wire. If they are the push-in type, use a flat-bladed screwdriver in the release slot of each wire to depress the retainer clip and release each wire.

5. Check which of the three-way switches you removed was wired directly to the light fixture. In the X10-controlled switch setup, this switch must be the master switch. The switch wired directly to the power supply must be the slave switch.

6. Install the slave switch first. Connect the black (hot) lead and the black traveler lead both to the black lead from the X10 slave switch. Secure the connection with a twist-on connector.

7. Connect the red traveler wire to the red lead from the X10 slave switch. Secure the connection with a twist-on connector.

8. Connect the two white (neutral) leads in the box to one another. Secure the connection with a twist-on connector.

9. Now install the master switch. Connect the two white leads in the box to one another. Secure the connection with a twist-on connector.

10. Connect the black traveler wire to the black lead of the X10 master switch. Secure the connection with a twist-on electrical connector.

11. Connect the red traveler wire to the red lead of the X10 master switch. Secure the connection with a twist-on electrical connector.

12. Connect the black wire going to the light to the blue lead of the X10 master switch. Secure the connection with a twist-on electrical connector.

13. Use a small screwdriver to remove the plate from the master switch and expose the setting wheels. Use the screwdriver to set the house code on the left set wheel of the switch to D. Set the unit code on the right set wheel to 3. Replace the plate by pressing it in place.

14. Install both switches in their respective boxes using the screws provided with them. Install the cover plates.

15. Turn the power in the circuit back on.

16. Manually turn the master switch to the on position to turn the controlled light on.

17. Plug the mini-controller into another receptacle on the same AC power system as the circuit you have been working on.

18. Send a command from the mini-controller to turn D3 off. The controlled light should go off.

19. Try a dimming command. Does it work? Why or why not? How about an All On command? All Off?

20. Test the master and slave switches manually to be sure they function correctly in that mode.

Case Projects

Case Project 11-1: Design Lighting Automation for a Garage and Driveway

A client wants you to install automated lighting in his garage and outside lighting to illuminate the driveway. The home's garage is attached to the house and is entered from the house through an interior door. The overhead garage doors each have electric openers on them that can be activated by switch buttons near the interior door or by an RF opener in each of the family cars (one opener in each car opens one garage door). The client wants the interior lights in the garage to turn on whenever someone enters the garage from the house or when either of the garage doors is raised to allow a car to enter. She also wants the outside driveway lights to turn on when either garage door opens, but not when someone enters the garage from inside the home. Diagram an automatic lighting solution that accomplishes the lighting the client wants as simply as possible. Note the hardware pieces required and how each is to be connected and set up so it performs as needed. For this exercise, assume there are no other automated lighting zones in the home.

Case Project 11-2: Design an Automated Room Lighting Zone

You are installing a home lighting system in a living room that has two overhead lights, both controlled by three-way switches at opposite ends of the room. It also has three lamps all plugged into split duplex outlets, and all controlled as a group by two three-way switches at opposite ends of the room. How would you design the lighting control for this room so both overhead lights and lamps are controlled as a unified zone, and so the overhead lights can also be operated manually from their switches? Draw a design using RadioRA control modules and other hardware to control the lighting and conform to UL electrical safety standards.

Case Project 11-3: Find a Programmable X10 Controller

One of your clients wants a programmed home lighting system, but does not want to run it from her computer. She wants a small controller that can be permanently located in the home's master bedroom from where she can direct the lighting system. She would like to be able to program the system from a graphic interface using a keyboard or a touch screen. The lighting system in her home has both X10 modules and wireless modules controlling some of the lights. The system has eight zones and each has up to four control modules in each zone. Visit some X10 Web sites (http://www.X10.com and http://www.smarthome.com are two to start with) and see what you can find that is available to meet her requirements. Compare the features of the controllers available and write a recommendation on the one you think is best.

Case Project 11-4: Design an Automated Lighting System for a Cabin

A client has a cabin located on a remote mountain lake. Lights there are powered by a generator with battery backup. The cabin can only be reached by boat across the lake. Its dock is in a little cove with a narrow rocky entrance that is dangerous to enter at night unless illuminated. The cabin's owner visits the cabin only about once a month and does not want to have lights come on nightly with a timer when he is not there. He would like a system that turns the dock and safety lights on only when his boat (or any other boat) approaches within 200 yards of the entrance to the cove after sundown. What can you suggest that meets his need and that does not require other visiting boats to have any special device on board to signal for the lights to come on?

Heating, Air Conditioning, and Water System Control

OBJECTIVES *After studying this chapter, you should be able to:*

- Describe the design and operation of zoned and non-zoned heating and air-conditioning systems.
- Identify and describe the function of heating and air-conditioning components and their controls.
- Describe how to install a heating and air-conditioning control system, set the control system, and program it for year-round operation.
- Describe the design and operation of automatic water system **controllers.**
- Identify the major components in automated water systems and describe their functions.
- Describe the installation and programming of an automated home water system.

OUTLINE Introduction

Design and Operation of Zoned and Non-zoned HVAC Systems

HVAC Components

HVAC Controller System Installation and Programming

Water Control System Design

Water Control System Components

Water Control System Installation and Programming

Introduction

Heating, ventilation, and air-conditioning (HVAC) systems and water systems are the home utilities most frequently automated. In this chapter, you will learn how HVAC systems operate to provide comfortable home environments.

You will find out how to identify the major components and control devices in HVAC systems and describe how each functions. You will also learn how to install **automated controls** for an HVAC system and discover how to set the controls of the system and program its control panel so it will provide a comfortable interior environment under a variety of conditions. You will also discover how to design a water control system that will provide maximum convenience and flexibility to the homeowner. You will find out about the major components in a water control system and how each functions. Finally, you will learn how to install an automated water control system and program it for automatic operation with a minimum of user input.

Designing and Operating Zoned and Non-Zoned HVAC Systems

Heating, ventilation, and air-conditioning (HVAC) systems in American homes vary as widely as the locations and climates where they are installed. Heating systems for low-altitude southern latitude homes may be little more than a few **space heaters** placed in strategic locations, while whole-house **air conditioning (A/C)** is a must. At the other extreme, a home in the mountains of Montana or almost anywhere in the northern tier of states may need a **central heating system** for 9 months of the year and no A/C at all. And then there are many areas such as Chicago, Omaha, and Denver (to name just three of hundreds) where central A/C and heating are both necessities for year-round home comfort.

Most automated control systems are designed to work with central heating and A/C rather than with room-size units. This chapter discusses the automated control of central heating and A/C systems only. Most homes have at least one of these and many have both.

Space Heaters, Room Air Conditioners, and Evaporative Coolers

Confining ourselves to central (whole-house) systems means making only this brief mention of space heaters (electric or gas), as well as window and other room air conditioners. All of these units can be automatically controlled with **thermostats** or timers, but none are large enough, or have a large enough impact on a home's interior climate, to warrant a whole-house temperature control system.

Also not covered is the class of air conditioners known as **evaporative or "swamp" coolers,** which are frequently designed to cool and humidify an entire house. These units are used mainly in dry desert climates where they can attain maximum cooling efficiency. They do not work well in high-**humidity** climates. Their effectiveness depends on moving large volumes (several thousand cubic feet per minute) of evaporative-cooled air through the home, drawing it in through the cooler and expelling it out through open windows or doors. Like space heaters and room air conditioners, evaporative coolers can be automatically controlled with a thermostat or timer, but nothing more than on or off commands are needed (or are possible) for them. Changes in their cooling efficiency are best made by adjusting the air exits (open windows or doors) rather than the operation of the cooler.

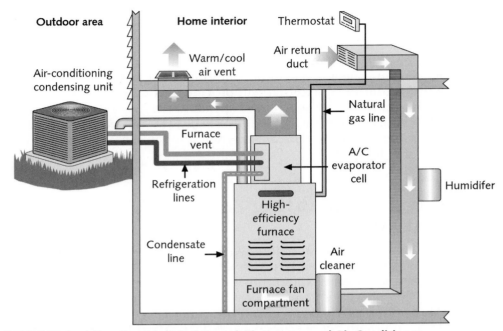

Outdoor area

Air-conditioning condensing unit

Home interior

Warm/cool air vent

Thermostat

Air return duct

Natural gas line

A/C evaporator cell

Furnace vent

Refrigeration lines

Condensate line

High-efficiency furnace

Air cleaner

Furnace fan compartment

Humidifer

FIGURE 12-1   A Basement-Located Forced-Air Furnace and Air Conditioner

Central Heating Systems

Central heating systems are divided into two types: **forced air** and **radiant.** Forced air systems all function by drawing air from the home's interior into a **furnace** where it is heated and then blown out again into the home through heating **vents** connected to the furnace by metal **ducts.** All forced-air heating systems rely on the continual transfer of air from the home's interior to the furnace (where it is heated) and then back again. They keep the air temperature in the room comfortable by heating the air itself.

A typical forced-air heating system is shown in Figure 12-1. The system is shown combined with a central A/C system, which uses the same ducts and **blower** as does the furnace. When heating is required, the furnace provides it, and the A/C unit is idle. When cooling is necessary, the furnace is idle, and the air conditioner takes over.

Radiant heating systems do not have any air movement system. They work by heating the air in the house through convection or by heating the house itself rather than the air inside it. Older radiant systems use circulating hot water or steam from a central **boiler** to heat standing radiators or baseboard radiators in each room of the house. These units warm the air around them, which then expands and becomes lighter. The heated lighter air rises away from the radiator and is replaced by cooler air that, in turn, is heated and rises away from the radiator. Radiator heating systems do not heat evenly and are not as efficient as other radiant or forced-air systems. Few, if any, of these systems are being installed in new homes, although millions still function in older homes.

Radiant heating systems now being used in newer homes consist of a large water heater connected to pipes that run at spaced intervals under the floors of the home. The heating system operates by heating water in the central boiler and then

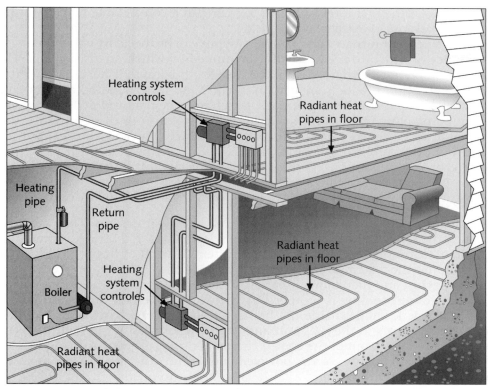

FIGURE 12-2   A Radiant Heating System with Piping in the Floor

circulating the water through all the pipes beneath all the floors of the house. Heat from the pipes is absorbed into the structure of the house and radiates from there to the home's interior air, warming it to a comfortable temperature. A section of a radiant heating system is shown in Figure 12-2.

Although this heating method sounds less efficient than forced-air heating, it is actually more energy efficient and provides more even heat distribution throughout the home than the other type. For these reasons, radiant heating systems, though more expensive to install initially, are growing in popularity.

Furnace Fuels

Fuel for the combustion chambers of nearly all modern furnaces is either fuel oil or natural gas. Some coal-burning furnaces remain in use, and some that were once coal-burning have been converted to burn oil or gas. Their numbers are declining, however, since all new residential furnaces use oil or gas. Oil is stored in a tank and piped from there to the furnace. Gas comes through a pipeline from a pressurized storage facility. The flow of either oil or gas fuel is not controlled as a means of regulating heat. Fuel settings are constant in furnaces, as is the amount of heat produced by the unit when running. The total amount of heat produced by the system to heat the home is controlled simply by turning the furnace on or off. Some newer furnaces for use in very cold climates have two operating levels; the first level is used for most heating and operates the furnace at up to 75% of its rated capacity. The second level, which operates the furnace at maximum capacity, is used only if the first level is unable to maintain a comfortable temperature in the home because of extreme cold outside.

Both forced air furnaces and radiant systems can be fueled by either oil or gas. The choice of which to use is usually determined by which one is available. Gas tends to be preferred wherever it is available, both because it burns cleaner (the usual combustion products are carbon dioxide and water) and because gas also leaves no ash or residue in the furnace, which must be removed periodically. Gas requires a pipeline connection for home delivery, but if lines are installed in a residential area, delivery is much simpler than pumping fuel oil from a large truck into an underground storage tank. Gas has historically been less expensive than oil, but the cost of both fuels is rising and coming closer to equality.

Central Air Conditioning (A/C)

Central air-conditioning (A/C) systems are all forced-air designs. They move air in much the same way as a furnace heating system. In fact, most A/C systems use the same ducts and vents as the heating system. Instead of heating the air that is drawn into the system from the home's interior, the air conditioner cools it by **refrigeration** and then blows it back into the home through the duct system.

Refrigeration cooling is provided by the A/C **condenser** and **compressor** unit, which is located outside the home. Chilled **refrigerant** from the **condenser unit** flows to the A/C unit and cools the air passing through it. Excess humidity condenses out of the cooled air and is carried away by a drain pipe.

A/C units are mainly electric powered and use no other fuel. The compressor in the condenser unit requires a fairly large motor to operate it and the blower unit that moves air over the **cooling coil** and into the home is also powered by a large electric motor. These motors make air conditioners heavy consumers of electric power. In most homes, the A/C system is the largest user of electric power when operating, because it requires substantial power and because it operates for long periods of time.

Single- and Multiple-Unit HVAC Systems

Most homes have only a single forced-air furnace, which serves to heat the entire house. The larger the home is, however, the more difficult it is to heat it adequately or evenly with a single furnace. Long duct runs from the centrally located furnace to the furthest rooms in the home tend to dissipate much of the heat in the air traveling through them so that by the time it reaches its destination, its temperature has dropped too low to raise the room temperature to an acceptable level. Similarly, the sheer size of many homes may exceed the heating capacity of a single furnace.

The effect of A/C systems is also limited by the distance their cooled air must travel and the size of the blower required to move it. Though blowers run more quietly than **fans,** they still make some noise and the noise level increases with the size of the blower. As with a single-unit furnace, a point is reached where the size of the home exceeds the ability of one A/C unit to effectively cool it. A second unit is the logical answer.

The Air Conditioning Contractors of America (ACCA) publishes Manual J, which contains the industry standard for calculating heating and air-conditioning loads in residential buildings. To meet or exceed its recommended furnace capacity, many homes have two or more furnaces and A/C systems located in different parts of the structure. Whenever a home has two or more furnaces and A/C systems, the HVAC system is by definition divided into **zones**, each of which is heated by a different furnace and cooled by a separate A/C system. One zone, with one HVAC system handling its heating and cooling, might be the main floor of the home, while another HVAC system may handle the lower floor or the second floor. The controls of each system must be programmed to work together or the efficiency of both is degraded and the home is poorly heated and cooled.

Radiant heating systems are less likely to require two or more separate heating units than forced air systems. This is because radiant home heating systems can be larger in overall capacity than forced air units, and because radiant systems are more efficient in carrying heat over long distances without significant loss. Finally, radiant heating systems can be more easily divided into zones and their heat distribution adjusted for distance than forced air systems. Few homes of less than 5,000 square feet floor area have more than one radiant heating unit, and many of these systems can go to even higher capacity.

Homes with radiant heating systems often have forced-air A/C systems for cooling. The two systems are completely separate, but they can be controlled by one set of thermostats and a single central controller.

Zoned and Non-Zoned Systems

Heating and cooling zones are not just the result of multiple furnaces or air conditioners. They can be set up in homes with only a single furnace and A/C unit as well. The purpose of heating zones is to more evenly and efficiently distribute heat throughout the home. The two factors that make zones necessary in many home HVAC systems are distances from the central furnace to some parts of the home and the effects of outside climate, mainly sunlight.

In an HVAC system, the further heated air travels from its source (the furnace), the more heat is lost in transit so that less remains when the air finally reaches its destination. The same is true for the cooled air from an A/C system: It warms as it moves through the ducts, absorbing heat from them as it moves. In a home with only one zone and relatively equal heated air distribution to all rooms, this results in the rooms closest to the HVAC system being the warmest, and those furthest away being the coldest. If colder rooms are brought up to normal temperature, warm ones are then too hot for comfort and vice versa.

Radiant heating systems with only one zone also have this problem. The heating water cools more by the time it reaches the most distant pipes than it does before reaching the nearby ones. The result again is uneven heating.

The A/C system suffers from a similar unequal heat distribution, only this time it is caused by cool air from the air conditioner warming more as it travels to the outer parts of the home than it does for those nearer. This causes the central core of the house to be cooler than those outer parts that are more removed from the air conditioner. The condition is made worse by the effects of summer sunlight, which can heat the southern exposed side of the home more than the northern side and thus make it even hotter. Once again, equally distributed cool air results in an unequally cooled home, some parts of which are comfortable and some not.

Another factor contributing to this condition is adjacent areas, which may be hotter or colder than the home's living area by design. Even with good insulation in the walls, some **heat transfer** occurs from the warmer area to the colder one. Figure 12-3 shows how outside climate, sunlight, and adjacent areas all contribute to unequal heating and cooling of a home with only a single zone.

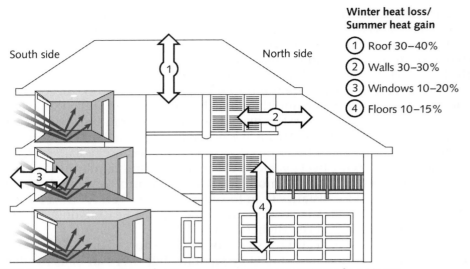

South side   North side

Winter heat loss/
Summer heat gain

1 Roof 30–40%

2 Walls 30–30%

3 Windows 10–20%

4 Floors 10–15%

FIGURE 12-3   How Heat Transfer Creates Uneven Temperatures in a House

Dividing the heating and A/C systems into zones with separate temperature controls in each allows the system to compensate for inequalities in heating and cooling of the home by sending more or less heated or cooled air to needed areas, independent of what is happening in other zones. Each zone is treated like a separate apartment with its own HVAC system and controls. If the controls of each zone had to be continually adjusted for changing conditions by hand, the system would be too cumbersome to use effectively. Automating the HVAC system allows it to be controlled by a preset program based on input from temperature sensors (thermostats) located in each zone.

Ventilation

In addition to heating, some HVAC systems may need to provide for **ventilation,** the exchange of air inside the home with outside air. Ventilation may be necessary even when the inside and outside temperatures are the same and no heating or cooling is required. If a home is well insulated and sealed against outside weather conditions, the air inside it remains cut off from outside air unless a door or window is opened. With occupants inside the home, this can result in depletion of the oxygen in the air, stagnant odors, a buildup of humidity, and accumulation of dust and bacteria.

Ventilation provides fresh air in the home. It can be accomplished in conjunction with the operation of the heating or cooling system or without changing the temperature of the new air brought in. In either case, the new air may be filtered or humidified so it does not add to the air-quality problems already present in the home.

For ventilation to be effective, the HVAC system must have both an intake vent on the outside of the home through which outside air is drawn in and an exhaust vent through which stale inside air is expelled. In a sealed home, an intake vent without an exhaust vent will not work, and neither will the reverse setup. Both are required. The intake vent may be a part of the HVAC system and controlled so it can be opened or closed as needed. The outside exhaust vent can also

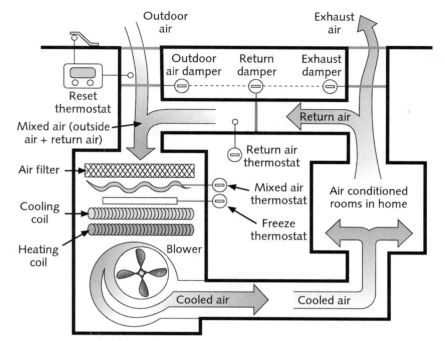

FIGURE 12-4   An Air-Conditioning System with Outside Ventilation

be part of the HVAC system's air return duct system, or it can be a separate vent or vents that exhaust air into the attic, where it will be expelled outside by an attic fan. Figure 12-4 shows a typical A/C system with built-in outside ventilation and the exhaust of room air. The amount of outside air entering the system is controlled by **dampers** (air duct doors).

HVAC Components

HVAC systems all include the same basic components. Some of these can vary greatly in size and capacity, but their functions are standard. This section describes all of the important components of HVAC systems except for the metal ducts of forced-air heating systems and A/C systems and the piping network of radiant heating systems. The reason for not including ducts or piping is that technicians who work on the automation controls for HVAC systems rarely, if ever, install or remodel metal ducts or pipes for radiant heating. The former are made and installed by sheet metal workers who specialize in building heating and A/C duct systems. The latter are installed by plumbers and pipefitters who specialize in heating systems.

Duct systems and pipes are nearly always installed when a home is built. The control technician may install control mechanisms at the time the HVAC system is installed, or he or she may be called in later to automate a system that was originally a manual one. By installing additional control components, an HVAC system that was originally a single-zone system can be converted to multi-zone in order to provide more even temperature control and better economy.

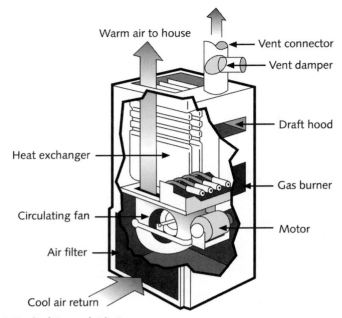

Warm air to house

Vent connector

Vent damper

Draft hood

Heat exchanger

Gas burner

Motor

Circulating fan

Air filter

Cool air return

FIGURE 12-5   A Typical Forced-Air Furnace

Furnaces

Furnaces provide the heat in an HVAC system. Forced air furnaces are controlled by a thermostat, which turns them off or on. Most furnaces do not have graduated control, only a simple on or off switch. When they are on, they produce heat at full capacity. When they are off, they produce none. Some furnaces may have two operating levels as an energy saving device; the furnace operates at the lower level and uses less energy unless extreme cold requires it to go to second-level operation to maintain a comfortable temperature in the home.

Figure 12-5 shows the main parts of a forced-air furnace and how it works. As was shown in Figure 12-1, furnaces and air conditioners are often built as combination units using the same blowers and duct work to transport the air they heat or cool.

Gas and oil furnaces, which are connected to a gas or oil line, have a safety device called a **thermocouple.** This device is a flexible metal tube, which produces an electric current when heated. In a furnace that has a continuously burning **pilot light** (that functions to ignite the gas or oil when the furnace is turned on), the thermocouple is placed in the flame of the pilot light. As long as the pilot light flame is burning, the thermocouple produces electricity, which in turn signals a **valve** on the gas or oil line to stay open so fuel can flow to the furnace. If the pilot light goes out, the thermocouple cools off and stops producing the electric current. This signals the gas or oil **safety control valve** to close, shutting off the gas or oil and preventing any buildup of unburned fuel, which could cause a fire or explosion.

Furnaces without continuously burning pilot lights are fired by a **spark igniter,** but they also have thermocouples as safety devices to control the flow of fuel. Just before the spark igniter fires the furnace, the fuel valve is opened and remains open for a few seconds while the furnace fires and the pilot light begins to burn.

When the pilot light starts burning, the thermocouple begins producing current and this signals the valve to stay open. If the pilot light does not come on, the thermocouple does not signal and the valve closes as soon as the delay period ends.

Radiant boilers also have thermocouples on their heating units to prevent any buildup of unburned fuel. Most times when a furnace stops working, the problem is a failed thermocouple. Other safety devices in furnaces and boilers include temperature sensors, which shut the units down if they get too hot. Heat buildup in a furnace can occur if a blower motor or drive belt fails or if a **circulating pump** on a radiant unit fails.

Refrigeration Air Conditioners

Air conditioners work like a refrigerator to cool air and circulate it. Air conditioners consist of a condensing/compression unit and a cooling coil through which air is blown by a fan or blower. The condensing/compression unit and cooling coil are connected by pipes or tubing so a refrigerant fluid can circulate between them. The condenser unit is located outside the home so the heat it extracts from the cooled air can be exhausted to the outside air. The cooling unit is located inside in the main air flow duct of the HVAC system where it can cool the air passing through it.

The refrigerant, which is a gas when it is at room temperature and pressure, is held inside a sealed circulation system in the A/C unit. As the refrigerant passes through the condenser/compressor unit, it is compressed by a compressor into a liquid under high pressure. Compressing the gas makes the liquid it forms hot and it is cooled by the fan of the condenser/compressor unit as it passes through the **condenser coil.** From the coil, the cooled liquid (still under high pressure) flows to the cooling coil in the air conditioner unit. There it is permitted to expand back into a gas at lower pressure and, as it does so, it absorbs a large amount of heat from the cooling coil and the surrounding air. This cools the air, which then is forced into the home by a blower. The expanded gas, which has absorbed heat from the air, flows back to the condenser/compressor unit where the processes of **condensation** and compression turn it back into liquid form and its heat is extracted in the condenser coil by the cooling fan of the condenser. Figure 12-6 shows how an air conditioner works.

Thermostats

A thermostat is the main control device for HVAC systems. It consists of a temperature sensor coupled to an adjustable switch. A thermostat can be set at a given temperature and whenever the temperature falls below the set temperature, the switch on the thermostat closes. This sends a signal to the furnace to turn on and raise the room temperature. When the temperature reaches the set level, the thermostat switch opens and this signals the furnace to turn off.

Thermostats can be dual-acting so that they control both heating and cooling equipment. In this case, when the room temperature falls too low and the thermostat switch closes, the furnace will turn on. When the temperature is above the set figure and the switch is open, the A/C unit turns on and cools the room until the switch closes, signaling the air conditioner to turn off. Dual-acting thermostats have a second switch, which sets them to either heat or cool mode. This prevents both the heating and A/C units from being active at the same time. The switch can be set by the user and determines which system is operational. Without such

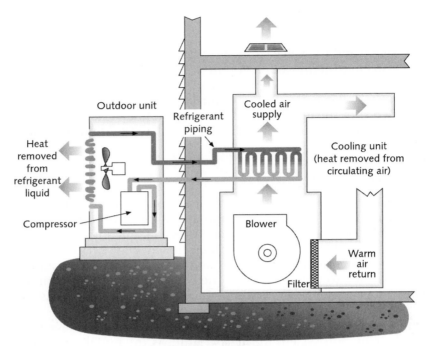

FIGURE 12-6   How an Air Conditioner Works

a switch, the two systems would constantly conflict with one another, the furnace continually raising the room temperature and the A/C lowering it.

In HVAC systems that have multiple zones, a thermostat normally is installed for each zone. This allows the temperature to be maintained at a set level in each zone. The actual temperature set may be the same in each zone, but the individual thermostats are still useful because they allow the system to act independently in each zone to maintain the desired temperature. Twice as much heat, for example, may be needed to maintain a temperature of 70 degrees Fahrenheit (F) in a room over an unheated garage on the north side of the house, compared with a sunny room on the south side. The individual thermostats can signal for the necessary heating in each zone.

Sensors

Thermostats are not thermometers. They do not measure the temperature. They only signal when one particular set temperature is reached. In automated HVAC systems, temperature sensors are sometimes used, which do signal the actual room temperature. These sensors do not signal the HVAC system to turn on or off; they simply signal the existing temperature so the system's central control can determine from that data what system to turn on and when. In a zoned system, a temperature sensor is placed in each zone so as to provide temperature information about the zone, enabling the HVAC system to act independently to maintain the correct temperature in each zone.

Another type of sensor used in HVAC systems is a **humidity detector.** This device measures the amount of moisture in the air and can be used to signal an HVAC system when to turn on or turn off a **humidifier** that adds moisture to the incoming air stream. Humidifiers are normally used only with furnaces. The process of heating air also dries it, and moisture often needs to be added for comfort in a heated room. This is especially true in normally dry climates where outside winter humidity levels can fall below 10%. Air-conditioning systems rarely use humidifiers because cooled air has a higher relative humidity than it did before cooling. Most A/C systems actually extract moisture from the air as they cool it. The extracted water is then drained out of the system and discarded.

HVAC systems also contain safety sensors to protect the equipment and the home. For the air conditioner, the safety mechanism is a **freeze sensor,** which signals if the cooled air is at or below the freezing point of water. This could happen if the A/C were inadvertently turned on in winter, or if the controlling thermostat fails to shut it off at the set temperature. The freeze sensor provides a safety check on the system that shuts it down before it freezes up and damages the cooling coils and dehumidifier. For the furnace, a high-temperature sensor is included that shuts the furnace down if its internal temperature rises too high. Such an event could occur if the blower motor or drive belt fails, or if the air vents were blocked so heated air could not exit from the system.

Air Handlers

Air handlers might better be named air movers. They are the mechanical devices that move air in an HVAC system. The two basic types of air handlers are fans and blowers, and each can be used in either of two ways: to force air forward from the air handler into an area by increasing pressure; or to exhaust air out of an area by decreasing the pressure.

Fans are sets of angled blades on a shaft spun by a motor. An airplane propeller is a fan, but one designed to pull or push the aircraft through the air rather than to move the air. A fan designed to move air has wider blades set at a shallow angle. Fans move air well and are widely used as cooling devices by themselves. As their size and speed increases, however, they become noisy. This is because the leading edges of the blades "bite" into the air as they turn and produce a small shock wave that can be heard as sound. In small fans, this sound is barely audible, but in large ones turning fast, it quickly becomes a roar that nobody wants to listen to for long. The sound problem for fans is made worse by the fact that the outer tips of the blades are moving much faster than the hub in the middle of the fan. In large fans rotating at high speed, the moving blade tips can actually approach the speed of sound. If they were to exceed the speed of sound, they would produce a continuous sonic boom that would be deafening to anyone nearby.

Because fans are noisy in large sizes, a second type of air handler was developed that greatly reduced this problem: blowers. A blower looks somewhat like the paddle wheel on the rear of a river boat and acts on air similar to how a paddle wheel works on water. The rotating wheel of a blower is closely fitted in a housing with one exit on the outside. As the wheel spins, air is drawn into its center and impelled outward by the blades of the wheel. When the air finds the exit opening in the housing, it moves into it, forced along by the moving blades.

Blowers can be made larger than fans and they move slower, rotating at 400 to 500 rpm in contrast to fans, which often spin at 3,600 or 7,200 rpm. Because of their slower speeds, blowers are quieter, and because of their size, they can move large amounts of air. Most HVAC systems use blowers rather than fans as their main air handlers.

Blowers usually push air forward, although they can be used to exhaust air as well. Fans can also push air as well as exhaust it, but most exhaust air handlers are fans. Forced-air handlers work by increasing the pressure of the air, causing it to move forward to where the pressure is less. Exhaust fans work by reducing the air pressure behind them as they push the air forward in front of them. The reduced pressure causes air further back in the system to move forward to where the pressure is less. HVAC systems use forced air handlers to move air through the system. If fresh air is taken into the system and used air exhausted, these functions are handled by an exhaust fan.

Dampers

Dampers control the flow of air in a duct by opening or closing. They act like a door that can be opened to allow the passage of air or closed to block it. Dampers are not hinged at the side like a door, however. One type is pivoted in the middle with half the damper door on one side of the pivot and half on the other. The door is fixed to the pivot rod and the rod extends outside the duct so it can be used to open or close the damper. This arrangement makes the pivot easy to open and close because the pressure of the moving air in the duct on it is balanced: There is as much air pushing to close the door as to open it.

Another type of damper is a set of louvers like those used to close the A/C vents on the dashboard of a car. The louvers are set in the duct and an activator rod that can open or close them extends outside the duct. Dampers can be automated by small motors called servomechanisms (servos for short) that are attached to the pivot rod or activator rod and open or close the damper on command. Motorized dampers usually operate on 24-volt AC power and are controlled by a controller. Figure 12-7 shows a louvered damper configured for installation in a rectangular duct. The damper's motor and control are mounted on the right side of the square casing.

Dampers are normally installed fully open. This means their normal position is open and they can only be closed by a signal from a controller. If the power fails or the system shuts down, all the dampers open. It should be noted that the function of dampers is limited to blocking the flow of air in a duct. They cannot increase the flow, only reduce or stop it.

Duct Boosters

If an area in a home has too little heated or cooled air reaching it, installing a damper will not help. To get more air into the room, an auxiliary fan is needed. These fans are called duct boosters because they are placed in the duct just as a damper is and they pull more warmed or cooled air through the duct and into the room than the HVAC blower alone can provide.

FIGURE 12-7   An Automated Damper for a Rectangular Duct

A duct booster has its own motor and can be controlled and operated independently of the main furnace. Duct boosters can only operate when the furnace is running because they do not have any source of heat other than the furnace. They are only fans that distribute the furnace's heat more efficiently to a specific area. If the room it serves is consistently underheated, a booster can be set to operate whenever the furnace is running in order to bring the room temperature up to normal. If a lower room temperature is only an occasional problem, the booster can be connected to its own temperature sensor and set to operate only when the room temperature is too low.

Controller Panels

Most HVAC systems are controlled by a thermostat or by multiple thermostats, if the system has zones. An HVAC system only needs a control panel if it is to be centrally controlled from one point. The control panel can be an independent multi-zone controller or an interface to a computer, which provides the control through a software program.

Forced-air systems typically use a controller such as an RCS brand six-zone HVAC control, which provides all the control features needed for a multi-zone system. The controller does all of the following:

- Communicates directly with up to six RCSLinc units, which control 24-volt A/C, normally open, motorized zone dampers and furnace thermostats
- Communicates with an outdoor RCSLinc unit, which controls the air-conditioning compressor/condenser unit
- Sets and maintains independent temperature control of up to six zones
- Automatically changes over from heat function to cooling based on time setting or temperature setting
- Provides a built-in short-cycle control function for the A/C compressor to prevent damage from too rapid starting and stopping cycles
- Shows visible LED status indicators on the panel for verification of the system's operations.

This unit can also be interfaced with a computer through an RS-232 or RS-485 connection so computer software can program its operation and provide more complete automation for the HVAC system. The control unit must be connected by low-voltage wires to each of the zone thermostats and to the zone dampers it controls. If it interfaces with a computer, the computer instructions are still implemented through the motor and temperature control connections of the controller rather than directly by the computer, so the same wired connections are needed in either case.

Multi-zone radiant heating systems are more easily controlled than forced-air systems because each zone can be turned off or on with a single valve control in the hot water circulation system. Since all the piping exits from the system's boiler, the entire control system can be wired there. No dampers are used in these systems, but thermostats are normally wired in each zone so the controller knows the room temperature and can adjust the hot water controls accordingly. Figure 12-8 shows an eight-zone radiant heating system control setup with an automated valve control on each zone pipe. The water storage tank is in the foreground of the picture and the boiler is in the background. The home in which this system is located is in Park City, Utah, at an altitude of about 8,000 feet, so no air conditioning is required for the system.

FIGURE 12-8   A Radiant Heating System with Automatic Controls

HVAC Controller System Installation

HVAC systems are nearly always installed at the time the home is built. Not only the physical equipment for the systems, but also the control devices (thermostats, temperature sensors, and damper controls) and wiring for them are built into the basic structure of the home. While it is possible to retrofit an automated control system to a previously installed HVAC system, this is unusual and is likely to be both difficult and expensive.

This section describes the controller wiring and devices that need to be in place for an HVAC controller system. Also described is how the wiring connects to termination points on each of the control devices as well as the central controller. The description assumes that the system is centrally controlled from a controller panel, although the wiring for thermostats would be similar if they were the only control mechanism in the system.

Controller Wiring

Wiring from HVAC controllers to the motorized control units on dampers and water valves, and to thermostats and temperature sensors, can all be done with Category 2 telephone wire. Thermostats send only power on and power off signals, and dampers receive only those same signals. Temperature sensors send more complex data, but still in relatively simple form and at slow speeds so the low capacity wire is adequate.

Dampers, valves, and thermostats generally run on 24-volt AC power. This is provided by a step-down transformer wired to the home's AC power supply. Two wires supply power to each unit. No ground is required because of the low voltage. Temperature sensors and thermometers can also be AC powered, but they require little power and many run on batteries that only need replacement every 3 to 5 years. The battery-operated type may be wireless and communicate to the HVAC controller by radio signals.

As with security system wiring, HVAC wiring can be stapled in place, bent around sharp corners, and carried near high-voltage power lines without any effect on its function. A home run wire pair must be run from each thermostat to the controller panel and from each damper to the controller panel. If the system is using temperature sensors in addition to thermostats, these must each have a home run wire or an RF connection to the control panel.

In a forced-air system, heated or cooled air to each zone is controlled by a damper installed in the zone ducts coming from the main blower. At the controller panel, the thermostat located in each zone is wired to the control panel's sensor connection for that zone, and the damper is wired to the controller's command out connection for the same zone. In a radiant system, the thermostat wiring is the same as for forced air, but the command out connections for each zone go to the respective zones' water control valves located near the boiler for the system. The starter control on the furnace or boiler is also connected to the control panel.

Air-conditioning control wires go to the condenser/compressor unit outside the house and also to the temperature sensor inside the cooling coil enclosure. If the system has outside air ventilation, the dampers for exhaust air and incoming air must also be wired to the control panel along with the return air thermostat or sensor that helps the system determine how much fresh air is needed.

Termination Points

HVAC wiring is all connected to screws or nut posts on the devices and should be attached firmly so it will remain in place for a long period of time. The wiring is AC, which means the current flowing in it reverses direction 60 times per second, so the order in which the two wires are attached to each device does not matter. AC motors and solenoids operate the same, regardless of the termination point to which the "hot" lead is connected.

When wiring the control panel, wires should be carefully labeled to identify the zone and device to which they are connected. Care should be taken to wire everything to the right connection point. A misplaced sensor or damper wire can cause the whole system to malfunction by turning on heat or cooling in the wrong room based on temperature data from another zone. Once wired, the HVAC system should be manually tested and each zone physically checked to be sure the system is actually functioning as the control panel says it is. Figure 12-9 shows the typical wiring for a multiple-zone HVAC controller.

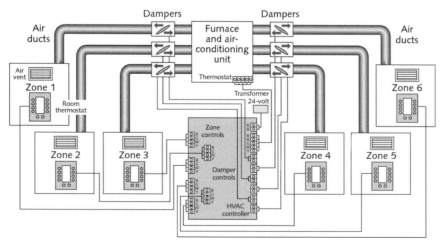

FIGURE 12-9   A Multiple-Zone HVAC Controller Wiring Diagram

Safety sensors in both the furnace and A/C are internally wired to shut each unit down if a safety problem arises. These shutdowns override the normal operation of the control panel, which does not need to be connected to the freeze or overheat sensors unless the system has an emergency notification feature that alerts the homeowner to the problem.

Computer connection for an HVAC control panel is through an RS-232 or USB port on the computer and the panel. The connection requires only four wires and can be changed with an adaptor to an RJ-11 or RJ-45 connector if the computer is separated from the control panel by more than a few feet. With the smaller connectors, a cable can be run between the computer and the panel to link them.

Setting and Programming HVAC Controllers and Sensors

Once installed, tested, and programmed, a good HVAC control system should function on its own with virtually no required input from the homeowner for long periods of time. Setting and adjusting the system to function on its own may take a little time, but it results in a comfortable environment over long stretches. This section describes setting the controllers and sensors to operate without additional human input.

Sensor and Thermostat Settings

Thermostats should be set to a comfortable room temperature, around 68 to 70 degrees in the winter and 75 to 78 degrees in the summer. Some people may prefer only one year-round temperature somewhere in between those given, but most feel better with two settings, and the dual range is more economical as well. Once set, thermostats can be left in the same position indefinitely.

Experience may show that some zone thermostats need to be set a few degrees higher or lower than the desired temperature in order to maintain that actual temperature in the room. This can be a result of the thermostat itself (for example, it reads slightly above or below the actual temperature) or its location (perhaps it has been placed in the room at a location where the temperature is consistently higher or lower than the room average). There is nothing wrong with making these adjustments and they will not affect the economy of the system as long as the actual room temperature level (as contrasted with the thermostat setting) is not increased substantially.

Zone Programming

The control panel, based on input from the thermostats, starts the HVAC system (furnace or A/C) as needed and opens or closes the dampers to each zone as required. Only those areas of the home that need heat or cooling receive it. In radiant systems, the controller opens or closes the appropriate valves on the piping system to control the flow of hot water.

Zones can be set to maintain different temperatures if desired. A garage or storage room, for example, might be kept 10 or 15 degrees colder than the rest of the home. The garage might be eliminated entirely from the A/C system. A sun room might be allowed to rise 10 degrees above the rest of the home's A/C level, and in winter to drop 10 degrees below the heated level of other rooms.

FIGURE 12-10   An Automated Thermostat Control

Time-of-Day Programming

Many people prefer lower heating temperatures at night, and a control panel or programmable thermostat in a single-zone HVAC system permits such time-of-day changes. The typical system allows for four temperature changes per day, so a day's program might look like the following:

- 5:30 am – Temperature up to 70 degrees prior to family awakening.
- 8:00 am – Temperature down to 55 degrees while family is at work or school.
- 4:30 pm – Temperature up to 70 degrees prior to family's arrival home.
- 10:30 pm – Temperature down to 65 degrees while family is asleep.

Weekend times and temperatures would probably be adjusted somewhat differently because the family's schedule would be different on those days. Figure 12-10 shows an automated thermostat for a single-zone HVAC system. A control panel has similar inputs for a zoned system.

Seasonal Presets

In addition to timed temperature settings on a daily basis, control panels allow for longer range settings as well. These seasonal adjustments can include temperature changes that vary depending on the time of year and time adjustments for away periods. If the home is not occupied year-round, the HVAC system can be programmed to reduce operation to the minimum needed to avoid freezing or overheating whenever the occupants are not in the home. This can be programmed automatically (using a motion sensor and a time delay—if no motion is detected in the home for 12 hours, the minimum operation cycle starts) or it can be integrated with the home's security system (thus, whenever the security system is armed because the family is away from home, the HVAC system goes to minimum operation status).

Remote Access

The minimal operation function of an HVAC system, whether timed or activated by a security system or sensor, is made better by a remote access feature for the system. Since no one likes coming home to a cold (or overheated) house, being able to contact the HVAC system by telephone or by computer on the Internet enables the homeowner to call ahead so the home temperature is normal when the family arrives. The radiant heating system pictured in Figure 12-8, for example, is in a home occupied only part of the year by a family who otherwise lives on the East Coast. When the owners are away, the HVAC system is on minimum operation. When they arrive at the Salt Lake International Airport, about an hour's drive from Park City, a telephone call and a few command inputs change the home's HVAC system to normal operation so that by the time the owners arrive at the front door, the home is heated normally and ready to receive them.

Water Control System Design

Automated home water systems are rare inside American homes, but millions are installed outside in their yards. This is actually the logical place for them because about 65% of the culinary water (drinking-quality water delivered to users through municipal systems) usage in the U.S. is for lawns and gardens, while only about 35% is used inside the home in bathrooms, laundries, and kitchens. Only a few inside uses of water can be automated. The rest usually require the presence of whoever is using the water, so remote operation is pointless.

Outside watering, however, is ideally suited for automation both because it is time consuming to do manually and because the hours when it is most economical and beneficial to do it are hours when most people prefer to be asleep.

This section describes the design of water control systems that can do all the work of watering the entire yard and do it during the hours between sundown and dawn, thus automating the water system to eliminate both the labor and the timing problem of having to do it with the garden hose.

Zoned Water Systems

Sprinkler systems are usually installed by plumbers who specialize in outdoor watering systems. Sprinkler systems were once almost all made of metal pipe and were both expensive and difficult to install properly. Most new systems are constructed of **PVC** (polyvinylchloride, a hard plastic) pipe that is less expensive than metal and far easier to install. PVC pipe is easily cut with a small saw or a pipe cutter and is flexible enough to be bent into gentle curves. The fittings that bend it around sharp corners and allow sprinkler heads and other attachments to be joined to it are all plastic castings that cost a few cents per piece. Equally important, PVC pipe can be joined by "welding" it with a type of liquid glue that dissolves the surfaces of two parts to be joined and fuses them together in a waterproof joint that is as durable as the pipe itself. Reduced materials expense and easier installation have brought the real cost of a sprinkler system down by half in recent years and made a fully automatic system affordable for almost every homeowner.

Many sprinkler systems are installed in the yards of new homes at the time of their construction, but many others are added later when the homeowner can better afford them or when changing the landscaping creates a convenient opportunity. In either case, many systems are originally installed with manual controls, a series of hand-operated valves that turn various segments of the system on and off. Each valve controls a zone in the system.

As with HVAC systems discussed earlier in this chapter, neither the design nor the installation of sprinkler systems is covered in depth; this chapter concentrates on the control systems that direct them. The design of sprinkler system zones is determined by the two main factors discussed next.

The Role of the Water Service Line

How large an area a sprinkler system zone can cover in a home installation is limited by how much water is available to the home through its **water service line.** The amount of water available is, in turn, determined by the size of the pipe used for the home water line pipe and the minimum amount of pressure forcing water through that pipe. Homes may have a half-inch (inside pipe diameter) water line, a ¾-inch, a 1-inch, or occasionally an even larger line. **Water pressure** in the lines also varies greatly, depending on the community's water system design, the availability of water in the area, and how the water system is pressurized. Home water systems can have water pressure from a few pounds per square inch up to 125 pounds per square inch or even more. If pressure is much above 125 pounds, a relief valve placed in the main incoming water line is a good idea to reduce the pressure to a level that protects water heaters from leaking or bursting.

Water Pressure Counts!

The sprinkler system designer should know the average water pressure for the area and design the system's zones so they are no larger than the maximum area that can be covered using the available water pressure. The maximum number of sprinkler heads per zone depends on the type of heads used and the area each covers, but the total water flow in each zone cannot exceed the capacity of the available water supply. Zones can be smaller than the maximum because the length of time the water is on in a zone can be reduced so it does not receive too much water, but there is no way to get enough water to an oversized zone. Lengthening the watering time will not do it because the water lines cannot carry enough water to make the sprinkler heads reach all parts of the zone. If the designer is in doubt, it is better to err on the side of making the zones smaller rather than too large. That way, coverage can be adjusted with the timer and the yard will not have dry spots.

Not all areas of a yard may require the same amount of water. If an area requires less, it should be placed in a separate zone so the watering time for that area can be adjusted to conserve water. If part of the yard, for example, needs a **drip irrigation system** for flowers and other decorative plants, while the rest requires regular sprinklers for grass, the water system should be zoned accordingly so each type of watering can be adjusted independently. Figure 12-11 shows a sprinkler design for a home with grass and garden areas. The sprinkler system has five zones, one of which uses only small sprinkler heads and waters the garden areas. One zone waters some garden area and some grass, while the other three zones use large sprinklers and water only grass areas.

Interior Zones

A home with a large **atrium** or other inside garden area may require an interior watering zone with the yard sprinkler system. This can be done by having the control valve outside with the other zone valves in the system and running a line

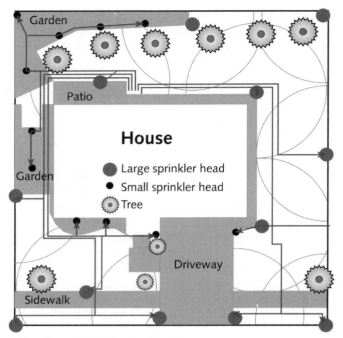

House

- Large sprinkler head
- Small sprinkler head
- Tree

Garden

Patio

Garden

Driveway

Sidewalk

FIGURE 12-11   A Five-Zone Sprinkler System Design

(usually smaller than the PVC pipe used for standard sprinkler lines) into the home through an exterior wall next to the garden area to be watered. Copper tubing is the preferred choice for piping in an interior line. A quarter-inch tube is usually large enough for interior watering, and the line should be carefully sealed where it passes through the outer wall so insects cannot enter through the opening.

Because any failure in the sprinkler system zone inside the home is likely to cause considerable damage, safety precautions should always be added to the control program of such a zone. These should include a default automatic shutoff within a few minutes after the normal shutdown time for the zone, and a **flow valve** or **pressure valve** that shuts off water flow automatically if the amount of water moving in the line exceeds a set level. These precautions can keep an interior watering zone safe from flooding that could damage the home.

Timed Water Systems

All automatic sprinkler systems are timed to regulate the amount of water for each watering zone. If the zones divide the home's yard into landscaped areas of different water requirements, then the system can be timed to correctly water each one. Most water control systems can adjust zone times to the minute so water requirements can be very precisely defined over a period of time and the system set to meet them exactly with no wasted water and no plants withering from insufficient water.

In addition to timing each zone independently for watering, many automated controls allow seasonal adjustments to those times for each zone. Some systems have only three or four seasonal periods to work with, while others allow for monthly or even more frequent adjustments. These seasonal programs allow watering periods for each zone to start short in the spring when heat is low and rain may provide much of the needed water, and then grow longer (in one or two steps or half a dozen) as the year advances to midsummer. Once the hottest days of the year are past, the watering times for the zones again decrease.

Each zone can be adjusted for time upward or downward by a set percentage or a number of minutes in some systems. Others require each program to be entered separately and set for a specific time to be implemented. These take longer to program than those that can be advanced or reduced as a system by a set percentage or number of minutes, but they work just as effectively once the program is entered.

The independent settings available for each zone mean that different watering patterns can also be accommodated. Drip irrigation systems typically do not need to increase their watering times as much in midsummer as sprinklers do because they lose less water to evaporation. Drip systems also require far less water pressure overall than sprinklers and can often be timed to run simultaneously with a sprinkler zone without having an adverse effect on either zone.

Watering Scenes

Drip irrigation zones and surface lawn sprinkler zones are two examples of watering scenes. Like lighting scenes defined as integrated units in a home, as discussed in Chapter 11, watering scenes are simply areas of landscaping outside that are intended to be viewed as a unit. These areas generally have similar water requirements, which can best be met by treating them as separate zones. Many landscaping designs include water features that require special water input, and these should always be set up as scenes with controlled water.

Water features in landscaped yards can be fountains, waterfalls, ponds, or combinations of these. Water features usually use recirculating water that is moved by a pump from the lowest point in the fountain or pond to the highest point and then allowed to flow down. As the water reaches the low point of the feature again, it is lifted back to the top and flows down once more. This recirculation minimizes the amount of water actually used in a water feature and eliminates the need for any continuous outlet or drain through which used water would flow.

Although recirculating water features use far less water than a continuous flow, they still use some, and they need to be "watered" regularly to replace the water they lose through evaporation or outflow. Some features also need to be flushed with a charge of fresh water so they do not become stagnant. Both replacement water and flushing with clean water can be automated as part of a watering system if the water feature is set up as a zone with its own timing and water flow.

Remote Access

Automatic water control systems run mainly on preset timed schedules that require little change over time. The controller for the system is usually located in or near the **valve box** where the wiring to the zone valves is connected. Locating the controller near the valve box minimizes the length of wiring that must be run to each of the valves. However, it is not always a convenient location for programming or accessing the system after installation. Remote access requires only a single cable to connect the system with a computer from which it can be controlled much more easily. Moving access to the computer also means that a distant connection to the system is possible either over the Internet or through a telephone link so that in an emergency the water system can be turned off or rescheduled while the family is away.

Water Control System Components

The main automatic sprinkler system components are the **solenoid valves** that control the watering zones and the central controller on which the watering schedule is programmed. Other components are accessories to these devices. They include sensors and pumps.

Solenoid Valves

Solenoid valves are the main operating device for all automated water systems. They are simple and very reliable valves that turn water off or on by means of a solenoid that can be energized or turned off from the controller. The solenoid is a round electromagnet with a hole in its center in which is placed a metal plunger. The plunger can move within the electromagnet, and its lower end functions as a small valve. Figure 12-12 shows a cross section of a solenoid valve in both the open and closed position.

As shown in Figure 12-12, the solenoid with its plunger (denoted as the solenoid on valve in the figure) is placed directly over the main valve so the pilot valve sits on top of the main valve itself. A small hole through the center of the main valve serves as a pilot orifice for opening and closing the main valve. The pilot orifice is opened and closed by the solenoid's pilot valve. The main valve's operation is started by the pilot orifice opening or closing, but the valve is actually forced open or closed by water pressure in the pipe, thus leveraging the effect of the solenoid, which does not have to be large or powerful to control even a high-pressure valve.

When the valve is closed, the solenoid pilot valve is held down in the pilot orifice by a spring above the plunger. Water from the side orifice pressurizes the area above the main valve and pushes it down into the closed position. This pressure seats the valve firmly and prevents it from leaking. As long as the pilot valve keeps the pilot orifice closed, the main valve remains off.

When the solenoid is energized by the automatic controller, its magnetic force pulls the plunger up into the solenoid's electromagnet center, as seen at the right

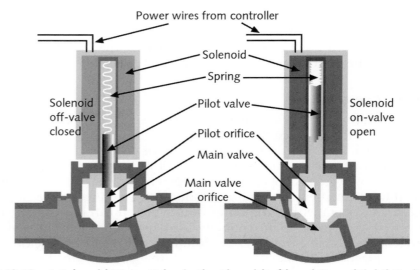

FIGURE 12-12   A Solenoid Water Valve in the Closed (Left) and Open (Right) Positions

in Figure 12-12. This allows water to flow through the pilot orifice to the other side of the main valve. As the water pressure is equalized on both sides of the valve, it opens slightly, and flowing water pushes it open completely, allowing the full pressure of water to flow through the valve. Water continues to flow from the side orifice into the area above the main valve and down through the pilot orifice, but the water pressure above and below the valve is equal and the flowing water keeps the valve open.

When watering is completed, the controller turns off the solenoid and the pilot valve closes off the pilot orifice. Afterward, the pressure above the main valve increases because water cannot flow out through the pilot orifice. The increasing pressure forces the main valve down until it closes and the water pressure above it seals it tightly in place. The valve remains closed until the activated solenoid releases water through the pilot orifice and again equalizes pressure on both sides of the valve.

A solenoid valve installed on each zone pipe of a sprinkler system enables a controller to open and close each valve in a programmed sequence. Solenoid valves are not adjustable for water flow rate. They have only two conditions possible: closed or fully open. Total water flow in each sprinkler system zone must be adjusted by the length of time water flows to the zone, or by a separate hand-set adjustable valve in the zone line that adjusts the amount of water flowing in the line when the solenoid valve opens it.

Controllers

The controller is the electronic device that controls the operation of the valves according to a timed schedule. It functions like an electric clock and calendar combination and signals each zone valve when to commence and end its watering cycle. Early controllers were electric clocks with **mechanical relays** built into them that could be adjusted for the day and time when they were to turn a water valve on or off. Today, all controllers are electronic and activate each zone's solenoid valve by means of a current triggered from a relay built into the unit's circuitry. Figure 12-13 shows a control unit installed on a sheltered exterior wall (on a covered patio) that controls the sprinkler system for the yard.

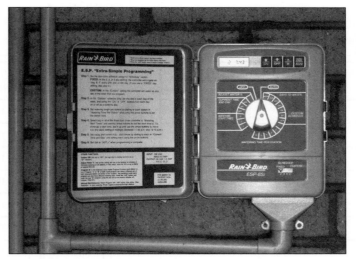

FIGURE 12-13   A Six-Zone Sprinkler Controller

The controller can be set to turn on the sprinkler system at a specified start time on specified watering days. On watering days, the sprinkler cycle begins with Zone 1 turning on at the start time. The controller then turns the zone off after the water has run for a specified number of minutes, and turns on the next zone in succession for its allotted time until all zones have been activated. The controller shown in Figure 12-13 allows two separate programs to be set for each zone so the watering schedule can be adjusted for the season. Two schedules are not really enough to adjust for the full range of watering needs during an entire year, so the controller also allows the watering days to be adjusted from a maximum of every day down to a minimum of once a week.

The controller also allows the timed schedule for each zone to be adjusted as a percentage of the full time. Thus, a zone set to water for an hour (60 minutes) can be set to run at 80% (48 minutes), 50% (30 minutes), or any percentage down to 10%. These additional adjustments provide plenty of flexibility in the watering schedule to meet any need.

If, for example, the yard needs a minimum of 40 minutes of watering per zone every week during the spring and fall of the year, and a maximum of 80 minutes of watering every week during the height of summer, then Schedule 1 could be set for a 10-minute watering time per zone, and Schedule 2 to a 20-minute watering time per zone. By varying the use of each schedule during each week, a wide range of total weekly watering times is possible over a 7-day period. An example of the schedules and total watering times that can be achieved is given in Table 12-1.

Additional variations are possible with this system by adjusting the schedule to run at a specified percentage of the set times. Other controllers have even more programmable schedules available and can be programmed on a monthly or daily calendar.

Standard automatic sprinkler controllers are designed to function independently of any networked home automation system. Once programmed, they function according to schedule without further input from the homeowner or other sensory data.

A number of systems are now available that connect directly to a computer and allow the sprinkler system to be programmed, controlled, and monitored from a PC.

TABLE 12-1
Graduated Watering Plan Using Two Schedules

Month	Day 1	Day 3	Day 5	Day 7	Total Watering Time
April	Schedule 1	Schedule 1	Schedule 1	Schedule 1	40 minutes
May	Schedule 1	Schedule 2	Schedule 1	Schedule 1	50 minutes
June	Schedule 1	Schedule 2	Schedule 1	Schedule 2	60 minutes
July	Schedule 2	Schedule 2	Schedule 1	Schedule 2	70 minutes
August	Schedule 2	Schedule 2	Schedule 2	Schedule 2	80 minutes
September	Schedule 2	Schedule 1	Schedule 2	Schedule 2	70 minutes
October	Schedule 2	Schedule 1	Schedule 2	Schedule 1	60 minutes

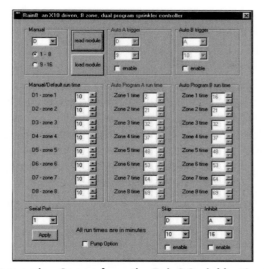

FIGURE 12-14 A Programming Screen from the Rain8 Sprinkler Control System Software

These systems use either a full X10 technology controller or an X10 interface unit that allows the computer to communicate with and program a stand-alone controller. Either of these alternatives provides a greater measure of convenience and control for the water system and also allows for additional input from other sensors that can improve the system's performance and economy.

Rain8 is an X10-based sprinkler control system that can control up to eight sprinkler zones as a single unit and can be networked in multiple units to control up to 256 zones. This system is manufactured by a Texas firm, WGL and Associates, and can be operated in dual modes with sprinkler zone times controlled directly from a PC or with times preset in the controller and triggered by simple X10 commands. The system provides for two schedules in each zone, all of which can be independently controlled and timed.

This system also provides several fail-safe features to eliminate malfunction due to interference or dropped signals in the X10 technology. These include default limits on a sprinkler's on time that the user can set, and a program block that prevents more than one sprinkler zone from being on at a time, which the user can override. Figure 12-14 shows a programming screen from the Rain8 system. Further information on the system is available at http://www.wgldesigns.com.

RCI Automation also offers other automated sprinkler control units that function using UPB signals, WiFi communication, and other technologies. These can control up to eight sprinkler zones and interface with a number of controllers. Such controllers are made by JDS Technologies, Applied Digital, and others. These RCI systems also have the fail-safe features of limiting the on time of any single zone valve (in this case to an 80-minute maximum) and blocking more than one valve from being on simultaneously (again, this feature can be bypassed, if necessary, for a watering program).

By interfacing with an X10, UPB, WiFi, or other controller, all the features of the controller itself become available for use with the watering system. These features include:

• Controls for sprinkler zones using standard X10 commands that are entered through the inside controller

- Control that can be extended to any wireless or remote wired unit with which the inside controller can communicate
- Timed schedules that can be started with a single command from the controller, and which can be entered manually or started at a preset time by the system program
- Watering schedules that can be temporarily suspended by use of a preset code from the controller or sent remotely
- Watering schedules that can be weather delayed by input from sensors connected to the controller

Smart Electronics Corporation offers a graphical interface software and hardware package for Windows-based PCs that allows for complete control of the sprinkler system, plus input from temperature, humidity, and **soil moisture sensors** that can further refine the watering schedules. Details about the system and prices for components are available at http://www.hometoys.com.

Sensors

Sensors for use with sprinkler system controllers are all weather-related devices that sense outdoor conditions and allow the computer/controller to delay or alter the sprinkler system's schedule accordingly.

Rain sensors detect the presence of water in a small holder and signal to the controller. The controller is usually set to respond to this signal by delaying or stopping a scheduled sprinkler cycle. The amount of rain that must be present in the holder to trigger a signal can be set by the user (usually from a minimum of one-eighth of an inch to a maximum of one inch). This is done by simply adjusting the float mechanism in the holder so it lifts only when the desired water level in the holder is reached. Rain sensors need to be placed where they will not be "watered" by the sprinkler system or acquire water from any source other than rain. They need to be in the open, however, so that falling rain can hit them directly. This often means that leaves and other debris can also fall into them. A screen over the collector helps prevent debris from accumulating, but the sensor should also be checked occasionally and cleaned as needed.

A soil moisture sensor is another type of water sensor. It works by measuring the electrical conductivity of the ground in which it is buried. Wet earth transmits electric current better than dry earth and so the sensor can detect the presence of water in the soil by measuring its conductivity. When the amount of moisture brings the current capacity up to a preset level, the sensor sends a signal to the controller. The moisture level can be adjusted on the sensor. It should be placed a few inches in the ground at a well-drained spot in the yard (a location high enough that water from the surrounding area does not drain into it and produce false high-moisture readings for the yard). Since a moisture sensor senses water in only one tiny spot and is used to represent the entire yard in the controller program, its placement should be carefully selected so it is a fair example of the landscaped area.

A freeze sensor is useful for sprinkler systems located in areas where sudden low temperatures during the growing season are possible. Freeze sensors send a signal when the temperature around them falls to near freezing (37 degrees normally, although some can be set to lower or higher temperatures). As with the rain and moisture sensors, when the freeze sensor signal is received by the system controller, the response is to delay or eliminate a sprinkler cycle until the temperature warms.

Finally, a **wind sensor** can be useful for sprinkler systems, especially where water conservation is important. A wind sensor can be set to signal the controller whenever the average wind speed rises to a preset level (adjustable from 12 to 35 miles per hour). If a high wind is blowing, the controller delays the sprinkler cycle to avoid wasting water. A wind sensor should be set in an open location where it can accurately measure wind speed and where unusual gusts of wind will not be directed at it by other obstructions nearby.

Pumps

The majority of automatic sprinkler systems do not include any pumps because the water pressure needed to operate them is provided entirely by the utility supplying water to the home. Some systems, however, may have inadequate pressure for a sprinkler system without the aid of a pump. In these systems, the pump must normally start before the sprinkler system begins its cycle in order to build up enough pressure to operate the system.

Figure 12-15 shows a typical home water system with a pressure pump and tank. For the internal needs of the home, a pressure switch in the main incoming line controls the operation of the pump. For the sprinkler system, another pressure switch can be used in the sprinkler line, or an electronic switch can be installed on the pump so it can be programmed to activate before the sprinkler cycle begins and then stay in operation until the cycle is completed. This assures adequate pressure throughout the sprinkling period.

In the configuration shown in Figure 12-15, the pressure switch turns the pump on whenever water pressure on the delivery side of the switch falls below a set level. But because the pressure switch is sensitive, it turns off as soon as normal pressure is restored. If water use is continued, the pressure again drops and the switch turns the pump on again. This rapid on-and-off cycling can continue indefinitely and is very hard on the pump and the pipe system. An automatic switch that keeps the pump on while the sprinklers are operating gives much smoother service and eliminates unnecessary wear on the pump and motor resulting from frequent starting and stopping.

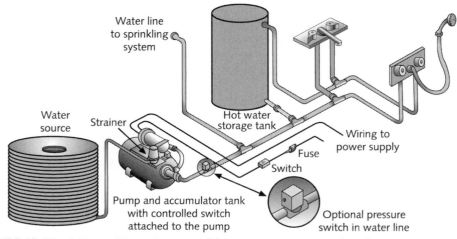

Water line to sprinkling system

Water source

Strainer

Hot water storage tank

Wiring to power supply

Fuse

Switch

Pump and accumulator tank with controlled switch attached to the pump

Optional pressure switch in water line

FIGURE 12-15   A Home Water System with a Pressure Pump and Tank

Water Control System Installation and Programming

Whether an automatic control system is installed as an original part of a new sprinkler system or added to an existing system, the installation procedure is almost identical. This section describes how to install the controller and wiring for the control system and how to program its operation.

Locating the Controller

The controller for the sprinkler system must be located where it can be easily wired to each of the zone valves. If the zone valves are located in one place, then the controller can be sited near them in a sheltered position and at a height that is convenient for working at the controller box to program it. Most controllers have lids that can be closed to protect the inside electronics from the weather, but most do not lock.

If a non-locking controller is used, it should be placed in a location where children or potential vandals will not find it easy to tamper with. This may mean placing it inside the house and running the control wiring outside to the valves in a conduit. At the least, the controller should be placed in the back yard, out of view from the street and in as sheltered a position as possible. If a locking box protects the controller from tampering, a sheltered location is still preferable, but it can be more openly positioned for convenience in wiring.

Because a wiring path from the controller to the valves is necessary, the controller should be placed so wiring from it to the ground will not have to pass through concrete patios or sidewalks, or other obstructions that will make the installation more difficult. Before installing the controller, check the path its wires must follow to the zone valves to be sure they can connect without major excavation or the moving of landscaping. If the valves are located in two or more places, both paths should be as convenient as possible. It is often easier to relocate the controller box than to overcome serious wiring obstacles.

The controller also requires a power source, usually 24-volt AC that can be wired from an outside power outlet or light. The controller comes equipped with a step-down transformer to reduce the voltage from 110 to 24, or one can be purchased as an accessory to go with it. Built-in transformers are part of the controller box, but separate ones may need an enclosure (covered junction box or similar) placed near the controller box and connected to it by a conduit for the wiring. The only power connection needed for a sprinkler system is the one to the controller box. Power to operate the solenoid valves is supplied through the wires connecting them to the controller.

If the controller is to be interfaced with a computer or inside control device (X10-type or wireless), then a wired or wireless path from the inside control device to the outside controller is also required. This will be a two-wire connection using telephone wire or a wireless transceiver connected to the outside controller by a similar two-wire connection.

Low-Voltage Wiring

Because it is outside and mostly underground, the low-voltage wiring from the controller to the valves should always run in conduit. Most jurisdictions allow PVC pipe to be used for this purpose and require that the system be grounded only

FIGURE 12-16   Sprinkler System Controller with Wiring Conduits to Power Supply (Left) and to Valve Box (Right)

at the controller. Some codes require metal conduit and an additional ground at the valves. Before installing the wiring, check what local requirements apply in your area.

The left photo in Figure 12-16 shows a controller box installed on a home's rear wall with wiring conduit running to it from a nearby outlet box and from the controller up to the underside of the overhead balcony. The right photo in Figure 12-16 shows the conduit extending down the spiral stairway support and underground to the nearby valve box. By going overhead to the outside edge of the patio, the wiring avoided the necessity of passing through the concrete slab or running on top of it.

Two wires connect each solenoid valve to the controller. All these wire pairs can be run through a single conduit, but they must be carefully color-coded or labeled so they can be correctly connected at both ends. The wires all carry AC current so there is no required order for wiring each pair, but the correct wires for each zone must connect to the zone's location in the controller or else the controller will be programmed for the wrong zones.

The power connections to the controller and to the remote access unit, if there is one, are also AC and can be wired with a single pair of wires connected in any order. All controllers have a built-in switch that turns power to the unit off so no separate power-line switch is necessary.

Once the wires have been run to the controller and from it to the valves, all the connections can be made according to the labels or color codes of the wires. The valves should always be located in a valve box, such as the one shown in Figure 12-17. This type of open-bottom box provides protection for the valves and a cover that keeps water away from the electrical connections. Figure 12-17 shows the valve box with its lid removed and the valve wiring connections made with twist-on connectors. These should only be used if the wiring will not be submerged in water. If there is any danger of the wiring getting wet, either from the sprinklers or from outside weather, the connections should be soldered and carefully taped to keep moisture out.

FIGURE 12-17   A Five-Zone Sprinkler System Valve Set in a Valve Box

Programming

After the sprinkler system is wired, it should be tested by manually activating each zone to be certain all zones are wired correctly. If the sprinkler system is a new installation, this testing can also extend to checking that all the sprinklers function correctly. The sprinklers can also be adjusted for area coverage. Once the system checks out, it can be programmed to function automatically.

Time-of-Day Settings

The first programming for the system is to set the time of day and the calendar date, if the controller has one. All the zone times key from the base time setting in the controller.

The next scheduling selection is to choose the overall watering schedule, the days on which the sprinkler system activates. Controllers can be set to activate the system every day, every other day, or another schedule custom-programmed by the user. Whatever schedule is selected, the final input for it is to select a "day one" from which the system times its schedule. If the schedule calls for watering every 3 days, for example, and Sunday is selected as day 1, the system activates on Sunday, then on Wednesday, then on Saturday, and the following week starts the schedule on Tuesday, the third day after the preceding Saturday.

The set schedule usually applies to all zones. Most controllers do not permit individual watering day schedules for zones. Those that have multiple schedules available, however, can be programmed so half of the zones function on one schedule and half on another, or a similar arrangement. All the zones will actually function on both schedules, but the watering times of those not wanted on a schedule can simply be set to zero for that schedule. The system then bypasses them and activates only those zones with set times. On days when the second schedule runs, the zones set to zero time in the first schedule are set to normal watering times, and those that are activated in the first schedule are set to zero and bypassed.

Seasonal Presets

Controllers that can be programmed with multiple schedules usually also have calendar timers that allow each schedule to be set for activation on a specific date. Schedule 1 might activate starting April 15, for example, and Schedule 2 on June 1.

Other schedules, if available, could have other activation dates. In most controllers, activating a second schedule does not deactivate the one already running. Each schedule needs an off date, as well as an on date, to avoid overlapping.

Zones

With the daily schedule and seasonal presets established, all that remains to complete the sprinkler system programming is to set times for the individual zones. This is a somewhat trial-and-error process. The system should provide enough water for all zones, but not too much. Finding that level for each zone may take several time adjustments during the first year of the system's operation.

Setting watering times can be assisted by testing how much water the sprinklers distribute in each zone. This can be done while the system itself is being manually tested for functionality. A flat-bottomed plastic bucket can be placed in the center area of each zone and allowed to fill from the sprinklers to a depth of one inch (using a translucent bucket with the 1-inch depth marked on the outside and a few drops of food coloring in the bucket to color the water makes the measurement easy).

Once you know how long the sprinklers require to distribute an inch of water, you can compare this amount with the weekly or monthly water needed for lawns in the local area. Then, time the sprinklers in each zone to stay on for the proper length of time to get the amount of water required. This procedure allows you to set the zone times close to the ideal. If dry spots show up as the season progresses, the times can be adjusted upward. If areas of the yard appear to stay too wet, some downward adjustment may be in order.

Summary

- Forced-air heating systems heat by moving warmed air from a furnace unit to the rooms of the home. Radiant heating systems heat by circulating hot water from a boiler to radiators or pipes beneath the floors of the home, from which warmth radiates to the interior air. Home heating systems are gas or oil fueled.

- Central air-conditioning systems are all forced-air designs. They move cooled air from a cooling unit to the rooms of a home to reduce the room air temperature. Most A/C systems are electric powered, and in addition to the inside cooling unit have an outside condenser/compressor unit that compresses the refrigerant and dissipates its heat.

- HVAC systems that are divided into zones can better heat and cool a home. The zone controls enable the system to overcome the effects of climate, sun heating, and uneven heat loss in a home, and also provide for varying temperatures in each zone, if desired.

- HVAC components consist of furnaces that provide heat, air conditioners that cool air, air handlers that move air, thermostats that signal temperature changes, sensors and dampers that control the operation of the system, and controllers that automate the operation of the HVAC system.

- HVAC systems are installed when a home is built. Control systems can be installed then or later. In either case, the thermostats and dampers are wired to the control panel, as are the other sensors and controls in the system.

- HVAC control panels can be programmed for automatic operation of all zones in the system. They also permit time-of-day programming, seasonal presets, and remote access to the system by telephone or Internet connection.

- Automatic water systems are usually installed by plumbers who specialize in sprinkler systems. They can be automated at the time of installation or later.

- Sprinkler systems are divided into zones, which are areas within a yard that require similar water amounts and are small enough to be watered by a group of sprinklers controlled by one valve.

- A sprinkler system can be programmed to time each zone independently. This allows for customized watering schedules for specialized zones such as water features and interior garden areas.

- Sprinkler system zones are controlled by solenoid valves that are electrically operated and are activated individually by signals from a central controller.

- A sprinkler controller is an electronic device that controls the operation of the solenoid valves according to a timed schedule. It functions like an electric clock and calendar combination and signals each zone valve when to commence and end its watering cycle.

- Controllers can be programmed to function automatically or they can be interfaced with a computer or remote control device for programming and control.

- Home water systems without sufficient water pressure may need a pressure pump installed as an accessory for a sprinkler system. Other system accessory devices include rain sensors, wind sensors, soil moisture sensors, and freeze sensors.

- Sprinkler systems are wired with single-pair telephone wire. Underground and outdoor portions of wiring should be placed in plastic or metal conduit according to local code requirements.

- Programming for controllers includes time-of-day and calendar-date settings, seasonal presets, zone times, start times, and multiple schedule settings (if these are used for the individual system).

Key Terms

Air conditioning (A/C) The part of an HVAC system that provides cool air to lower the indoor air temperature.

Air handlers The mechanical devices that move air in an HVAC system—mainly blowers and fans.

Atrium An area within a home that is open to the outside or walled in by glass, where plants can grow as in the outside yard.

Automated control A method for controlling HVAC or water systems by preset programming rather than the manual setting of controls.

Blower A device that uses a rotating bladed wheel to move air in an HVAC system. Not the same thing as a fan.

Boiler A device for heating water to be used in heating a building. Not the same as a water heater.

Central heating system The part of an HVAC system that provides warm air from a central furnace to raise the indoor air temperature.

Circulating pump A device in a radiant heating system that continuously circulates the heated water through the boiler and heating pipes.

Compressor The part of an A/C system that squeezes the refrigeration gas under pressure into a smaller volume until it converts into a liquid.

Condensation The process in an A/C system by which the refrigeration gas changes back into a liquid as it cools. Also, the process by which water vapor is removed from cooled air and deposited as a liquid.

Condenser coil A coil through which hot refrigerant passes and is cooled by flowing air, similar to how an automobile radiator cools engine coolant.

Condenser unit The part of an A/C system in which the refrigerant is converted back into a liquid from a gaseous state.

Controller The electronic device that controls the operation of sprinkler system zone valves through a preset timed schedule.

Cooling coil The part of an A/C system that cools the surrounding air by the expansion of the refrigerant in the coil, which absorbs heat from the air.

Dampers Devices that act like doors in ducts. They can be closed to block air passage or opened to allow it.

Drip irrigation system A type of irrigation system in which small tubes deliver water flow directly to individual plants or plots of ground without spraying. The system conserves water compared to spraying sprinkler systems.

Duct A metal tube or open-ended box though which air can flow. Can be rigid or flexible and of any size.

Evaporative or "swamp" cooler A type of air cooler that works by evaporation of water into air, which is thereby cooled as it passes through the cooler.

Fan A device consisting of a balanced set of angled blades on a shaft spun by a motor. Fans move air and are one type of air handler.

Flow valve A type of valve that measures the amount of water flowing through it and can adjust that flow automatically.

Forced air A type of heating or A/C system that works by blowing heated or cooled air into a home through ducts and air vents.

Freeze sensor A device that sends a signal when the temperature around it falls to near freezing (37 degrees normally, although some can be set to lower or higher temperatures) so that action can be taken to prevent freezing.

Furnace A device for heating air so it can be blown into the interior of a home.

Heat transfer The process by which heat radiates from warmer objects to cooler ones.

Heating, ventilation, and air-conditioning (HVAC) systems Systems that provide controlled heating, cooling, and ventilation in buildings.

Humidifier A device that adds moisture to an incoming air stream.

Humidity Moisture contained in air. All air contains some water vapor (gaseous water). The amount can vary with the pressure and temperature of the air.

Humidity detector A device that measures the amount of moisture in the air around it and signals that data to a controller or computer.

Mechanical relay An electrical device activated by an electromagnet being energized that functions as a switch to turn a control current on or off; an automatic remote-controllable switch.

Pilot light A small flame that burns continuously in a furnace to ignite the furnace's main burner when it is turned on.

Pressure valve A type of valve that shuts off water flow automatically if pressure in the pipe drops below a set minimum, indicating there is a leak or break in the pipe.

PVC Polyvinylchloride; a hard strong plastic used to make pipes and many other products.

Radiant A type of heating system that works by warming a home with hot water that flows to radiators in each room or in pipes beneath the floors. The radiators or floors then warm the interior air.

Rain sensor A device that detects the presence of rain water in a small holder and signals to the controller when a preset level is reached.

Refrigerant fluid A gas/liquid that circulates in an A/C system, continuously extracting heat from the surrounding air as it changes from a liquid to a gas and back again.

Refrigeration A method of cooling air that uses the heat absorption of expanding gas to extract heat from the surrounding air.

Safety control valve A specialized valve on the fuel line of a furnace or other appliance that must receive a continuous electric current from a thermocouple in order to stay open. If the current stops, the valve closes, shutting off the fuel flow.

Soil moisture sensor A device that can measure the electrical conductivity of the ground where it is buried. This measurement allows it to detect the amount of water in the soil.

Solenoid valve An automatic valve operated by an electromagnet that can be energized or turned off to open or close the valve. It is the main control device in automated water systems.

Space heater A small heater, usually electric, but sometimes gas fired, designed to heat a small area or room. Manually controlled.

Spark igniter An electrical device that ignites the fuel in an oil or gas furnace when it starts.

Temperature sensor A device that signals the actual temperature of the air surrounding it to an HVAC controller or a computer.

Thermocouple A safety device that produces electric current when heated. Used to shut valves on gas or oil lines if the furnace fails to start or run properly.

Thermostat A special type of temperature sensor that sends a signal when the air temperature reaches a preset level.

Valve   A device for controlling the flow of fluids (gases or liquids) in a pipe. Can be manually or automatically controlled.

Valve box   An open-bottom box sunk to its lid in the ground where the zone valves and sometimes the controller of a sprinkler system are located.

Vent   An opening in a wall or floor through which air flows from a connected duct into a room.

Ventilation   The process of exchanging air in a confined room or space. Ventilation exhausts the old air out of a room and brings new air in.

Water pressure   The force that impels water through a pipe. Water pressure is produced by gravity (the weight of water pushing down from a higher elevation

to a lower one) or by artificially pressurizing a sealed water system, usually with compressed air.

Water service line   The pipe that supplies water to a residence from a public utility. It enters the home and connects to the interior plumbing and exterior water system.

Wind sensor   A device that can be set to signal a controller whenever the average wind speed rises to a preset level (adjustable from 12 to 35 miles per hour). Used to shut off sprinkler systems in high-wind conditions.

Zone   In sprinkler systems, multiple sprinkler heads in an area that are controlled by one valve. In HVAC systems, a part of a home separately heated, cooled, and controlled by one thermostat.

Review Questions

1. A space heater is a device that _____.

   a. Heats water in a shallow container to provide humidity

   b. Heats a small area or a single room

   c. Fits into a small space

   d. Heats objects placed in it

2. The two main types of central heating system are _____ and _____.

3. A radiant heating system heats using _____.

   a. Reflected sunlight

   b. Hot water pipes beneath the floors

   c. Large windows that trap heat from the sun

   d. Heated bricks placed around the home

4. The two main types of air handlers are _____ and _____.

5. How does refrigeration work?

6. If a home has more than one furnace or A/C unit, it also has more than one _____.

7. What is heat transfer?

8. A thermostat and a thermometer are the same thing. True or false?

9. The function of a damper is to open or close a _____ so air can pass or be blocked.

10. Furnaces that do not have pilot lights are started by a(n) _____.

11. A boiler and a circulating pump are both devices that would be found in what type of system?

    a. Air conditioning

    b. Ventilating

    c. Radiant heating

    d. Forced air furnace

12. What does the compressor in a condenser unit compress and why?

13. A thermostat consists of a temperature sensor and a(n) _____.

14. Blowers are used in large HVAC systems rather than fans because fans make more _____.

15. An HVAC control panel is connected by home run wiring to which of the following?

    a. Thermostats

    b. Dampers

    c. Both of the above

    d. Neither of the above

16. Home sprinkler system zones should be sized according to how large an area the home water system can water at one time. True or false?

17. The total amount of water flowing to a water system zone is usually adjusted by changing the _____.

    a. Area of the zone

    b. Length of time that water flows

c. Size of the pipe to the zone

d. Time of day for watering

18. A solenoid valve is operated by _____.

a. A motor

b. A spring

c. An electromagnet

d. Air pressure

19. To reduce the amount of water going to a zone, the solenoid valve for that zone can be set to 50% open instead of fully open. True or false?

20. How many zones in a controlled sprinkler system can be on simultaneously?

a. One

b. Two

c. Three

d. Four

21. A(n) _____ sensor could be used to help control a standard sprinkler system, but would not be helpful with a drip irrigation system.

22. How would a freeze sensor function in an automatic sprinkler system?

23. Most automatic sprinkler controllers operate on _____ power.

24. Wiring for sprinkler system components requires what type of wire?

a. 110-volt AC cable

b. Cat5 cable

c. Single-pair telephone wire

d. Armored cable

25. The first item that should be programmed on a water system controller is the _____ setting, while the second item is the _____ setting, if the controller has one.

Hands-On Projects

Project 12-1: Wire a Thermostat

In this project, you wire a thermostat to an electric light that serves as a simulated furnace. To complete the task, you will need a thermostat (standard or automated), a 9-volt electric light in a socket with connector screws, a 9-volt battery in a holder with connector screws, and some single-strand insulated wire to make the connections. You also need a hair dryer to test the project when you have finished wiring it.

1. Remove the battery from the holder until you have finished wiring the thermostat.

2. Strip a short length of wire at both ends and connect the positive lead of the battery holder to one connector on the light socket.

3. Strip the ends of a second length of wire and connect the negative lead of the battery holder to one connector screw on the thermostat.

4. Strip the ends of a third length of wire and connect the other connector screw on the thermostat to the open connector on the light socket.

5. Set the thermostat to 70 degrees (or just above the present room temperature) so it is on. Place the battery back in the holder. The light should go on because the thermostat is on.

6. Plug in the hair dryer and turn it on low heat. Direct the dryer's air stream on the thermostat until it opens and the light goes off.

7. Immediately turn off the dryer and see how long it takes for the thermostat to close and turn on the light.

Project 12-2: Program an Automatic Thermostat

In this project, you program four time-of-day settings with corresponding temperatures into an automatic thermostat. To complete your task, you will need an automatic thermostat (Dayton Fuel Trimmer Model T-110 or similar).

1. Open the back of the thermostat's case and insert a 9-volt battery to power it. The thermostat can be programmed without being connected to an HVAC system.

2. Set the Heat/Cool switch to Heat. Set the Master Switch to On.

3. Press the Set Day button. Use the Time Forward or Time Back button to set the day to the number (1 through 7) of today. Monday is 1, Sunday is 7.

4. Press the Set Clock button. Use the Time Forward or Time Back button to set the correct hour of the time. Be sure you get the correct setting for AM or PM.

5. Press the Set Clock button again. Use the Time Forward or Time Back button to set the correct minutes of the time.

6. Press the View Program button. The program opens to the first setting.

7. Advance or back up the time to 5:30 am using the Time Forward or Time Backward buttons.

8. Use the Up Arrow and Down Arrow buttons to set the temperature to 70 degrees.

9. Press the View Program button. The program opens to the second setting.

10. Advance the time to 8:00 am using the Time Forward button.

11. Use the Down Arrow button to set the temperature to 55 degrees.

12. Press the View Program button. The program opens to the third setting.

13. Advance the time to 4:30 pm using the Time Forward button.

14. Use the Up Arrow button to set the temperature to 70 degrees.

15. Press the View Program button. The program opens to the fourth setting.

16. Advance the time to 10:30 pm using the Time Forward button.

17. Use the Down Arrow buttons to set the temperature to 60 degrees.

18. Press the Run Program button to close the program. You can view the program by pressing the View Program button repeatedly to advance through the settings.

Project 12-3: Wire a Thermostat to a Damper

In this project, you wire a thermostat to a motorized damper that could be used as a zone control in an HVAC system. To accomplish this, you will need a thermostat (standard or automated), a 24-volt motorized damper, a 24-volt output AC transformer with connector screws, and some single-strand insulated wire to make the connections. You also need a hair dryer to test the project when you have finished wiring it.

1. Unplug the transformer until you have finished wiring the thermostat and damper.

2. Strip a short length of wire at both ends and connect one lead of the transformer to one connector on the damper.

3. Strip the ends of a second length of wire and connect the other lead of the transformer to one connector screw on the thermostat.

4. Strip the ends of a third length of wire and connect the other connector screw on the thermostat to the open connector on the damper.

5. Set the thermostat to 70 degrees (or just above the present room temperature) so it is on. Plug in the transformer. The damper should open because the thermostat is on.

6. Plug in the hair dryer and turn it on low heat. Direct the dryer's air stream on the thermostat until it opens and the damper closes.

7. Immediately turn off the dryer and see how long it takes for the thermostat to close and open the damper.

Project 12-4: Wire a Controller for a Zone Sprinkler System

In this project, you wire an automatic sprinkler controller to control a four-zone sprinkler system. To complete the task, you must use a light to simulate each of the sprinkler system zone valves so you can wire the system in a classroom rather than outside in a yard. For this project, you will need an automatic sprinkler controller (Rain Bird ESP 6Si or similar) with a 24-volt transformer, a standard 110-volt AC plug, four 24-volt lights in sockets that can be wired individually, 20 small twist-on wire connectors, and a roll of telephone wire (single-pair).

1. Connect the pair of wire leads for Zone 1 that extend out of the bottom of the controller to the connector posts or screws on one of the lights. Add a length of telephone wire to both of the colored leads before connecting them to the light so the wires are long enough to place the light 2 or 3 feet from the controller. Use twist-on connectors to connect the new length of wire to each lead from the controller, and then connect the new wires to the light.

2. Repeat Step 1 to connect another light to the Zone 2 leads, then Zone 3 and Zone 4. Finish one zone completely before starting the next so you do not get the wire leads mixed up.

3. Attach the two power leads from the controller to an 8-foot length of telephone wire. Connect the other end of the wire to the 24-volt transformer's output wires.

4. Connect the input wires of the transformer to a standard AC plug that can be plugged into a 110-volt AC outlet.

5. Set the controller and the connected lights on a table and plug the power cord of the transformer into a wall outlet. When the controller is plugged into its power source, the display window should begin to blink.

6. Test the wiring by performing the following steps:

   a. Set the twist dial to the Auto position.

   b. Set the Schedule switch at the bottom of the controller to the 2 position.

   c. Press the Manual Start (Man Start) button in the top-right corner of the controller once. The Zone 1 light should go on. If it does, proceed with the test. If it does not, unplug the controller and check the wiring to the lights.

   d. Press the Manual Start button again. The Zone 2 light should come on and the Zone 1 light should go off.

   e. Press the Manual Start button again. The Zone 3 light should come on and the Zone 2 light should go off.

   f. Press the Manual Start button again. The Zone 4 light should come on and the Zone 3 light should go off.

Project 12-5: Program a Controller for Date, Time, and Schedule

In this project, you program an automatic sprinkler controller for the correct date and time. You then program start times for a watering cycle and test the program. For this project, use the same sprinkler controller and wired lights you used in Project 12-1.

1. Plug the controller's 24-volt transformer into a power outlet.

2. Set the Schedule Switch to the Fixed/2 position. This sets the watering schedule to water every 2 days.

3. Check the display window on the upper-left side of the controller. The display should say PGM A (Program A) to the left of the time and day display. If it says PGM B, press the A/B button once to change it to PGM A.

4. Turn the controller dial to the Current Time and Day/Day position.

5. Press the Up Arrow or Down Arrow button at the top of the controller to set the day. This controller does not have a calendar function, so the day is either Day 1 or Day 2. Set the timer to Day 1.

6. Turn the controller dial to the Current Time and Day/HR (Hour) position.

7. Press the Up Arrow or Down Arrow buttons at the top of the controller to set the correct hour (be sure to include AM or PM in your setting).

8. Turn the controller dial to the Current Time and Day/MIN (Minute) position.

9. Press the Up Arrow or Down Arrow button at the top of the controller to set the correct minutes.

10. Turn the dial to the Watering Time Per Station/1 position. Use the Up Arrow button to set the watering time at 1 minute.

11. Repeat Step 10 for each of the other three stations. Set all the watering times for 1 minute.

12. Turn the dial to the Watering Start Times/1 position. Use the Up Arrow button to set the watering start time a few minutes ahead of the current time. The start times can only be adjusted in 15-minute increments, so set the start time to at least 3 minutes, but not more than 18 minutes, ahead of the current time.

13. Turn the dial to the Auto position and wait to see if the controller turns on each of the lights at 1-minute intervals when the time reaches the start time.

14. If the test does not work correctly, follow the preceding steps again to reset the program. Check carefully to be sure you are setting each function correctly. Test the program again.

Project 12-6: Program a Controller for a Dual Schedule

In this project, you program an automatic sprinkler controller for a dual schedule. You then test the dual program to see if it functions correctly. For this project, use the same sprinkler controller and wired lights you used in Project 12-2.

1. If the Program A you entered in the controller in Project 13-3 is still not in place, repeat the Project 13-3 steps to enter that Program A into the controller, then continue with this project.

2. Verify that the Schedule Switch is still set to the Fixed/2 position. This sets the watering schedule to water every 2 days.

3. Check the display window on the upper-left side of the controller. If the display does not say PGM B to the left of the time-and-day display, press the A/B button once to change it to PGM B.

4. Turn the dial to the Watering Time Per Station/1 position. Use the Up Arrow button to set the watering time at 1 minute.

5. Repeat Step 4 for Station 3. Set its watering time for 1 minute.

6. Repeat Step 4 for Stations 2 and 4. Set their watering times for 2 minutes.

7. Press the A/B button once to change back to Schedule A.

8. Repeat Step 4 for Stations 2 and 4 in Program A. Set their watering times for 0 minutes.

9. Turn the dial to the Watering Start Times/1 position (this sets the start time for Program A). Use the Up Arrow button to set the watering start time a few minutes ahead of the current time. The start times can only be adjusted in 15-minute increments, so set the start time to a time at least 3 minutes, but not more than 18 minutes, ahead of the current time.

10. Turn the dial to the Watering Start Times/2 position (this sets the start time for Program B). Use the Up Arrow button to set the watering start time to be 15 minutes ahead of the time you set for Program A to start. Both Program A and B should start on Day 1, 15 minutes apart. This enables you to test both programs one right after the other. In an actual sprinkler system, the two programs would run on different days.

11. Turn the dial to the Auto position and wait to see if the controller turns on each of the lights at 1-minute intervals when the timer reaches the start time.

12. If the test does not work correctly, follow the preceding steps again to reset the programs. Check carefully to be sure you are setting each function correctly. Test the program again.

Case Projects

Case Project 12-1: Design Zones for an HVAC System

You have been asked to design the zones in a home HVAC system. The home is a two-story house with an attached double garage. All four bedrooms are upstairs, two of them over the garage. The living room, family room, and kitchen are downstairs. The owner would like heat available to the garage, but not air conditioning. The rest of the home should have heat and A/C. A maximum of four zones are available in the HVAC system. Explain how you would configure the HVAC zones so as to provide even heating and A/C in the home and heat only in the garage.

Case Project 12-2: Diagnose a Malfunctioning Heating System

In a large home with a six-zone HVAC system, Zone 3, which covers three upper-floor bedrooms, is consistently cold during cold nights. Setting the thermostat for this zone to a higher level does not help. The rooms' temperature does not rise at all. Some heat is getting to these rooms because they are definitely warmer than the outside air, but not enough to make them comfortable. When you check the radiant heating system in the utility room, you find that the water pipe leading to Zone 3 is cool to the touch, while all the other zone pipes are hot. From what you know about heat transfer and zoned heating systems, which piece of HVAC equipment would you check first to solve this problem? Which piece would you check next and why?

Case Project 12-3: Reprogram a Sprinkler System to Meet Restrictions

The town where a home with a sprinkler system is located is suffering from a drought. To conserve water, the town council has restricted lawn watering to a maximum of 2 hours on each of 4 days of the week only (Monday, Wednesday, Friday, Sunday). The home's eight-zone sprinkler system is set to operate every third day, but it requires 4 hours to complete its full eight-zone cycle (30 minutes per zone). How would you reprogram the sprinkler system so it would be in compliance with the water restrictions, none of the zone watering times would be shortened, and the system would make maximum use of the available watering times?

Case Project 12-4: Solve a Water Deficiency Problem

A house that was recently purchased has a sprinkler system installed in the yard, but the new owner finds that it does not work very well. There is not enough water pressure to operate the six sprinkler heads included in each of the system's four zones. The low pressure is caused by the small (half-inch) water service pipe that comes into the home. It just cannot carry enough water for six sprinkler heads. The homeowner finds that the size of the incoming pipe cannot be enlarged because a local ordinance specifies a half-inch line as the maximum allowed for a residence. The homeowner wants to put all new landscaping in the yard, but is reluctant to do so without the sprinkling system. Hand-watering would be very time consuming. Can you suggest a way that the sprinkling system could be altered so it would work with the available water pressure? (Hint: A pressure pump will not help because the small water line cannot supply it with any more water than it does the sprinkler system. If there is no way to get more water, you must use less at one time.)

Power Supply and Conditioning

After studying this chapter, you should be able to:

Introduction

In this chapter, you will learn about the various disturbances that can affect the **alternating electric current** that powers a home network, as well as all the other electric lights and appliances in the home. You will also discover how **direct current (DC)** disturbances can affect the data transmission circuits in computers, networks, and other types of electronic equipment. You will learn to use test equipment to determine the nature of **power disturbances** and interference

417

in network circuits and identify the causes of these disturbances. Finally, you will find out how to install **power-conditioning** components that block or suppress power disturbances and how to find other solutions to power-conditioning problems when hardware devices are not a viable option.

Types of AC Electric Power Disturbances

Alternating current or AC, the type of electric power in almost all homes, is the most useful and most vulnerable utility service we rely upon. The average American home has more than 150 electric devices within its walls, the majority of them operating on the home's AC current (the remainder operate on battery power, which is direct current or DC, discussed in the next section of this chapter).

AC power comes into our homes from generating facilities over a grid of power transmission lines built up during more than a century. The grid includes high-voltage lines that transmit power from the generation point to **transformers** near the point of use. These reduce or "step-down" the voltage of the current, usually in several phases, until it is at the 220-volt level required for home service.

Most AC-powered devices are designed specifically to operate on the U.S. standard for home electric power, which is 60 **cycles** or **hertz (Hz)** at 110 volts. A few devices, such as electric stoves, air-conditioning condenser/compressor units, and electric clothes dryers, require more power than can be conveniently delivered on a 110-volt AC circuit. These devices use 220-volt, 60Hz power. Conversely, some devices in the home do not need electric current as high as 110 volts to operate at peak efficiency. These devices, such as doorbells, thermostats, electronic controllers, and low-voltage lighting, typically operate on 24-volt AC power.

Since some home appliances require 220-volt power, the home's **electric service** must include incoming power at that voltage. The 220-volt service is brought into the home as two 110-volt "legs" of power. Both legs are combined to power the 220-volt circuits for the appliances that need this level of current. The 110-volt circuits for the rest of the home's electric power use only one leg of the service line and thus have 110 volts of potential. 24-volt appliances and lights must use a **step-down transformer** that reduces the 110-volt power to the lower voltage. A step-down transformer can be built into a device or it can be installed in a 110-volt circuit so low voltage devices can be connected to it.

Most power disturbances that affect home electric service result from a single fact: AC electric current cannot be stored. It cannot be stored at the generating facility where it is created and it cannot be stored in the home to await future use. All the electricity being used in any home at this moment was generated less than a second ago and transmitted from its source to the point of use. The use in any home can vary from a few hundred watts during the late night hours to several thousand watts on a hot evening when all the electric appliances and air conditioning in the home, as well as its lights, are operating.

But whatever the need, when a switch is thrown to activate any circuit, the current for that circuit must be generated and sent to it at that moment, neither later nor sooner. Ideally, the AC current must be at exactly 110 volts potential and exactly 60-Hz **frequency,** regardless of the load on the circuit or the distance traveled by the current to reach it. Imagine for a moment the variations in demand for electricity that must be met from second to second in each home and collectively throughout the nation's **electric grid** that serves over 125 million homes, plus millions of additional business and industrial users.

TABLE 13-1
Power Interruptions and Their Duration

Type	Duration (Cycles)	Duration (Seconds)
Instantaneous	0.5 to 30 cycles	0.008 to 0.5 seconds
Momentary	30 to 120 cycles	0.5 to 2 seconds
Temporary	120 to 12,120 cycles	2 to 120 seconds
Sustained	More than 12,120 cycles	More than 120 seconds (2 minutes)

How, you may well ask, is it possible to maintain the exact electric standard at all times on the total electric grid throughout the country and in each individual user home or business? The short answer to that question is that it is not possible 100 percent of the time. And therein is the source of most power disturbances. The six major types of disturbance are discussed in the following sections.

Power Interruptions

A **power interruption** is a loss of current flow: the voltage in the circuit drops from 110 volts to 0 and electricity stops moving through it. Interruptions can be short or long and are divided into four categories by the electric industry: instantaneous, momentary, temporary, and sustained. Table 13-1 shows the length of each type of interruption in current cycles (60 per second for U.S. AC) and in actual time.

Interruptions, regardless of length, are usually caused by a problem in the generation or transmission of the electric power. Since AC power cannot be stored, any interruption of its generation or transmission instantly results in a loss of power in the user circuit.

Damage to the transmission grid is the most common cause of interruptions. The damage can be caused by high winds, falling trees, vehicle accidents, lightning strikes, equipment failure, or a host of other events. Utility companies try hard to prevent these interruptions by keeping the grid infrastructure as strong and robust as possible, and by making it **redundant** (having more than one transmission route over the grid to every user, so if one fails, the other will still function). Nevertheless, interruptions do occur. Their length is usually determined by the cause and how resilient the grid is where the problem exists.

If the grid is openly redundant at the point of damage, there will not be an interruption because the current simply continues to flow in the undamaged part of the grid and still reaches the user. If the redundant line must be switched into service when the primary line is damaged, the utility's automatic switching equipment may still be able to make the move in less than a half second, making the interruption instantaneous, but nothing more. This is important for most networks and electronics equipment that operates on low-voltage DC current stepped down from higher-voltage AC. Most computer power supplies are designed to bridge an instantaneous interruption without losing their 5-volt DC power output. This means they will not lose any data in transmission or in memory. The instantaneous interruption may result only in a wavering of the computer monitor (which is AC powered at 110 volts and will react to even a half-second power loss).

Momentary or longer interruptions will shut down computers and electronic equipment causing the loss of data in transmission or data stored in memory. Data being recorded on magnetic media or laser devices (CD and DVD) will be corrupted or lost. To avoid such losses, as well as possible damage to software and hardware not properly shut down, power interruptions that cannot be prevented must be overcome by other means.

Spikes, Surges, and Other Transients

A **spike** in an electrical circuit is a sudden increase in the voltage level and the amount of current flowing. A spike happens very fast (from its start to the peak voltage of the spike may be only 5 **nanoseconds**) and is of a very short duration (typically about 50 nanoseconds from start to end). A slightly slower and longer-lasting voltage increase is often referred to as a **surge.** Both spikes and surges are actually **impulsive transients**—sudden short bursts of electrical current that find their way into an electrical circuit. Figure 13-1 shows how a spike looks on the normal sine wave of AC current.

Impulsive transients are the most damaging type of power disturbance so far as electronic equipment is concerned. Though they are short events, spikes and surges can load a circuit briefly with an extra 10,000 volts of electricity for a fraction of a second. That kind of power amplification (100 times the 110-volt circuit's normal voltage, or more) can damage or destroy components by literally burning them up or melting them into useless blobs.

Spikes and surges have many causes. Among them are lightning strikes, poor grounding of electronic equipment, and **electrostatic discharge (ESD).** Lightning is the most damaging cause of spikes because it can easily put 50,000 volts into a spike and keep it at that level for a second or more. No electronic hardware can endure that amount of current without serious damage or total destruction.

ESDs can also produce voltage spikes of 50,000 volts or more, but they are of shorter duration than lightning-produced spikes. Even so, they can do heavy damage when introduced into a poorly **grounded** device. ESDs are caused by the buildup of a static charge on the surface of any conductor. When the charge has developed to a high enough voltage, it will **arc** across a nonconducting gap

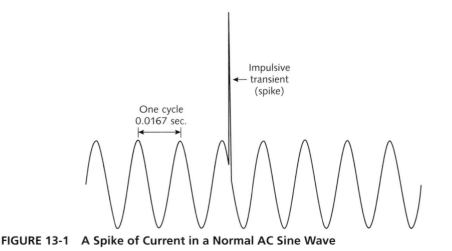

FIGURE 13-1   A Spike of Current in a Normal AC Sine Wave

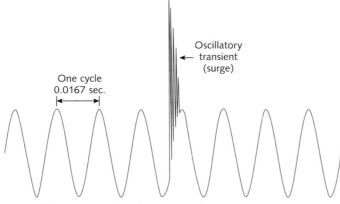

One cycle
0.0167 sec.

Oscillatory
← transient
(surge)

FIGURE 13-2   An Oscillatory Transient in a Normal AC Wave

to reach a ground. The arc is an ESD that travels from the object on which the charge exists to a grounded object that discharges the voltage. If the properly grounded object is an electronic device, the ESD will pass through it to ground by the shortest least-resistant available path—the ground connection—usually doing no damage to the equipment as it does so. If the ground connection is poor, however, the ESD will invade any conducting path in the equipment circuitry, seeking a better route to ground, and will instantly damage or destroy any components it encounters that cannot tolerate its voltage peak.

Electrostatic charges can be produced by feet walking on carpet, hands rubbing over smooth furniture surfaces, a balloon rubbed on a wall, wind passing over an object, a spinning fan blade, or many other sources. An electrostatic charge is harmless; it is the discharge of that charge when the object holding the charge comes near an uncharged object that does the damage. To protect electronic equipment, ESDs must be prevented or they must be grounded in a manner that keeps their power away from all electronic circuits.

Another type of transient power disturbance is an **oscillatory transient.** This type of transient occurs when an electric motor or similar device that is connected to a circuit is turned off. When power to the motor is switched off, it slows down and finally stops rotating, but during the slowdown it acts as a **generator,** temporarily creating electric current and pushing additional voltage into the circuit. That added current produces an oscillatory transient, as shown in Figure 13-2.

Another cause of oscillatory transients is a change in the current flow of the circuit. Flowing electricity acts somewhat like water flowing in a pipe. If a faucet is suddenly turned on, the water pressure in the pipe drops momentarily because a new outlet for it has been created. Similarly, if a faucet is suddenly turned off, you may hear a hammering sound caused by the water pressure suddenly rising in the closed pipe. In either case, the water pressure will quickly return to normal and no damage will result, but an oscillation has occurred.

Electricity also resists changes in a circuit's current, either greater or reduced flow. Greater flow momentarily reduces the voltage (electrical pressure), while reduced flow momentarily increases it. The result is an oscillatory transient of increased or decreased voltage. These transients are usually short, 10 **microseconds** or less, and they typically are not as high-voltage as impulsive transients, but they can still do considerable damage to electronic circuits.

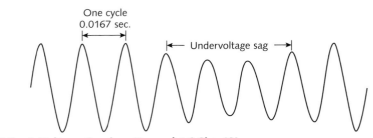

FIGURE 13-3   A Voltage Sag in a Normal AC Sine Wave

Undervoltage Sags

A **sag** is a period of reduced voltage in a circuit that lasts from 0.008 second to 1 minute. It is caused by turning on equipment that requires a large starting current. A large electric motor, for example, may draw five or even six times as much current when starting as it does when running at speed. Starting a home air-conditioning compressor motor can easily produce a sag in the circuit it is connected to and also on other circuits supplied by the same service line. So much power goes to the starting motor that the voltage available to the rest of the home's circuits drops. Again, electricity acts like water in a pipe: if all the faucets in the house are turned on at the same time, the pressure in the pipe will drop. Figure 13-3 shows how a sag affects AC current.

A sag can also be caused by a **fault** outside the home electrical circuits. Such faults may include heavy equipment starting up elsewhere in the electric grid, a grounded line that drains away current from the grid, or heavy current loads that are switched onto the grid. Whether the source of the sag is in the home or comes from outside, it can cause damage in electronic equipment if not addressed.

A sag that lasts longer than a minute is often referred to as a **"brownout"** and can be even more damaging than a shorter sag. Electric motors and transformers are designed for the specific voltage they operate on. If the voltage stays too low for a significant time, these devices draw more current (amperes) as they try to maintain speed or output. The increased current flow causes them to overheat and burn out quickly.

Overvoltage Swells

A **swell** is the opposite of a sag: an increase in a circuit's voltage that lasts from .008 second to 1 minute. Swells are caused mainly by sudden large load reductions in a circuit, such as when a heating device (hair dryer, space heater, stove, etc.) is turned off. As current demand drops, the electricity flowing in the circuit resists the change and the voltage swells upward.

Swells can damage electric motors and transformers by causing them to overheat just as sags do. In the case of a swell, the increased voltage forces more current through these devices than they were designed to carry, causing them to overheat and burn out rapidly. Swells have the same overloading and overheating effect on electrical contacts, semiconductors, and electrical insulation. A swell will quickly burn out most electronic circuits and may also create a fire hazard from degraded insulation. This is especially true if the swell lasts longer than a minute or becomes a constant **overvoltage** condition.

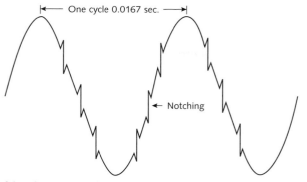

FIGURE 13-4   Notching in a Normal AC Sine Wave

Waveform Distortion

A graphic representation of normal AC current flow shows a smooth sine wave oscillating at 60 cycles each second. No bumps or variations occur in the wave form and no changes arise in the shape of the wave, its frequency, or its strength. **Waveform distortion** disturbs the characteristic wave form by adding or subtracting bits of current from outside sources. Several kinds of waveform distortion can disturb AC current, but the two most likely to affect electronic equipment using the AC power are **notching** and **noise.**

Notching gets its name from the fact that it produces small sharp breaks or notches in the wave form of an AC current. The notches are tiny increases or decreases in voltage that break the smooth flow of the wave, as shown in Figure 13-4. They are extremely short, usually less than 1 microsecond (.001 seconds) in length, and the voltage change they cause is generally less than 10 volts, but they occur frequently (often 300 or more times per second) and can cause data corruption or loss in electronic equipment. Notching in a home environment is usually caused by light dimmers and similar devices connected to the AC power.

Noise in an AC circuit is caused by unwanted outside voltage being added to the current normally flowing in the circuit. The noise voltage is a higher frequency than the AC current and rides along on top of the AC waveform making the smooth wave look instead like a jagged series of sharp peaks and valleys. Figure 13-5 shows what noise looks like compared to a normal AC sine wave.

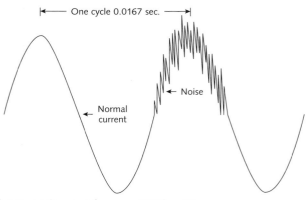

FIGURE 13-5   Noise Superimposed on an AC Sine Wave

Sources of noise include radio transmitters of all kinds, arc welders, and electronic control circuits that function intermittently. Noise in the AC power system can cause distorted video displays, hard disk drive failure, and data errors in both memory and transmission.

Other types of wave distortion include **DC offset,** in which a constant DC current is added to the AC current in the circuit by induction from an outside source, and **harmonic AC currents,** which are also induced in the circuit by some power supplies, electric motors, and similar equipment. The harmonic currents are at frequencies that are multiples of the regular AC current frequency (120Hz, 240Hz, 480Hz, etc.) and they distort each cycle of the AC sine wave by adding voltage to it at regular intervals.

DC offset and harmonic wave distortion can both damage electronic equipment or corrupt data, but neither is likely to be a problem for the home network or its connected nodes simply because the equipment that produces these distortions is not likely to be found in a home environment.

Voltage and Frequency Variations

Voltage variation in an AC circuit means a continuing random series of voltage changes in the current up to 5% above or below the normal voltage. It is caused by rapid and random load changes on the circuit, such as those produced by arc welders, arc furnaces, arc lights, and similar devices. Voltage variation is rare in home electrical service unless the residence is located near industrial facilities with the type of equipment that produces it.

Even rarer is **frequency variation** in AC circuits. This is usually caused by variations in the AC generator speed that, in turn, are caused by large rapid changes in the load placed on the generator. Unless a home is operating on an emergency generator or a relatively small utility system, frequency variation is not a significant risk. Even if this is the case, most electronic equipment is somewhat tolerant of frequency variation and will not be adversely affected. Only electric motors and devices that rely on the steady AC frequency for their timing (speed controls, clocks, timers, etc.) will overheat or run inefficiently due to frequency variation.

Types of Electromagnetic Interference

Even more widespread than AC power disturbances are several forms of **electromagnetic interference (EMI)** that can adversely affect the performance of networks and electronic equipment. EMI can originate anywhere within the whole range of electrical signal frequencies, from analog audio signals (20-Hz to 20-kHz) through radio frequency (RF) signals, to television signals, all the way up to infrared signals at the beginning of the visible light spectrum. EMI that occurs in radio frequency wavelengths is referred to as **radio frequency interference (RFI).** Other wavelengths of EMI signals include those below the AM radio band (over about 1,000 feet in length) and those above the television broadcast bands (less than an inch in length). This wide range of wavelengths makes EMI easy to receive, conduct, and transmit over numberless paths in the world of electronic signals. EMI has multiple sources and travels by both **conduction** and radiation at the speed of light.

Sources of Radio Frequency Interference (RFI)

The three basic sources or causes of Radio Frequency Interference (RFI) are:

- Spurious emissions from a transmitter
- The effects of fundamental **overload** in a device
- Externally generated spurious signals or intermodulation

These RFI sources fall into two general categories: intentional and unintentional. The intentional sources produce RF signals that are intended for reception by specific users who have antennas and receivers matched to their signals. These intentional signals become RFI, however, when they are also received by other antennas and devices that are tuned to them by chance rather than design.

Spurious Transmitter Emissions

Intentional RF interference is produced by AM and FM radio stations, as well as shortwave, CB transceivers, amateur transmitters, and any other communication device that transmits a radio frequency signal.

All these transmitters produce broadcast signals intended for reception by specific tuned receivers, but these intentional signals can also be received by other devices close to the transmitter or that are tuned in some manner to the transmitter's frequency.

All transmitters also generate some signals outside their primary frequency. These signal transmissions are called spurious emissions. Like the transmitter's primary frequency broadcast, they can be picked up not only by the intended receivers, but by any device near enough to the transmitter or that is tuned in some manner to one (or many) of the transmitter's spurious emissions.

Spurious emissions can be harmonics (signals that are a multiple of the transmitter's primary broadcast frequency) or mixing frequencies caused by the process used in most transmitters to generate the primary transmission frequency. Mixing frequencies are usually close to the primary frequency or a harmonic frequency, but are not as strong as either of these. Finally, transmitters can produce noise—random-generated signals at any frequency—that is of a lower strength than the primary or harmonic frequencies. Noise can be caused by several factors in a transmitter.

Manufacturers of transmitters are required to reduce (but not entirely eliminate) spurious transmissions from their products, and most do a creditable job of doing so. But some transmitters may develop RFI problems after years of use, while others, even though they comply with required RFI reduction standards, may still produce enough interference at some frequencies to cause problems for nearby networks.

Fundamental Overload RFI

Most RFI is not caused by spurious signals but rather by the primary or fundamental broadcast frequencies of transmitters that simply overload networks or other electronic equipment. These fundamental frequency broadcasts are intentional outputs of the transmitters, but are intended for reception only by radio receivers tuned to their frequency. Properly designed electronic equipment should be able to select only a desired signal and reject others (or reject all RF signals), but today literally thousands of transmitted signals are constantly being propagated across the frequency spectrum and not all equipment is up to the task.

A strong fundamental transmitter signal may simply overpower the filtering or **shielding** capacity of network equipment and force its way into the system. If blocked by the shielding of the equipment on the network, RF signals may be picked up by the network wiring (if not properly shielded), by the AC wiring that powers the equipment, by a telephone line connected to the equipment, or by a coaxial cable that connects the network to outside Internet, video, or audio services. Even if partially filtered, enough of a strong fundamental signal may find its way onto a network to seriously interfere with data transmission.

RF Noise

Although it is an unintentional form of RFI, almost all RF transmitters produce some noise as a byproduct of their signal output. Harmonic and mixing RFI are not generally regarded as noise because they occur on specific frequencies related to the transmitter's fundamental frequency. Noise, on the other hand, can be generated on any frequency, and at almost any strength, although it is usually far below the fundamental or harmonic signal strengths. Noise can result from defects in transmitter components, from poor wiring connections, or from the function of amplifiers.

Noise is generally low power but can be produced on any frequency and by almost any transmitter. This means it can be mobile, moving into and out of proximity to a network, causing interference when the source is close, but none when it moves away. Modern technology has resulted in hundreds of transmitters being placed within or near most home networks. The following list of RF transmitters gives an idea of their number, but it is not complete. Any of these is capable of producing RFI.

- Police and other emergency transceivers
- Business, taxi, and other two-way radios
- Military communications
- Amateur radio
- Cellular phones
- Land mobile transmitters and repeaters
- Pagers
- Marine radios
- Aircraft radios and radar systems
- Wireless motion detectors and remote control devices

Other sources include wireless phones, cell phones, commercial and government radio systems, microwave ovens, aircraft and marine radios, radar systems, motion detectors, remote controls, and literally thousands of other commercial, industrial, and medical devices that generate radio waves.

Unintentional Sources of EMI

In addition to transmitters, unintentional sources of EMI that may be within or outside the radio frequency spectrum, are most commonly those electrical devices that produce a spark when operating. A spark is produced whenever any two contacts with different electrical potential come close enough to one another that the current can jump the gap between them and equalize the voltage potential.

This happens whenever a switch is closed as the two contacts in it come together. Just before they touch, the gap between them narrows to the point where the voltage difference between the contacts forces the current across the still open distance, creating a tiny spark for perhaps a microsecond. That spark is a potent generator of EMI in an array of frequencies and strengths that scatter out from the source, conducted through any wiring connected to it and also radiated into the surrounding space from any conductor that can serve as an antenna.

Besides ordinary light and appliance switches wired into every home, common spark generators of EMI include electric welders, motors with brush contacts, relays, spark plugs in internal combustion engines, and neon or fluorescent lights that are not properly functioning. Other devices with switches that produce EMI include light dimmers, TV and computer CRT displays, and switching power supplies (used in most electronic equipment). This means the source of EMI in a home LAN may be in the same room or building as the LAN. Indeed, it may originate in one or more of the LAN's nodes.

In addition to the AC power disturbances they cause, as discussed in the previous section, arcing electrical equipment (such as arc welders, arc lights, and similar devices) also produces EMI. Any conductor that develops a sufficient electrostatic charge that it arcs to discharge its potential to ground produces EMI as it does so. This type of unintentional EMI source includes power-line insulators, static discharge protectors, and lightning rods. Lightning, because of its immense power, produces huge EMI over long distances whenever it strikes the ground or connects one cloud to another in the sky.

Listed next are some of the more common sources of EMI. This list is intended to show the scope of EMI sources and to help in identifying these sources. It is by no means complete.

- Doorbell and low-voltage light transformers
- Toaster ovens
- Electric blankets and heating pads
- Fans
- Light dimmers
- Appliance switches
- Water bed and aquarium heaters
- Furnace and air-conditioning system controls
- Ultrasonic pest control devices
- Fluorescent, neon, and mercury vapor lights
- Electric fences and ground sensors
- Lightning arrestors
- Alarm systems and smoke detectors
- Electrical toys such as model trains, especially if radio-controlled

Finally, EMI can originate in space, from the sun, from some of the other planets in the solar system, from more distant celestial objects that generate RF waves, and even from the background radiation that permeates all space.

Of these space-based EMI generators, the sun is by far the most potent. Electromagnetic storms on its surface throw off enormous amounts of radiant energy, some of which strikes the earth. Its effects, in addition to exciting the earth's atmosphere into displays of the aurora borealis, include the disruption of radio and television broadcasts and the dispersion of RFI around the globe.

EMI is more disruptive to digital transmission on a network than most forms of AC power disturbances. This is because RFI is composed of DC energy, which is the same as the low-voltage DC current that carries network data signals. When EMI invades a network or piece of electronic equipment, it mimics the legitimate signals being transmitted on the net or through the circuits of the equipment. The result is corrupted data, lost data, slower performance by the network as it is forced to resend lost or damaged data packets, and in extreme cases, the complete shutdown of the system because it becomes confused by the spurious signals being received and does not know how to correct the problem.

Power-Conditioning Components

Residential power-conditioning components can address almost all of the power disturbances described in the previous section. A number of companies manufacture power-conditioning equipment and every component can be purchased in a variety of price ranges and quality standards. Because the risk of damage to a home LAN and its associated electronic equipment is relatively high (about one-third of home networks and 31% of home computers have been damaged at least once by power disturbances of electromagnetic interference), it is important to install high-quality power-conditioning equipment that will adequately protect the expensive equipment in a home network and a digital home automation system.

The two fundamental quality considerations in power-conditioning equipment are conditioning efficiency and speed. The component must correct power disturbances to well within the tolerances of the equipment it is protecting and it must act rapidly, often very rapidly, to make the corrections before the protected equipment is damaged. As noted previously, some power disturbances rise to full strength within a 5-nanosecond window of time and almost all are at or near their peak within 200 microseconds. High-quality conditioning components act within those time constraints to bring the AC power back within acceptable parameters. Most other features in such equipment are conveniences to enhance their ease of use and keep the user informed of their status, but they do not materially affect the unit's performance.

The three basic types of power conditioners are **whole house surge suppressors, single-circuit surge suppressors,** and **voltage regulators.** A typical example of each is described in the following sections, including their major specifications, performance levels, and added features. The descriptions are based on units manufactured by American Power Conversion (APC) Corporation, but several other companies manufacture similar devices.

Whole House Surge Suppressors

A residential whole house surge suppressor is the first line of defense against damaging electrical surges and spikes that originate outside the home and travel into it through the home's AC service line. The suppressor is large enough in capacity to accommodate the total peak load of electricity coming into the home, but it blocks power disturbances that are outside normal AC parameters.

Whole house surge suppressors must be installed at the home's service/circuit breaker panel and only by a qualified electrician. The suppressor safely reduces the severity of power transients caused by utility accidents, power interruptions, and lightning.

Whole house surge suppressors are equipped with **fuses** and **thermal disconnects** that react quickly to protect against dangerous high- and low-current wiring faults, as well as other catastrophic surge events. The panel-mount suppressor is a passive unit. This means it continues to allow normal AC power through to the home if any of the surge protection components are damaged, but the unit will no longer protect against new power disturbances until the damaged components are replaced.

Residential suppressors have typical maximum surge current ratings of from 120 **kiloAmps (kA)** to 160kA per phase of home AC power. These maximum surge current ratings provide protection from all but the largest (and rarest) surges and spikes. Units with larger surge current ratings are available, but are not cost effective for most home installations.

The suppressor also provides noise filtering that attenuates EMI and RFI line noise that may be present in the AC wiring. It does not address EMI or RFI that enters the home through other wiring or by broadcast transmission or noise that originates inside the home.

The APC suppressors have several convenience features including remote monitoring capability that allows for configuration of a remote alarm or indicator via terminal block connections, LED status indicators for each phase of power the suppressor is controlling, and an audible alarm to provide immediate failure notification when suppression circuitry has been damaged by a heavy strike or surge. The units also have a surge event counter that counts the power transients the suppressor has kept from reaching the home.

Single-Circuit Surge Protectors

A single-circuit surge protector, or line protector as it is also known, provides reliable surge protection for a home LAN or home office computers and electronic equipment that are connected to AC power through a single outlet. The protector is plugged into the AC outlet and computers and other network devices are plugged into the protector's multiple outlets.

Single-circuit protectors should be used to protect network and other electronic equipment, whether or not the home AC service includes a whole house surge suppressor. The whole house suppressor protects only against power disturbances originating outside the home and reduces or suppresses these only to a level that is within the tolerance of the home wiring system. Its suppression is not necessarily sufficient to entirely protect sensitive electronic circuits from damage. The single-circuit protector provides additional protection against the residual effects of outside power disturbances, as well as first-line protection against disturbances originating inside the home.

These smaller everyday power disturbances can degrade the performance of electronic equipment and decrease its useful life. The single-circuit protector provides a high level of protection for computers, electronics, and connected devices, and in addition also provides surge protection for phone lines and incoming coaxial cables. Its circuitry offers protection against large power disturbances and multiple surge events.

A series of surge suppressors manufactured by American Power Conversion Corporation include a number of features in addition to those noted earlier. Many other companies also manufacture surge suppressors, some of which have these additional features, while others do not. The APC suppressors have instantaneous reaction to lightning strikes and wiring faults. If a protector's surge components are damaged due to a power spike or overvoltage, excess power is still prevented from reaching connected equipment. Once the circuit has been compromised, the unit disconnects equipment from the power supply. Some surge suppressors do not have this feature and will continue to let power through even after the circuit has been damaged, thus leaving equipment exposed to other damaging surges.

Other features of the APC single circuit protector include an overload indicator that illuminates whenever connected equipment draws more power than the unit's rated capacity, noise filtering circuitry that attenuates EMI/RFI line noise, eight or more widely spaced outlets that allow block power supply units to be connected without covering adjacent outlets, a phone line splitter, and status indicator LEDs.

Voltage Regulators

A voltage regulator provides automatic voltage regulation for protection against **undervoltage** sags and overvoltage swells. These **voltage fluctuations,** if uncorrected, gradually degrade the life of electronic components, potentially causing their premature failure. The voltage regulator not only adjusts voltages to safe levels, but also provides surge protection against electrical surges and spikes, including lightning. It automatically steps up low voltage and steps down high voltage to levels that are suitable for connected equipment and has a **resettable circuit breaker** that prevents heavy power surges and spikes from reaching connected equipment. Unlike suppressors, which typically have thermal fuses that must be replaced after an overload event damages them, the regulator's resettable circuit breaker allows easy recovery from overloads at the flip of a switch.

The regulator unit is connected to a single circuit, usually in place of a suppressor, although both can be placed on the same circuit for additional protection if needed. The average home-use regulator has a 10-amp capacity, but larger units are available. It also has status indicator LEDs that display input voltage levels (low, normal, or high) and widely spaced outlets that permit bulky transformer block plugs to be connected without covering up other outlets.

Uninterruptible Power Supplies (UPS)

While surge suppressors and voltage regulators address most of the AC power disturbance problems that may affect a home network, neither can provide protection against the most common and most potentially damaging disturbance: power interruptions. For protection from power interruptions, instantaneous or sustained, an **uninterruptible power supply (UPS)** is required.

UPS units are battery-powered backup power supplies designed to sense an AC power interruption and quickly and smoothly switch the electronic equipment connected to them from the interrupted AC circuit to the UPS's battery backup power. The switchover permits the network and its connected equipment to be shut down in an orderly manner without data loss. Alternatively, if the UPS has the **run-time** capacity, the network can be kept running on the backup battery power for up to 24 hours, or until normal AC power is restored.

All UPS units use batteries to supply their backup power. Unlike AC power, which cannot be stored, DC power (or the potential for producing DC power) can be stored in limited quantities in batteries. All batteries are DC devices that provide direct current output and must be charged or recharged with direct current input.

Since the UPS unit functions on an AC power circuit and must supply AC power as output to its connected equipment, two **power conversions** are required in the UPS. Incoming AC power is first converted to DC current that is used to charge the UPS battery pack. When the UPS detects a power interruption, it reverses the process and converts its stored DC power back to AC power and then outputs it to keep its connected equipment running.

UPS units are available from several companies and come in a variety of sizes. The smaller units provide 200 to 300 watts of 110-volt AC power output for a typical run time of perhaps 3 minutes. This is just enough power to run a single computer and one or two peripherals, and just enough time to quickly close any running software, save open files, and shut the equipment down in a normal manner.

These small UPS units provide the minimum backup power, and minimum time, to save a computer and its network from damage or data loss due to a power interruption. They also typically provide surge protection and voltage regulation for the AC power they are connected to, up to the limit of their rated capacity. If a computer is connected to a UPS, even a small one, it probably does not need any additional surge protection or voltage protection, although a whole house surge protector is still a good investment in this situation.

Larger (and more expensive) UPS systems have the same basic functions and protections as the small units, but offer larger AC outputs and longer running times. The increased output is achieved by adding more and larger batteries to the UPS configuration. Some UPS units, like the APC example described next, can have their output expanded by the addition of external battery packs. These can be purchased after the original unit as required to protect a growing system.

The APC Smart-UPS XL 750VA is a midsize UPS that provides power protection with scalable runtime for computers and data networks. The base unit provides up to 600 watts of AC power for a run-time of 50 minutes at 300 watts and 17 minutes at 600 watts. With up to ten additional battery packs, run-time can be extended to 24 hours or more and optimized for the specific application. With included management software for servers and workstations, users can provide safe system shutdown and advanced UPS management. The unit connects to a network computer through a serial or USB port. Other standard features on this unit include automatic self test, automatic voltage regulation, wiring fault indicator, hot swap batteries, load meter, network-grade line conditioning, overload indicator, replace battery indicator, and LED status displays.

Power Conditioning Installation and Setup

Before installing power-conditioning equipment, it is a good idea to evaluate the risk of power disturbances in the home. The higher the risk of power problems, the more robust should be the conditioning equipment used to combat its effects on the home LAN.

According to a study by IBM, an average of about 120 power disturbances hit a typical computer each month. Most are minor events and have little or no adverse effect. A few cause data loss, expensive repair, or both. Among the factors affecting the likelihood of power disturbances in a particular home environment are the following:

- How frequent are power interruptions (any length) to the home's AC service? More than three outages per year is above average, while 12 or more constitutes a severe risk.

- Older homes with aged wiring are more prone to power disturbances than newer ones. On average, homes more than 10 years old are slightly more at risk; those more than 30 years old are moderately at risk; and those more than 50 are severely at risk.

- Overloaded wiring is also a risk factor for power disturbances. If wiring has not been upgraded in homes over 25 years old, the risk of overloading is high. It gets higher with increasing age over 25 years.

- Power distributed on poles above ground is about ten times more likely to have disturbances than underground power distribution because the latter is not as subject to weather hazards, car accidents, fires, and other natural events.

- Industrial equipment that uses large amounts of electricity can cause power disturbances for nearby residential users. The closer a home is to industrial facilities, the greater the risk of disturbances from these sources.

- As much as half of power disturbances are caused by the power environment within the home. Shop equipment, welders, copy machines, compressors, vacuum systems, and many other electric appliances can contribute to the problem. The more electrical appliances a home has, especially those that are heavy power users, the greater the risk of power disturbances.

- The further power must travel to the home from the generation point or regional distribution center, the more likely power disturbances become. Rural homes are most at risk from this factor, but the risk is proportional to distance from the power source for all homes.

- Lightning is the most apparent cause of power disturbances. The number of lightning storms that occur each year in areas of the U.S. varies widely. Check Figure 13-6 to see the risk in your area.

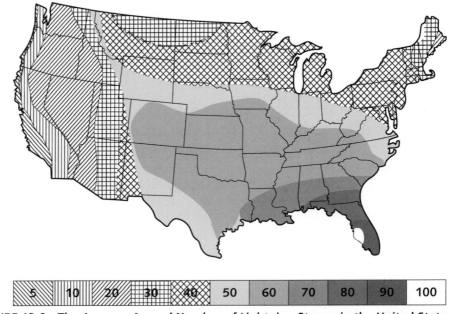

FIGURE 13-6   The Average Annual Number of Lightning Storms in the United States

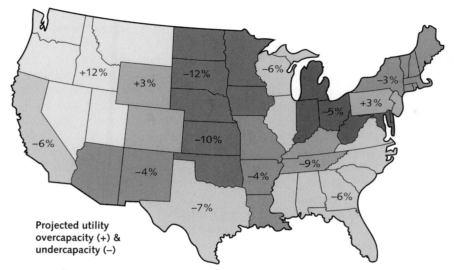

FIGURE 13-7 The Estimated Power-Generating Undercapacity in the United States

- Electric generating capacity is presently strained to meet the power demand in the United States, a fact that increases the risk of power disturbances. The undercapacity varies in different parts of the country. Check Figure 13-7 to see how severe the problem is in your area.

Installing a Whole House Surge Suppressor

If a home is located in an area of the country subject to 30 or more storms per year, a whole house surge suppressor is a must for protecting the network and computer equipment. These suppressors cost from $90 to $250, depending on their maximum suppression capability. Installation will require at least 2 hours and must be done by a qualified electrician. Installation may take more time and cost more if service panel changes or wiring upgrades are needed, but the suppressor is still a good investment for the homeowner. It is the single most important power-conditioning device for areas that have significant lightning or other outside power disturbance risks.

Installing a UPS

Once the whole house surge suppressor is installed, evaluate what additional power conditioning is needed. If the risk of power interruptions is moderate or higher (more than three interruptions per year), a UPS unit should be considered to provide backup power for the network and critical equipment.

Unless the home network is a business as well as a personal site, backup power to keep it running throughout a sustained power interruption probably is not necessary. A UPS unit that will provide backup power long enough to shut down the network and other equipment in an orderly manner is probably sufficient.

To determine the size of the UPS needed for the network, calculate the total amount of watts of power needed to keep critical equipment functioning to allow an orderly network shutdown. Usually, this equipment will be only the main network computer, the network modem, and powered bridges or switches, if any. If the network has a server separate from the computer's hard drives, it may need to be included, too, but only if it is needed for data storage in the event of an emergency shutdown. Peripherals such as printers, scanners, and so on do not need to be connected to the UPS.

When the required UPS output has been determined, a unit can be selected that will provide the backup power needed for the amount of time necessary to shut down the network. The time needed for a manual shutdown can be determined by the user. Start with a worst case scenario that includes several software applications running and files open in memory. Time the shutdown procedure and add a safety factor of 50% or at least 3 minutes, whichever is longer. Use this manual shutdown time calculation to determine the amount of run-time the UPS should have. Use the same figure even if the UPS has automatic shutdown software included. The automatic shutdown time will not be longer than the manual, and the manual procedure should always be available as a backup for the automated one.

If the home network is left running unattended for significant periods of time, the UPS unit should have a connection to the network's main computer on which software is installed that will initiate an automatic shutdown and save any at-risk data in the event of a power interruption. Without such management software, a UPS offers far less protection because, in the event of an interruption, its backup power run-time may expire before anyone can manually shut down the network. The result will be a network crash with the same data loss and risk of damage to equipment as would have occurred without the UPS.

When installed, a UPS unit becomes part of the home's AC electric system, which is a high-voltage system. It is important to set up the UPS correctly and be certain the home's AC wiring is properly connected to accept it. To install a UPS unit on a residential circuit, perform the following steps:

- Unpack the UPS unit and connect its batteries to the unit's internal wiring. Most UPS units are shipped with their batteries disconnected so they will not discharge during shipment.
- Place the UPS unit so it is at least 18 inches from any display screen, which will ensure that its magnetic field will not affect them. Connect the UPS unit to the AC circuit via a wall outlet.
- Check to see that the wiring of the AC circuit does not have any faults. Some UPS units have a built-in wiring fault indicator that illuminates if the unit is connected to a faulty circuit. Faults in the wiring could be lack of a ground, reversed polarity (hot and neutral wires reversed), or an overloaded neutral. Any fault should be corrected before proceeding further.
- Connect all the AC hardware units that are to be backup powered by the UPS to its AC outlets. If the unit has outlets that are surge-protected only (no backup power), connect only noncritical hardware to these outlets. Do not exceed the load capacity of the UPS.
- If the UPS has a phone or modem connection, connect this hardware to the unit.
- If the UPS has a ground connection, attach any additional ground wires from other hardware or protection devices on the circuit to the ground.
- If the UPS has a data line, connect it to the LAN's main computer and install the control software that comes with the unit on the computer.
- Follow the instructions that come with the UPS unit to power it up and configure it for the individual LAN system. Batteries in the UPS will need to be charged for several hours before it is capable of providing full backup power.

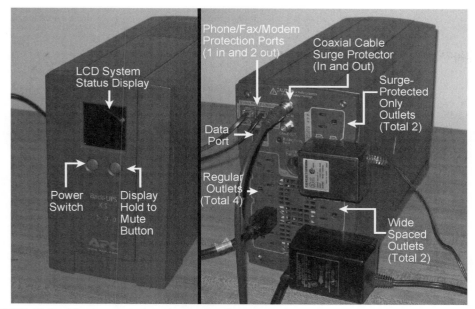

FIGURE 13-8   The Front and Back Panels of a UPS Unit

Figure 13-8 shows the front and back panels of a UPS unit with hardware cables partially installed. Individual connections and control features of the unit are labeled.

It is not practical to connect a UPS to equipment in more than one room, so if there is a risk of data loss on other computers on the home LAN, each may need its own UPS unit. Again, only critical hardware needs to be connected through the UPS. If the automatic shutdown software on the main computer has the capability to control other computers and UPS units over the network, it should be programmed to shut down each of them prior to shutting down the network and the main computer.

Installing Single-Circuit Surge Protectors

When a UPS has been installed on a computer system, no other power conditioning is necessary. UPS units provide surge suppression and voltage regulation in addition to their power backup function.

On computers that do not have backup power protection, a single-circuit surge protector is a good idea for protection against power disturbances originating inside the home. A second reason for using a protector is that high-quality suppressors also filter EMI and RFI from the AC power line. This further reduces the risk of data loss or corruption from another source.

To get the full benefit of its protection, all of a node's hardware (computer, printer, scanner, etc.) should be connected to AC power through the protector, up to the limit of its amp rating. Protectors should not be overloaded, even if they have sufficient outlets to allow it, because they have circuit breakers that will trip on overload whether or not it is caused by a power disturbance.

Single-circuit surge protectors are not expensive, but the cheapest ones should be avoided because they do not act fast enough or have a large enough suppression capability to protect electronic equipment well. Fast protectors with a suppression

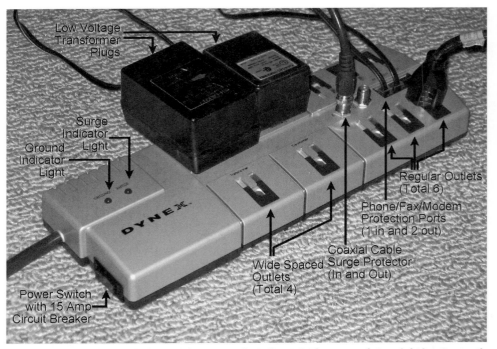

FIGURE 13-9   A Surge Protector with Connections for AC, Phone, and Coaxial Line Protection

capability of 1000 **joules** or more cost from $35 to $50 each. Figure 13-9 shows a 15-amp capacity surge protector with multiple AC outlets, a phone/fax/modem connection and a coaxial cable connection.

Installing Voltage Regulators

Voltage regulators are not generally needed on home LAN equipment, unless the risk of sags or swells is high. This will only be the case if the home's AC power is subject to voltage drops due to inadequate generating capacity in the area, or from the periodic effects of heavy-use electrical equipment inside the home or nearby.

If a voltage regulator is needed, it can be used in place of a surge protector because it will provide surge protection in addition to its main function. Voltage regulators do not have any battery power and so will not be helpful in the event of a power interruption. They simply regulate the power that comes into them on the AC line by increasing or reducing its voltage to within normal parameters.

Regulators can be obtained in a variety of sizes, from a 3-amp output capacity up to 15 amps or more. Like protectors, they are not expensive and are well worth the investment for circuits at high risk for sags or swells. Small units cost around $40 and are manufactured by several companies.

Sourcing and Reducing EMI and RFI

EMI becomes evident on a LAN when it impedes the flow of data by slowing down the network, or on occasion even stopping its function entirely. Just knowing the EMI is affecting the network is no help in correcting the problem, however. To reduce or eliminate EMI from a network, a technician must know three things about the interference: how it is affecting the network; what path it is taking to reach the network; and what source is creating it. Armed with these answers, the technician can generally find a workable solution to reduce or eliminate the EMI.

The Effects of EMI on a LAN

The most common effect of EMI on a network is to slow down or stop the transmission of data by interfering with the data packets on the network. As individual packets are corrupted or destroyed by spurious signals traveling on the network, these packets must be sent again and their correct receipt confirmed. This process slows down the network's transmission speed in direct proportion to the number of packets corrupted.

Other effects of EMI on a LAN usually arise in the individual nodes of the system and can be severe. Magnetic storage devices may lose data or store it in a corrupted version that cannot be later retrieved. If an EMI signal is strong enough to approach the operating voltage of electronic circuit boards, it may cause components to malfunction or even burn them out. A malfunctioning processor chip will not perform its intended functions correctly and the result is usually a system lockup or a crash since the computer is unable to determine what has gone wrong or where it is supposed to go next.

Once EMI is suspected of affecting the LAN, the next step is to determine if the problem is continuous or intermittent. This helps determine the source of the EMI since some emitters of EMI operate all the time, while others produce interference only at specific intervals.

Finding EMI Sources

As noted in an earlier section of this chapter, the possible sources of EMI, both inside the home and outside, are numerous and varied. Finding them is as much art as science, often with a bit of luck added to the mix.

If the EMI in the LAN is constant, search for continuously operating sources within the home. If a continuous source is found, a filter can be attached to reduce or eliminate it. If the continuous source is from outside, its path to the LAN should be determined and blocked, if possible, to eliminate the problem.

If the EMI source is intermittent, time its occurrences as accurately as possible, and look for matching operating cycles for sources inside the home. If the LAN slows down whenever the furnace activates, good places to look for the source of the EMI are the furnace motor circuit, or the low-voltage control circuit that governs the heating system. Timing EMI episodes is the best clue to finding their source.

Do not be disappointed if you cannot find an intermittent EMI source inside the home. Many outside intermittent sources exist and most of these are beyond the control of the home LAN user. In such cases, the solution is the same as for a continuous outside EMI source: Find a means of blocking the EMI path to the LAN.

Even without clues from the behavior of the LAN, you can often find sources of EMI in the home. Use a portable radio with both AM and FM reception to search for EMI sources. Place the radio as near as possible to a suspected source, turn the volume control to a medium level on any station, then slowly sweep the tuner through the entire AM and FM frequency bands, listening for any high-volume EMI as you do so. If you find a "hot spot" on the radio tuner, leave the set tuned to that frequency and shut off the suspected EMI device. If the EMI decreases or stops, you have found a source.

EMI Paths

EMI can travel from its source to a network or other susceptible device in three ways: **electromagnetic radiation,** conduction, and **induction.** In most cases, EMI may travel using more than one of these methods to reach the network or device it contaminates. The path from the source to the **receptor** (the **"victim"** of EMI) may be complex, but discovering it is key to finding a workable solution to block it.

Radiated EMI is, in effect, the same as transmitted RFI: It is broadcast into the air or space in all directions from its source. Because it is going out in all directions, the power of radiated EMI decreases rapidly with distance from the source. It weakens as the volume of space it travels through increases. This gives a clue as to how it can be most easily blocked.

Conducted EMI travels over wires connected to both the source of the EMI and the receptor. The wires may be AC power lines, antenna feed lines, telephone lines, data transmission lines, or ground-connect lines. Conducted EMI also weakens with the distance it travels, but not nearly so much as radiated EMI. The conductor wire resistance is the only factor weakening the EMI signal, so it can persist over a greater distance and at a higher strength than a similar-strength radiated EMI traveling through space.

Induction of EMI occurs when a circuit containing EMI comes in close enough proximity to another circuit that the two are magnetically coupled. When that happens, the current passing through the first circuit will induct a current into the second circuit, even though the two circuits have no actual contact with one another. The induced current in the second current is an EMI signal, although the current in the original circuit may have been part of its normal function.

As noted, a given EMI signal may travel by more than one of these methods to reach a receptor. For example, a power transformer on a pole produces an EMI signal by induction in a telephone wire passing near it. The EMI signal may infuse the telephone line with noise that disrupts voice or data transmission on it, or it may be at a frequency that does not affect analog telephone communication. Instead it may be conducted through the telephone line into the home and onto the home LAN via the modem that connects the LAN to its ISP. The frequency of the EMI may be such as will affect the LAN's data transmission or not. In either case, it may travel still further, this time by radiation from the unshielded telephone line to a television or radio antenna. From the antenna, it will be conducted as EMI into the receiver and perhaps disrupt video or audio programming.

Most EMI travels by conduction on at least part of its path, but rarely by conduction alone. When an EMI signal is radiated from a source, it will usually be picked up by a conductor wire that is connected to the receptor equipment. Similarly, inducted EMI is usually inducted from a source into a conductor that then carries it to the victim which it then affects.

Eliminating EMI by Movement

The easiest, and sometimes the most effective, means of eliminating EMI is by movement. Either the source of the EMI can be moved so its signal is out of effective range of the network, or the network node can be moved so it is out of range of the EMI source. The latter may seem impractical, but it really is not. Radiated EMI attenuates quickly with distance from the source, approximately as the square of the increasing distance from the source. Thus, if you double the distance of a victim node from the source of an EMI signal, the EMI will be only one-fourth as strong; at three times the distance, only one-ninth as strong. This is often enough attenuation that the EMI is no longer a problem.

EMI is also not uniform in strength in all directions. Moving a node even a foot or two in one direction or another may be enough to reduce EMI to acceptable levels. Likewise, turning the equipment even a few degrees may block the EMI signal enough from entering that it is no longer a threat.

EMI sources can often also be moved, especially if they are portable equipment. Most of the EMI sources listed earlier in this chapter can be moved some distance to decrease their proximity to LAN hardware, and most can also be repositioned so their EMI radiation is redirected away from sensitive receptors. Before any other solutions are tried, including grounding, a little movement is an inexpensive and often effective solution. You can (and should) still do the grounding afterward, however.

Grounding

Many EMI problems can be solved with proper grounding of network components. Grounding often supplies a better (lower resistance) path for EMI currents to travel than going into the network or connected equipment. Grounding is not a cure-all for EMI problems, but it is definitely a good place to start.

All network equipment should be grounded, if possible, to the AC wiring's ground wire. This is the third (green) wire in an AC cable or cord and is the D-shaped hole in a modern AC outlet. Most AC-powered hardware has a three-wire cord that includes a ground connection, but you should check to be sure all components are actually connected to this ground.

If the AC ground is available, do not ground equipment separately to water pipes or other structural elements as a backup safety ground. Multiple ground connections frequently have different voltage potentials and this can actually produce EMI in some parts of the system. Ground all equipment on the network to the same ground to avoid voltage differentials and use surge protectors to ground telephone lines and cable connections to the same ground circuit.

In older homes, the AC outlets do not have the third D-shaped hole for a ground wire. In these circuits, the ground wire is attached to the metal box that contains the outlet. The center screw of the outlet face plate connects to the grounded box through the frame of the outlet. If the LAN is in a home with this old-style AC ground system, a surge protector should always be used for the network equipment connections and the protector should be grounded to the outlet's center screw by an adaptor or with a direct wire connection. Before using an old-style outlet for a network, it is a good idea to remove the face plate on the outlet and be sure you can actually see the ground wire connected to the metal box.

Shielding

Shielding is used to isolate a radiating source of EMI. It is only a practical solution if the source is fairly small and can be surrounded by a grounded metal shield. If the frequency of the source's EMI is low (below 30KHz), use an iron or steel shield that will absorb the EMI and carry it to ground. If the EMI frequency is higher, it will not absorb as easily, but a copper or aluminum shield will reflect it back inside the shield repeatedly until it absorbs and grounds.

Shielded cables such as Cat5 are essential for network wiring to reduce the risk of EMI signals entering the LAN. Cable shielding is usually a metal wire mesh to preserve flexibility in the cable, but this is not as effective as a solid metal shield. If a network cable is known to pass near an EMI source, or has been established as having EMI in it, running the cable in a grounded metal conduit will provide better shielding and usually eliminate the EMI problem.

Other forms of shielding include thin metalized film and spray-on conductive coatings that can be attached to hardware that is an EMI source, or to units that are victims of EMI. These shields do not absorb EMI, but they do reflect it, and this may be enough to keep it away from sensitive LAN equipment.

Filters

Filters are the usual defense against conducted EMI and should be used on all LAN wiring whenever EMI is a threat. AC surge protectors reduce EMI by the use of filter circuits, and the better units also have filtering for telephone and coaxial circuits.

For specific EMI problems, either a high-frequency or low-frequency filter can be used as required. Filters work by passing through some frequencies while attenuating others. The attenuation is accomplished either by absorption or reflection or both. Filters for general network use reduce the most common frequencies of EMI, such as those produced by lights, motors, RF equipment, and so on, but they do not cover all frequencies of EMI nor do they eliminate EMI completely. If EMI persists despite the presence of a general filter, a more frequency-specific unit may be the solution.

Summary

- Because AC electric power cannot be stored but must be created at the moment it is used, several types of power disturbance occur frequently in AC power circuits.

- The six main types of power disturbances are: power interruptions, transients, undervoltage sags, overvoltage swells, waveform distortions, and frequency and voltage variations.

- AC circuits are also affected by electromagnetic interference (EMI) and radio frequency interference (RMI), which enter the wiring from a variety of sources.

- Power conditioning is the process of eliminating disturbances from electric current flow so the power supply for a home LAN and related equipment is smooth and clean.

- Power-conditioning hardware includes surge and spike (overvoltage) suppressors, voltage regulators, and uninterruptible power supplies (UPS) that provide backup battery power for LANs and computer equipment.

- Whole house surge suppressors must be installed by a licensed electrician. Other power-conditioning equipment can be installed by a technician following procedures to insure the equipment is compatible with the circuit.

- EMI and RMI can be reduced or eliminated from circuits by eliminating the source of the interference or moving the source onto a different or more distant power supply line. EMI and RMI can also be controlled by filtering and good grounding practices.

Key Terms

Alternating current (AC)   Electric current that alternates its direction of flow several times each second. In the U.S. AC current is 60 cycles (or reversals) per second.

Arc   An electric current that travels through the air from one conductor to another, making a visible spark or small lightning bolt as it does so.

Brownout   A term for reduced voltage in a standard AC power grid. It occurs when the load on the circuit exceeds the capacity of the generator making the electricity.

Conduction   A means of EMI travel in which the EMI current flows in one or more wires.

Cycle One complete sine wave form in AC current; the wave form of a cycle goes from 0 voltage to 110 volts positive, to 0 volts, to 110 volts negative, and back to 0.

DC Offset A power disturbance in which a continuous DC current is added to an AC current causing the AC current sine wave to be offset to higher positive or negative voltage by the amount of the DC current's voltage

Direct current (DC) Electric current that flows in one direction only, from a positive potential to a negative potential, as in a battery.

Electric grid All the components that make up an electric power system; generators, transmission wires, switches, transformers, and others.

Electric service line The power line that connects the electric grid to an individual residence or building, supplying electric power to the home.

Electromagnetic interference (EMI) Any form of electrical energy that is not an intentional electric current on a circuit but comes into it from an outside source.

Electromagnetic radiation A method of EMI travel in which EMI signals are radiated into the air or space in all directions from a source.

Electrostatic discharge (ESD) An arc produced by a surface charge on an object discharging to another object, creating a spark or arc across the gap of space between them.

Fault Any defect in a circuit that prevents current from flowing or that causes too much current to flow.

Frequency The number of cycles per second that an AC current sine wave reverses direction; in the U.S. system, 60 cycles per second is the standard.

Frequency variation A power disturbance in which the frequency of AC current changes rapidly over time; caused by rapid changes in the load on the circuit.

Fuse A device to limit current flow in a circuit; the fuse "blows" or burns out when the circuit overloads, stopping the flow of current through it.

Generator The mechanical device that creates alternating electricity; it is similar to an electric motor but is turned by steam or an internal combustion engine and generates electricity.

Grounded Being connected to the earth by an electrical conductor so electric current can travel into the ground to complete a circuit.

Harmonic AC current A power disturbance in which AC current is produced in a circuit at a multiple frequency of the normal current frequency.

Hertz (Hz) The term for cycles of frequency; AC current is 60Hz, an AM radio station may be 1280KHz (kilohertz), and the lowest frequency the human ear can hear is 20Hz.

Impulsive transient A sharp voltage increase in a current that lasts .008 seconds or longer; transients can be instantaneous, momentary, temporary, or sustained.

Induction The process by which current flowing in a circuit produces a corresponding current in another nearby circuit. A means by which EMI travels from one circuit to another.

Joule A unit of electrical energy equal to the work done when a current of one ampere is passed through a resistance of 1 ohm for 1 second.

KiloAmp 1,000 amperes of current.

Line protector A device for protecting an electrical circuit from spikes, surges, and noise.

Microsecond One thousandth of a second.

Nanosecond One millionth of a second.

Noise A form of power disturbance that causes a smooth AC sine wave to look like a series of sharp peaks and valleys.

Notching A form of power disturbance where tiny voltage variations cause an AC current's sine wave to have notches in it.

Oscillatory transient A power surge that changes voltage several times before fading to 0; an alternating current surge.

Overload More current flowing in a circuit or its components than they were designed to carry.

Overvoltage A power disturbance in which the voltage of a circuit remains consistently higher than normal.

Power conditioning The process of maintaining electric current at its correct voltage and frequency and keeping it free of noise and other EMI.

Power conversion Changing AC current to DC or the reverse.

Power disturbance Any interruption or abnormal alteration in the voltage, frequency, or sine wave of an AC current.

Power interruption A complete loss of voltage and current flow in a circuit for any period of time.

Radio frequency interference (RFI) Any unwanted electrical energy at a radio frequency entering an electrical circuit where it is not wanted.

Receptor Any device that receives EMI, whether intentional or unintentional.

Redundant line A design feature in an electric grid in which all users are connected to the grid by at least two paths so the failure of one will not cause an interruption.

Resettable circuit breaker A device that prevents an overload in a circuit by cutting off current flow when it exceeds the circuit capacity; an overload switch that is not damaged by an overload and can be reset to function again.

Run-time The amount of time a UPS will keep normal power supplied to the devices connected to it.

Sag A momentary drop in the voltage of an AC current.

Shielding A form of EMI suppression that involves surrounding an EMI source or victim with metal to absorb or reflect EMI.

Single-circuit surge suppressor A device for blocking spikes, surges, and some noise from an AC circuit.

Spike A sudden short increase in voltage in an AC circuit.

Step-down transformer A device for decreasing the voltage of an AC current branch from the standard 110 volts to a lower figure, such as 24 volts, for use by a low-voltage device.

Surge A momentary increase in the voltage of an AC circuit.

Swell A short-duration low-value increase in the voltage of an AC circuit.

Thermal disconnect A device that protects an electrical circuit by melting or burning out in order to stop current flow when the load exceeds capacity.

Transformer A device for increasing or decreasing the voltage of an AC circuit.

Transient A sudden sharp increase in the voltage of a circuit.

Undervoltage A power disturbance where the voltage in a circuit is consistently lower than the designed voltage.

Uninterruptible power supply (UPS) A battery-powered device that keeps an AC circuit functioning by supplying it with temporary power converted from DC batteries.

Victim A receptor of EMI.

Voltage fluctuations A power disturbance where the voltage of an AC circuit changes rapidly over a short time period.

Voltage regulator A device for maintaining the voltage of an AC circuit at a set level with little variation.

Voltage variation A power disturbance in which the voltage of an AC circuit is consistently lower than its designed level.

Waveform distortion Any form of power disturbance that alters the normal smooth sine wave of AC current over a period of time; examples are DC offset, notching, and noise.

Whole house surge suppressor A device installed on the service line of a residence that blocks transients that originate outside the home.

Review Questions

1. The standard for AC residential electric current in the U.S. is _____ volts and _____ volts at _____ Hertz.

2. AC devices that operate using lower voltage than the standard residential circuit can be powered through a _____.

3. How long can AC electric power be stored before it is used?

4. What are the four categories of power interruptions and what is the major difference between them?

5. What is the most common cause of power interruptions?

6. What is an impulsive transient and how long does it last?

7. Which of the following describes a spike impulsive transient?

   a. A 10-volt increase in current for 1 minute

   b. A 100-volt decrease in current for 10 seconds

   c. A 1-volt decrease in current for 1 microsecond

   d. A 10,000-volt increase in current for 5 nanoseconds

8. When an electrostatic discharge (ESD) occurs, it produces a spark or arc from the charged item to ground. True or false?

9. An electrostatic charge can be produced by which of the following?

   a. Feet walking on nylon carpet

   b. Wind blowing over an exposed surface

   c. Rotating fan blades

   d. All of the above

10. An oscillatory transient is most likely to be caused by

    a. A light bulb being turned off

    b. A television or radio being turned on

    c. An electric motor being turned off

    d. A light bulb burning out

11. A sag is a temporary decrease in voltage in a circuit, and a swell is a temporary increase in voltage. True or false?

12. Which of the following are wave form distortions in AC current?

    a. Noise

    b. Surge

    c. Notching

    d. Spark

13. What is EMI?

14. What are the two types of RMI?

15. What lighting or lighting control devices can produce EMI?

16. Where is a whole house surge suppressor installed in a residential electric system?

17. What are the two most important features of a single-circuit surge protector?

a. The number of outlets it has and the length of its connecting cord

b. Its action speed and capacity

c. The length of its connecting cord and the capacity of its circuit breaker

d. The amount of RFI it suppresses and its DC power rating

18. The function of a voltage regulator is to:

    a. Decrease DC voltage in an AC circuit

    b. Increase AC voltage in a DC circuit

    c. Maintain a preset standard voltage in an AC circuit

    d. Limit DC voltage in a circuit to a present maximum

19. How do UPS units store electric power that they can output as AC current?

20. To increase the run-time of a UPS unit, the load on its output must be reduced, or what must be increased?

21. Which home is likely to have the greatest risk of power disturbances?

    a. A 5-year-old home with underground electrical service

    b. A 35-year-old home with above-ground electrical service

    c. A 75-year-old home with above-ground electrical service

    d. A 5-year-old home with above-ground electrical service

22. When a UPS unit is installed on a circuit, an additional voltage regulator for the circuit is not needed. True or false?

Hands-On Projects

Project 13-1: Evaluate the Risk of Power Interruptions in a Home

In this project, you will evaluate the risk of a power interruption occurring in the electric service of your residence or a residence of a client or friend. After doing research and making a list of the risk factors affecting the home, you will assign a numerical value to each factor and add the risk numbers together to obtain the total risk. Finally, you will determine whether the risk factors warrant the installation of a UPS unit to provide emergency power to the home LAN.

1. Find answers to the following questions about the residence and write down what you discover:

   a. Does the home have underground or above-ground electric service?

   b. Is the home's electric service served by redundant lines (you may need to contact the local power company to determine this)?

   c. What is the average number of storms per year in the area where the home is located? (See Figure 13-6.)

   d. What is the electric generating undercapacity where the home is located? (See Figure 13-7.)

   e. How old is the home?

   f. How many power interruptions have occurred at the home during the past year?

2. Assign risk numbers for each of the answers to the questions in Step 1. Rate each factor as specified below:

   a. If the electric service is underground, rate it 1; if above ground, rate it 10.

   b. If the home's electric service is served by redundant lines, rate it 1; if not, rate it 6. If you do not know, rate it 3.

   c. If the average number of storms per year in the area where the home is located is 10 or less, rate it 1; if 20, rate it 2; 30, rate it 3; between 30 and 100, rate it 10.

   d. For each percentage point the electric generating is undercapacity where the home is located, rate 1 point (1% under equals 1 point, 2% under equals 2 points, etc., up to 12% under equals 12 points).

   e. If the home is 10 years old or less, rate it 1, 25 years or less, rate it 2; 50 years or less, rate it 4; 75 years or less, rate it 8; older than 75 years, rate it 10.

   f. Rate the home 1 point for each power interruption that has occurred at the home during the past year up to a maximum of 12.

3. Add up the total points you assigned to the home in each of the questions in Step 2. The maximum possible points are 60, which represents the highest risk of a power interruption. The lowest risk is 4 points. Where does your home fall on this scale?

4. Based on the risk factors you evaluated, should this home's LAN have a UPS?

Project 13-2: Determine Average and Peak Loads on a Circuit

In this project, you will check the average load and the peak load on a home LAN's electric circuit and determine if there is any danger of overloading the circuit.

1. Shut down the LAN and any peripheral equipment connected on the same circuit.

2. Identify the residence circuit breaker that protects the LAN circuit. To do this, plug a lamp into the LAN's surge protector and turn off one circuit breaker at a time until the lamp goes off. Use the same procedure to check which other outlets are connected to the LAN circuit.

3. Note the amp capacity of the breaker. (It will be a number embossed on the breaker face or on its switch.) Then, turn the breaker back on.

4. Determine the power requirement in watts of each piece of hardware connected to the LAN circuit. It will be listed on the label of the hardware in its user's guide.

5. Add up the total watt requirement of all electrical items connected to the LAN circuit.

6. Divide the total load in watts by 110 (the voltage of the circuit) to get the amp load on the circuit; amps equals watts divided by volts.

7. If the result of Step 6 is a number larger than the amp rating of the breaker, the circuit is overloaded. This condition must be corrected by removing some devices from the circuit.

8. If the calculated load is within the breaker's amp rating, check the devices in the circuit to be sure none contain motors or other components that have higher startup power loads than when running.

9. If you find any, factor their startup load requirements into the total load calculation and see if the breaker will be temporarily overloaded when some equipment starts. If it will be, some equipment should be moved off the circuit.

Project 13-3: Identify a UPS Unit for a LAN

In this project, you will identify a UPS unit that is suitable for use as backup power for a LAN that has the following equipment: a Dell-brand PC computer with an 18-inch flat LCD screen, an HP-brand LaserJet 1000 printer, an HP 6100 Scanner/printer/copier/fax, a Dell server, a Macintosh G-5 video editing computer with two 18-inch flat LED screens, and two Lacey 250 gigabyte hard drives.

1. Determine which pieces of hardware on the LAN need to have backup power (not all of them do).

2. On the Internet, check the Web sites for the manufacturers of the equipment that needs backup power in order to find out the amount of power each piece of hardware draws.

3. Add the total power requirements of the hardware together to get the output requirement for the UPS you need to purchase.

4. Assume the minimum run-time to shut down the LAN is 13 minutes. This is quite a long shutdown period, but the video editing computer on the LAN takes that much time to save large video files to the hard drives.

5. On the Internet, find a UPS system that has the power output and run-time (at the output load you need) for the LAN. Find the lowest price unit you can, but be sure it meets the minimum requirements with at least 10% spare capacity.

Project 13-4: Eliminate an Electric Motor EMI Source

In this project, you will determine the most cost-effective way to eliminate an EMI source that originates in a ¾-horsepower AC electric motor in a home furnace. The motor is producing a low frequency (7,200Hz) EMI signal that is affecting the home LAN.

1. On the Internet, find the cost of a filter that can be inserted in the source AC circuit to suppress 7,200Hz EMI coming from the motor.

2. On the Internet, find the cost of a filter that can be inserted in the LAN's AC circuit to suppress 7,200Hz EMI coming from any outside circuit.

3. On the Internet, find the cost of a new ¾-horsepower AC electric motor that runs at 7,200 rpm.

4. Assume the cost of installing the motor or either of the filters will be equal to one and a half times the cost of the hardware (motor or filter).

5. Determine which of the three alternatives is most cost effective: replacing the motor, installing a source filter, or installing a receptor filter. Which is likely to be the more permanent solution?

Case Projects

Case Project 13-1: Determine the Source of EMI in a Home LAN

A client asks you to help him with an EMI problem he's having on his LAN. Every fall, usually in October, the LAN is affected by severe EMI that continues for up to half an hour, then stops for a time, and resumes again. The EMI continues intermittently through the winter months, but stops in the spring and does not begin again until the fall. The client thought the EMI was caused by cold dry wind producing electrostatic discharges on the power line insulators near the 20-year-old home. He installed a whole house surge suppressor, but it did not help. What other possibility can you suggest as the source for this intermittent EMI? How would you prove your suggestion? How would you correct the problem?

Case Project 13-2: Analyze the Cause of Slow LAN Performance

A client asks you to help her improve the performance of her LAN. The LAN functions at a high rate of data transmission as long as only one data stream is being sent on it. The client says that whenever she sends data from her computer to the combination printer/fax/copier that is also a node on the LAN, she cannot access the Internet or send data to any other node until the printer finishes its task and ceases operation. The LAN has plenty of capacity for multiple users, but the printer seems to require its entire capacity. Do you think the LAN could be overloaded by the computer sending data to the printer, or is something else slowing down the network? If so, what could it be?

Case Project 13-3: Evaluate the Need for Power Conditioning Equipment in a Home

You are installing a LAN with a connected security system and an Internet connection at a residence. Because the home has above-ground electric service and is located in an area where lightning storms are frequent, you suggest the owner install a whole house surge suppressor and a UPS unit to protect the LAN from power disturbances. The owner responds that he does not need this protection because the home has several grounded lightning rods on its roof. Is he right, or could power disturbances still reach the LAN?

Case Project 13-4: Determine the Cause of Recurring Computer Crashes

A home owner with a LAN calls you to help him with an interference problem. He says his main network computer crashes every time the home's air-conditioning compressor/condenser starts up. He has put a filter on the LAN circuit and another one on the A/C compressor's circuit to stop the interference, but the problem persists. When you check the network, you confirm it does crash as the compressor starts and all the lights in the room go dim for several seconds simultaneously. What do you believe is actually causing the problem and how can it be corrected?

Troubleshooting Integrated Technology Systems

OBJECTIVES *After studying this chapter, you should be able to:*

- Describe and Analyze Integrated System Problems
- Read Integrated System Wiring Diagrams and Schematics
- Use Testing Tools and Equipment to Trace Faults to Their Source
- Identify Major Faults in Integrated Systems
- Correct or Repair Identified Faults to Make an Integrated System Function Normally

OUTLINE Introduction

Analyzing Integrated System Problems

Reading Wiring Diagrams and Schematics

Using Test Equipment

Identifying Integrated System Faults

Troubleshooting and Repairing Integrated Systems

Introduction

In this chapter, you will learn how to analyze network and integrated system problems to find their causes. Problem analysis is not the same as **troubleshooting** problems, although troubleshooting may form part of the analytical process. You will learn how to read **schematics** and system **wiring diagrams** so you can determine whether **open circuits** or **short circuits** exist in a network. You will also find out how to use basic test equipment to test the voltage, amperage, and resistance of circuits. Finally, you will discover how to troubleshoot an integrated system and take the necessary corrective action to resolve problems and bring the system on-line at full operating capacity.

Analyzing Integrated System Problems

Ideally, an integrated home network, if its wiring and other infrastructure is installed correctly, will function at full capacity the first time it is turned on and continue to do so for an extended period of time. Unfortunately, this does not happen in many cases, either because the network installation and setup were not done correctly, or because existing factors affecting the network were not foreseen. Many networks also develop problems after initial installation due to the cumulative effect of "wear" factors on network components and infrastructure, or due to new problems and defects that develop during the network's operation that reduce or block its operation.

Solving a network problem always involves three steps:

1. Analyzing the problem situation and circumstances to determine what is really wrong, as opposed to what may appear to be wrong at first glance.

2. Problem solving. Determining what is causing the network malfunction by tracing backward from the evident symptoms of the problem to its root cause.

3. Implementing a workable solution that will resolve the actual problem, not just the symptoms of the problem.

Analyzing Network Problems

Analyzing an integrated home network problem means determining what is actually wrong with a network rather than just the easily observable symptom that tells you something is wrong. The first symptom may turn out to be the real problem, but many times it is not and you should always analyze the whole network operation before implementing a solution.

For example, if an open circuit develops in the AC power supply of a piece of equipment, the problem may be observed to be a burned-through segment in a power cord, or a nonfunctioning power **transformer** inside the hardware. Either of these conditions is a symptom of a problem. Putting in a new power cord or transformer will probably get the network functioning again, at least temporarily, but it may not solve the underlying problem that the troubleshooting technician should ask: Why did the power cord burn through, or why did the power transformer stop functioning? The answer to that question may simply be a defective part, but if the network functioned normally for a while with that part, something else must have caused it to become defective. A good technician will use analysis to find that deeper problem and correct it as well.

Analysis begins with observations and questions. Among the observations you should make and the questions you should ask are the following:

- How was the network functioning before the problem developed?

- How did the problem occur? Suddenly or slowly over a period of time?

- What else was happening on the network, or around it, when the problem developed?

- Was any new hardware or software installed shortly before the problem arose?

- Could the problem be an operator error rather than a system malfunction?

- Is the problem intermittent or continuous? If intermittent, does it occur randomly or at specific times? Are the occurrences related to any other events on or around the network?

This list is not exhaustive by any means, but it gives you an idea of some things to think about—analyze, that is—in relation to a problem. Other questions and observations to make will come to you through your own experience in troubleshooting. Try to think of similarities between this problem and others you have seen or read about.

Do not be afraid to ask questions of others who are familiar with the network on which you are working. Also, describe the problem to others who have experience in troubleshooting and so benefit from their experience as you develop your own. You will not encounter many new problems in a network—even the more unusual difficulties have occurred before—and a good troubleshooter will seek the solutions others have developed to remedy such problems.

Problem Solving

Once you have analyzed the network and its operation to determine what is actually malfunctioning, the next step in the troubleshooting process is to problem solve the malfunction. Your objective is to bring the network back up to full operation and, if possible, prevent the same problem from happening again. Analysis has told you what is wrong with the network—now you must find the specific cause of the problem and develop a solution that is effective and economical. If a network data cable has failed, for example, you may want to replace the cable, but you may also want to follow a different path for the new cable so the same failure does not recur. Or you may want to install a wireless connection that will not be affected by whatever caused the cable failure.

In problem solving, helpful steps to follow include:

- Describing the problem as specifically as you can. It is a good idea to write down your description so the details are clearly understood.
- Describing all the known symptoms of the problem, not only what is malfunctioning on the network, but any noises, smells, color changes, changes in non-network equipment, and other factors you observe.
- Testing to find out which components and infrastructure elements are working correctly and which ones are not. Are the nonworking items totally dead, or just reduced in function?
- Trying to trace the problem backward from the last symptom that occurred (the one that brought the network down or caused the user to realize something was wrong) to earlier ones that will lead you to the cause of the problem.
- Bypassing nonworking elements or substituting a working device for a nonworking one to determine if such procedures solve the problem.
- Being cautious of substituting working devices for nonworking ones until you know what caused the initial failure.
- Looking for evidence of conditions and substances (described in the next section) that cause most electrical failures.

Implementing a Solution to the Problem

Once you have determined the cause or causes of a problem, you can develop a solution that will repair the existing problem and hopefully prevent its recurrence. You should also evaluate your solution to be sure it does not cause another problem while curing the first. Installing a cooling fan, for example, to reduce the heat level in a network equipment closet may not be a good solution. Fans produce EMI and the fan's motor probably has a high startup current flow. Either of these factors could adversely affect the network even as the fan rectifies the original heat problem.

The previous example demonstrates what is called the law of unintended consequences. It is not really a law, only an expression of the fact that any change in a system may produce effects in addition to the ones the technician intended. You should be aware of this fact and evaluate whether a solution you are adopting for a problem will have any unintended consequences for the network.

Your solution should also be cost effective. This does not necessarily mean you should adopt the cheapest possible solution, but rather the one that will be the best value for the network owner. If a power supply fails on a 3-year-old computer, for example, it may be more cost effective to replace the entire central processing unit than to put a new power supply into an old (and obsolete) piece of hardware. This may be true even if the new computer costs three times as much as the replacement power supply. It will probably give longer and better service than the old repaired equipment that could fail from a different cause at any time.

Finally, network problem solutions should look to the future whenever possible, just as the initial installation of a network should provide for future expansion. In many cases, the minimum solution to a problem will also be the least expensive, but may leave no capacity for newer equipment to be installed later, or for the network to grow. Where possible, it is better to leave some unused excess capacity in circuits, cables, breakers, current loads, and other components so the system has room for future growth.

In most electronic repairs, the labor of the technician is the most expensive element. Be aware of this as you develop problem solutions. Your time spent in analyzing the malfunction and problem solving the cause should constitute the majority of the time you devote to troubleshooting the project. Repairing the problem should involve only as much of your time as is required to install, set up, and test whichever new devices or infrastructure is needed to bring the network back up to full operating level.

Do not spend 2 hours repairing an item that can be replaced for the cost of 1 hour of your time. Generally, do not repair electronic equipment that is more than 3 years old. It will be obsolete by then and the customer will be better served by replacing it with new and up-to-date gear.

Reading Wiring Diagrams and Schematics

AC wiring installed in homes is typically described in wiring diagrams that form part of the house plans or **blueprints** (so named because they are reproduced with blue printing on a white background; or the reverse, white printing on a blue background). Blueprints will note all of the wiring that forms part of the home's original infrastructure. If extensive electrical systems, such as a home theater or a home LAN, are installed after the original constructions, additional wiring diagrams will be prepared showing the electrical connections and equipment needed.

The internal wiring of electronic equipment is described in diagrams called schematics that show each component in an electronic device and how the components are connected to create the device.

Schematics and wiring diagrams may be complex, but they are not difficult to read and interpret once you understand a few basic rules about how they are composed. This section briefly explains how to read wiring diagrams and schematics.

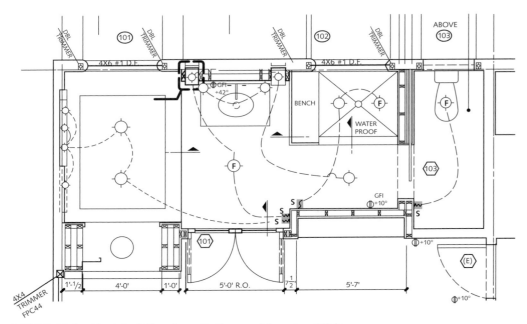

FIGURE 14-1   Electrical Plan for a Bathroom Showing Wiring

AC Wiring Diagrams

Some AC wiring diagrams are composed of dashed lines that represent wires and **electronic symbols** that represent electrical fixtures. These are plotted on scale floor and wall plans for the house or other building being wired. A simple example of an AC wiring diagram that includes wire lines is shown in Figure 14-1.

Note that the electrical lines are dashed and curved. They do not run in straight lines, and where two electrical lines cross one another, one jumps over the other with a half-circle loop, as shown at the right center of Figure 14-1. This indicates the wires are not connected at the point where they cross. The wire lines in this diagram show how the lights and fixtures in the room, designated by symbols, are connected to one another and to the switches that control them.

In the construction of the house, the wires do not have to follow the exact path shown in the plan. The electrician who installs the wires may elect to follow a different path that is shorter or more convenient, but the fixtures must all be placed in position and connected in the manner indicated by the wiring plan.

Some electrical plans may not show the electric wires at all, leaving the paths that they will follow entirely to the discretion of the installing electrician. The plan simply shows the placement and location of each fixture and, where necessary, designates its function so the electrician will know how the device must be wired. Figure 14-2 shows a larger bathroom electrical plan whose wire paths are not shown, but the function of various devices is nonetheless noted to indicate the required wiring.

The plan in Figure 14-2 is more complex than that in Figure 14-1 and also more specific. By designating the function and placement of every fixture and device, the designer has taken most options away from the installing electrician. The lighting system in this plan is designed for automatic control and maximum user convenience, as are the other electrical fixtures in the room.

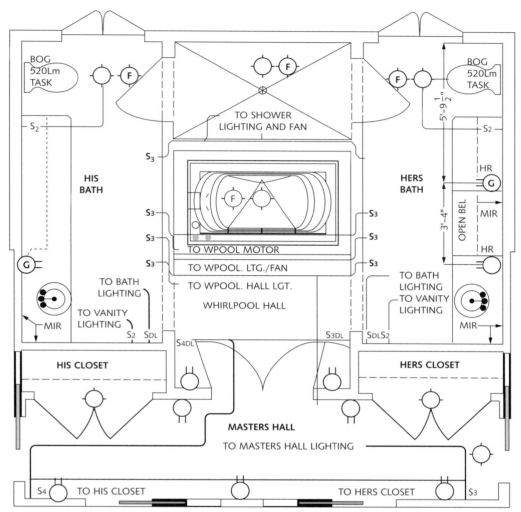

FIGURE 14-2   A Large Bathroom Electrical Plan Showing Device Functions

The AC electrical symbols used in both types of electrical plans provide a convenient shorthand method of designating fixtures and other devices without writing out the name of each piece every time it is specified in a plan. Table 14-1 lists the common AC wiring symbols and their meanings. This list is not complete. A number of additional, though less commonly used, symbols exist, but you should at least know the symbols shown in the table in order to work with lighting control and other AC automation systems.

Schematics

Schematics are the written plans of electronic equipment. They describe in detail exactly which components are contained in the device and exactly how they are wired together. Unlike AC wiring diagrams or plans, schematics always show all the wiring, as well as all the components, in a device. The symbols used in schematics are somewhat different from those used for AC diagrams, and much more numerous. This is because many different types of components reside in electronic

TABLE 14-1
Common AC Electrical Symbols and Their Meanings

Symbol	Meaning	Symbol	Meaning
	Ceiling fixture	S	Single pole switch
	Drop cord fixture	S_2	Double pole switch
	Fan hanger fixture	S_3	Three-way switch
	Junction box	S_4	Four-way switch
	Pull switch fixture	S_{WP}	Weatherproof switch
	Special fixture	S_L	Low-voltage switch
	Recessed fixture	S_D	Dimmer switch
	Single outlet	S_A	Special switch
	Duplex outlet		Fluorescent fixture
	Triple outlet		Telephone port
	Quad outlet		Intercom port

TABLE 14-1
(*continued*)

Symbol	Meaning	Symbol	Meaning
	Split-wired outlet	T	Thermostat
WP	Weatherproof outlet		Flush-mounted panel box
S	Duplex outlet with switch	•	Push button
A	Special duplex outlet	CH	Chimes
	230 volt outlet	TV	Television antenna port

devices and each has its own symbol. Most electronic hardware operates on DC current (usually converted from AC power through a power converter), and most, though not all, of the components in them are different from AC components.

In modern electronic hardware, many, if not most, of the wires noted in a schematic are not wires at all but metal pathways on a printed circuit board. These printed circuit boards take the place of individual insulated wires used in earlier electronic hardware. The metal circuit wires printed on the boards are not insulated because they are firmly attached to the rigid board and are not likely to come in contact with any outside current that could damage them. The circuit boards plug into ports on larger circuit boards or on the frame (chassis) of the device.

For cost effectiveness as well as performance, electronic hardware contains as few hand-wired parts as possible. Circuit boards with smaller components attached to them through high-tech production methods (robotic part insertion, wave soldering, etc.) are far less expensive than a hand-wired circuit of equal complexity. They are also faster (because they are smaller, meaning the electric current has less distance to travel), and more reliable (because they do not contain the possible errors of human manufacturing processes).

The result of circuit board construction and automated production techniques in the manufacture of electronic hardware is that there are usually few easily repairable parts in these devices. Repair of electronic hardware now consists mainly of diagnosing which of its circuit boards has a defective part on it and then replacing the entire board with a new plug-in board. The labor required to do so makes repair of individual components on a circuit board too costly in most cases.

Technicians should, however, still know how to read schematics so they can determine correct hookup procedures for hardware and so they can diagnose

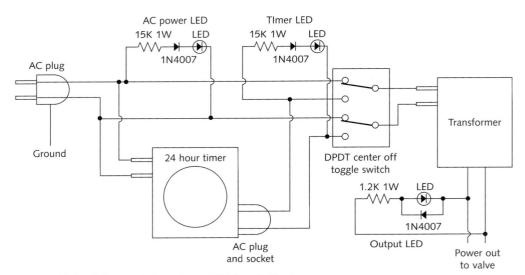

FIGURE 14-3   **Schematic for a Timed Valve Activator**

internal hardware problems that may affect the network they are working on. A typical simple schematic for a timed activator circuit on a sprinkler valve is shown in Figure 14-3. This device can be set to turn a valve on and then off at specific times of the day by sending a current through its output wires to the valve to hold it open, and then stopping the current to allow the valve to close.

Note that this schematic is for a low-voltage output AC device. It has some components wired at 110-volt AC current and some wired at low-voltage AC. The transformer is the component that reduces the high voltage to low for output to the valve, which is also low voltage.

Each component in the schematic is designated with a symbol, and its electronic properties are also described, with text written beside it. This is the usual method for describing components in a schematic.

All the wires in the device are also shown in the schematic. Where wires are connected to one another, a large dot is placed over the point where they intersect in the schematic. In this schematic, if wires cross but are not connected, the same half-circle loop used in AC wiring diagrams is placed in the line, representing one of the wires so as to show the two are not connected. All schematics do not use this convention. Some may simply show the wires crossing without any dot to indicate they are connected. The user should assume that crossing wires without a dot are not connected, even if the half-circle loop is missing.

Points where wires terminate—in other words, where they are attached to a component by a screw or other removable terminator—are designated with a small circle. Points where a wire is connected directly by a soldered joint to a component are not removable and do not show any connecting dot or circle.

Figure 14-4 shows some of the common symbols used in schematics for electronic components. Only a few of the most common symbols are shown. Many others exist for different components.

Correctly reading wiring diagrams and schematics helps you troubleshoot the circuits and devices those documents describe. When you know how a circuit's wiring is connected and how it should function, you can identify a malfunction by testing each part of the circuit to find out whether it is actually functioning as

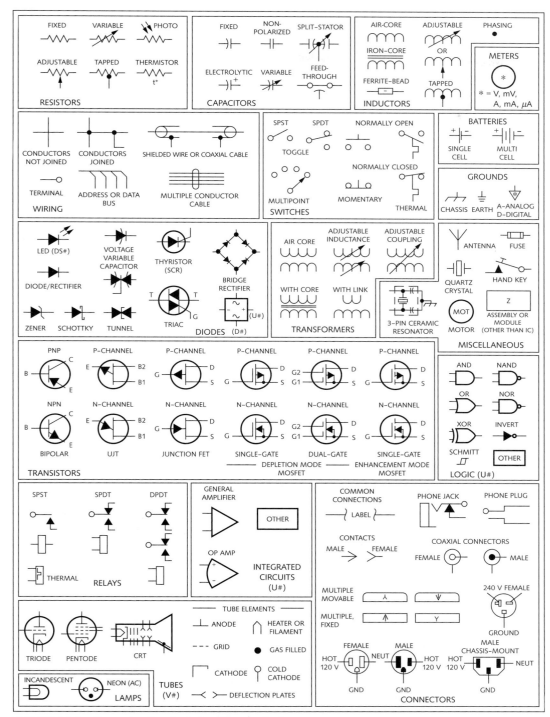

FIGURE 14-4 Common Symbols for Electronic Components

it should. Once you have isolated the nonfunctioning part or component in the circuit, you can replace it or take other corrective action to get the network running again.

Using Test Equipment

Literally dozens of specialized electronic test devices are available to assist in troubleshooting and repairing electric and electronic systems. This section describes only a few of the most popular and most essential test instruments and how to use them. Every troubleshooter should have these instruments, at a minimum, before attempting to diagnose system problems. A partial list of other test devices (and their functions) that the technician may want to acquire is given at the end of this section.

Multimeters

The **volt-ohm-milliammeter (VOM)** is popularly known as a **multimeter.** It is an analog device and has been in use for many years as an electronic diagnostic device. As its name suggests, it can measure resistance, voltage in an AC or DC circuit, and small amounts of current flow or amperage. It can measure voltage up to 1,000 volts and resistance up to 15,000 ohms, but is limited to .25 amps of current flow. Because it is an analog device that displays readings on a constantly active needle dial, the VOM multimeter can measure variations in voltage, resistance, and current flow in a circuit as they occur. This is often very helpful in diagnosing problems in the circuit. One disadvantage of a VOM multimeter is that the **impedance** of the needle meter itself can affect the accuracy of voltage measurements in low-voltage circuits. A VOM multimeter is shown in Figure 14-5.

FIGURE 14-5   A VOM Multimeter

Another type of multimeter is the digital multimeter (DMM), which is a newer, more expensive, and sometimes more accurate instrument than the VOM multimeter. It is favored over the VOM multimeter when servicing digital equipment where extreme accuracy of voltage measurement is often required. It also does not have the problem of affecting voltage readings due to the impedance of the instrument.

For most testing purposes, the VOM multimeter is adequate, and using it is easy. The first step is to be certain the multimeter has good batteries installed. Batteries are only needed for checking resistance, but it is a good idea to keep good batteries in the multimeter at all times. You can check this by turning the multimeter's **function selector** to the highest setting for measuring resistance and then adjusting the meter needle to zero. Note that the resistance scale on the multimeter reads from right to left. Zero is on the far right side of the scale. If you cannot adjust the needle to zero, new batteries are needed.

To install batteries in the multimeter, do the following:

1. Set the function selector to OFF and unplug the test leads.
2. Use a screwdriver to loosen the screws on the meter's back cover. Remove the back cover.
3. Install fresh batteries according to the **polarity** markings (+ and −) in the battery compartment.
4. Replace the meter's back cover and secure it with the screws.
5. Replace the test **probes** in their proper sockets. The black probe goes in the common (ground) socket, while the red lead goes in the other (V.R.A.) socket.

If the pointer on the multimeter display does not rest exactly over the 0 on the left side of the AC/DC scale, use a small screwdriver to move the adjusting screw in the window of the meter to set the pointer to 0. When you adjust the meter and when you take readings from it, keep the meter lying flat on a nonmetallic surface. For accurate measurements, the meter should be used at room temperature (60 to 80 degrees F). For all measurements, use a range setting that results in a reading in the upper third of the meter scale. If you do not know what range the measurement will be in, start with a higher range setting than you think you will need, then reset the meter to a lower range if needed. To read the meter precisely, look at the scale from a viewing angle where the needle and its reflection on the mirror on the scale behind it are aligned one on top of the other so you cannot see any reflection of the needle in the mirror.

To measure voltage, do the following:

1. Set the function selector to one of the DCV positions (if you are measuring DC voltage) or one of the ACV positions (if you are measuring AC voltage).
2. Connect the test probes to the circuit.
3. Read the voltage on the voltage scale.
4. If the reading is very low on the setting you are using, adjust to a lower range and read the voltage again.

To measure DC current (not AC) up to 250 milliamps (.25 amps), do the following:

1. Set the function selector to the 250 **MiliAmp** (mA) DCA position.
2. Break the circuit in which you want to measure current.

3. Connect the black probe to the positive side of the break, and the red probe to the negative side of the break.

4. Read the results on the scale.

5. If the reading is in the lower part of the scale, set the function selector to the lower DCA position and try again.

To measure resistance, perform the following steps:

1. Disconnect power from the circuit you are testing.

2. Set the function selector to one of the OHMS positions.

3. Touch the probes together and turn the OHMS adjustment to bring the pointer to the 0 mark on the right side of the top scale.

4. Touch the probes across the circuit or component you are testing and read the results on the scale.

5. If necessary, reset the meter to a different range to get a reading in the upper half of the scale.

Oscilloscopes

An **oscilloscope** is essentially a **voltmeter** with a visual display. It is more useful than a multimeter or voltmeter in some testing situations because it shows what the waveform being measured actually looks like and can reveal any abnormalities in the current flow. In addition to measuring voltage, the oscilloscope can also measure frequency, time periods, and wave slopes (the shape of the wave). The face of an analog oscilloscope is shown in Figure 14-6.

An oscilloscope has many controls and adjustment knobs that often confuse or intimidate the novice user, but like the multimeter, it is not difficult to use once you understand the controls. Some of the display controls are much like those on a television set, while the others are mainly for measurement adjustment. The main operating controls on an oscilloscope are:

- **Focus**   Adjusts the sharpness of the display
- **Intensity**   Adjusts the brightness of the display

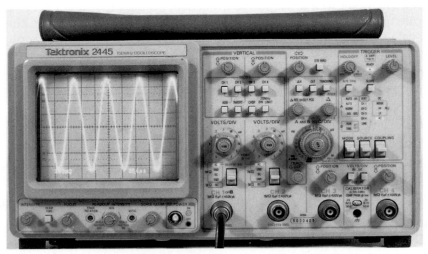

FIGURE 14-6   Front View of an Oscilloscope Showing Adjustable Controls

- **Vertical center** Adjusts the vertical positioning of the display
- **Horizontal center** Adjusts the horizontal positioning of the display
- **Vertical gain** Adjusts the height of the display
- **Horizontal gain** Adjusts the width of the display
- **Sweep** Adjusts the frequency of the display (not the frequency of the tested wave)
- **Sync selector** Permits internal or external synchronization
- **Z-axis** Adjusts the intensity of modulation of the trace
- **Calibration scale** Allows voltage waveforms to be measured on a scale

Before each use, an oscilloscope must be set up and **calibrated** so its test readings and display are accurate. Many oscilloscopes have internal set up and calibration functions that activate when the unit is turned on. If the oscilloscope has this capability, it should be allowed to warm up and calibrate itself before being used. If it does not have a self-calibration function, perform the following steps to set it up and calibrate it for testing:

1. Before turning the oscilloscope on, turn the focus intensity and sync controls all the way down (counterclockwise).
2. Turn the vertical and horizontal controls to their midpoints.
3. Turn on the oscilloscope and turn the intensity up to minimum brightness.
4. Wait a couple of minutes until the unit warms up and then adjust the focus to get the sharpest possible image.
5. Center the trace image using the vertical and horizontal controls.
6. Connect an AC power source of 6.3 volts (9 volts peak) to the vertical input of the unit.
7. Use the vertical gain control to adjust the display so the height of the 9-volt peaks (positive and negative together) of the wave equal 1.8 divisions of the calibration scale on the display.
8. Use the sync control to adjust the wave form until it is stationary on the display.
9. Use the horizontal gain to adjust the wave form so three full sine waves are displayed.

The oscilloscope is now ready to use. As it is calibrated, each division on the scale is equal to 10 volts. The vertical attenuator can be set to multiply this calibrated 20-volt range by .1 (2 volts), 1 (20 volts), 10 (100 volts), or 100 (1,000 volts) as needed for test measurements.

Three types of probes are commonly used with an oscilloscope. A **low-capacitance probe** reduces the loading effect of the probe and is more accurate when measuring high-frequency or high-impedance circuits.

A **voltage-divider probe** steps down high-voltage circuits so their measurement will be within the range of the oscilloscope. The step-down ratio can be 10:1 or 100:1.

A **demodulation probe** is used for measuring radio frequency signals that must be detected before being displayed on the oscilloscope.

Analog oscilloscopes have now been replaced in many applications by digital oscilloscopes, which are more expensive and more accurate for some

FIGURE 14-7   A Handheld Mini-Oscilloscope

testing procedures. Digital oscilloscopes are also available which use a computer for their display function. These units consist of a hardware device that has the probes and some of the adjustments of a standard oscilloscope and that connects to a computer through a port. Software loaded on the computer interprets the incoming signal from the hardware unit to display the wave form on the screen. All the adjustments of a normal oscilloscope are possible with these units, but the adjustments are made using electronic controls on the screen.

Analog oscilloscopes are still adequate for most troubleshooting work on networks and for most electronic equipment. When buying an oscilloscope, analog or digital, it is important to get one with as wide a bandwidth (range of frequencies it can measure) as possible.

You may also want to consider a mini-oscilloscope such as the one shown in Figure 14-7. This is a digital unit, but costs less than a typical analog oscilloscope and still provides accurate measurements over a wide frequency range. It is also a convenient unit to carry on location and faster to set up and use than a standard size unit.

Voltage Testers and Test Lamps

Two other simple test devices should be part of every technician's troubleshooting kit. They are a **neon voltage tester** and a **test lamp**.

A neon voltage tester is shown in Figure 14-8. It consists of two lead wires connected to a neon bulb and is used to test for the presence of voltage in a circuit.

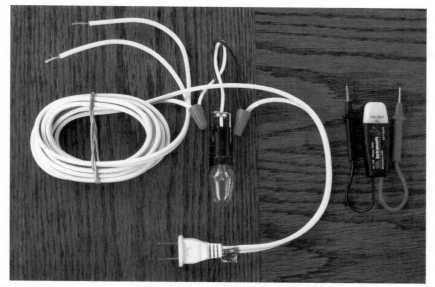

FIGURE 14-8   A Test Lamp (Left) and Neon Voltage Tester (Right)

To test an AC outlet in a home, the two probes are inserted in the outlet. If the bulb lights, the circuit has voltage. The voltage tester is inexpensive and small enough to keep in your pocket. It should be used to check that power is off before working on any AC home wiring.

A test lamp consists of an AC plug and two lead wires with attached alligator clips. A lamp is wired in series in one of the lead wires. A test lamp is also shown in Figure 14-8. It is used to test continuity in a circuit. When the test lamp is plugged into an AC outlet and the clips are connected to the two power wires of a device, the lamp will illuminate if there is continuity in the circuit of the device. The clips can be moved to any part of the circuit to determine if there is continuity between the points where the clips are attached. Defective parts of a circuit can thus be easily identified.

Specialized Test Equipment

Literally dozens of specialized test instruments are available for electronic troubleshooting and testing. The four units described earlier will probably be sufficient for most common applications, but technicians who perform extensive tests and do major repairs may want to purchase some of the additional instruments that make such work easier and faster. Among the commonly used specialized instruments are:

- **Signal generator**   To generate test signals and wave forms for testing various video and audio hardware devices
- **Ammeter**   To measure current flow in circuits and components
- **Transistor checker**   To test transistors for continuity and performance parameters
- **Capacitor checker**   To test capacitors for continuity and performance parameters

- **Megohmmeter**   To test electrical insulation for resistance to current leakage
- **Ohmmeter**   To test resistance in circuits and components
- **Signal level meter**   To measure the signal strength of video and audio signals

Identifying Integrated System Faults

To identify faults in an integrated network system or one of its nodes, you must first locate the segment of the network that is not functioning correctly, then carefully compare how the network is functioning with how it *should* function. Nearly all network problems are localized—that is, they are caused by a problem in one node or one segment of the network. This is true because, even though the cause of the trouble is likely to be found in one specific place, the fault may affect the entire network. Finding the source of the problem is a big first step toward correcting it.

Most breakdowns in electronic equipment or failures in network performance have some very basic causes. These include:

- EMI or RFI invading the network or a node
- Power disturbances
- Heat buildup
- Moisture or condensation
- Dirt or contaminants
- Improper installation of equipment or infrastructure
- Manufacturing defects or damage
- Wear caused by excessive movement
- Damage by weather or animals

This section discusses how to identify some of these causes and outlines steps to correct them.

Troubleshooting EMI and RFI

Troubleshooting EMI and RFI can be a frustrating job because interference has many causes and can affect various parts of a network. Some causes of interference are discussed in Chapter 13. As with most network problems, the first step in finding a solution to an interference fault is to determine where it is happening in the network. Since interference entering the network at any point is likely to infect the whole network system, the easiest way to localize the problem is to find the source of the interference.

To test devices within the home for interference, use a portable AM/FM radio with an external antenna. Tune the radio to a point on the FM band where you cannot hear a broadcast station with the antenna fully extended. Without moving the tuner control, change to the AM band and see if you hear the broadcast of any AM station. If you do, move the tuner to a clear point on the band and flip back to the FM band to see if the tuned location is clear on that band as well. Continue this procedure until you find a tuned location on the radio that has no AM or FM station. Turn up the volume of the radio.

With the radio at high volume and its antenna extended, tour the house room by room listening for interference that will be audible as noise (static) on one or both of the radio frequency bands. If you detect interference in a room, move the radio around until you get the strongest signal. Look for an operating electrical device that could be producing the interference. Tag any suspect devices you find, but complete your testing tour before attempting to correct any problems you find.

If there is no interference in a room when you enter it, turn on every electrical device in the room, one by one, and check to see if any interference develops as each device is activated. Do not skip anything electrical, including battery-operated devices. Tag each one that produces an interference signal on the radio.

Test for interference in every room of the house, including the garage, basement, storage areas, and outside patios and balconies, if these contain electrical devices. Be sure to check the furnace and the air conditioner when they are operating. If the home has a sump pump, check it when it starts and runs. In the garage, start the cars to see if their ignition systems produce interference. Also start any other gas or diesel engines on equipment in the garage and check whether they produce interference when starting or running.

When your EMI/RFI inspection tour is complete, make a list of the devices you found that are producing interference. The number of problem devices you are dealing with in a home may determine the solution you adopt.

You can deal with devices that produce interference in three ways:

- Repair, replace, or eliminate the device.
- Move the device to a location where its EMI does not affect the network.
- Filter the device to reduce EMI, or the network to reduce EMI reception.

If the device is inexpensive and well used, such as an old hair dryer, replacement may be the best option. More expensive devices, such as furnace and A/C motors, may warrant repair. If a device is replaced, do not assume the new model will be interference-free. Test it to be sure, and return it if it is not.

Interference decreases **(attenuates)** as the square of the distance from the source to the network entry point increases. This means that doubling the distance from the source to the network will reduce the interference to one-fourth its previous level; tripling the distance reduces it to one-ninth. This is usually enough to eliminate an interference problem by simply reducing its level to the point where network data packets are not adversely affected.

Try moving portable devices, especially low-power ones, to another location to reduce their interference potential. For lawnmowers and other gas-powered equipment, do not start them in the garage or near the house. Even moving a device to another spot in the same room may place it far enough away from network nodes or wiring that its interference will cease to be significant.

If moving a device or repairing it does not help, and elimination is not possible, filtering may be the best answer. A filter applied to the source of interference will prevent the noise from entering the AC wiring and passing to the network. Filters on the network nodes will prevent the noise in the AC lines from entering the network. If several sources of interference are present, filtering at the network nodes may be the most cost-effective answer. If only one source of noise is evident, a filter on the source may suffice. In truth, filters on both the source and the network may be required to suppress interference conducted through the home.

Radiated interference that cannot be eliminated by movement, repair, or replacement requires **shielding.** Again the shielding can be applied at the source, or on the network nodes and wiring, or both. Network hardware is generally shielded to some extent and network cables can be shielded if necessary, but shielding is usually easier to install on a serious interference source than on all the potential receivers.

Shielding consists of a grounded metal enclosure around the interference source. The metal enclosure receives the interference before any other receiver and grounds it harmlessly into the earth. The more completely a metal shield encloses an interference source, the more completely it will block interference.

If radiated interference is coming into the home network from an outside source, it usually cannot be eliminated by any of the methods used for inside interference. Often, outside interference sources cannot even be located, and if they are, the home owner usually has no means of influencing whoever controls them to deal with them effectively. Filtering the network and shielding its components are the only reliable methods of dealing with this type of problem. If the direction of outside interference can be determined, it can sometimes be blocked by shielding (aluminum foil or wire mesh) on the wall of the home facing the interference source.

Troubleshooting Power Disturbances

Power disturbances in a home's AC wiring can be detected by tests with an oscilloscope or a multimeter. Disturbances are actually a more common source of network problems than is interference, and power disturbances are growing in frequency and severity. As the nation's power grid becomes more complex and its capacity more strained (or overloaded), this trend will continue.

As discussed in Chapter 13, these disturbances can be best corrected with power-conditioning devices such as whole house surge suppressors, single-circuit protectors, and UPS units. Circuit protectors, which include interference protection for telephone and data lines as well as AC wiring, are adequate protection for video, audio, telephone, and automation control lines in an integrated network because the devices connected to these lines will not lose data in the event of a power interruption. Computer network and security system data lines, on the other hand, can lose data or be prevented from transmitting data at critical moments in the event of a power interruption. For this reason, both of these systems should be connected through a UPS system or otherwise provided with battery backup power for use in an emergency.

Heat Buildup

All electrical activity—from current flowing in a wire to the complex functions of transformers, motors, cathode ray tubes, and resistors—produces heat. The resistance to the movement of electricity that is present in any device through which current passes is transformed into heat within the device. This heat must be dissipated (radiated or conducted) away from the device to prevent it from overheating and melting or burning out. The three methods of transferring heat away from a device are:

- **Conduction** Placing a cooler object next to a heat source so that excess heat is conducted into the object.
- **Convection** Moving cooler air past a heat source with a fan so excess heat is dissipated by the moving air.
- **Radiation** Increasing the surface area of a heat source and/or cooling the air around it so heat radiates from the surface to the surrounding cooler air.

Heat damages devices by increasing the resistance of circuits, which in turn increases the current flowing and produces still more heat. Heat causes materials to dry out, crack, blister, and finally burn. When this happens, the device fails.

Well-designed electronic devices have built-in methods of dealing with heat to prevent buildup. These include vent fans that blow cooler air through the unit to remove heat from it, **heat fins** (rows of metal spines that increase the surface area to more quickly radiate heat away from an object), and **heat sinks** (large pieces of metal that absorb heat from a smaller component, thus preventing overheating in the smaller component). Heat fins and heat sinks are often built together in a device.

In many cases, the overheating of a device is caused by its placement, or its placement in relation to other hardware. Equipment that has a cooling fan should not be placed so the intake or exit vents of the fan are obstructed. Fins should always be kept open and available to the coolest air in the room. No electronic device should be placed in an enclosure or cabinet that restricts the free flow of air around it. Likewise, all electronic devices should be kept away from other heat-producing sources (home heat vents, radiators, large motors and transformers, fireplaces, etc.) that will add to their heat buildup.

Generally speaking, the cooler you can keep electronic equipment (down to 50 degrees F), the better it will operate and the less chance overheating will occur in its components. Remember, while the overall unit may be adequately cooled, overheating can still occur inside the device, especially if ventilation is blocked or increased heat has built up in one component.

When inspecting a device, look for signs of heat damage such as discoloration, cracking insulation, burned surface marks, bits of ash, and so on. If you find any of these, a heat problem is almost sure to exist in the component or near it.

To correct overheating defects, you must replace any damaged or heat-stressed parts that are contributing to the heat buildup, and eliminate any problems affecting the system's built-in heat control technology. If this does not solve the problem, additional cooling devices (fans, heat sinks, etc.) may be needed.

Moisture and Condensation

Moisture also causes circuits to draw more current and produce more heat. This speeds the breakdown of sensitive components. Water with salt or other contaminants in it becomes a conductor and can short circuit a system quickly. Water causes expansion of materials, as well as warping, swelling, and dissolving—any of which can lead to the rapid failure of a device.

Heat and moisture are connected and often work in tandem to destroy a device. When the equipment gets hot, it heats the air immediately around it, which enables that warm air to retain more moisture. When the device cools down, it also cools the surrounding air, causing it to hold less moisture. The excess moisture turns to water droplets on the cool component. When this process is repeated over time, the moisture eventually causes damage to, or failure in, the device.

Good ventilation is the best protection against moisture. To avoid condensation, keep room temperatures fairly even and do not run evaporative coolers in rooms with electronic equipment. Keep liquids away from electronic gear. Spilled liquids always cause damage, but even the evaporation from a hot cup of coffee can condense on cold metal parts, producing water droplets that can run onto circuit boards and short them out.

Dirt and Contaminants

Foreign material in electronic equipment can easily cause circuit failures. The most common contaminant problem is dust and dirt building up in the cooling vents of the equipment from air circulated by the fan. This dirt should be cleaned periodically, both to prevent it from being pulled onto circuit boards or other components and to prevent it from obstructing the flow of cooling air to the device.

Other contaminants that can be drawn in through the ventilating system and cause shorts include moisture and small insects. The latter can blow out a circuit board if one happens to land on its surface and bridge one of the circuits.

When diagnosing network problems, look for signs of any liquid spilled, splashed, or dropped on the equipment. Even if no liquid remains when you do your inspection, if evidence exists of liquid previously in the system, it probably did damage before evaporating.

Installation, Weather, and Animal Problems

Installation, weather, and animal problems are less common than some of the others discussed earlier, but they occur frequently enough to warrant some attention. Improper installation is usually a result of carelessness or not reading and following the manufacturer's instructions.

When diagnosing equipment faults, read the instructions that came with the device, if possible. Check to see that the device is properly set up. Are connections secure and tightened? Is the device correctly grounded (and not grounded twice, creating a loop for EMI)? Are any incorrect substitutions made for parts or cables (e.g., crossover cable for a straight through, or a high-impedance antenna hookup for a required low one)?

Is the device built to perform as the client thinks it will, or is he or she mistaking its capability? Digital receivers are required for digital signals, video or audio; analog devices will not work. Not all DVD players will play all types of DVDs. Checking the device and the instructions that came with it will help resolve these and similar problems.

Weather can affect electronic networks and devices in ways other than how lightning and other electrical disturbances might. Weather wears on anything exposed to it over time. Insulation deteriorates, plastic cases become brittle and crack, metals corrode. Any wiring or device exposed to outside weather will age more rapidly than inside equipment and be subject to failure sooner. Watch for weather-related damage on exterior wires and devices, but also on interior wiring and sensors placed in windows, on doors, or in attics and crawlspaces. Weather may penetrate these places and affect equipment, especially if the installer used interior-grade materials.

Finally, remember that animals like to chew on electronic wires and devices. Dogs and cats will chew wires inside and outside. So will rodents (rats, mice, squirrels, chipmunks, etc.), and they can often access wires in places where larger animals cannot. Broken or shorted wiring should be routinely checked for teeth marks. If they are found, replace the entire line: You cannot tell where all the damage is. Rarely, a woodpecker or other bird will also damage or break a wire. If small animals and other wildlife are common around a home, be aware that these creatures can bring down a network or audio visual system.

Repairing Circuit Faults

The most common circuit faults that result from the causes previously mentioned are:

- Short circuit
- Open circuit
- **Ground fault**

Short circuits are the result of current traveling a shorter direct path across its source. For example, if the two wires in a frayed or chewed AC power cord touch one another, the current flows from one wire directly to the other without ever passing through the device the cord is attached to, creating a short circuit.

Signs of a short circuit are a tripped breaker or a blown fuse, increased heat, low voltage, high amperage, and burning or smoke in the circuit. Short circuits must be located and opened so current flows only in the prescribed path.

An open circuit occurs when the circuit wiring and connections are incomplete. The current cannot flow because there is no complete path from source to neutral. An open circuit has no current flow and infinite resistance, so the lack of current leaves any device on the circuit dead and inoperative. The open point in a circuit must be found and closed before the circuit will function normally.

A ground fault is similar to a short circuit. It occurs when the current finds a path to ground before reaching the neutral side of the power source. For example, if an antenna wire touches the metal frame of a video tuner, the signal it carries will go into the frame and from there to ground, bypassing the connection to the tuner circuitry entirely.

Grounded AC circuits can be dangerous because the grounded device may keep on functioning (although poorly) and the user may not notice the ground fault immediately. If the user touches the metal to which the circuit is grounded, he or she may become part of the ground path and receive a shock.

In addition to their shock potential, ground faults produce abnormal amperage readings for a device and often abnormal voltage and resistance readings as well. Sometimes a ground fault will trip a circuit breaker or blow a fuse, but not always. Abnormal performance and current readings are the most reliable indicators of ground faults. Whatever the cause, a ground fault must be corrected to make a circuit function normally and keep it safe.

Troubleshooting Circuit Faults

Common methods used to find circuit faults are:

- Measuring the circuit for abnormal voltage, amperage, and resistance
- **Substitution** of a known functional component for a suspected defective component
- **Bridging** across a component in a circuit to determine if the bridged component is defective
- Heating or freezing a component to determine if it fails under adverse conditions
- **Signal tracing**
- Testing components
- **Bypassing** components and cable segments

Voltage, amperage, and resistance measurements, using a multimeter or an oscilloscope, can determine open circuits, short circuits, and ground circuits. The technique for making measurements with these instruments is described earlier in this chapter. Remember that the multimeter or oscilloscope must always be connected in parallel when measuring voltage, and in series when measuring current. Also remember that the power must always be shut off when measuring resistance.

Substitution simply means putting a known working part into a circuit in place of one you suspect is defective. If the substitution cures the problem, then the repair is already made. Remember, however, substitution may not provide the answer as to why the original part failed. If that is not determined, the newly substituted part may fail prematurely as well. It is wise to check for other root causes before substituting expensive parts into a circuit.

Bridging is the technique of substitution without removing the suspected part. Instead, another functioning component is simply bridged across the suspected component and connected at the same points in the circuit. If the problem is corrected, the defective component is later removed. If the problem remains, the suspected part is probably not defective. Bridging can only be done safely on a component in an open circuit, not one that is short-circuited. In the latter case, bridging may have no effect, or it may damage the new component.

Heating or freezing a suspected component or connection will often cause its condition to become more evident. Heating, usually with a hair blower, is a good technique for making a weak component fail entirely, thus revealing it as the source of a circuit fault. Freezing, by contrast, may restore a component temporarily to normal working order. Cold is transferred to the component by touching it with a chilled (nonconducting) object.

Signal tracing is also done using a multimeter or oscilloscope as a detector. A suspected transmission line can be probed at different points along its length to determine if a signal is present. If the signal stops at any point, then an open circuit must exist in the preceding segment of line.

If a trace detects a point in a line where a signal fails, the fault can be confirmed by bypassing the line with another laid temporarily along another path. If the bypass line restores the circuit to normal (or even subnormal) functioning, then the fault is definitely in the original line.

Summary

- Troubleshooting network problems should begin by making observations and asking questions, including how the network was functioning before a problem developed, the timing of the problem, what was happening simultaneously on the network or nearby, and whether the problem is continuous or intermittent.

- Problem solving in network troubleshooting means developing a solution that will correct the existing problem with a network and cure the underlying condition that caused it so the same malfunction does not occur again.

- Problem-solving steps include writing down the specific malfunction, noting all symptoms that have affected the network, testing components and infrastructure, tracing the problem backward from its latest symptom, substituting or bypassing components to determine their status, and detecting evidence of common conditions that cause network failures.

- AC wiring diagrams and plans use lines and symbols to denote wiring and fixtures used in home wiring systems. Electronic schematics also use lines and symbols to denote the composition of electronic devices, but the components and their symbols are different than those used in AC diagrams.

- Multimeters measure voltage, resistance, and current using two probes and various settings. Resistance is measured with current shut off; voltage is measured with a circuit connection in parallel; and current is measured with a circuit connection in series.

- Oscilloscopes also measure voltage resistance and current, but also provide a visual output of the sine wave they are measuring. An oscilloscope can measure a much wider range of frequencies than a multimeter.

- Many other test devices are commonly used to test circuits in electric systems and electronic devices. Among these are voltage testers, test lamps, signal generators, component testers, megohmmeters, and signal level testers.

- The common causes of integrated system faults are EMI or RFI invading the network or a node, power disturbances, heat buildup, moisture or condensation, dirt or contaminants, improper installation of equipment or infrastructure, manufacturing defects or damage, wear caused by excessive movement, and damage due to weather or animals.

- The presence of EMI and RFI in a home can be tested using a radio tuned to an open AM and FM frequency and placed near suspected sources. EMI and RFI can be cured by repairing or replacing a source device, moving it, or filtering the interference.

- Power disturbances can be corrected with power-conditioning equipment. These include whole house surge suppressors, UPS units, and single circuit protectors.

- Most common circuit faults are short circuits, open circuits, and ground faults. Each can be detected and corrected by testing and several other techniques, including substitution, bridging, heating and cooling, signal tracing and bypassing.

Key Terms

Ammeter A specialized test instrument for testing current flow (amperage) in a circuit.

Attenuate Weaken; radiated electrical energy weakens (attenuates) as its distance from the source increases.

Blueprint A schematic drawing of a building or device produced on blue paper with white text and graphics or the reverse.

Bridging A test method in electronic troubleshooting that involves using a replacement part to bridge (pass over) a suspected defective part.

Bypassing A test method in electronic troubleshooting that involves eliminating one part or segment of a circuit to see if it functions without the missing part.

Calibrated Set to an established standard scale or level.

Capacitor checker A specialized test instrument for checking capacitors in electronic circuits.

Demodulation probe A probe used with an oscilloscope to measure radio frequency signals.

Electronic symbol A graphic figure used to represent an electric fixture or electronic component in wiring diagrams and schematics.

Function selector A control on a multimeter that sets the test function it will perform.

Ground fault A circuit fault that involves the circuit being connected to a ground in addition to the normal negative-to-neutral circuit connection.

Heat fins Metal ribs placed on electronic devices to increase surface area and help dissipate heat produced by the devices.

Heat sink A large metal plate placed on electronic devices to absorb heat produced by the same. Often used with heat fins.

Impedance Resistance to the flow of current in a device or wire.

Low-capacitance probe A probe used with an oscilloscope to measure radio frequency and high-impedance signals.

Megohmmeter A specialized test instrument for testing the resistance of insulation material to current flow (amperage) in a circuit.

Multimeter A combination test instrument for testing voltage, resistance, and current flow (amperage) in a circuit.

MiliAmp A thousandth of an ampere (.001 amps)

Neon voltage tester A specialized test instrument for testing the presence of voltage in a circuit.

Ohmmeter A specialized test instrument for testing resistance in a circuit.

Open circuit An incomplete circuit that does not have a continuous path from current source to neutral.

Oscilloscope A combination test instrument for testing voltage, resistance, and current flow (amperage) in a circuit. It also displays the sine wave of the circuit being tested.

Polarity The negative (−) and positive (+) sides or ends of a circuit.

Probes Metal leads used with electronic test instruments to connect to circuits for testing them.

Schematic A graphic representation of the wiring and components of an electronic device.

Shielding Metal plating or mesh placed around a source of EMI to block interference or around a receiver of EMI to block the reception.

Signal generator A specialized test instrument for testing the path of a signal in a circuit.

Signal level meter A specialized test instrument for testing the strength of a signal at various points in a circuit.

Signal tracing A troubleshooting technique that involves determining the path and strength of a signal in a circuit to determine the location and source of a fault.

Short circuit A circuit with too little resistance that permits too much current to flow. Caused by a connection in the circuit providing a shorter path to neutral than normal.

Substitution A troubleshooting technique that involves substituting a known working part for a suspected defective part to see if doing so repairs the circuit.

Test lamp A specialized test instrument for testing whether a circuit is complete and working.

Transformer An electronic device that increases (steps up) or decreases (steps down) voltage.

Transistor checker A specialized test instrument for testing transistors for correct functioning.

Troubleshooting The common name for analyzing, testing, and correcting electrical and electronic problems in all types of systems.

Volt-ohm-milliammeter (VOM) A combination test instrument for testing voltage, resistance, and current in a circuit. Commonly known as a multimeter.

Voltage-divider probe A probe used with an oscilloscope to measure high voltage in a circuit.

Voltage tester A specialized test instrument for testing voltage in a circuit.

Voltmeter A voltage tester.

Wiring diagram A graphic representation of the wiring and fixtures of an AC electric system in a home or building.

Review Questions

1. Why is it important to analyze a network problem before attempting to correct obvious faults?

2. What are some questions to ask when analyzing network and system problems?

3. How does the law of unintended consequences relate to network and system repairs?

4. What does cost effectiveness mean when troubleshooting a network and recommending or implementing repairs?

5. House plans, including wiring diagrams, are usually called _____ because of their color.

6. In some AC wiring diagrams, the wires are represented by _____.

7. If wire lines that meet or cross in a schematic are connected, this will be indicated by a _____ at the meeting point.

8. Some AC wiring diagrams do not show the wires at all, only the labeled fixtures, so the installer can determine the paths of the wires. True or false?

9. What is a circuit board?

10. A multimeter measures _____, ____ _____, and _____.

11. When measuring resistance, the power in the circuit must be turned on. True or false?

12. What does the function selector on a multimeter determine?

13. How does the resistance scale on the multimeter differ from the other scales on the instrument?

14. The black probe of the multimeter connects to which port?

15. How can an oscilloscope reveal power disturbances in a circuit?

16. Before use, an oscilloscope must first be warmed up and _____ so its measurements will be accurate.

17. The focus and intensity adjustments on an oscilloscope adjust the _____ and _____ of the instrument.

18. A voltage divider probe is used to _____ the voltage in a circuit so it will be within the range of the oscilloscope.

19. A neon voltage tester is used to test for the presence of voltage in a circuit. True or false?

20. Two important causes of breakdown in electronic equipment are _____ and _____.

21. What are the three methods of dealing with a device that produces EMI or RFI?

22. What is shielding and what is it used for?

23. Frequent power interruptions that disrupt a network can be overcome by the use of a _____.

24. The three most common faults in circuits are _____, _____, and _____.

25. What do bypassing and substitution mean in electronic repair?

Hands-On Projects

Project 14-1: Use a Multimeter to Measure Resistance in a Circuit

For this project, you will need an extension cord and a lamp, a good battery, and a multimeter to measure the resistance in two circuits.

1. Turn the function selector on the multimeter to the lowest setting on the resistance (ohms) scale. Touch the probes together and adjust the meter needle to zero using the ohms adjust control.

2. Place the extension cord on a table with both the plug end and the outlet end pointing toward you. Do not plug the cord into an outlet. Resistance readings are always taken with the power off.

3. Put the probes into the outlet end of the cord, one on each side as if you were plugging the meter into the cord.

4. Read the resistance on the top dial of the meter. Why is the resistance so high?

5. Try changing the function selector to the highest resistance scale. Does the needle move? Why or why not?

6. Change the function selector back to the lowest resistance scale.

7. Place a piece of metal (a paper clip or a small screwdriver blade) firmly across the two blades of the extension cord plug.

8. Read the meter on the resistance scale. Why did the resistance drop?

9. What is the resistance for the cord you are measuring? Note it for comparison with the lamp you test next.

10. Turn on the lamp you will measure for resistance and unplug it while it is still turned on.

11. Touch the probes to the blades of the lamp plug. What is the resistance? Is it more or less than the cord? Why do you think that is?

12. Turn the lamp switch off. What happens to the resistance reading? Why?

Project 14-2: Measure Voltage Using a Multimeter

For this project, you will measure the AC voltage in a wall outlet. You will need a multimeter and access to a working wall outlet. Remember that the power is on in the outlet. Do not touch the ends of the probes when measuring the outlet voltage.

1. Turn the function selector on the multimeter to the 250-volt mark on the ACV scale. There is no need to adjust the meter. It should read zero volts.

2. Insert the black probe in the common side (smaller slot) of the outlet you are testing. Insert the red probe into the negative side (larger slot) of the outlet.

3. Read the meter on the AC scale. What is the voltage reading?

4. The entire AC scale on the meter represents 250 volts, so each large mark on the scale denotes 25 volts (one-tenth of the 250 volt total for the whole scale). What is the voltage of the outlet?

5. Remove the probes and change the function selector to the 1,000 (1K) volt setting on the ACV scale.

6. Test the outlet again. What is the voltage?

7. At this setting, the scale represents 1,000 volts total, so each large mark on it represents 100 volts. What is the voltage reading for the outlet using this scale? Is it the same voltage as when you used the lower 250-volt scale?

Project 14-3: Locate EMI in a Home Environment

For this project, you will use a portable battery-powered radio to locate sources of EMI in a home. You will need a working radio, preferably one with an external antenna.

1. Tune the radio to a point on the FM band where you cannot hear a broadcast station with the antenna fully extended.

2. Without moving the tuner control, change over to the AM band and see if you hear any broadcast AM station. If you do, move the tuner to a clear point on the AM band and flip back to the FM band to see if the tuned location is clear on that band as well.

3. Continue this procedure until you find a tuned location on the radio that has no AM or FM station. Turn the volume of the radio up high.

4. With the radio at high volume and its antenna extended, tour the room you are testing, listening for interference that will be audible as noise (static) on one or both of the radio frequency bands.

5. If you detect interference in the room, move the radio around until you get the strongest signal. Look for an electrical device that is operating and could be producing the interference.

6. Tag any suspect devices you find, but complete your testing tour before attempting to correct any problems you find.

7. If there is no interference in a room when you first test it, turn on every electrical device in the room, one by one, and check to see if any interference develops as each device is activated.

8. Tag each device that produces an interference signal on the radio. Test as many rooms in the house as you wish.

9. How many sources of EMI did you find? How would you reduce or eliminate EMI in the house?

Project 14-4: Make a Test Lamp

For this project, you will make a test lamp, a device used to test continuity in a powered circuit. To make the lamp, you will need a simple light socket with lead wires and a short extension cord. You will find the test lamp very useful, but you cannot easily buy this device in a store. When you are done making it, keep it as part of your troubleshooting toolkit.

1. Separate the two wires of the extension cord by about 3 inches approximately 2 feet from the plug end.

2. Cut the wire connected to the wide blade of the plug in the middle of the separated section. Do not cut the other wire.

3. Strip both ends of the cut wire and insert the light socket into the cord circuit using twist connectors to join the wires of the socket to the ends of the cut cord wire. You can connect either socket wire to either end of the cut cord wire.

4. Cut off the plug end of the cord and separate the two wires by a length of 6 inches.

5. Strip the insulation off both wire ends by one-half inch and twist the wire strands in each wire tightly so they will not fray out.

6. When completed, test your lamp. Tape one end of the cord to the table and keep the other end separate from it.

7. Put a light globe in the socket and plug the cord into an outlet.

8. Touch the loose wire lead to the taped one. If the test lamp was correctly made, the lamp will light. This shows it is a complete circuit.

9. You can test any circuit by touching the wire leads of the test lamp to points on the two sides of the circuit. If the lamp lights (even dimly) the circuit is complete (has continuity) and current is flowing through it. The test light is wired in series with the rest of the circuit, so it may light brightly if resistance in the circuit is low, or dimly if resistance in the circuit is high.

Case Projects

Case Project 14-1: Troubleshoot an Electrical Problem

A client asks you to solve a mysterious problem with his home network. Almost every evening between 7 and 9 pm, the circuit breaker on his network trips, shutting down the system. If the breaker is reset, it trips again a few minutes later. After 9 pm, the problem disappears and the network functions normally until the next evening. You observe what else is happening in the home during the evening hours and find that the two children are usually doing homework on their computers, playing video games, or watching television (none of which are affected by the tripped breaker). The client's wife cleans up the dinner dishes and puts them in the dishwasher. She then reads, watches TV, or helps the children with homework. When the dishwasher finishes, she puts the dishes away. The client has not been able to find any electrical equipment that could be causing the breaker to trip. The network's central computer is in the family room next to the kitchen, but it is on its own circuit and when you shut down that circuit during the day, nothing else except the network appears to go off. When you test the network circuit during the evening, you find a sudden voltage drop just before the breaker trips. The drop occurs before the breaker trips every time. About the time the client's wife finishes putting away the dishes from the dishwasher, the voltage drops and the circuit functions normally. Where or what would you test to find what is causing this intermittent problem?

Case Project 14-2: Reconnect a Lost Network Node

A node on a home network has lost its connection to the network. The problem began when the homeowner had some remodeling work done in the kitchen. She assumed the cutoff was only temporary due to the construction process, but when the work was finished, the bedroom node on the network remained dead. When you test the network wiring, you confirm that the data circuit to that computer is open, probably because the cable in the kitchen wall was inadvertently cut by the workmen remodeling the area. Repairing the cable now that the remodeling is complete is impossible without great expense. Pulling a new cable will also be very difficult. For security reasons, the homeowner does not want to put in a wireless hub. How can the lost node be reconnected to the network without needing a new cable?

DHTI Acronyms

Acronym	Meaning
1080i	1,080 interlaced lines
1080p	1,080 progressive-scanned lines
16:9	16 units wide by 9 units high
4:3	4 units wide by 3 units high
480i	480 interlaced lines
480p	480 progressive-scanned lines
720p	720 progressive-scanned lines
A/C	Air conditioning
A/V	Audio/visual
AC	Alternating current
ADSL	Asymmetric digital subscriber line
AGC	Automatic Gain Control
AM	Amplitude modulation
Amp	Ampere
AP	Access Point

ARP	Address Resolution Protocol
ASCII	American Standard Code for Information Interchange
ATM	Asynchronous transfer mode
ATSC	Advanced Television Standards Committee
ATSC	Advanced Television Systems Committee
AVI	Audio Video Interleaved
AWG	American Wire Gage
BC	Bare Copper
BNC	Bayonet Neill Concelman connector
Cat3	Category 3 wire
CAT5	Category 5
Cat5	Category 5 cable
Cat5e	Category 5e cable
Cat6	Category 6 cable
CATV	Community Antenna TV/Cable Television
CCS	Copper Clad Steel
CCTV	Closed Circuit TV
CD	Compact Disc
CDMA	Call Centre Management Association
CD-R	Compact Disc-Recordable
CD-RW	Compact Disc-Rewritable
CEBus	Consumer Electronics Bus
CMS	Content Management System
COAX	Coaxial
CompTIA	Computer Technology Industry Association
CRT	Cathode ray tube
D/C	Direct current
dB	Decibel
DC	Direct current
DD	Dolby Digital
DDEX	Digital Data Exchange
DDNS	Dynamic Domain Name Servers
DDP	Datagram Delivery Protocol

DES	Data Encryption Standard
DHCP	Dynamic Host Configuration Protocol
DHTI	Digital Home Technology Integration
DID	Direct Inward Dialing
DLP	Digital Light Processing
DMM	Digital Multi-Meter
DMZ	Demilitarized Zone
DNS	Domain Name System or Domain Name Server
DOCSIS	Data over Cable Service Interface Specification
DRM	Digital Rights Management
DSL	Digital Subscriber Link
DSL	Digital Subscriber Line
DSS	Direct Station Select
DSSS	Direct Sequence Spread Spectrum
DTCP	Digital Transmission Content Protection
DTS	Digital Theater Sound
DTSES	Digital Theater Sound-Extended Surround
DTV	Digital television
DVB-S	Digital Video Broadcasting Satellite
DVB-T	Digital Video Broadcasting Terrestrial
DVD	Digital Versatile Disc
DVD	Digital video disc
DVDA	Digital Versatile Disc Audio
DVI	Digital Video Interface
DVR	Digital Video Recorder
EDTV	Enhanced-definition television
ELV	Electronic Low Voltage
EMF	Electromotive Force
EMI	Electromagnetic interference
ESD	Electrostatic discharge
FDDI	Fiber Distributed Data Interface
FM	Frequency modulation
FTP	File Transfer Protocol

GFCI	Ground Fault Circuit Interrupter
GPRS	General Packet Radio Service
GPS	Global Positioning System
HAVi	Home Audio Video interoperability
HD	High Definition
HDCP	High Definition Copy Protocol
HDCP	High Definition Content Protection
HDMI	High-Definition Multimedia Interface
HDTV	High-definition television
HomeRF	Home Radio Frequency
HPCC	HomePlug Command and Control
HPNA 2.0	HomePNA 2.0
HTTP	HyperText Transfer Protocol
HVAC	Heating Ventilation and Air Conditioning
Hz	Hertz
ICMP	Internet Control Message Protocol
IEEE	Institute of Electrical and Electronics Engineers
IHA	Internet Home Alliance
IKE	Internet Key Exchange
IP	Internet Protocol
IR	Infrared
IrDA	Infrared Data Association
ISO	International Standards Organization
ISP	Internet service provider
KSU	Key Service Unit
LAN	Local area network
LCD	Liquid Crystal Display
LCoS	Liquid Crystal on Silicon
LED	Light-emitting diode
LFE	Low Frequency Effects
LNB	Low Noise Block Down Converter
MAC	Media Access Control
MAU	Multistation Access Unit

MIDI	Musical Instrument Digital Interface
MLV	Magnetic Low Voltage
MP3	Motion Picture Standards Group Layer 3
MPEG	Moving Picture Experts Group
MRAV	Multiroom Audio/Video
MTU	Maximum Transmission Unit
NAS	Network Attached Storage
NAT	Network Address Translation
NEC	National Electric Code
NEXT	Near End Cross Talk
NFPA	National Fire Protection Association
NIC	Network Interface Card
NID	Network Interface Device
NM	Nonmetallic cable
NTSC	National Television Standards Committee
NTSC	National Television System Committee
OGG	Ogg Vorbis
OSGi	Open Services Gateway initiative
OSI	Open Systems Interconnection Model
OTA	Over the Air
P2P	Peer to peer
PAL	Phase Alternative Line
PBX	Private Branch Exchange
PC	Personal computer
PDA	Personal Digital Assistant
PIR	Passive Infrared Sensor
PLC	Power-line carrier
PoE	Power over Ethernet
POTS	Plain Old Telephone Service
PPPoE	Point-to-Point Protocol over Ethernet
PSTN	Public switched telephone network
PTZ	Pan Tilt Zoom
PVC	Polyvinylchloride

QAM	Quadrature Amplitude Modulation
QoS	Quality of Service
QS	Quad Shield
RADSL	Rate-adaptive digital subscriber line
RCA	The Radio Corporation of America
REN	Ringer Equivalence Number
RF	Radio Frequency
RFI	Radio Frequency Interference
RGB	Red Green Blue
S/N	Signal-to-noise
SACD	Super Audio Compact Disc
SDSL	Symmetrical digital subscriber line
SDTV	Standard-definition television
SIA	Security Industry Association
SIP	Session Initiation Protocol
SKIP	Simple Key Management for Internet Protocol
SMDR	Station Message Detail Reporting
SMTP	Simple Mail Transfer Protocol
SSID	Security set identifier
STP	Shielded Twisted Pair
STP	Shielded Twisted Pair wire
SVGA	Super Video Graphics Array
TCP/IP	Transmission Control Protocol/Internet Protocol
TDR	Time Domain Reflectometer
TIA	Telecommunications Industry Association
UDP	Universal Datagram Protocol or User Datagram Protocol
UPB	Universal Powerline Bus
UPnP	Universal Plug-and-Play
UPS	Uninterruptible power supply
URL	Universal Resource Locator
USB	Universal Serial Bus
USB	Universal Service Bus
UTP	Unshielded Twisted Pair

UV	Ultraviolet
VBR	Variable Bit Rate
VGA	Video Graphics Array
VHS	Video Home Standard
VoIP	Voice over IP
VPN	Virtual private network
VSB	Vestigial Side Band
WAN	Wide area network
WAP	Wireless Access Point
WEP	Wired Equivalent Privacy
WiFi	Wireless Fidelity
WMA	Windows Media Audio
WPA	WiFi Protected Access
WPA2	WiFi Protected Access 2
WWW	World Wide Web

APPENDIX B

CompTIA DHTI+ Examination Objectives

Objective	Chapter and Section(s)
1.0 Networking	
1.1 Identify basic networking protocols and their uses and know when/how to apply them.	Chapter 1: Introduction to DHTI (Basic Components of DHTI Systems) Chapter 2: Network Basics (Network Architecture, Receiving and Sending External Information on a LAN, Major Hardware Components of a Home Network) Chapter 3: Home Network Design and Configuration (Wired Network Types, Network Configuration and Settings for a Home Network)
1.2 Recognize and implement methods of network security.	Chapter 2: Network Basics (Network Architecture, Receiving and Sending External Information on a LAN, Major Hardware Components of a Home Network) Chapter 3: Home Network Design and Configuration (Wired Network Types, Network Configuration and Settings for a Home Network)
1.3 Configure, set up, and maintain a residential LAN (local area network).	Chapter 3: Home Network Design and Configuration (Wired Network Types, Network Configuration, and Settings for a Home Network) Chapter 4: Installing Network Components and Low-Voltage Wiring (Planning a Home Network, Basic Central LAN Components, Wiring a Home LAN, Configuring the Hardware and Cables)
1.4 Configure, set up, and maintain a secure wireless network.	Chapter 3: Home Network Design and Configuration (Wired Network Types, Wireless Protocols and Standards, Network Configuration and Settings for a Home Network)

1.5 Identify and define network cabling characteristics and performance.	Chapter 4: Installing Network Components and Low-Voltage Wiring (Planning a Home Network, Basic Central LAN Components, Wiring a Home LAN, Configuring the Hardware and Cables)
2.0 Audio/Video	
2.1 Implement, maintain, and trouble-shoot multiroom audio systems. Identify common interference sources.	Chapter 6: Video and Audio Fundamentals (From Analog Recording to Digital Transmission, Audio Recording and Broadcasting, Sources of Audio and Video Services, Connected Audio/Video System Design) Chapter 7: Audio and Video Installation and Setup (Audio and Video System Components, Installation and Setup of Audio and Visual Systems, Configuration and Settings for External Audio and Video, Configuration for Internal Video/Audio, Monitoring and Maintaining Video/Audio Systems)
2.2 Install, configure, and maintain a residential home theater system.	Chapter 6: Video and Audio Fundamentals (From Analog Recording to Digital Transmission, Television Recording and Broadcasting, Sources of Audio and Video Services, Video Storage Files and Formats, Connected Audio/Video System Design) Chapter 7: Audio and Video Installation and Setup (Audio and Video System Components, Installation and Setup of Audio and Visual Systems, Configuration and Settings for External Audio and Video, Configuration for Internal Video/Audio, Monitoring and Maintaining Video/Audio Systems)
2.3 Assess, install, and configure content management systems and describe their applications in a residential environment.	Chapter 6: Video and Audio Fundamentals (Audio Recording and Broadcasting, Television Recording and Broadcasting, Sources of Audio and Video Services Video Storage Files and Formats, Connected Audio/Video System Design) Chapter 7: Audio and Video Installation and Setup (Audio and Video System Components, Installation and Setup of Audio and Visual Systems, Configuration and Settings for External Audio and Video, Configuration for Internal Video/Audio)
2.4 Implement, maintain, and trouble-shoot multiroom video systems.	Chapter 6: Video and Audio Fundamentals (Television Recording and Broadcasting, Sources of Audio and Video Services, Video Storage Files and Formats, Connected Audio/Video System Design) Chapter 7: Audio and Video Installation and Setup (Audio and Video System Components, Installation and Setup of Audio and Visual Systems, Configuration and Settings for External Audio and Video, Configuration for Internal Video/Audio)
3.0 Telephony/VoIP	
3.1 Differentiate and describe POTS vs. VoIP delivery. Identify and trouble-shoot common issues.	Chapter 10: Telecommunications Fundamentals and Installation (Telecommunications System Types and Characteristics, Local Telephone Systems, Telephone Components and Features, Telephone Installation and Configuration, External Services)
3.2 Describe and define the fundamentals of telephone systems.	Chapter 10: Telecommunications Fundamentals and Installation (Telecommunications System Types and Characteristics, Local Telephone Systems, Telephone Components and Features, Telephone Installation and Configuration, External Services)

4.0 Security and Surveillance Systems	
4.1 Maintain, configure and trouble-shoot basic security systems and applications.	Chapter 8: Security and Access System Fundamentals (Security Design and Installation Factors, Security System Types, Security System Equipment Locations, Security System Components) Chapter 9: Security System Installation and Setup (Security System Installation, Component Installation and Setup, Accessory Security Components, Uninterruptible Power Supply, Backups and Fail-Safe Systems, Security System Programming and Settings, Maintenance, Servicing, and Recovery)
4.2 Describe basic security terminology and apply installation procedures and methodologies	Chapter 8: Security and Access System Fundamentals (Security Design and Installation Factors, Security System Types, Security System Equipment Locations, Security System Components) Chapter 9: Security System Installation and Setup (Security System Installation, Component Installation and Setup, Accessory Security Components, Uninterruptible Power Supply, Backups and Fail-Safe Systems, Security System Programming and Settings, Maintenance, Servicing, and Recovery)
4.3 Identify, configure, install, maintain, and troubleshoot security and surveillance cameras.	Chapter 9: Security System Installation and Setup (Security System Installation, Component Installation and Setup, Accessory Security Components, Security System Programming and Settings, Maintenance, Servicing, and Recovery)
5.0 Home Control and Management	
5.1 Identify user interfaces and their appropriate applications.	Chapter 3: Home Network Design and Configuration (Wired Network Types, Wireless Protocols and Standards, Network Configuration and Settings for a Home Network) Chapter 4: Installing Network Components and Low-Voltage Wiring (Planning a Home Network, Basic Central LAN Components, Wiring a Home LAN, Configuring the Hardware and Cables) Chapter 11: Home Lighting Control (Automated Home Lighting Design, Types of Automated Lighting Systems, Lighting Control Components, Programming the Automated Lighting System)
5.2 Define and recognize control systems that integrate subsystems in the home.	Chapter 3: Home Network Design and Configuration (Wired Network Types, Wireless Protocols and Standards, Network Configuration and Settings for a Home Network) Chapter 4: Installing Network Components and Low-Voltage Wiring (Planning a Home Network, Basic Central LAN Components, Wiring a Home LAN, Configuring the Hardware and Cables) Chapter 11: Home Lighting Control (Automated Home Lighting Design, Types of Automated Lighting Systems, Lighting Control Components, System Installation and Setup, Programming the Automated Lighting System, Troubleshooting the Automated Lighting System)
5.3 Identify commonly used communication protocols and their application.	Chapter 3: Home Network Design and Configuration (Wired Network Types, Wireless Protocols and Standards, Network Configuration and Settings for a Home Network)

	Chapter 4: Installing Network Components and Low-Voltage Wiring (Planning a Home Network, Basic Central LAN Components, Wiring a Home LAN, Configuring the Hardware and Cables
5.4 Describe basic HVAC (Heating Ventilation and Air Conditioning) terminology and install peripheral control devices.	Chapter 12: Heating, Air Conditioning, and Water System Control (Design and Operation of Zoned and Non-Zoned HVAC Systems, HVAC Components, HVAC Controller System Installation and Programming)
5.5 Describe basic lighting terminology and install peripheral control devices.	Chapter 3: Home Network Design and Configuration (Wired Network Types, Wireless Protocols and Standards, Network Configuration and Settings for a Home Network) Chapter 4: Installing Network Components and Low-Voltage Wiring (Planning a Home Network, Basic Central LAN Components, Wiring a Home LAN, Configuring the Hardware and Cables) Chapter 11: Home Lighting Control (Automated Home Lighting Design, Types of Automated Lighting Systems, Lighting Control Components, System Installation and Setup, Programming the Automated Lighting System, Troubleshooting the Automated Lighting System)
5.6 Identify and install component power protection devices.	Chapter 13: Power Supply and Conditioning (Types of AC Electric Power Disturbances, Types of Electromagnetic Interference, Power-Conditioning Components, Power-Conditioning Installation and Setup, Sourcing and Reducing EMI and RFI)
6.1 Identify and apply the fundamentals of troubleshooting and diagnostics.	Chapter 14: Troubleshooting Integrated Technology Systems (Analyzing Integrated System Problems, Reading Wiring Diagrams and Schematics, Using Test Equipment Identifying Integrated System Faults, Troubleshooting and Repairing Integrated Systems)
6.2 Given a scenario, demonstrate how to apply troubleshooting skills to integrate subsystems.	Chapter 14: Troubleshooting Integrated Technology Systems (Analyzing Integrated System Problems, Reading Wiring Diagrams and Schematics, Using Test Equipment Identifying Integrated System Faults, Troubleshooting and Repairing Integrated Systems)
6.3 List and describe the benefits of verification of installation.	Chapter 4: Installing Network Components and Low-Voltage Wiring (Planning a Home Network, Basic Central LAN Components, Wiring a Home LAN, Configuring the Hardware and Cables Chapter 13: Power Supply and Conditioning (Types of AC Electric Power Disturbances, Types of Electromagnetic Interference, Power Conditioning Components, Power Conditioning Installation and Setup, Sourcing and Reducing EMI and RFI) Chapter 14: Troubleshooting Integrated Technology Systems (Analyzing Integrated System Problems, Reading Wiring Diagrams and Schematics, Using Test Equipment, Identifying Integrated System Faults, Troubleshooting the Repairing Integrated Systems)
6.4 Deliver appropriate manuals and documentation to the end user upon completion of installation	Chapter 14: Troubleshooting Integrated Technology Systems (Analyzing Integrated System Problems, Reading Wiring Diagrams and Schematics, Using Test Equipment Identifying Integrated System Faults, Troubleshooting and Repairing Integrated Systems)

Glossary

1080i Screen resolution of 1,080 interlaced lines; resolution standard for HDTV screens.

1080p Screen resolution of 1,080 progressive-scanned lines; top resolution standard for HDTV screens. Not yet in common use, but products that can deliver it are in development.

16:9 Wide-screen aspect ratio for HDTV and SDTV screens; 16 units wide by 9 units high.

2-conductor wire A high-voltage cable containing a hot wire, a neutral wire, and a ground, which is not counted as a conductor.

3-way switch An arrangement by which either of two switches can control power to a device.

4:3 The standard aspect ratio for SDTV, EDTV, and NTSC analog television; 4 units wide by 3 units high.

480i Screen resolution of 480 interlaced lines; minimum resolution standard for SDTV screens.

480p Screen resolution of 480 progressive scanned-lines; minimum resolution standard for EDTV screens.

720p Screen resolution of 720 progressive-scanned lines; minimum resolution standard for HDTV screens.

A/V receiver A digital television system component that serves as the central connecting unit of the system. May include a digital tuner, amplifiers, distribution connections, and other devices.

AC cable Armored cable; type of cable sheathed in metal. Also, any cable used for AC wiring.

Access control Restriction on who has the right to use a computer system or to enter a physical location.

AC power Alternating current electric power; the standard electric service in homes.

Adapters Devices that change audio and video connectors from one form to another, including gender changes.

Administrator A person designated to maintain and control a computer system or subsystem.

Advanced Television Systems Committee (ATSC) The non-profit electronics industry organization that sets standards for the television industry and which developed the U.S. digital television standard (DTV); also known by the acronym ATSC.

Air conditioning (A/C) The part of an HVAC system that provides cool air to lower the indoor air temperature.

Air handlers The mechanical devices that move air in an HVAC system. They are mainly blowers and fans.

Alarm An audible warning of fire, illegal entry, or other security breach. Usually a siren, a ringing bell, or similar sound.

Alternating current (AC) Electric current that alternates its direction of flow several times each second. In the U.S., AC current is 60 cycles or reversals per second.

Always-live duplex outlets An AC electrical outlet that is not controlled by a switch and always has current available.

American Wire Gage (AWG) The standard for electrical wire sizes.

Ammeter A specialized test instrument for testing current flow (amperage) in a circuit.

Ampere The unit used to measure electric current flow.

Amplifier A device that strengthens the power of an RF signal it receives from an antenna or pre-amplifier.

Amplitude modulation (AM) A method of converting sound waves to radio signals by varying the amplitude (strength) of the signal, but not its frequency.

Analog Sound or video signals that are analogous to (have patterns similar to) the actual sounds or images.

Analog data Data in nonnumeric form, such as radio waves, sound waves, and others.

Antenna The component of a radio or television receiver that receives a broadcast signal from a distant transmitter.

Appliance Any piece of equipment that performs a specified task.

Arc An electric current that travels through the air from one conductor to another, making a visible spark or small lightning bolt as it does so.

Arm To activate a security system so its sensors are functioning.

Aspect ratio The ratio of a video screen's width to its height. NTSC screens have a 4:3 ratio; HDTV screens a 16:9 ratio.

Atrium An area within a home that is open to the outside or walled in by glass, where plants can grow just as in the outside yard.

Attenuate Weaken; radiated electrical energy weakens (attenuates) as its distance from the source increases; used to describe radio and television signals.

Automated control A method for controlling HVAC or water systems by preset programming rather than the manual setting of controls.

Automation The process of controlling a lighting or other system remotely, either with manual commands from a controller or by a programmed set of instructions from a computer without the need for human supervision.

Average use The amount of electricity used in a home over a period of time, usually a day.

Backdoor A means of accessing a software program other than the normal one (front door).

Bandwidth The amount of data that can travel over a communication line or wireless connection in a given length of time; the data transmission capacity in a line.

Barn doors A term that describes the effect of viewing a 4:3 aspect ratio image on a 16:9 aspect ratio screen; the viewer sees black bars on either side of the screen, which are referred to as barn doors.

Baud One complete cycle or wave in an analog transmission signal. A baud starts at zero voltage, goes up to maximum positive voltage, comes back to zero, goes to maximum negative voltage, and finally returns to zero voltage.

Biometric identifiers Unique distinguishing physical features of an individual that can be used as a means of identification. These include fingerprints, eye retinas, voice prints, and facial characteristics.

Blower A device that uses a rotating bladed wheel to move air in an HVAC system. Not the same as a fan.

Blueprint A schematic drawing of a building or device produced on blue paper with white text and graphics, or the reverse.

Bluetooth A short-range (100 meters) wireless connection technology currently used for networking.

BNC connector One of the standard connector types used for audio and some video system connections.

Boiler A device for heating water when heating a building. Not the same as a water heater.

Breaker A safety device wired into a circuit to cut off current flow if the circuit becomes overloaded.

Break switch A sensor that monitors a closed switch and signals if the switch is opened and the circuit broken. Used on doors, windows, and containers whose normal state should be closed.

Bridge An intelligent switch that limits data flow on a LAN.

Bridging A test method in electronic troubleshooting that involves using a replacement part to bridge (pass over) a suspected defective part.

Broadband Any method of transmitting large amounts of data in a short time span. Usually accomplished by using multiple frequencies or data streams; a large (or wide) bandwidth technology.

Brownout A term for reduced voltage in a standard AC power grid. It occurs when the load on the circuit exceeds the capacity of the generator making the electricity.

Bus topology One form of network architecture for Ethernet.

Bypass Any means of evading a security device by making it appear to be functioning in a normal state when, in fact, conditions have changed. A window sensor, for example, could be bypassed by slipping a loose plate onto its contact points to make the switch continue to signal normal (closed) even when the window is opened.

Bypassing A test method in electronic troubleshooting that involves eliminating one part or segment of a circuit to see if it functions without the missing part.

Cable modem A device that converts digital data to analog signals and connects a LAN to an ISP via the cable television connection.

Cable run A cable installed between two connecting points such as a patch panel and jack.

Calibrated   Set to an established standard scale or level.

Caller ID   A system in which a signal sent by the telephone company with an incoming call identifies the calling number and its registered owner. Telephones with electronics capable of reading this signal display the caller ID information on a small LCD screen.

Call restriction   A telephone system feature that allows certain calls to be blocked from incoming or outgoing lines.

Camera   A device that creates a video image in digital or analog form. Cameras can produce still images or moving images.

Capacitor checker   A specialized test instrument for testing capacitors in electronic circuits.

Carrier frequency   In radio transmission, a signal at high frequency on which a signal of lower frequency is carried.

Carrier Sense Multiple Access/Collision Avoidance (CSMA/CA)   A method of data transmission in which nodes avoid data packet collisions through use of a token or other device controlling the movement of data.

Carrier Sense Multiple Access/Collision Detection (CSMA/CD)   A method of data transmission in which data packets contend for space on the network, and nodes sense packet collisions that require resending.

Category 2 (Cat2) telephone wire   A low-speed data wire containing two pairs of solid core insulated wires. Named for its widespread use in wiring analog telephone lines.

Category 3 (Cat3) wire   Low-speed data wire with the same composition as Category 2 wire but having either four or six pairs of wires.

Category 5 (Cat5) cable   An eight-wire UTP cable that permits high-speed data transmission; an advanced version is Category 6 (Cat6) cable.

Cathode ray tube (CRT)   The display tube in a television set or monitor on which the picture appears.

CatX cable stripper   A tool for cutting and removing the outer jacket from Category 5 and other similar types of cable.

CD burner   A device for recording compact discs (CDs). A CD burner can also play a CD, but CD players are not always burners.

CD player   A device for playing recorded compact discs.

CD-R (CD-Recordable format)   A compact disc that can only be recorded on once; after which the disc is fixed and unchangeable.

CD-RW (CD-Rewritable format)   A CD that can be recorded on multiple times, with succeeding recordings added to or replacing older ones on the disc.

Central heating system   The part of an HVAC system that provides warm air from a central furnace to raise the indoor air temperature.

Central processor   The main computer in a DHTI system or, alternately, the component in a computer that performs calculations on data.

Centrex   Central exchange, the main switching point in the telephone company where calls are connected by switching them to the line called.

Certification   The process of attesting to the qualification or competence of a person to perform certain services.

Checksum   A mathematical method for a receiver to determine if a data packet has been corrupted.

Cipher lock   A door lock that opens only when a numeric code is entered on a keypad mounted near it. The code can also be placed on a memory card that is swiped through a reader to open the lock. Some locks can have many access codes, some only one.

Circuit   A conducting "circle" in which electricity flows from a source to a device, through the device, and back to the source.

Circuit-switched lines   The normal method of completing an analog call in which the calling line is physically connected to the called line by a defined path, and the path is kept open for the duration of the call.

Circulating pump   A device in a radiant heating system that continuously circulates the heated water through the boiler and heating pipes.

Clean power   AC current that does not contain noise, interference, surges, or spikes.

Cliff effect   A description of the sudden termination of the range of digital video and audio transmissions, which can end as if they "fell off a cliff."

Codec   Coding/decoding: a software program or hardware device that encodes and/or decodes digital transmissions.

Command module   A power-line or wireless device that sends commands (manually generated or programmed) to control modules that control lights and perform other automated functions.

Compact disc (CD)   A plastic disc on which audio files are encoded using the pulse code modulation method.

Component video   A method of transmitting video signals using three separate connector cables, usually terminated with RCA plugs.

Compression Any technique that uses math algorithms to reduce the size of digital files for storage or transmission.

Compressor The part of an A/C system that squeezes the refrigeration gas under pressure into a smaller volume until it converts into a liquid.

CompTIA Computer Technology Industry Association, a global information technology (IT) trade association with more than 13,000 members in 89 countries. It works to advance the IT industry, promote IT public policy, and develop standards for training professionals in the industry.

Concealed wiring Retrofitted wiring installed within the finished walls of a structure.

Condensation The process in an A/C system by which the refrigeration gas changes back into a liquid as it cools. Also the process by which water vapor is removed from cooled air and is deposited as a liquid.

Condenser coil A coil through which hot refrigerant passes and is cooled by flowing air in the same manner as an automobile radiator cools engine coolant.

Condenser unit The part of an A/C system in which the refrigerant is converted back into a liquid from a gaseous state.

Conduction A means of EMI travel in which the EMI current flows in one or more wires.

Cone The part of a speaker that vibrates in the air producing sound.

Connector Any cable with end plugs designed to transmit electronic signals from one device to another. Includes such systems as RCA, BND, S-video, FireWire, DVI, HDMI, and others.

Control device Any device that initiates a process, directs equipment to perform a function, or responds in a preset manner to remote commands.

Control module In lighting control systems, any of the devices that receive lighting control instructions from the controller and implement them by adjusting the lights.

Controller A lighting control device that sends lighting commands to control modules in an automated lighting system. Same as command module. In water management, the electronic device that controls the operation of sprinkler system zone valves through a preset timed schedule.

Converter A device that renders digital data into analog signals, or vice versa, in audio and video systems. *See* decoder.

Cooling coil The part of an A/C system that cools the surrounding air by the expansion of the refrigerant in the coil, which absorbs heat from the air.

Cordless phones Phones with transceivers (transmitters and receivers) that operate on radio frequencies. They have a limited range from the base unit, which contains another transceiver.

Credential A written certification that an individual has specified qualifications, abilities, or expertise.

Crossover cable A connector cable in which two pairs of wires are connected to different pins (crossed over) at each end of the cable.

Crossover network A wiring circuit in speaker systems that permits several speakers of different ranges to function seamlessly together.

Crown molding Angled decorative slats installed at the junction of interior walls and ceiling. It can conceal cable.

Current The flow of electricity, measured in amps.

Cycle One complete sine wave form in AC current; the wave form of a cycle goes from 0 voltage to 110 volts positive, to 0 volts, to 110 volts negative, and back to 0.

Daisy-chain wiring Wiring light fixtures in parallel one to another with only the first light being connected directly to a power supply and each of the others receiving power in succession down the chain.

Dampers Devices that act like doors in ducts. They can be closed to block the passage of air or opened to allow it.

Datacasting Providing additional program material or non-program-related resources with a digital television or audio broadcast signal (video, audio, text, graphics, maps, services, etc.) to specially equipped computers, cache boxes, set-top boxes, or DTV receivers.

DC Offset A power disturbance in which a continuous DC current is added to an AC current causing the AC current sine wave to be offset to higher positive or negative voltage by the amount of the DC current's voltage.

Decibel (dB) A unit for measuring sound level. Used in audio engineering.

Decoder A device that renders digital data into analog signals, or vice versa, in audio and video systems. *See* converter.

Demarcation point The physical place where the incoming telephone lines from the telephone company connect to the internal wiring of the home; also known as demarc.

Demodulation probe A probe used with an oscilloscope to measure radio frequency signals.

Denial of Service A method of attacking a LAN or server by sending thousands of unanswerable requests to it.

Detection devices   Sensors that discern changes in conditions indicating the presence or passage of a person or persons.

Detector   A sensor designed to react to a certain event such as pressure, motion, heat, or light.

Dial tone   A continuous tone sent on a telephone line by the PBX or the central exchange to indicate it is available for a call.

Digital   Any kind of data that is recorded in numerical (discrete) form rather than analog (continuously varying).

Digital audio file   A recorded music or voice audio segment stored on a computer hard drive or other similar media.

Digital audio tape (DAT)   The standard for recording uncompressed digital audio on tape at the same quality level as a CD.

Digital cable ready   A government designation for television sets denoting their capability to play digital cable programs.

Digital data   Data in the form used by computers that consists only of the binary numeric digits 1 and 0. Nearly all forms of information can be rendered into digital form for processing or transmission and then converted back into a form that can be displayed or understood by people.

Digital Home Technology Integration (DHTI)   A connected home environment in which a computer system manages data and controls subsystems in the home.

Digital subscriber line (DSL)   An advanced form of ISDN line that makes possible the transmission of voice and high-speed digital data on the same line at the same time.

Digital television (DTV)   Television signals that are in numerical format and create a picture in pixels rather than rasters like analog TV.

Digital Transmission Content Protection (DTCP)   A copy protection system used to prevent unlicensed copying of HDTV programs.

Digital tuner (digital converter, digital encoder/ decoder)   The electronic device that decodes digital audio or television signals so they can be played on radio or television receivers; a tuner can be an internal part of the receiver or contained in a separate set-top box.

Digital video disc (DVD)   Media for recording digital video. *See also* DVD.

Digital Video Interface (DVI)   A connector system for distribution video (but not audio) HDTV signals. It includes HDCP technology.

Digital video/audio switcher   A device that receives digital signals from multiple sources and routes them as instructed to multiple outputs.

Direct current (DC)   Electric current that flows in one direction only, from a positive potential to a negative potential, as in a battery.

Direct Sequence Spread Spectrum (DSSS)   A method of signal hopping or rapidly changing frequencies in a specified sequence to transfer data at high speed.

Disarm   To deactivate a security system so its sensors are not functioning.

Dish antenna   A parabolic-shaped antenna that receives satellite broadcasts.

Distortion   Changes in reproduced sound waves caused by excessive power or other interference in the recording and playback processes.

Distribution   The process of transmitting data throughout a system so it is available to all components.

DNS server   Special computers on the Internet that keep databases of IP addresses and their corresponding domain names.

Dolby Digital 5.1   A digital surround-sound technology that is broadcast as the audio part of HDTV signals and used in home theater systems to enhance audio. It includes five separate channels of audio, each connecting to a separate speaker, plus a low-frequency subwoofer-connected channel that together produces true-to-life audio.

Domain name   A unique name assigned to a network and registered with ICANN.

Domain Name Service (DNS)   A part of the TCP/IP protocol that translates domain names into their corresponding IP addresses.

Dot matrix printer   A printer that prints using a matrix of small pins that strike the paper through a ribbon and combine to form characters.

Double protection   In security systems, at least two devices monitoring a means of entry into a home so that if one is disabled or fails, the other can still detect an intruder.

Downlink   Digital data being transferred into a ground-based system from a satellite.

Drip irrigation system   A type of irrigation system in which small tubes deliver water flow directly to individual plants or plots of ground without spraying. The system conserves water compared to spraying sprinkler systems.

Drive ring   A metal ring that can be hammered into walls to carry cable on the exterior of buildings.

Drywall   The paper-covered gypsum board that is fastened to studs to finish the walls in most homes.

DSL splitter   A device attached to the NID, which splits an incoming DSL data line, isolates other telephone lines and equipment, and allows a home run to be installed to the modem.

Duct   A metal tube or open-ended box though which air can flow. Can be rigid or flexible and of any size.

Duress code   An access code (to be used in the event a person is accosted or threatened at the door) that opens the door, but also summons help.

DVD (digital video disc)   A small plastic record on which digital video programs are recorded using a laser process.

Dynamic Host Configuration Protocol (DHCP)   A method of automatically assigning IP addresses to nodes on a LAN.

Eave   The underside of a roof that overlaps the outer walls of a building.

Electric grid   All the components that make up an electric power system; generators, transmission wires, switches, transformers, and so on.

Electric service   Electric power purchased from a utility, also the cable bringing the electricity to a home from the utility.

Electric service line   The power line that connects the electric grid to an individual residence or building, supplying electric power to the home.

Electromagnet   A magnet created by passing a current through a coil of wire. Used in audio speakers and many other devices.

Electromagnetic field   Force field of electrons generated by high-voltage equipment and wires.

Electromagnetic induction   The process whereby a high-voltage current flowing in a circuit produces (induces) a current in an adjacent circuit even though the circuits are not connected to one another.

Electromagnetic interference (EMI)   Any form of electrical energy that is not an intentional electric current on a circuit but comes into it from an outside source.

Electromagnetic radiation   A method of EMI travel in which EMI signals are radiated out into the air or space in all directions from a source.

Electronic symbol   A graphic figure used to represent an electric fixture or electronic component in wiring diagrams and schematics.

Electrostatic discharge (ESD)   An arc produced by a surface charge on an object discharging to another object, creating a spark or arc across the gap of space between them.

Emergency response   Any assistance delivered as a result of a security system call-in. Fire department, police department, and medical teams are all emergency responses.

Encoder   A device or software program for changing analog audio or video signals into digital format. Video versions are also called coders.

Enhanced-definition television (EDTV)   An ATSC-defined television that can play both analog and DTV digital programs and has a minimum resolution of 480p by 704 pixels wide in 4:3 or 16:9 format.

Equalizer   An audio device for balancing the levels at which sound frequencies are recorded or played back.

Ethernet   The most common form of LAN architecture. It uses bus or star topology and employs CSMA/CD to manage the flow of data on the network.

Evaporative or "swamp" cooler   A type of air cooler that works by evaporation of water into air, which is thereby cooled as it passes through the cooler.

Extensions   In telephone systems, an additional phone wired in parallel into a single line.

Face plate   The decorative cover on wall boxes that contain data jacks or AC power outlets.

Fail-safe   A theoretical term for a security system that can't be disabled or bypassed. Not possible to achieve in actual systems.

False alarms   Any security breach not produced by a genuine security threat.

Fan   A device consisting of a balanced set of angled blades on a shaft spun by a motor. Fans move air and are one type of air handler.

Fault   Any defect in a circuit that prevents current from flowing or that cause too much current to flow.

Fax machine   A digital device that operates on an analog telephone line. It sends and receives printed documents by transmitting them as bit-mapped images.

Fiber Distributed Data Interface (FDDI)   Large fast networks constructed almost entirely of fiber-optic cable.

Fiber-optic cable   A very high-speed means of transmitting data using light beams through glass or plastic threads or fibers.

Field   One scan of an interlaced television frame that refreshes one half the frame.

Filter   A device to remove interference from an electric circuit.

Filtering   A method of analyzing data by comparing it to preset parameters; data meeting the parameters can then be blocked, approved, or otherwise directed.

Fire-stop   A wooden crosspiece set between studs in interior walls to retard fire.

Firewall A software program or a hardware device that controls or filters the information passing from the Internet onto a LAN and from a LAN onto the Internet. It is used to maintain the security of the private system.

Firewire An IEEE 1394 data transmission protocol.

FireWire IEEE 1394 standard data cable and connectors for transmitting digital data at high speed between two devices or nodes on a network.

Flat panel screen An LCD monitor or television. *See* liquid crystal display (LCD).

Flow meter A device that measures the amount of water flowing in a pipe and signals its measurement to a recording device. Used to detect leaks or breaks in pipes or sprinkling systems.

Flow valve A type of valve that measures the amount of water flowing through it and can adjust that flow automatically.

Fluorescent lighting Lighting tubes that contain phosphorescent material and glow when a high-voltage current is passed through them. The tubes are made in various lengths, up to 8 feet.

Forced air A type of heating or A/C system that works by blowing heated or cooled air into a home through ducts and air vents.

Frame One complete refresh of a television screen, which can be two scans (fields) if interlaced, or one scan if progressive.

Freeze sensor A device that sends a signal when the temperature around it falls to near freezing (37 degrees normally, although some can be set to lower or higher temperatures) so that action can be taken to prevent freezing.

Frequency The number of cycles per second that an AC current sine wave reverses direction; in the U.S. system, 60 cycles per second is the standard.

Frequency division multiplexing An analog technology for carrying multiple data streams, voice or data, on the same wires.

Frequency Hopping Spread Spectrum (FHSS) A method of signal hopping or rapidly changing frequencies in a random sequence to transfer data at high speed.

Frequency modulation (FM) A method of broadcasting sound by varying the frequency of the carrier wave but not its strength.

Frequency variation A power disturbance in which the frequency of AC current changes rapidly over time; caused by rapid changes in the load on the circuit.

Function selector A control on a multimeter that sets the test function it will perform.

Furnace A device for heating air so it can be blown into the interior of a home.

Fuse A device to limit current flow in a circuit; the fuse "blows" or burns out when the circuit overloads, stopping the flow of current through it.

Gas detector A device that senses the presence of natural gas, carbon monoxide, or other poisonous gas.

Generator The mechanical device that creates alternating electricity; it is similar to an electric motor but is turned by steam or an internal combustion engine and generates electricity.

Ground A wire that connects an electric device to the earth so excess current can flow to the ground rather than elsewhere.

Ground fault A circuit fault that involves the circuit being connected to a ground in addition to the normal negative-to-neutral circuit connection.

Ground start system Local telephone system that seizes a telephone line as soon as a receiver is lifted so no other call can transmit on the line.

Grounded Connected to the earth by an electrical conductor so electric current can travel into the ground to complete a circuit.

Hacker A person who attempts to gain unauthorized access to a computer or network.

Harmonic AC current A power disturbance in which AC current is produced in a circuit at a multiple frequency of the normal current frequency.

HDTV tuner A digital television tuner that receives full resolution HDTV signals in 16:9 format.

Header Data at the beginning of a data packet identifying its source and destination.

Heat fins Metal ribs placed on electronic devices to increase surface area and help dissipate heat produced by the devices.

Heating, ventilation, and air-conditioning (HVAC) systems Systems that provide controlled heating, cooling, and ventilation in buildings.

Heat sensor A device that monitors air temperature and signals if it rises above or below a set range. Used in fire detection alarms (high temperatures) and freeze detectors (low temperatures).

Heat sink A relatively large piece of metal or other material placed near a heat source to absorb heat from it and dissipate it into the surrounding air. Often used with heat fins.

Heat transfer The process by which heat radiates from warmer objects to cooler ones.

Hertz (Hz) The term for cycles of frequency; AC current is 60Hz, an AM radio station may be 1280 KHz (kilohertz), and the lowest frequency the human ear can hear is 20 Hz.

High-Definition Content Protection (HDCP) A copy-protection system used in DVI and HDMI connector systems to prevent unlicensed copying of HDTV or DVD programs.

High-Definition Multimedia Interface (HDMI) A connector system for distribution video and audio HDTV signals in a single cable. It includes HDCP technology.

High-definition television (HDTV) A digital television format with a 16:9 aspect ratio and a minimum resolution of 720p or 1080i.

Home Audio Video interoperability (HAVi) A standard for audio and video connectors and cables designed to allow multiple technologies to interface with one another.

Home lighting system The entire electrical light configuration in a home, including all interior and exterior lights and the controls (manual or automated) that direct them.

HomePlug Command and Control (HPCC) A power-line control system recently introduced by the HomePlug Alliance. Few products using this technology are available yet.

HomePNA The original standard of HomePNA technology currently in use for networks using telephone lines for connectivity.

HomeRF (Home Radio Frequency) A wireless network technology for home LANs.

Home run cable A data cable running direct from a DSL splitter to the network modem assuring a clean incoming signal.

Home run connection A wired connection that links a light directly to its power supply without the current to it passing through any other fixture except a switch.

Home security system A DHTI subsystem of hardware and software designed to prevent unauthorized persons from entering a home or yard, using any of its data systems, or removing anything from them.

Home theater A DHTI subsystem for displaying television programs and recorded video programs. Similar to a home entertainment center, but usually specialized primarily for video viewing.

Host Another name for a computer on a LAN.

Hot wire The wire in a cable that is connected to the source of electric potential; usually colored black.

HPNA 2.0 The network standard currently in use for HomePNA networks.

Hub A device that connects nodes on a LAN and broadcasts data received from any node to all other nodes.

Huffyuv A lossless codec for digital video compression.

Humidifier A device that adds moisture to an incoming air stream.

Humidity Moisture contained in air. All air contains some water vapor (gaseous water). The amount can vary with the pressure and temperature of the air.

Humidity detector A device that measures the amount of moisture in the air around it and signals that data to a controller or computer.

IEEE 1394 A fast serial protocol that transmits data at speeds from 100 to 400 Mbps.

IEEE 802.11b A WiFi protocol based on the IEEE 802.11 wireless network standard.

IEEE 802.11g An advanced fast protocol based on the IEEE 802.11 standard.

Impedance Resistance to the flow of current in a device or wire.

Impulsive transient A sharp voltage increase in a current that lasts .008 seconds or longer; transients can be instantaneous, momentary, temporary, or sustained.

Induction The process by which current flowing in a circuit produces a corresponding current in another nearby circuit. A means by which EMI travels from one circuit to another.

Infrared (IR) device A device that functions as a controller by sending command signals using infrared light, which is invisible to human eyes. Also, a device that detects heat (infrared-wavelength energy) or changes in the heat level of its surroundings.

Infrastructure Any of the wiring, conduit, connectors, wireless hubs, switches, routers, and other hardware that enable the subsystems of a DHTI system to communicate with one another and the outside world.

Inside arming Using a code or command to activate perimeter devices in the security system, which may activate some interior devices, but usually not all. Inside armed is the normal status for the security system at night while the family is asleep.

Integrated Services Digital Network (ISDN) A technology that uses a telephone line to transmit digital data at high speed.

Intercom A line in key systems that allows any extension to connect to any other for an internal call without accessing an outside line.

Interference Any signal that corrupts or blocks an analog or digital data signal.

Interlaced A method of refreshing a video screen in which only odd-numbered lines are scanned on one pass and even-numbered lines on the next.

Internet A worldwide web of interconnected, but independent, networks over which data travels from source to destination by various routes.

Internet Corporation for Assigned Names and Numbers (ICANN) The group that assigns and regulates domain names and IP addresses through accredited registrars.

Internet Home Alliance (IHA) An industry group whose objective is to develop the market for home technologies that require a broadband or persistent connection to the Internet.

Internet service provider (ISP) A company that provides Internet connections to home and business LANs.

Internetwork Packet Exchange/Sequenced Packet Exchange (IPX/SPX) The protocol used by Novell NetWare networks.

Intruder Any unauthorized person trying to enter a secured home or yard.

IP address Internet Protocol address, a 32-bit address consisting of four numbers separated by periods, used to uniquely identify a device on a network.

IPCONFIG A TCP/IP utility that displays the computer's adapter address, IP address, subnet mask, and default gateway, and allows the DHCP to be renewed or released by the user.

Jack A connecting device terminating a cable, into which a plug is mated to connect a node.

Joist A horizontal support beam in a floor.

Joule A unit of electrical energy equal to the work done when a current of 1 ampere is passed through a resistance of 1 ohm for 1 second.

Keyboard An alphanumeric data input device for a computer. It may contain additional keys that input specific commands or run sequences of data.

Keypad A numeric pad similar to a telephone dial pad that is used to input passwords to cipher locks and other security devices.

Key system A local telephone system in which multiple phones are connected to multiple lines by switching buttons so any telephone can use any line.

KiloAmp A unit denoting 1,000 amperes of current.

Laser light Coherent light waves transmitted in parallel beams so they maintain their intensity over long distances. Used in light beam sensors and many other devices.

Laser printer A printer that works by picking up toner on a charged image on the surface of a rotating drum and depositing it on paper where the image is fused in place.

Latchkey function A command or code that programs the call-in device to notify a set telephone number when a specific door of the home is opened during set hours.

Letterbox A term that describes the effect of viewing a 16:9 aspect ratio image on a 4:3 aspect ratio screen; the viewer sees black bars above and below the image that are referred to as the letterbox effect.

Light beam sensor A device in two pieces, one of which transmits a beam of light to the other across a distance. If the light beam is obstructed, the sensor signals its failure to arrive.

Light scenes Illuminated rooms or areas within a room that not only provide light but beautify a setting or invoke a mood.

Light sensor A device that senses the presence of light and sends a signal if there is an absence of t.

Light-emitting diode (LED) An electronic device, similar to a vacuum tube, which emits red light or infrared radiation. Arrays of tiny LEDs are used to make television display screens.

Line of sight A term used to describe the location of wireless transmitters and receivers. It means that the receiver must be visible when viewed from the transmitter for the signal to be received.

Line protector A device for protecting an electrical circuit from spikes, surges and noise.

Line speed The amount of data that can travel over a communication line or wireless connection in a given length of time.

Liquid crystal display (LCD) A thin lightweight type of video display that uses liquid-crystal material sandwiched between two layers of electrodes to create a color image.

Load In electrical systems, the amount of current (amps) flowing in the wiring at a given time; measured in amps or watts.

Local area network (LAN) A regionally confined network consisting of computers that communicate and share data and services.

Lock-down feature A program in some security systems that allows all door locks to be deactivated during specified times so that they cannot be opened from the outside at all, even with a valid access code.

LonWorks A power-line control system common in Europe that has recently been introduced in the United States.

Loop start system Local telephone system in which a Centrex-supplied dial tone indicates when a line is available for a call, but no line is seized until a call is actually placed.

Lossless Type of video compression in which no data is lost. Huffyuv is an example.

Lossy Type of video compression in which some data is lost, but files can be made smaller than with lossless compression.

Low-capacitance probe A probe used with an oscilloscope to measure radio frequency and high impedance signals.

Macro A programmed sequence of commands that direct a computer to perform certain functions; sometimes used by hackers to gain control of a computer or LAN.

Magnetic key A key similar to a plastic credit card with its access information coded on a magnetic stripe that can be read by an access device.

Magnetic media Any media with a ferrous coating capable of storing analog or digital data recorded on it by a magnetic head.

Magnetic recording Any analog or digital data recorded on magnetic media.

Magnetic tape Plastic tape coated with ferrous material for recording data.

Make switch A sensor that monitors a switch that should remain open and signals if it is closed. Most commonly used as pressure pads, which are closed by weight passing over them.

MC cable Metal-clad cable used for AC wiring.

Mechanical relay An electrical device activated by an electromagnet being energized that functions as a switch to turn a control current on or off; an automatic remote-controllable switch.

Media Any means of storing or recording audio or video information: magnetic tape, CDs, DVDs, and phonograph records are some examples.

Megohmmeter A specialized test instrument for testing the resistance of insulation material to current flow (amperage) in a circuit.

Microphone A device that converts sound waves into an electronic analog signal that can be converted to digital form, transmitted over distances, and replayed or stored.

Microsecond One-thousandth of a second.

MIDI (Musical Instrument Digital Interface) file A type of digital music recording in which the file stores an actual musical score. MIDI files can be played on any electronic instrument or software-equipped computer.

MiliAmp A thousandth of an ampere (.001 amps).

MiniDV cassettes A small-format recording medium used in many digital video cameras.

Modem An electronic device that converts digital data into a form that can be sent over a telephone line. Modems are the most common method of connecting a home computer to the Internet through a telephone line or cable connection to a service provider.

Monaural Having only one track or sequence; not stereo. Used to describe single-track audio recording.

Monitor A screen output display for a computer or TV signal. Can be CRT or LCD. Also, a sensor that monitors one or more environmental conditions.

Monitored In security systems, watched either by a person at a control console or by the electronic security panel itself so any change in status can be responded to.

Monitoring service A commercial service in which a company's staff continually watches a home security system via a telephone line linked to the home.

Motion sensor A device that detects any object that radiates heat moving into its range and signals the change in its surroundings.

MP3 file A file created with an audio compression algorithm with the same name.

MP3PRO A file created with an audio compression algorithm with the same name; a higher compression than MP3, but equal in quality.

MPEG-1 A file created with a video compression algorithm with the same name.

MPEG-2 A video compression algorithm used for DVD video recording.

Multimeter A combination test instrument for testing voltage, resistance, and current flow (amperage) in a circuit.

Multiplexing Combining different types of data from multiple sources on a single transmission path.

Multistation Access Unit (MAU) Device used in a token ring star design to which all nodes are connected.

Multisystem VCR A video player capable of playing NTSC, PAL, or SECAM videocassettes on any analog television.

Nanosecond One-millionth of a second.

Narrow band In telecommunications systems, bandwidth of 128 Kbps or less.

***National Electrical Code* (NEC®)** A safety standard for electrical wiring and installation developed by the NFPA.

National Fire Protection Association (NFPA) Publisher of the *NEC®*.

National Television Standards Committee The government and industry group that established the NTSC analog television standard in 1953.

Neon voltage tester A specialized test instrument for testing the presence of voltage in a circuit.

Network A group of computers, information sources, and peripheral devices connected by cable or radio so they can share data and communicate with one another.

Network Addressing Translation (NAT) A method of assigning temporary IP addresses to overcome the shortage of available unique IP addresses.

Network BIOS Extended User Interface (NetBEUI) A Microsoft proprietary protocol commonly used for LANs.

Network interface card (NIC) A device for connecting a node to a LAN.

Network Interface Device (NID) A device that connects the ISP service line to a home's inside telephone and data wiring.

Network printer A printer of any type connected to a LAN as a node with its own IP address.

NM cable Nonmetallic cable used for interior AC wiring.

Node A computer or other device connected to a LAN by an NIC.

Noise A form of power disturbance that causes a smooth AC sine wave to look like a series of sharp peaks and valleys.

Non-interlaced A type of screen refresh that renews each line of the screen in order. This method of refreshing is also called progressive.

Notching A form of power disturbance in which tiny voltage variations cause an AC current's sine wave to have notches in it.

NTSC format The standard analog U.S. TV format with 525 scan lines and a 4:3 aspect ratio.

Numeric code A group of digits that serves as a password and must be entered into a cipher lock to open it.

Obstruction detector A safety device that detects anything unusual under a descending garage door and stops the door's downward movement.

Ogg Vorbis (OGG) An open source video compression format, and hence free of any patents.

Ohmmeter A specialized test instrument for testing resistance in a circuit.

Ohm's Law Describes the relationship among amperes (amps), the amount of current flowing; volts, the electrical potential between the two ends of a circuit; and resistance within the circuit to the flow of current.

Open circuit An incomplete circuit that does not have a continuous path from current source to neutral.

Open Services Gateway initiative (OSGi) A standard for audio and video connectors and cables designed to allow multiple technologies to interface with one another.

Oscillatory transient A power surge that changes voltage several times before fading to 0; an alternating current surge.

Oscilloscope A combination test instrument for testing voltage, resistance, and current flow (amperage) in a circuit. It also displays the sine wave of the circuit being tested.

Outlet box A wall fixture containing a data jack or AC power receptacle.

Outside arming A code or command that activates all security sensors to function while the home is unoccupied.

Overload Current flow greater than a circuit or its components can carry without danger of burning out.

Overloading Placing a current demand on an electrical circuit above its capacity to carry safely.

Overvoltage A power disturbance in which the voltage of a circuit remains consistently higher than normal.

Packet A small segment into which data is divided and packaged with a header and trailer for transmission on a network.

PAL format The European analog TV standard equivalent to the U.S. NTSC standard.

Panning motor An accessory device for a surveillance video camera. The panning motor slowly swings the camera right and left allowing it to cover more area than a stationary camera could.

Passport Obsolete discontinued type of power-line technology for networking.

Password A group of letters and numbers that must be entered into a security system in a home or into a computer system to gain access.

Patch cable An eight-wire straight-through cable in which all wires are connected to the same numbered pin at each end.

Patch panel A device consisting of a row or block of jacks, used for connecting all components of a network.

PATHPING An improved version of PING.

Peak use The maximum load of electrical consumption in a home.

Peripheral Any input or output device connected to a computer that sends or receives data from the processor. Examples of input peripherals are floppy drives and microphones. Examples of output peripherals are printers and control devices.

Phonograph A machine that plays analog recordings from a plastic disc embossed with grooves bearing the sound wave impressions.

Pilot hole A small hole drilled to locate a position or guide a larger drill bit.

Pilot light A small flame that burns continuously in a furnace to ignite the furnace's main burner when it is turned on.

PING A TCP/IP utility that enables a user at one computer to determine if that node can communicate with another computer connected to a network.

Pixel A picture element in digital television; the unit of color and brightness that forms the picture in digital television.

Plain Old Telephone Service (POTS) The most common method of home Internet connection.

Plenum An enclosed passage for moving air in a building. Also sometimes used as a channel for electronic wiring.

Plug A terminator on the end of a cable that mates with a jack to make a connection.

Polarity The negative (−) and positive (+) sides or ends of a circuit.

Potential The flow force of electric current, measured in volts.

Power conditioning The process of maintaining electric current at its correct voltage and frequency and keeping it free of noise and other EMI.

Power conversion Changing AC current to DC or the reverse.

Power disturbance Any interruption or abnormal alteration in the voltage, frequency, or sine wave of an AC current.

Power failure warning sensor A device that detects the presence of voltage (electrical pressure) in a circuit and signals its absence to a monitor.

Power interruption A complete loss of voltage and current flow in a circuit for any period of time.

Power-line carrier (PLC) The general term for control systems that operate by sending command signals over AC wiring.

Power-line network Network technology that transmits data over a home's AC power lines at the same time high-voltage power is running on the lines at a different frequency.

Power-line technology A network data transmission method in which data signals are sent on AC wiring using a different frequency and voltage than the regular current flowing in the circuits.

PowerPacket A high-speed power-line technology for networking.

Pre-amplifier A device that strengthens an original signal from an antenna or other source before it goes to an amplifier.

Pressure pads A make switch that is closed when a set weight is placed on it. The sensor then signals its changed state.

Pressure sensor A device that monitors gas or liquid pressure in a pipe or tank. If pressure drops too low (or, in a steam pipe, rises too high) the sensor signals the change.

Pressure valve A type of valve that shuts off water flow automatically if pressure in the pipe drops below a set minimum, indicating there is a leak or break in the pipe.

Private branch exchange (PBX) A local switching point where phone lines within an organization can be connected to one another or to outside lines.

Probes Metal leads used with electronic test instruments to connect to circuits for testing them.

Program A set of digital instructions that a computer executes in sequence to perform functions.

Programmable Capable of storing instruction sets for later execution in sequence.

Progressive A type of screen refresh that renews each line of the screen in order. The other type of refresh scheme is interlaced, which renews every other line.

Protocol A set of rules and standards that a network uses to communicate among its nodes.

Proxy server A LAN node isolated from the rest of the LAN that services all outside requests for data from the LAN.

Public switched telephone network (PSTN) The commercial network of telephone lines and transmission facilities over which most telephone calls are made.

Pulling cable The process of drawing a cable through an existing structure from one connector to another.

Pulse code modulation (PCM) The method used to record compact discs using MP-2 compression. Not a magnetic process. Uses light diffraction to record data.

Pulseworx The UPB-based control system developed by Powerline Control Systems.

Punchdown block   The usual means of centrally terminating telephone lines and connecting them to the trunk lines.

PVC   Polyvinylchloride; a hard strong plastic used to make pipes and many other products.

PVC conduit   Plastic pipe commonly used to transport water, but also a tube through which electronic wires and cables can be run.

QuickTime   An Internet and computer video file format that uses compression and is widely used by PC and Apple computers.

Raceway   An enclosed track in which to run cable; the track is attached to the surface of walls.

Radiant   A type of heating system that works by warming a home with hot water that flows to radiators in each room or in pipes beneath the floors. The radiators or floors then warm the interior air.

Radio frequency (RF)   Any electronic wave with a frequency in the radio band of the electromagnetic spectrum. Includes all radio and TV frequencies.

Radio frequency interference (RFI)   Any unwanted electrical energy at a radio frequency entering an electrical circuit where it is not wanted.

RadioRA   A proprietary wireless lighting control technology developed by Lutron and used for home lighting control.

Rain sensor   A device that detects the presence of rain water in a small holder and signals to the controller when a preset level is reached.

Rasters   The individual horizontal scan lines that make up an analog television picture.

RCA connector   One of the standard connector types used for audio and video system connections.

RealPlayer   A computer and Internet video compression algorithm.

Receptacle   A device into which AC-powered appliances can be plugged to obtain power.

Receptor   Any device that receives EMI, intentionally or unintentionally.

Recorded emergency message   A message recorded on a security system call-in device; it plays after a device calls a preset telephone number and gets a connection. Some systems dial multiple numbers in succession and some can play multiple messages.

Redundant line   A design feature in an electric grid in which all users are connected to the grid by at least two paths so the failure of one will not cause an interruption.

Redundant protection   In a security system, having two or more sensors or detectors monitoring one area

or entrance. Provides additional security to high-risk areas.

Refresh rate   The number of times per second a television picture is "painted" on the screen; usually 15 times per second for interlaced broadcasts or 30 times per second for progressive-scanned broadcasts.

Refrigerant fluid   A gas/liquid that circulates in an A/C system, continuously extracting heat from the surrounding air as it changes from a liquid to a gas and back again.

Refrigeration   A method of cooling air that uses the heat absorption of expanding gas to extract heat from the surrounding air.

Regenerator   A digital data amplifier that reads weakened data signals and re-creates them at full strength and without noise on the transmission line.

Relay   A remotely operated electric switch that is activated by a small current, but that controls a large current flow.

Remote access   A security system monitored by an outside commercial firm through a telephone or radio connection to the home control panel.

Remote-control   A hardware device or software program that allows a person or a computer to direct the operation of a subsystem from a distance.

Remote login   Connecting to and taking control of a computer from a remote location.

Repeater   A device in a LAN that receives and strengthens the data signal to offset its attenuation over distance.

Resettable circuit breaker   A device that prevents overload in a circuit by cutting off current flow when it exceeds the circuit capacity; an overload switch that is not damaged by an overload and can be reset to function again.

Residential gateway   A device that connects a LAN with a WAN that is part of the Internet; the gateway controls the data coming into or going out of the LAN.

Resistance   The force inhibiting the flow of electricity in a circuit, measured in ohms.

Resolution   The amount of clarity, detail, and sharpness in a television picture; usually measured in scan lines (interlaced or progressive) or pixels (e.g., 480 by 720). Also, the number of pixels on a computer monitor that are individually addressable by software.

Response devices   Security devices that can perform an action when commanded to do so by the security panel. They may sound alarms, call for assistance, activate systems, or perform other tasks.

Retrofit   To add new wiring or other infrastructure to an existing building.

RJ-11 jack   The standard single-line telephone connector wired with two wires.

RJ-14 jack   The standard two-line telephone connector wired with four wires.

RJ-45 crimping tool   A tool used to attach an RJ-45 plug as a terminator on the end of a cable.

RJ-45 jack   A terminator device on a cable into which a plug is mated to make a network connection.

RJ-45 punchdown tool   A tool used to attach an RJ-45 jack as a terminator on the end of a cable.

Rough-in   To install the outlet boxes and cable runs for a network.

Routable   A protocol that allows data to be sent to interconnected networks on the Internet.

Router   A device that connects two or more networks and directs the data traffic passing between them.

Run-time   The amount of time a UPS will keep normal power supplied to the devices connected to it.

Safety control valve   A specialized valve on the fuel line of a furnace or other appliance that must receive a continuous electric current from a thermocouple in order to stay open. If the current stops, the valve closes, shutting off the fuel flow.

Sag   A momentary drop in the voltage of an AC current.

Sampling   A technique for converting analog signals into digital form by taking quantified samples of the analog data.

Satellite link   An Internet connection to an ISP via a satellite through a receiver dish antenna. Also, a digital television connection that beams signals to receivers on the ground from a stationary satellite orbiting Earth in space.

Scan   Survey of a data set, space, or sensors to determine if particular data or a set of conditions is present.

Scan line   One horizontal line of picture data on a television screen.

Scan rate   The number of times per second that the electron beam in a CRT repaints the screen from top to bottom. Same as the refresh rate in progressive screens; equal to twice the refresh rate in interlaced screens.

Schematic   A graphic representation of the wiring and components of an electronic device.

SDTV television   A television that receives and displays all ATSC digital formats, but not necessarily at full high resolution.

SDTV tuner   An RF receiver that receives ATSC terrestrial digital television signals and decodes all Table 3 video formats. It can play to an analog or digital television set.

SECAM   A French-created European equivalent of the NTSC analog standard.

Security breach   An event that occurs any time a sensor signals a change of status to a security panel.

Security panel   The control center of a security system to which all of the system's sensors and surveillance devices report their status. The panel may also direct responses, record data, and sound alarms.

Security zone   An area in a home or its yard that is monitored by a specific group of sensors in the security system. Zones are set up with defined monitoring devices so the location and nature of any security breach can be quickly identified.

Sensor   Any device that detects or measures human activity or environmental conditions and sends data regarding its measurement to a processor. Also, a device that monitors an object (door, window, floor area) or condition in or around the home and signals the security panel if its status varies significantly from normal.

Server   A computer or device on a network that provides network services or manages network resources.

Service line   The cable that brings electric power into a home from the utility.

Service panel   The wall box containing a home's circuit breakers to which the service line is connected.

Service provider   A company that provides data transmission service or other utility services to consumers.

Shielding   Metal plating or mesh placed around a source of EMI to block interference, or situated around a receiver of EMI to block the reception. Also, metal webbing around a data line that grounds noise and interference before it can reach the data line.

Shock   Electric current flowing through a person.

Short circuit   A circuit with too little resistance that permits too much current to flow. Caused by a connection in the circuit providing a shorter path to neutral than normal.

Siamese cable   Any cable that consists of two or more types of cable (Cat5, coaxial, fiber optic, etc.) bundled together in a single sleeve.

Signal generator   A specialized test instrument for testing the path of a signal in a circuit.

Signal level meter   A specialized test instrument for testing the strength of a signal at various points in a circuit.

Signal-to-noise ratio (S/N ratio)   The difference in sound level between the recorded audio and the background noise on any type of audio recording.

Signal tracing A troubleshooting technique that involves determining the path and strength of a signal in a circuit to determine the location and source of a fault.

Single-circuit surge suppressor A device for blocking spikes, surges, and some noise from an AC circuit.

Single-pole switch An AC basic switch that opens or closes a circuit to control power to a fixture.

Smoke detector A device that signals the presence of smoke in the air around it. It does not react to heat or flame, only to the presence of smoke particles.

Softphones Software programs that display phone features (hold button, caller ID, message waiting) on the computer screen and route the calls through a handset or an earphone and microphone wired to the computer,

Soil moisture sensor A device that can measure the electrical conductivity of the ground in which it's buried. This measurement allows it to detect the amount of water in the soil.

Solenoid valve An automatic valve operated by an electromagnet that can be energized or turned off to open or close the valve. It is the main control device in automated water systems.

Space heater A small heater, usually electric, but sometimes gas-fired, designed to heat a small area or room. Manually controlled.

Spam The e-mail equivalent of junk mail; unwanted and unsolicited e-mail.

Spark igniter An electrical device that ignites the fuel in an oil or gas furnace when it starts.

Spike A sudden short increase in voltage in an AC circuit.

Split duplex outlet An AC electrical outlet in which one receptacle is always live and the other is controlled by a switch.

Splitters Devices that allow two or more telephones or peripherals to be connected to a single wall jack.

Spread spectrum Spread spectrum signals constantly change frequency, a process known as hopping, to reduce the power requirements for transmission.

Standard-definition television (SDTV) The approved minimum format for U.S. digital television with a 4:3 aspect ratio and an interlaced 480 by 720 pixel screen.

Star topology One type of network topology in which nodes are arranged in a star pattern.

Static Noise or interference on a telephone line that is typically heard as a background crackling sound.

Static electric discharge Also called electrostatic discharge; a current flow caused when two objects that have different static charges touch or come near enough for the discharge current to arc between them.

Step-down transformer A device for decreasing the voltage of an AC current branch from the standard 110 volts to a lower figure, such as 24 volts for use by a low-voltage device.

Stereo Two soundtracks recorded from the left and right side of a musical performance to give balance and depth to the recording.

STP cable Shielded Twisted Pair wire. Consists of up to four pairs of wires twisted in pairs and wrapped in a metal shield with an insulated outer sleeve.

Straight-through cables Cable wiring connected to the same terminator pins at both ends of the cable.

Studs The 2 × 4 or 2 × 6 wood uprights in the walls of homes.

Substitution A troubleshooting technique that involves substituting a known working part for a suspected defective part to see if doing so repairs the circuit.

Subsystem A group of hardware components and software set up to perform a specific task or function within a larger multiple-function DHTI system.

Suppressor A device for blocking surges and spikes in a circuit.

Surface wiring Wiring run along the outside of a wall, usually in a raceway, although it can be bare.

Surge A momentary increase in the voltage or current flow of an AC circuit.

Surveillance video camera A small video camera used in a home security system to monitor a yard or part of the home's interior. The camera's output can be displayed or recorded or both.

S-video connector One of the standard connector types used for video and audio system connections.

Swell A short-duration low-value increase in the voltage of an AC circuit.

Swipe slots An input device for plastic magnetic-striped keys. The plastic key is passed (swiped) through the slot in the device, which reads the key code and grants access.

Switch A device used in a LAN to direct data traffic among the nodes. Also, an electrical device that completes (turns on) or opens (turns off) a circuit.

Switchboard A device to which a number of telephone lines are connected and which can switch any line to connect to any other.

Switched duplex outlet An AC electrical outlet controlled by a switch that turns current to it on or off.

Switcher   A multiple video input panel with a single output to a monitor and a switch that allows the monitor to display any video input selected.

Synchronized   Operating or playing together at the same rate. Stereo sound tracks are synchronized, as are the picture and sound of a video program.

T-1 line   A high-capacity telephone trunk line capable of simultaneously handling up to 24 voice lines or 1.5 Mbps of data.

Telecommunications   The general name for all communication and data functions carried on telephone lines or radio signals, or performed by telephone hardware and software.

TELNET   A TCP/IP utility that allows a user in one location to access a computer in a remote location as if the user were physically sitting in front of the remote machine.

Temperature sensor   A device that measures the ambient air temperature and signals it to a security system monitor, HVAC controller, or computer that can compare it to programmed instructions and take appropriate action.

Terminator   A device on an Ethernet that ends the data flow in a bus topology.

Test lamp   A specialized test instrument for testing whether a circuit is complete and working.

Thermal disconnect   A device that protects an electrical circuit by melting or burning out to stop current flow when the load exceeds capacity.

Thermocouple   A device that transforms heat into an electrical current. Used as a safety device in gas appliances to monitor the pilot light. If the light goes out, the thermocouple stops generating electricity, thereby signaling a problem.

Thermostat   A special type of temperature sensor that sends a signal when the air temperature reaches a preset level.

Three-way switch   A type of AC switch used in pairs to turn a circuit on or off from two locations.

Token ring   A type of network in which data flows in a circular pattern and is controlled by a token.

TRACERT   A TCP/IP utility that shows the complete path data packets are taking from the computer to reach any given destination.

Trailer   Data attached to the end of a data packet.

Transformer   An electrical device that increases (steps up) or decreases (steps down) voltage in a circuit with an inverse increase or decrease in amperage.

Transient   A sudden sharp increase in the voltage of a circuit.

Transistor checker   A specialized test instrument for testing transistors for correct functioning.

Translator   Another term for a decoder that converts encrypted data into a readable form for display by an output device.

Transmission Control Protocol/Internet Protocol (TCP/IP)   The most common protocol used to connect networks.

Transmission   The movement of data from one location to another. The data can be digital or analog and the locations close together or distant.

Triad   A set of three dots in a color monitor which in combination can produce all colors.

Troubleshooting   The common name for analyzing, testing, and correcting electrical and electronic problems in all types of systems.

Trunk line   A telephone line from a PBX to the telephone company that can be used by any of the telephones connected to the PBX.

Truss   A triangular structure that supports a roof.

Tuner   The device in a radio or television that sets one station's frequency to be received and excludes all others.

Twisted-pair wires   A set of two wires twisted around one another in a specific manner to improve data transmission in a high-speed cable.

Uncompressed original video footage   Digital video recorded without use of any algorithm to reduce file size. Produces large files that contain all digital data in sequential format.

Undervoltage   A power disturbance in which the voltage in a circuit is consistently lower than the designed voltage.

Unified messaging   A computer-based system for storing messages from multiple sources. Storage is located on a PC's hard drive and the messages include voice mail, fax, and e-mail.

Uninterruptible power supply (UPS)   A battery-powered device that keeps an AC circuit functioning by supplying it with temporary power converted from DC batteries.

Universal Powerline Bus (UPB)   A power-line technology for controlling lighting that uses signals measuring from 4 Khz to 40 Khz. UPB control devices are made primarily by Pulseworx.

Universal Serial Bus (USB)   A bidirectional, isochronous, dynamically attachable serial interface for adding devices on a single bus.

USB port   Universal Service Bus port. One of the standard connector types used for data connections; includes video, but rarely audio.

UTP telephone wire Unshielded Twisted Pair wire. Consists of up to four pairs of wires twisted in pairs and wrapped in an insulated sleeve. Typically used to wire telephones.

Valve A device for controlling the flow of fluids (gases or liquids) in a pipe. Can be manually or automatically controlled.

Valve box An open-bottom box sunk to its lid in the ground where the zone valves and sometimes the controller of a sprinkler system are located.

Vent An opening in a wall or floor through which air flows from a connected duct into a room.

Ventilation The process of exchanging air in a confined room or space. Ventilation exhausts the old air out of a room and brings new air in.

Victim A receptor of EMI.

Video A digital display of pictures such as television programs or computer screens.

Videoconferencing A technology for sending full-motion picture images with a voice telephone call so callers can both hear and see one another.

Video distribution system A wired system for distributing video programs from multiple sources to multiple viewing sites in the home.

Video Home Standard (VHS) The U.S. standard analog video recording format used in videocassettes.

Video image Any image, analog or digital, displayed on a CRT or other type of screen.

Video server A computer set up as a storage location for video programming on a network, from which the video footage can be requested by other network nodes.

Virus A small self-propagating software program that spreads through networks by copying itself onto other computers where it remains resident in storage; it may then do damage to stored data and application programs, or allow hackers to access the system.

Voice mail A digital system that allows telephone callers to record messages that the system stores, retrieves, and plays on demand.

Voice over Internet Protocol (VoIP) A system whereby analog phone calls are digitized and sent in packets over the Internet. At the destination, the digital data is decoded to analog form and delivered to the receiver.

Volt The unit used to measure electric potential or flow pressure.

Voltage-divider probe A probe used with an oscilloscope to measure high voltage in a circuit.

Voltage fluctuations A power disturbance in which the voltage of an AC circuit changes rapidly over a short time period.

Voltage regulator A device for maintaining the voltage of an AC circuit at a set level with little variation.

Voltage tester A specialized test instrument for testing voltage in a circuit.

Voltage variation A power disturbance in which the voltage of an AC circuit is consistently lower than its designed level.

Voltmeter A voltage tester

Volt-ohm-milliammeter (VOM) A combination test instrument for testing voltage, resistance, and current in a circuit. Commonly known as a multimeter.

Wall-mounted controllers In video and audio systems, a device for controlling distributed programming in a room or for the whole system.

Water pressure The force that impels water through a pipe. Water pressure is produced by gravity (the weight of water pushing down from a higher elevation to a lower one) or by artificially pressurizing a sealed water system, usually with compressed air.

Water sensor A device that senses the presence of water and signals it to a security monitor or panel.

Water service line The pipe that supplies water to a residence from a public utility. It enters the home and connects to the interior plumbing and exterior water system.

Watt The unit of electric power, often calculated for a circuit by multiplying the number of volts by the number of amps.

WAV file A compression algorithm for sound files.

Waveform distortion Any form of power disturbance that alters the normal smooth sine wave of AC current over a period of time; examples are DC offset, notching, and noise.

Whole house surge suppressor A device installed on the service line of a residence that blocks transients that originate outside the home.

WiFi (Wireless Fidelity) IEEE 802.11-based wireless transmission protocol with an 11-Mbps transmission rate. It is presently the most popular wireless standard. Also known as WiFi/b.

WiFi/g Higher-speed version of WiFi/b that is backward-compatible with it.

WiFi5 IEEE 802.11-based wireless transmission protocol; a very high-speed technology.

Wind sensor A device that can be set to signal a controller whenever the average wind speed rises to a preset level (adjustable from 12 to 35 miles per hour). Used to shut off sprinkler systems in high-wind conditions.

Windows Media Audio (WMA) An MPEG-4 audio compressed file.

WINIPCFG   Windows 9*x* version of IPCONFIG; a utility for displaying a computer's adapter address, IP address, subnet mask, and default gateway, and for renewing or releasing its DHCP.

Wireless audio transmitter   A wireless transmitter that sends analog or digital audio signals to a computer.

Wireless hub   A device to which nodes in a wireless LAN can connect using radio waves.

Wireless NIC   A device in a wireless node that connects it to a hub using radio waves.

Wireless technology   Any of several methods of communicating digital data by means of radio waves without the need for any wires connecting the sender and receiver.

Wiring diagram   A graphic representation of the wiring and fixtures of an AC electric system in a home or building.

World Wide Web (WWW)   The global system of interconnected networks over which users can share data through the use of common protocols.

X10   A wired technology that transmits data on existing high-voltage AC power lines in the home, and thus requires no new wires for installation.

Zone   A group of lights controlled together to provide a specific scene or accomplish a specific task such as security lighting. Also, in sprinkler systems, multiple sprinkler heads in an area that are controlled by one valve. Also, in HVAC systems, a part of a home separately heated and cooled, and controlled by one thermostat. Also, the area around a wireless hub which its transmission reaches and from which it can receive data from wireless nodes.

Z-Wave   A wireless lighting control system developed in Denmark and now available in the U.S.

Index

Note: Page numbers referencing figures are italicized and followed by an "*f*". Page numbers referencing tables are italicized and followed by a "*t*".